STRUCTURAL CONDITION ASSESSMENT

STRUCTURAL CONDITION ASSESSMENT

Edited by

Robert T. Ratay, Ph.D., P.E.

JOHN WILEY & SONS, INC.

Published by John Wiley & Sons, Inc., Hoboken, New Jersey
Published simultaneously in Canada.

For general information on our other products and services or for technical support, please contact our Customer Care Department within the United States at (800) 762-2974, outside the United States at (317) 572-3993 or fax (317) 572-4002.

Wiley also publishes its books in a variety of electronic formats. Some content that appears in print may not be available in electronic books. For more information about Wiley products, visit our web site at www.wiley.com.

Library of Congress Cataloging-in-Publication Data:
Structural condition assessment / edited by Robert T. Ratay.
    p. cm.
  Includes bibliographical references and index.
  ISBN 0-471-64719-5 (cloth)
    1. Structural analysis (Engineering)   I. Ratay, Robert T.
  TA645.S75   2005
  624.1'71—dc22

                                                                    2004009416

10 9 8 7 6 5 4 3 2 1

■ CONTENTS

CONTRIBUTING AUTHORS

Howard W. Ashcraft, Jr., Esq., *Senior Partner, Hanson, Bridgett, Marcus, Vlahos & Rudy, LLP, 333 Market Street, San Francisco, CA 94105; hashcraft@ hansonbridgett.com*
> Mr. Ashcraft is a Fellow and board member of the American College of Construction Lawyers and a construction arbitrator in the American Arbitration Association's Large and Complex Case Program; he specializes in design and engineering issues.

Kimball James Beasley, P.E., *Manager, New York Office, Wiss, Janney, Elstner Associates, Inc., 1350 Broadway, New York, NY 10018; kbeasley@wje.com*
> Mr. Beasley is a structural engineer with over 30 years of experience investigating failures and designing repairs for both historic and contemporary buildings.

Cynthia Chabot, P.E., *Principal, Structural Engineer, Chabot Engineering, 15 Damon Avenue, Melrose, MA 02176; cchabot@chabotengineering.com*
> Ms Chabot has been a structural engineer for 15 years working on bridges and tunnels as well as small commercial and residential buildings; she is a member of the NCSEA Code Advisory/General Engineering Committee.

Robert W. Day, P.E., G.E., *Chief Engineer, American Geotechnical, 5764 Pacific Center Blvd., San Diego, CA 92121, dayr@amgt.com*
> Mr. Day is a leading geotechnical engineer and the author of over 200 published technical papers and five textbooks on geotechnical and foundation engineering.

James P. Donnelly, P.E., S.E., *Consultant, Wiss, Janney, Elstner Associates, Inc., 330 Pfingsten Road, Northbrook, IL 60062; jdonnelly@wje.com*
> Mr. Donnelly specializes in the evaluation and repair of existing structures, particularly post-tensioned, precast, and reinforced concrete buildings such as parking structures, and he is a member of the ACI Committee on Parking Structures.

Donald O. Dusenberry, P.E., *Principal, Simpson Gumpertz & Heger Inc., 41 Seyon Street, Waltham, Massachusetts 02453; dodusenberry@sgh.com*
> Mr. Dusenberry has 30 years of experience in the evaluation of structural performance and forensic engineering. He served for several years as the supervisor of the Simpson Gumpertz & Heger Inc. testing laboratory.

Allen C. Estes, P.E., Ph.D., *Colonel, Associate Professor, U.S. Army: West Point, Department of Civil and Mechanical Engineering, United States Military Academy, West Point, NY 10996; allen.estes@usma.edu*
> Col. Estes is the Civil Engineering Program Head at the United States Military Academy and has authored over 60 papers, articles, and book chapters on the subjects of reliability, life-cycle cost, and structural optimization.

Dan M. Frangopol, D.Sc., Dr. h.c., *Professor of Civil Engineering, Department of Civil, Environmental and Architectural Engineering, University of Colorado, Boulder, CO 80309; dan.frangopol@colorado.edu*
> Dr. Frangopol is an expert in the areas of structural reliability, structural optimization, bridge engineering, and life-cycle analysis, design, maintenance, and man-

agement of structures; he is the founding president of the International Association for Bridge Maintenance and Safety, and founding editor-in-chief of *Structure and Infrastructure Engineering: Maintenance, Management, Life-Cycle Design and Performance.*

Tian-Fang Jing, P.E., *Principal, Weidlinger Associates, Inc., 375 Hudson Street, New York, NY 10014; jing@wai.com*
Mr. Jing is an expert in the design of long-span fabric roof structures.

Leonard M. Joseph, P.E. S.E., *Senior Vice President and Principal, Thornton-Tomasetti Group, 2415 Campus Drive, Irvine CA 92612; ljoseph@thettgroup.com*
Mr. Joseph has 30 years of experience designing and investigating long-span, high-rise, and special structures across the United States and around the world.

Paul F. Mlakar, Ph.D., P.E., *Senior Research Scientist, U.S. Army Engineer Research and Development Center, 3909 Halls Ferry Road. Vicksburg, MS 39180; Paul.F.Mlakar@erdc.usace.army.mil*
Dr. Mlakar currently serves as the U.S. Army senior expert on weapons effects and structural dynamics.

Donald W. Neal, P.E., *President, Neal Engineering Associates Ltd., 16645 SW Pleasant Valley Road, Beaverton, OR 97007; don@nealengr.com*
Mr. Neal is a consulting structural engineer in practice, with special expertise in the design and evaluation of structural timber.

David H. Nicastro, P.E., *Founder and Chief Executive Officer, Engineering Diagnostics, Inc., 106 E. 6th Street, Austin, TX 78701; DNicastro@EngineeringDiagnostics.com*
Mr. Nicastro specializes in the investigation of problems with existing buildings, designing remedies for those problems, and resolving disputes that arise from them.

Antranig M. Ouzoonian, P.E., *Weidlinger Associates, Inc., Consulting Engineers, 375 Hudson Street, New York, NY 10014; ouzoonian@wai.com*
Mr. Ouzoonian has 50 years of consulting engineering experience in the design and investigation of distressed structures.

David B. Peraza, P.E., *Vice President, Thornton-Tomasetti Group, LZA Technology Division, 641 Avenue of the Americas, NY 10011; dperaza@LZATechnology.com*
Mr. Peraza, a recognized expert in the structural and forensic engineering industry, has extensive experience in building investigations, analysis, rehabilitation, and remediation.

Predrag L. Popovic, P.E., S.E., *Principal and Vice President, Wiss, Janney, Elstner Associates, Inc., 330 Pfingsten Road, Northbrook, IL 60062; ppopovic@wje.com*
Mr. Popovic, member of the ACI and PCI Parking Garage Committees, has evaluated over 1500 existing structures, including over 100 parking structures.

Brian E. Pulver, P.E., *Senior Associate, Wiss, Janney, Elstner Associates, Inc., 330 Pfingsten Road, Northbrook, IL; bpulver@wje.com*
Mr. Pulver specializes in the evaluation and repair of existing post-tensioned, pre-stressed, precast, and conventionally reinforced concrete structures such as parking garages.

Robert T. Ratay, Ph.D., P.E., *Consulting Engineer, 198 Rockwood Road, Manhasset, NY 11030, and Adjunct Professor at Columbia University, New York; rtratay@cs.com, Robert@RatayGroup.com*
Dr. Ratay has 40 years of structural design, analysis, condition assessment, forensic investigation, and university teaching experience; he is also the editor of books and author of articles on structural engineering.

Thomas Z. Scarangello, P.E., *Managing Principal, Thornton-Tomasetti Group, 641 Avenue of the Americas, NY 10011; tscarangello@thettgroup.com*
Mr. Scarangello has designed, investigated, and retrofitted numerous covered arenas and stadiums, open-air ballparks, and other buildings subject to earthquakes, hurricanes, and snow loads over the past 25 years.

Mark K. Schmidt, S.E., *Structures Group Unit Manager, Wiss, Janney, Elstner Associates, Inc., 330 Pfingsten Road, Northbrook, IL 60062; mschmidt@wje.com*
Mr. Schmidt has focused on the assessment, preservation, remedial design, and implementation of restoration programs for a wide variety of building envelopes for over 20 years.

Robert Smilowitz, Ph.D., P.E., *Principal, Weidlinger Associates, Inc., Hudson Street, New York, NY 10014; smilowitz@wai.com*
Dr. Smilowitz provides protective design services for public and private sector projects and participates in research and development programs to develop innovative protective design solutions

R. Wayne Stocks, P.E., *Vice President, Thornton-Tomasetti Group, Thornton-Tomasetti-Cutts Division, 2000 L Street, NW, Washington, DC 20036; wstocks@ttcutts.com*
Mr. Stocks specializes in the renovation and modernization of historic landmark structures, such as occupied federal buildings and national monuments, with cost-effective solutions that respect existing structure.

Eric C. Stovner, P.E., S.E., *Senior Associate, Miyamoto International, 2102 Business Center Drive, Irvine, CA 92612 ; EStovner@MiyamotoInternational.com*
Mr. Stovner has extensive experience in the design and investigation of structures, which includes earthquake engineering applications and the preservation of historic structures.

Andrea E. Surovek, P.E., Ph.D., *Assistant Professor, Civil and Environmental Engineering, South Dakota School of Mines and Technology, 501 East St. Joseph Street, Rapid City, SD 57701; surovek@sdsmt.edu*
Dr. Surovek has performed numerous condition assessments of industrial and commercial structures; her current area of expertise is the stability of steel building frames, with emphasis on advanced analysis methods for design.

Mark J. Tamaro, P.E., *Senior Associate, Thornton-Tomasetti Group, Thornton-Tomasetti-Cutts Division, 2000 L Street, NW, Washington, DC 20036; mtamaro@ttcutts.com*
Mr. Tamaro has completed numerous structural assessments of historic buildings and monuments for government, institutional, and private sector clients, and has lectured and published technical articles on many of these prestigious historic projects.

Wesley R. Terry, P.E., *Vice President of Engineering, Birdair Inc., 65 Lawrence Bell Drive, Amherst, NY 14221; WesT@birdair.com*
Mr. Terry has been involved in the design and construction of fabric membrane structures for 23 years.

Glenn G. Thater, P.E., *Associate, Thornton-Tomasetti Group, LZA Technology Division, 641 Avenue of the Americas, New York, NY 10011; GThater@LZATechnology.com*
Mr. Thater has extensive experience in condition assessment, remedial design, and structural alterations of buildings and special structures, and he has served as an expert in a wide variety of forensic investigations.

David Transue, P.E., *Project Engineer, Atkinson-Noland & Associates, Inc., 2619 Spruce Street, Boulder CO 80302; dtransue@ana-usa.com*
David Transue is a consulting engineer who specializes in masonry structures, with areas of interest including condition assessment, forensic engineering, materials research, design, and historic masonry.

Robert S. Vecchio, Ph.D., P.E., *Managing Principal, Lucius Pitkin, Inc., 50 Hudson Street, New York, NY 10013; rvecchio@lpiny.com*
Dr. Vecchio has been with Lucius Pitkin, Inc., since 1985; currently he is managing principal (partner) specializing in the fitness-for-service, fatigue, and fracture mechanics analyses of structures.

John L. Windle, P.E., *Consulting Engineer, John L. Windle, P.E., 604 South Concord Road, West Chester, PA 19382; jamwind@comcast.com*
Mr. Windle has been involved with all aspects of engineering and quality assurance in the communications tower industry for 50 years.

Scott F. Wolter, P.G., *President/Geologist/Petrographer, American Petrographic Services Inc., 6850 Chanhassen Road, Chanhassen, MN 55317; swolter@amengtest.com*
Mr. Wolter is a senior forensic material scientist specializing in concrete, aggregate, and rock.

Bojidar S. Yanev, Eng. Sc. D., P.E., *Executive Director, Bridge Inspection & Bridge Management, Department of Transportation, City of New York, 2 Rector St., New York, NY 1006; and Adjunct Professor, Department of Civil Engineering, Columbia University, New York; byanev@dot.nyc.gov*
Dr. Yanev heads the Bridge Inspection/Management Unit, which he established at the New York City Department of Transportation; he teaches a graduate course on Bridge Design at Columbia University.

Structures, like people, never get younger. Structures, like people, can maintain their good health with age, if properly cared for, examined, and treated when needed. One may view this book as a Structural Physician's Reference.

It may be said that a structure that has withstood the combined effects of use, abuse, loads, and environmental conditions over time has, in effect, proven itself. However, buildings, bridges, parking structures, stadiums, and all other structures do deteriorate with time as the result of repeated loadings, exposure to the elements, aging of materials, wear and tear from normal use, abuse, inadequate maintenance, and other factors. The deterioration may progress to the point or one or more individual debilitating events may have the consequence of compromising the structure's strength, stability, or serviceability.

Evaluating the condition of structures is an area of professional practice within the field of structural engineering. It is an increasingly active business driven in part by the need for maintenance, repair, and rehabilitation of buildings, bridges, and other deteriorating infrastructure; by change in ownership; by the choice of adaptive reuse of facilities; by the necessity of retrofitting for ever-changing code compliance; and by the need for increased physical security of corporate and public buildings.

While working knowledge of structural analysis and design, and understanding of structural behavior are indispensable for the engineer doing structural condition assessment, experience in field observation of structural problems is imperative. We analyze and design structures for strength, stability, and deformation, but the most common problems are lack of serviceability.

In simplified terms, structural condition assessment consists of visual observation, measuring, photographing, probing and sampling, field and laboratory testing, engineering analyses, record keeping, documentation, and report preparation. It is not to be confused with forensic investigation of structures, which is the determination of the causes and modes of failure (where failure is not only collapse but also the unacceptable difference between intended and actual performance).

While sophistication and accuracy of the field measurements, testing, and analyses are important, the reliability of the condition assessment lies in the interpretation of these data and in the judgments converting them to accurate conclusions and recommendations.

The purpose of this book is to fill the need for a comprehensive and authoritative reference in the practice of structural condition assessment, addressing technical as well as business and legal matters. In addition, by its nature, the book is also valuable reference on structural and material performance.

The editor's and the contributing authors' intention is to instruct the novice and guide the experienced engineer in performing the condition assessment of structures: how to conduct the assessment; how to inspect the structure; how to recognize various conditions; what to look for; what methods of field examination, laboratory testing, and analytical evaluations are available; how to report the findings; and what to recommend. This book has the aggregate components of the subject in one volume. In both its format and content it is a "first" rather than "another" book on the subject.

This is not a "what happened" or "lessons learned" book; instead, it is a "what I need to know," "what I need to do," and "how I need to do it" book. It is, at this time, the most complete book on the subject, suitable for use by the novice as a text to learn the necessary basics, by the expert as an authoritative reference, and as a guide to all to assist in the day-to-day conduct of structural condition assessment. It is hoped that this book will stimulate the study of structural condition assessment and will be embraced as the text in academic and continuing education courses on the subject.

No one single author would have the wide-ranging expertise and credibility to write on all the topics in this comprehensive reference book; therefore each chapter was prepared by a contributing author, or co-authors, with extensive experience and demonstrated expertise in its subject matter.

Twenty-one chapters of the book are grouped into four logical parts:

Part I—DEGRADATION, SAFETY AND RELIABILITY OF STRUCTURES. This first part, consisting of two chapters, provides background material for the understanding of the conditions and events that cause structures to deteriorate, creating aesthetic, serviceability, and safety problems. It also discusses in depth the theoretical aspects and reliability of structural condition assessment.

Part II—BUSINESS, CODES AND LEGAL ASPECTS OF CONDITION ASSESSMENT. This rather unique part, consisting of three chapters, addresses the subjects of how to manage the engineering practice of structural condition assessment, including agreements, compensation and other engineer-client relations. Hard-to-find information on past editions of building codes and material design standards is given with references to and contents of the ones often used in structural design. The nature and management of risk to both the engineer and owner are discussed in easily understandable legal terms.

Part III—SURVEY AND ASSESSMENT OF STRUCTURAL CONDITIONS. The appropriate condition assessment procedures as well as types, causes, signs, and consequences of degradation and defects for nine types of structures are discussed in separate chapters titled *Buildings, Historic Buildings and Monuments, Building Facades, Parking Structures, Stadiums and Arenas, Bridges, Tensile/Fabric Structures, Broadcast* and *Transmission Towers,* and *Foundations and Retaining Walls.* The assessment of vulnerability to terrorist action is discussed in a separate chapter, titled *Vulnerability to Malevolent Explosions.*

Part IV—EVALUATION AND TESTING OF STRUCTURAL MATERIALS AND ASSEMBLIES. This part includes five chapters, each on the inspection, testing, and engineering evaluation of a different structural material: Concrete, Steel, Masonry, Timber, and Fabric. The last chapter is a practical guide to field load testing as part of structural condition assessment.

Structural condition assessment is incomplete and is of questionable value without a written report. The last section of the book, APPENDIX A, gives succinct pointers and reminders for the preparation of reports.

I am grateful to the contributing authors for accepting my invitation to participate in the preparation of this book, and I thank them for their effort and cooperation. Without their contributions there would be no book.

I also owe thanks to people at John Wiley & Son, Inc., especially to Jim Harper, Editor, for his valuable assistance and willing cooperation throughout the preparation of the book; to Scott Amerman, Senior Production Editor, for his able supervision of the copyediting and further production; and to the copyeditor, illustrator, compositor, typesetter, and others whose professional work has been so important.

Robert T. Ratay, Ph.D., P.E.
Manhasset, New York

PART I

DEGRADATION, SAFETY, AND RELIABILITY OF STRUCTURES

CHAPTER 1

Defects, Deterioration, and Durability

DAVID H. NICASTRO, P.E., and ANDREA E. SUROVEK, Ph.D., P.E.

INTRODUCTION

In any condition assessment, it is important to recognize two fundamental categories of existing or potential performance failure within buildings and other structures: defects and deterioration. A *defect,* as used here, is the nonconformity of a component with a standard or specified characteristic. Defects may be introduced through poor design, manufacturing, fabrication, or construction before a structure begins its service life and (less frequently) by inappropriate operations and maintenance during its service life. *Deterioration,* as used here, is the gradual adverse loss of desired material properties. Eventual deterioration is normal for most construction materials owing to aging and weathering processes, and it must be addressed through strategic maintenance, repair, or replacement to avoid unwanted distress.

Defects may influence the rate of deterioration or may initiate premature deterioration for materials; hence the two are often involved in a cause-and-effect relationship (Fig. 1.1). This chapter provides a summary of common defects and deterioration

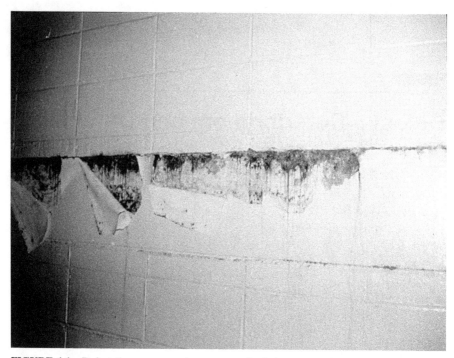

FIGURE 1.1   Defect (improper exterior waterproofing) that caused deterioration (corrosion of steel box beam).

mechanisms observed in existing construction. Identifying and understanding defects and deterioration are critical in extending the service life of existing structures. Without detection through condition assessments and subsequent remedy, defects and deterioration may lead to failure of components, systems, or even entire structures.

During a condition assessment, determining whether deterioration is natural or premature may help to identify defects within the system. However, deterioration can only be identified as "premature" if the expectations for normal service life are understood. Therefore, this chapter also discusses *durability,* the relationship between the expected service life (design life) of building materials and their actual service life in the absence of uncontrolled defects and deterioration. Finally, a glossary is provided for terms used commonly in structural condition assessments.

DEFECTS AND DETERIORATION

A condition assessment of an existing structure may be performed for a number of reasons, including (but not limited to) determining whether a structure is safe for public occupancy; documenting the condition of existing structures prior to performing construction nearby; evaluating the existing condition as part of a "due diligence" survey prior to purchase of a building; determining whether a structure is suitable for a change in usage; establishing code compliance or determining areas of noncompliance to be remedied; and determining the cause and appropriate remedy of problems that elicit tenant complaints, such as leaks, excessive floor vibrations, or trip hazards. In these instances, the purpose of a condition assessment is to identify the pervasiveness and severity of distress, defects, and deterioration, if any. Identifying the causes of such conditions is usually necessary to design proper remedies.

Defects

Defects are introduced most commonly during design, manufacture, fabrication, or initial construction; therefore, they are typically present at the beginning of the service life of a structure. *Latent* defects may be present during initial construction and lie dormant prior to manifesting themselves. Specifically, this section addresses three categories of defects: design defects, product defects, and construction defects.

Design Defects. In professional engineering design offices, it is common practice for designs to be peer reviewed, and many design errors are prevented by this process. However, defects in design often are created by poor detailing, insufficient detailing, or a lack of attention to constructability of details.

There is also occasional abdication of design responsibility that transcends errors or omissions. For example, it is common engineering practice to show on structural details the size and position of shelf angles to support brick masonry veneer but only to indicate the masonry by a phantom line, implying that the masonry itself is trivial to the engineer. A leading cause of masonry distress, including cracking, spalling, and collapse, is the lack of coordination between the structural steel and the supported masonry. A lack of expansion joints below shelf angles may lead to masonry crushing (Fig. 1.2) or a systemic failure such as bowing of the entire wall. These failures can

FIGURE 1.2 Masonry crushing due to lack of expansion joint below shelf angle.

be prevented by the engineer providing not only the traditional structural details but also the details pertaining to the interface between structural and architectural elements.

Product Defects. Poor product performance may be attributed to the manufacturer, the fabricator, and/or the designer. It is the responsibility of the designer to research and specify the appropriate product for a given application. Manufacturers' representatives often work closely with designers to assist in the selection of products for construction applications. Reputable manufacturers have vast experience with the performance of their products in different applications and have quality assurance programs that include product testing. However, defects still may be introduced at the factory; while some sources of defects are material-dependent, others may be the result of poor quality control, such as the introduction of contaminants in manufacturing or improper material storage. Errors in fabrication also may qualify as product defects. For example, if a steel column is fabricated with an out-of-straightness that exceeds the AISC-specified fabrication tolerances,[1] forces may be introduced in the member under loading that were not accounted for in the initial design process.

Construction Defects. Construction defects commonly are introduced when there is a breakdown in communication between the various groups of professionals and technicians that work separately to produce a structure. Poor communication between design professionals can lead to numerous problems, such as conflicting drawings and specifications, interference between structural elements and nonstructural systems, or misalignment of systems between stories.

Many defects are attributable to errors that occur during construction. For example, in concrete construction, defects that arise during the construction process may be due to improper placement, consolidation, curing, or finishing. Improper location of reinforcing and/or insufficient concrete cover are common sources of deterioration and subsequent distress in parking structures. Fabrication errors, such as misaligned bolt holes, are common defects in steel construction. With proper preparation and review of shop drawings, as well as field monitoring by the design team, many of these types of defects can be prevented.

Deterioration

Unlike defects that are typically present at the beginning of the service life of a structure, deterioration of a material or system is time-dependent. While some forms of deterioration may develop early in the service life of a structure, others are a matter of the aging of a material or system. Deterioration often is initiated or accelerated by the presence of a defect or the introduction of a catalyst. Water is the most significant catalyst for deterioration; approximately half of all failure mechanisms identified in construction involve water, including some subtle interactions.[2] For example, overstress may not be directly dependent on water, but the strength of wood and masonry may be affected by the presence of water, thus exacerbating deterioration due to loading. In weathering and biological or chemical attack, water is the primary catalyst for deterioration. Water-induced mechanisms include (but are not limited to) decay, corrosion, freeze-thaw action, soil erosion, settling, and upheaval. The deleterious presence of water is not limited to naturally occurring events, such as rainfall, but may be attributed to infiltration of water due to construction defects.

Several types of deterioration catalysts or degradation factors may be classified. ASTM E632,[3] "Standard Practice for Developing Accelerated Tests to Aid Prediction of the Service Life of Building Components and Materials," categorizes degradation factors that initiate or accelerate the deterioration of building materials and components as follows:

Weathering factors. Deterioration from exposure to water, temperature, wind, radiation, air, and air contaminants.

Biologic factors. Attack of a material by a living organism, such as bacteria or insect infestation.

Stress factors. Loads on a system, either sustained or periodic. For example, stress may result from gravity loads, thermal loads, shrinkage, swelling, or settlement.

Incompatibility factors. Chemical reactions, such as chloride attack of steel, or physical interactions between materials, such as abrasion.

Use factors. Wear and tear associated with construction and service or application of loads that exceed the strength of the system.

Table 1.1 illustrates the variety of factors that can influence the service life of building components. In many instances, degradation factors work in tandem; for example, corrosion of steel is a chemical attack (incompatibility factor), but it usually occurs in the presence of water (a weathering factor).

The time dependence of deterioration mechanisms is variable. In general, the development rate may increase, decrease, or remain roughly constant over time. Decreasing-rate mechanisms, such as shrinkage of concrete, will diminish in rate of deterioration over time, even though the cumulative deterioration is always increasing in magnitude. Increasing-rate mechanisms will develop faster over time, typically because the deterioration "feeds" on itself; for example, the erosion of the concrete shown in Fig. 1.3 facilitates ponding of additional water, thereby accelerating the erosion. Constant-rate mechanisms, such as carbonation of concrete, generally proceed linearly over time.

Obviously, it is easiest to predict the future behavior of systems affected by constant-rate mechanisms. Also, decreasing-rate mechanisms tend to be fairly straightforward to deal with in condition assessments because at any time that the structure is studied the rate of future deterioration will, by definition, be diminishing. Increasing-rate mechanisms pose the hardest problem in predicting future behavior; for instance, once freeze-thaw damage is evident, there may be little time remaining to arrest it before complete disintegration occurs.

Below is a brief overview of the defects and types of deterioration specific to the four most common building materials: concrete, steel, wood, and masonry. More comprehensive catalogs of defects and deterioration mechanisms are provided by SEI/ASCE 11-99[4] and Nicastro,[2] as well as in other chapters of this book.

Concrete

Defects. Concrete is a composite material; the interaction of its components leads to many of the design and construction defects common to the material. One of the principal design defects that causes distress in concrete members is improper detailing of reinforcement. The lack of proper detailing in concrete can lead to distress as minor as spalling and as catastrophic as collapse. At one point, the American Concrete Institute (ACI) estimated that repair and rehabilitation accounted for approximately 70% of the U.S. construction market.[5]

Other sources of concrete defects stem from errors in concrete mix design based on the numerous possible combinations of admixtures, aggregates, and cement types. Susceptibility to cracking, freeze-thaw damage, and chemical attack all can arise from poor mix design. In addition, poor quality aggregates and impurities in the mix water can affect quality adversely.

Most construction defects in cast-in-place concrete come from improper placement, curing, and finishing. A common consequence of these defects is uncontrolled crack-

TABLE 1.1   Degradation Factors Affecting the Service Life of Building Components[3]

Weathering factors	Radiation	Solar
		Nuclear
		Thermal
	Temperature	Elevated
		Depressed
		Cycles
	Water	Solid (such as snow, ice)
		Liquid (such as rain, condensation, standing water)
		Vapor (such as high relative humidity)
	Normal air constituents	Oxygen and ozone
		Carbon dioxide
	Air contaminants	Gases (such as oxides of nitrogen and sulfur)
		Mists (such as aerosols, salt, acids, and alkalis dissolved in water)
		Particulates (such as sand, dust, dirt)
	Freeze-thaw	
	Wind	
Biologic factors	Microorganisms	
	Fungi	
	Bacteria	
Stress factors	Stress, sustained	
	Stress, periodic	Physical action of water as rain, hail, sleet, and snow
		Physical action of wind
		Combination of physical action of water and wind
		Movement due to other factors, such as settlement or vehicles
Incompatibility factors	Chemical	
	Physical	
Use factors	Design of system	
	Installation and maintenance procedures	
	Normal wear-and-tear	
	Abuse by the user	

ing, particularly of slabs. While concrete cracking is inevitable owing to the nature of the material, it can be limited. In addition, crack locations can be controlled when the design is detailed properly and construction is monitored. Placement and finishing, adequate control joints and expansion joints, proper curing, and adequate concrete cover and/or slab depth all are required for the prevention of unwanted cracks in concrete slabs and members.

Additional problems often arise owing to quality control during concrete placement, the most infamous of which is the on-site addition of water to the concrete mix to increase workability. Improper vibration or consolidation during placement can lead to honeycombing, delamination, stratification, and spalling, as well as a reduction in structural capacity.

FIGURE 1.3   Erosion of concrete, an increasing-rate failure mechanism.

Most concrete defects can be prevented through proper design and construction monitoring. A comprehensive resource for design and construction of concrete structures is the *ACI Manual of Concrete Practice.*[6]

Deterioration. Much of the deterioration found in concrete stems from either the defects just discussed or the deleterious effects of water. Water is the primary catalyst for concrete weathering, scaling, erosion, bleeding, leaching, freeze-thaw damage, and chemical reactions. In some instances, incompatibility factors will not manifest distress without the addition of water (e.g., corrosion of steel reinforcement). Concrete is particularly susceptible to chemical incompatibility, including alkali-silica reactions, alkali-carbonate reactions, carbonation, and sulfate attack.[2,7]

Steel

Defects. Of the materials commonly used in construction, steel is manufactured under the most controlled conditions, and consequently, material defects are not particularly common. Most defects in steel members and structures arise from either improper design and detailing or errors in fabrication and erection. Steel defects that occur during construction typically stem from errors in erection, including misalignment of members and connections. Misalignment of members can lead to eccentric loads not accounted for in the original design. A dominant source of construction errors (and the subsequent structural distress) is improper installation of connections and/or poor-quality welding.

The cross-sectional shapes used in steel assemblages are particularly susceptible to stability failures. These failures may be on the local level (e.g., web buckling), at the member level (e.g., lateral torsional buckling), or on a system level (e.g., story buck-ling).[8] System instability is typically catastrophic in nature. This phenomenon is not limited to fully constructed facilities; it occurs more often as a result of insufficient lateral bracing during construction.

Deterioration. Deterioration of steel at the material level stems primarily from in-compatibility factors, such as chemical attack, that lead to corrosion. There are a large number of ways that steel can corrode. A partial list of corrosion types includes chloride-accelerated, concentration cell, crevice, deposit, electrochemical, electrolytic, galvanic, pinpoint, and thermogalvanic.[2] Section loss may occur due to corrosion, reducing the strength or stability of a steel member or system. Proper surface treatment of steel, such as corrosion resistant paint, often can reduce or prevent the occurrence of corrosion.

Deterioration in steel may stem from overstress conditions owing to underdesign, overload, unanticipated eccentric loads, or fatigue. These conditions may manifest plastic deformations or cracking in steel elements. Misalignment of steel members, particularly at connections, is a common cause of unanticipated eccentric loads on steel structures, which can lead to deterioration.

Wood

Defects. Wood is an organic material, and some defects are a natural part of the material; these are easily detectible during structural surveys. Wood defects generally result from growth or drying. Knots develop from branch growth, shakes are grain separations that develop between growth cycles, and checks are cracks resulting from the differential shrinkage rates of wood parallel and transverse to the grain that result from improper cutting and drying of dimension lumber. Splitting occurs in wood as a result of the same mechanism as checks.[2,9,10] Knots, shakes, checks, and splits create local stress concentrations in structural lumber and may lead to failure.

Other defects affecting the structural integrity, as well as the aesthetics, of wood components include deformations, such as bowing, twisting, crooking, and cupping.[11] These deformations may result from improper drying in fabrication, improper material storage, and one-sided coating applications.

Deterioration. Wood deterioration often can be attributed to deficiencies in mainte-nance, such as exposure to weathering factors without proper protective coatings. Leak-ing building envelopes, peeling paint, and damp conditions often lead to the accelerated deterioration of structural wood components.

As an organic building material, wood is especially vulnerable to biological deg-radation factors. Wood fibers act as a food source for a variety of organisms, such as fungus and bacteria, as well as several types of insect.

Wood rot results from a combination of conditions within a structure and functions as deterioration by consumption. A fungus breaks down the cellular structure and leaves the wood with little or no structural integrity. The various species of fungi, such as dry rot, wet rot, soft rot, and white rot, require a food source (cellulose), oxygen, water, and particular temperature, light, and pH ranges to thrive. Wood typically does not rot if it is continuously submerged in water or if it is well protected from water; however, the intermittent wetting and drying of wood provides ideal conditions for fungal attack and resulting deterioration.

In addition, softened, damp wood provides an ideal environment for several bur-rowing insects. Termites, carpenter ants, carpenter bees, marine borers, and various beetles consume wood fibers; loss of structural integrity and premature failure may result. Non-consumption-based degradation factors also affect wood, such as elevated humidity and moisture levels, which may reduce the strength and stiffness of wood.[12]

Masonry

Defects. Masonry units may consist of brick, concrete (CMU), terracotta, and natural stone. Each material possesses unique criteria for design, fabrication, and installation. However, there are some characteristic defects common to all types of masonry systems.

Like concrete, masonry structures are composites; many defects in masonry systems arise from improper attention to the interface of the components. The most common structural application for masonry is exterior wall systems composed of masonry units, mortar, and reinforcement. For an exterior wall system to function, both structural (load-bearing) and environmental (weatherproofing) aspects must be addressed in the design.

Defects in the structural systems of a wall often result in increased stress factors. For example, deficiencies in the masonry units themselves, such as underfired brick, make the units more susceptible to cracking, spalling, and crushing. Improperly selected mortar, for either initial construction or repointing, may lead to crushing of the surrounding masonry due to differences in expansion. Deficient detailing of control and expansion joints may restrain normal movement of masonry and result in elevated internal stresses within the wall. Similarly, the improper detailing or installation of soft joints below shelf angles often results in unintended compressive forces and related distress within a wall. The design and placement of reinforcement and lateral anchorage are also critical in controlling the load paths within a wall system because defects can result in instability or excessive stress (Fig. 1.4). If mortar selected for pointing or repointing is more impervious to water than the masonry units, water does not evaporate as readily, which may lead to accelerated weathering.

FIGURE 1.4   Defective masonry wall tie installation (pintles, not engaging eyes).

Masonry systems are particularly susceptible to distress owing to defects in detailing of the structural and architectural elements. Defects in the weather-resisting systems often result in accelerating or initiating weathering factors and premature failure. Deficiencies in the weather resistance of the masonry units may include fabrication defects such as face-bedding sedimentary stone, porous brick masonry, or deficient terracotta glazing, all of which lead to increased water absorption. Improper detailing and installation of flashing are notorious masonry defects that result in trapped water within the wall system. Additional defects commonly found in masonry walls include improperly designed or missing flashing end dams, insufficient weep holes, mortar droppings that block weep holes, and improperly terminated flashing.

Deterioration. Deterioration in exterior masonry walls usually propagates at an increasing rate owing to weathering factors. Specific weather-related mechanisms include freeze-thaw damage, salt recrystallization, and embedded steel corrosion. Stress factors and incompatibility factors also contribute to accelerating the deterioration of masonry, such as cracking or crushing of masonry at lintels.

Various types of masonry units share similar deterioration mechanisms that include spalling, cracking, and delamination; these often result in increased water infiltration, as well as a reduction in structural integrity. For example, embedded steel elements may corrode in a moist environment, and the corrosion by-products result in material expansion in the form of pack rust. Pack rust may expand several times the thickness of the original steel, and the swelling induces stress in the surrounding masonry units, an effect known as *rust jacking*. In addition, the corrosion of embedded steel structural components results in a reduction in the structural capacity of the exterior wall. In extreme conditions, these deterioration mechanisms working in concert result in failure of the wall and tremendous economic and safety hazards. Cracking and spalling on the face of a brick can result in progressive deterioration from increased water absorption and freeze-thaw damage.

Some mortar deterioration is expected. Mortar may be considered a maintainable component with a design life of approximately 30 years, and it is expected to require condition-based maintenance such as repointing. Mortar acting as a sacrificial element accommodates most of the structural movement and absorbs most of the water. Eventually, the binders within the mortar erode or break down. Mortar deterioration is readily verifiable visually or by using a probe. Generally, loose, friable mortar or mortar that is cracking or separated from the surrounding masonry has reached the end of its service life, but the same conditions also can result from defective original construction.

DURABILITY

Durability is the quality of maintaining satisfactory aesthetic, economic, and functional performance for the design life of a material or system. To evaluate the design life of a material or system, it is first necessary to classify the design life of whole buildings. Table 1.2 describes five building categories based on design life. A selected design life for a building may be self-fulfilling because its materials and details are chosen to have commensurate durability. For example, designers select very different products for a modular retail building with a design life of 20 years than for a high-rise office building with a design life of over 50 years.

Design Life

ASTM E2018-99, "Standard Guide for Property Condition Assessments: Baseline Property Condition Assessment Process," [13] defines the term *expected useful life*

TABLE 1.2 Categories of Building Life[18]

Category Description	Building Life	Examples
1. Temporary accommodation	Up to 5 years	Facilities used during construction, temporary exhibition buildings
2. Short-life building	5 to 30 years	Temporary classrooms, buildings for short-life industrial processes, modular buildings
3. Medium-life building	30 to 60 years	Most industrial buildings, retail and warehouse buildings
4. Normal-life building	At least 60 years	New health and educational buildings, new residential structures
5. Long-life building	60 to 120 years	Major theaters, courthouses, government and institutional high-quality buildings

(equivalent to *expected service life* and *design life* used in this book) as the average amount of time that an item is estimated to function when installed new and assuming that routine maintenance is practiced.

Not every component within a structure is designed to last as long as the structure itself. Economic factors and design constraints dictate that some components and materials act as sacrificial elements (e.g., zinc coatings), some are exposed to more severe weathering (e.g., roofing), and some cannot be manufactured at a reasonable cost to last long term (e.g., paint). Table 1.3 presents examples of the design life of construction components broken into three simple categories. Components with design lives much shorter than the building's life should be identified as *replaceable,* components with design lives equal to or greater than the building's life should be identified as *permanent,* and components that may last with proper treatment on a scheduled or condition-based cycle for the life of the building should be identified as *maintainable.*

It would be useful if all building components were classified according to this type of system so that failures could be understood as "premature" versus "normal." Currently, designers, manufacturers, and owners disagree, sometimes using all three classifications for one component. Categorizing all building products and materials affects a variety of special interests and remains a complex undertaking.

Determining the influence of a product or system on the operation of the entire structure may be useful in estimating its economically appropriate service life. White[14]

TABLE 1.3 Categories of Component Life[18]

Category Description	Life	Examples
1. Replaceable	Shorter life than building life; replacement can be envisioned during initial design	Floor finishes, mechanical components, roof membranes
2. Maintainable	Will last, with periodic maintenance, for the life of the building	Exterior wall systems, doors, and windows
3. Permanent	Will last for the entire life of the building with little or no maintenance	Foundations and main structural elements

developed an approach using four levels of operational influence of a system to evaluate its service life:

Highly critical. Failure would cause the use of the facility to stop during repairs.

Critical. Failure would reduce efficiency, and repairs would have to be performed outside of normal operating hours.

Not critical. Failure would require remedial work, but it may be scheduled for convenience.

Not affecting the structure. Failure (typically of ancillary components) would not affect operations.

Schmalz and Steimer[15] presented a classification method (after Stillman[16]) in which failure modes are classified into six levels according to safety and economic factors:

1. Danger to life
2. Danger to health
3. Costly repair
4. Frequent repair repeat
5. Interruption of building use
6. No exceptional problems

To both of these classifications one could add another important criterion—whether deferring repairs would result in greater cost later, not due to inflation but rather due to an increasing-rate mechanism. For example, postponing repair of freeze-thaw damage can greatly expand the scope of future remedial work. Postponing remedy of the ponding condition shown in Fig. 1.5 will lead to premature failure of the roof membrane, requiring expensive roof replacement.

Most building components require at least maintenance, if not replacement, during the service life of a building under normal environmental loads. The "weakest link in the chain" may compromise all others if not addressed properly. For example, neglecting to replace deteriorated sealant joints may lead to the premature failure of adjacent components, such as masonry freeze-thaw damage, wood rot, and structural steel corrosion.

Systems whose failure would pose a threat to public safety or building use require a longer design life than those that pose little safety or economic threat. Table 1.4 lists the typical design life of representative building components, compiled from several publications.[15,17,25]

Service Life

Determining the durability of components exposed to weather, pollution, natural disasters, human error, and the general wear and tear from use is a complex task. The service life of a structure also depends on human factors, such as the changing needs of property owners and communities, changing aesthetic values, and changing economic and political climates. Consequently, the design life of a structure and its actual service life both vary and may not be defined completely.

Whether a component is designed and classified to last only a short time or for the entire life of the building, *premature failure* occurs when the actual service life is shorter than reasonably expected. Premature failures typically are caused by defects or improper maintenance that allows for uncontrolled deterioration.

ASTM E2018-99, "Standard Guide for Property Condition Assessments: Baseline Property Condition Assessment Process,"[13] defines the term *remaining useful life* as

FIGURE 1.5 Severe ponding on a roof, which inevitably will lead to membrane failure if not remedied.

TABLE 1.4 Examples of Building Component Design Life[15,17,25]

Building System	Typical Design Life (in years)
Foundation	>100
Structure	>100
Exposed parking decks	30
Brick units in masonry	>100
Mortar in masonry	25
Wood siding	20–40
Doors	25
Windows	20–40
Asphalt shingles	15–30
Roof membrane	10–30
Finishes	7–20
Floor covering	5–10
Suspended ceilings	10–20
Concealed flashing	40

a subjective estimate based on observations, average estimates of similar items, or a combination thereof, of the remaining time that an item, component, or system is able to function in accordance with its intended purpose before needing replacement. Such period of time is affected by the initial quality of the item, the quality of the initial installation, the quality and amount of preventive maintenance exercised, climatic conditions, extent of use, etc.

Estimating Durability and Design Life

At present, there is no standardized measure of the actual durability of most construction products; therefore, a professional performing a condition assessment must use a combination of personal experience and published literature to evaluate material failures in the field. Information regarding the design life and durability of building products can be obtained from a variety of sources, including professional and technical society guidelines, manufacturers' technical literature, published test results from recognized laboratories, published case histories and journal articles, and experience of the designer and his or her colleagues.

Standards. Professional and technical societies often provide unbiased industry standards written by committee and published with the consensus of designers, manufacturers, and contractors. Such standards improve communication between producers and consumers regarding the suitability and durability of building components. Using published standards, different claims made by manufacturers may be evaluated with more confidence.

Recent research regarding the service life of buildings indicates the need for more standards regarding the durability of materials, components, and systems. Manufacturers often provide conflicting and unclear information regarding the durability of the materials and components they market. Further, the more complex interaction of several products within a system also must be considered and possibly standardized to ensure reliable performance and to facilitate identification of deficiencies.

In 1988, the European Union stated the goal of increasing public health through better building practices. This EU directive motivated the British Standards Institute (BSI) to draft a standard, "Durability Requirements for Products,"[18] presenting categories of design life for various construction systems. Some results of the BSI study are included in the tables of this chapter.

Similarly, ASTM International has published several diverse standards in recent years regarding the durability of building materials and products, including

ASTM E2136[19]: "Standard Guide for Specifying and Evaluating Performance of Single Family Attached and Detached Dwellings—Durability."

ASTM E2094[20]: "Standard Practice for Evaluating the Service Life of Chromogenic Glazings."

ASTM E774[21]: "Standard Specification for the Classification of the Durability of Sealed Insulating Glass Units."

ASTM F793[22]: "Standard Classification of Wallcovering by Durability Characteristics."

Manufacturer's Technical Literature. Unfortunately, many building products continue to lack clear standards for durability. Information provided by manufacturers is a blend of marketing and science, often with no clear distinction between the two. Many manufacturers do not address product design life directly in their literature; they may only mention how long the product line has been on the market. Sometimes the only indicator of design life is the guarantee period provided by the manufacturer.

However, guarantees usually limit the liability of the manufacturer in the case of premature failure of the component. Liability is often difficult to isolate when products fail because proper product selection by the designer and proper installation by the contractor are beyond the control of the manufacturer. This sharing of liability often reduces the direct accountability of the manufacturer, and the written guarantees reflect these limits.

Special attention should be paid to the language in the manufacturer's literature to ensure that the expectations of the owner can be met. Some owners may assume mistakenly that a guarantee is a maintenance agreement or insurance policy. In fact, most guarantees severely restrict the legal remedies available to the owner and do not cover incidental or consequential damages, loss of profits, or inconvenience.

Manufacturers' guarantee periods alone do not always provide reliable information regarding product durability. They are established by manufacturers based on several factors, including the length of time a product has been on the market (longer with proven success), longer durability from product improvements, the guarantee periods set by competitors, and the product's performance under accelerated wear and weathering tests.

Accelerated Testing. Durability often is quantified by the performance of a material under an accelerated weathering test. In such a test, the material of interest is subjected to magnified environmental loads. These harsh conditions applied over a relatively short test period are extrapolated to predict the behavior of the material under normal weathering for a longer period of time. The *acceleration factor* is the ratio between the test period and the associated real service time.[3] Therefore, the exposure time until material failure during an accelerated weathering test multiplied by the acceleration factor yields the theoretical design life of the material.

While such data may be very useful in predicting the durability of components, accelerated testing remains highly subjective. Testing may not reproduce all the ancillary environmental loads contributing to deterioration of a material, such as the effects of atmospheric pollution or biologic attack. In addition, there is no industry consensus on the true acceleration factor for the most commonly used artificial weathering devices.

Even given reliable acceleration factors for such equipment, there remains variation in the repeatability and reproducibility of accelerated weathering tests. In addition, many standardized test methods do not set criteria to determine when the test subject has "failed"; they only recommend a standard practice for conducting the testing. Therefore, several precautions are necessary to produce meaningful results using accelerated weathering techniques.

A reliable accelerated testing program would correlate the results of laboratory testing to conditions observed in the field or to a control specimen exposed to non-accelerated weathering.[3] It is also important to define the scope of the testing, with more limited or targeted testing likely to achieve more meaningful results. For example, some chemical product manufacturers will perform testing to determine whether their product is likely to stain specific adjacent materials at the site.

An example of a broader and more difficult scope of testing may be to determine the durability of a specific paint for use on the exterior of a building. The substrate, environmental exposure, and watertightness of the walls comprising the painted surfaces will affect the performance of the product in service. These additional factors must be considered in developing a meaningful accelerated weathering test.

The various combinations of different degradation factors must be taken into account so that laboratory test results can represent adequately the environmental loads that govern the in-service performance of a material. Identifying the factors most likely to affect a building component for a given installation may serve to exclude certain products from use prior to testing. The nature of accelerated testing is such that only one or two environmental loads are magnified at a time. Therefore, complex interactions affecting a material may not be captured adequately in a laboratory test alone.

According to ASTM E632,[3] the following steps are necessary to produce meaningful results from accelerated testing:

1. Identify the attributes and properties critical to the service life of a material or component.
2. Identify the type and range of degradation factors affecting the test subject in service.
3. Identify likely failure mechanisms.
4. Select various test methods.
5. Define performance requirements and the scope of the testing program.
6. Perform pretesting to refine the test methodology.
7. Perform the predictive service life test.
8. Correlate accelerated test data with in-service test data or observations.

Human Perceptions of Durability

Another complication to predicting durability is that there is no industry consensus on what constitutes the end of service, or failure point, for many construction products. For example, some exterior seals begin degrading owing to environmental exposure soon after they are installed; even with proper design and installation, they degrade continuously until at some arbitrary point it is determined that they must be replaced. Before replacement is implemented, some owners will allow the seals to degrade further than other owners, perhaps allowing water infiltration to become pervasive before replacing the seals. An owner's definition of the end of service life may depend on internal policies and tolerance for deterioration.

In addition, expectations for durability vary between communities, and certain prevailing weather conditions affect durability as well. For example, solar radiation causes more rapid deterioration of components in the southern regions of the United States than in the northern regions; therefore, faded paint may be more accepted in some southern communities, whereas such deterioration may be dubbed a failure in the North. The quality of "average" and "acceptable" workmanship and building condition varies between communities as well.

Economics play a critical role in the durability of structures. Most owners consider the design life in planning capital improvement programs, which factor into the calculation of life-cycle costs. On purely economic criteria, the lowest life-cycle cost is usually the best choice among multiple options. Owners often choose to implement capital projects with short life expectancies simply because they lack one important piece of information: They do not have a basis for evaluating actual product durability.

Improvements in Durability

History has proven that if a building is designed, constructed, and maintained properly, it may remain in service for centuries or longer. Unfortunately, deferred maintenance and defects in design, fabrication, and construction often lead to premature failures and shortened service lives for many buildings throughout the United States.

According to the U.S. Census Bureau,[23] over $13.6 billion was spent in 1992 to repair private, nonresidential buildings less than 25 years old, including $7.3 billion spent on buildings less than 15 years old. An additional $1.11 billion was spent in demolition. When comparing these figures with the reported total of $90.6 billion spent on new construction, demolition and repair of relatively young buildings represent a significant portion.

The quality of modern American design and construction was addressed in a report published by the U.S. government in November 1995 entitled, *National Planning for Construction and Building R&D.*[24] According to this report, building defects and uncontrolled deterioration often correlate with poor indoor air quality and occupant discomfort. In addition to health consequences, the U.S. report estimated that half the world's energy is spent in the construction and operation of buildings; therefore, improving the durability and service life of structures requires attention. Several of the cited goals focus on the elimination of design and construction defects that often lead to poor serviceability, poor indoor air quality, and premature failure of materials, components, and systems. In addition, proper maintenance, repair, and rehabilitation of existing construction (as opposed to demolition) were identified in the report as a viable means of achieving less pollution and waste. The report noted that construction waste from demolition and new construction occupies 20% to 30% of all landfill space.

MAINTENANCE

Even without the presence of design, fabrication, or construction defects, some components will require maintenance to control natural deterioration. The term *maintenance* has been used historically to describe vastly different procedures, including repairing concrete spalls on precast concrete cladding panels, replacing sealant joints, and underpinning a foundation. Of these examples, only repairing concrete spalls is the type of intervention that is envisioned when a component is termed *maintainable*.

It is reasonable that it will be necessary during the design life of a concrete facade to perform some condition-based maintenance. Underpinning, however, is necessary only when the foundation, a "permanent" component, suffers premature failure. Sealants are not maintainable components and instead are categorized as "replaceable" at the end of their service life. These different operations imply different uses of the word *maintenance*. Therefore, it is useful to define categories of reasonable maintenance types. BSI[18] defined three categories of maintenance, as shown in Table 1.5.

When maintenance is not performed as required by the condition of components, it is commonly termed *deferred* maintenance. Uncontrolled deterioration and premature failure often result from such deferment, which then becomes neglect.

ACCEPTING UNDESIRABLE EXISTING CONDITIONS

In the design of new structures, prescriptive codes and specifications set minimum standards for design. When the condition of existing structures is assessed it is not always feasible (or even technically sound) to apply design equations or design stan-

TABLE 1.5 Maintenance Levels[18]

Level Description	Scope	Examples
1. Corrective maintenance	Maintenance restricted to restoring items to their original function after a failure	Reglazing broken windows, remedying sources of active leaks
2. Scheduled maintenance	Maintenance at predetermined time intervals, regular cycle	Repainting, sealant replacement, recoating roof membrane with solar reflective paint
3. Condition-based maintenance	Maintenance as a result of distress	Tuck-pointing masonry, replacing roof membrane

dards directly. Because building codes generally do not provide guidance on the analysis of existing structures, it is often necessary to apply engineering judgment to determine whether an existing condition is acceptable, safe, or serviceable.

When engineers are asked about comfort level regarding a calculated overload condition on an existing structure, many consider 5% to 10% overstress acceptable. They understand that a factor of safety was built into the original design, that there are practical limits to the accuracy of such calculations, that the actual live and dead loads may be well defined (as opposed to the estimates used before construction), and that being overly conservative could result in substantial economic waste (unnecessarily repairing structures that are expected to function satisfactorily). However, no published code recognizes this distinction between design calculations for new construction and analysis of existing components; an engineer may be accused of negligence for accepting such an overstress condition owing to a lack of published support for the rationale, despite sound engineering judgment.

A new code currently in development, the International Existing Building Code (IEBC), was intended originally to become the primary code for work in existing buildings, and provisions regulating repair, alterations, additions, and change of occupancy in the International Building Code (IBC) were to be deleted. It remains to be seen whether the final edition will recognize that the criteria for new construction may be overly conservative for evaluating existing construction.

It is even more difficult to standardize acceptance of distressed conditions. Design rules are based on the behavior of new materials in their original configuration. For example, while concrete design requires the engineer to consider time-dependent effects such as creep, the code equations do not take into account reduction in strength or service life owing to deterioration or delamination.

There are many questions involved in accepting undesirable existing conditions. Is the public served by remedying an existing condition to strict compliance with design standards and building codes? Does it pose an unreasonable financial burden to improve a structure that may be "good enough" but is not in strict compliance? Does the current condition of the structural element or system fall within the assumptions used to develop the code equation? In particular, the last question requires that the engineer not only understand the basis for the code equation but also use substantial judgment in determining whether the existing conditions are compatible with those assumptions.

GLOSSARY OF TERMS

Biologic degradation factor. A degradation factor directly associated with living organisms, including microorganisms, fungi, bacteria, and insects.[3]

Catastrophic. Description of an unintended event resulting in loss of life, severe personal injury, or substantial property damage.

Collapse. A structural failure resulting in the total destruction of all or a portion of a structural system, with consequential damage to the nonstructural systems.

Damage. Distress to property (including building components) caused by a failure.

Deficient. Lacking some desirable element or characteristic. Deficient is used generally to describe design, fabrication, installation, and/or deterioration resulting in a defect.

Defect. The nonconformity of a component with a standard or specified characteristic. Defect is used sometimes as a synonym for *failure,* but the preferred meaning is to indicate only a deviation from some (perceived) standard that may, but will not necessarily, result in a failure. See also *Latent defect.*

Degradation. The lowering of a material's characteristics (such as strength or integrity). Similar to decomposition but not necessarily implying an organic process. Also similar to deterioration but not necessarily time-dependent.

Degradation factor. An external factor that adversely affects the performance of building components and materials, including weathering, biologic, stress, incompatibility, and use factors.[3] See also *Environmental load*.

Design life. The period of time after installation during which all properties of a material, component, or system *are intended or expected* to exceed the minimum acceptable values when maintained routinely.

Deterioration. The gradual adverse loss of physical or chemical properties of a material.

Distress. The individual or collective physical manifestations of a failure as perceivable problems, such as cracks, spalls, staining, or leakage.

Durability. The quality of maintaining satisfactory aesthetic, economic, and functional performance for the design life of a material or system.

Failure. An unacceptable difference between expected and observed performance; also, the termination of the ability of an item or system to perform an intended or required function. Most failures are not catastrophic.

Failure mechanism. An identifiable phenomenon that describes the process or defect by which an item or system suffers a particular type of failure.

Failure mode. A description of the general type of failure experienced by a system. A broader term than failure mechanism, encompassing fundamental behavior such as shear, tension, etc.

Flaw. A relatively small imperfection in a material or component. Note, however, that even small flaws can cause catastrophes if they occur in critical areas.

Hazard. An attribute of a component or system that presents a threat of harm, injury, damage, or loss to person or property.

Incompatibility factor. A degradation factor resulting from detrimental chemical and physical interactions between materials and components.

Maintainable. A material or component that may last with proper treatment on a scheduled or condition-based cycle for the service life of a building or structure.

Natural disaster. One of the following generally recognized phenomena of nature resulting in destruction of property and/or personal injury: fire, flood, hurricane, major storm, tornado, hail, earthquake, avalanche, or blizzard.

Permanent. Material or component with a design service life equal to that of the building or structure.

Premature failure. Failure of a material, component, or system prior to the end of its design life.

Progressive collapse. The collapse of multiple bays or floors of a structure resulting from an isolated structural failure owing to a chain reaction or "domino effect."

Rehabilitate. Extensive maintenance intended to bring a property or building up to current acceptable condition, often involving improvements.

Renovate. Make new; remodel.

Repair. To restore an item to an acceptable condition by the renewal, replacement, or mending of distressed parts.

Replaceable. Easily exchanged building components or equipment. Usually, replaceable components have a design life that is shorter than the life of the structure they occupy.

Restore. To bring an item back to its original appearance or state.

Service life. The period of time after installation during which all properties of a material, component, or system *actually* exceed the minimum acceptable values when maintained routinely.

Sound. Good structural condition.

Stable. In a state of stable equilibrium, i.e., resistant to buckling or stability failure. Stable should not be used as a synonym for *good condition.*

Stabilize. The act or process of returning a material, component, system, or structure to a stable condition.

Stress factor. A degradation factor resulting from loads on a system, either sustained or periodic.[3]

Use factor. A degradation factor resulting from design, installation, maintenance, wear and tear, and user abuse.[3]

Weathering factor. A degradation factor associated with the natural environment; see also *Environmental load.*[3]

REFERENCES

1. AISC (2000). *Code of Standard Practice for Steel Buildings and Bridges,* American Institute of Steel Construction, Inc., Chicago.

2. Nicastro, David H., ed. (1998). *Failure Mechanisms in Building Construction,* ASCE Press, New York.

3. ASTM E632 (1996). "Standard Practice for Developing Accelerated Tests to Aid Prediction of the Service Life of Building Components and Materials," in *Annual Book of ASTM Standards,* ASTM International, West Conshohocken, PA.

4. SEI/ASCE (1999). ASCE 11-99, *Guideline for Structural Condition Assessment of Existing Buildings,* ASCE Press, New York.

5. Busel, J. P., and Barno, D. (1996). "Composites Extend Life of Concrete Structures," *Composites Design and Application,* **Winter,** pp. 12–14.

6. ACI (2003). *Manual of Concrete Practice,* American Concrete Institute, Farmington Hills, MI.

7. Mindless, Sidney, J. Francis Young, and David Darwin (2003). "Durability," in *Concrete,* 2d ed., Prentice-Hall, Upper Saddle River, NJ, pp. 477–514.

8. AISC (1999). *Commentary on the Load and Resistance Factor Design Specification for Steel Buildings,* 3d ed., American Institute of Steel Construction, Inc., Chicago.

9. Breyer, Donald E. (1993). *Design of Wood Structures,* 3d ed., McGraw-Hill, New York.

10. Marotta, Theodore W., and Charles A. Herubin (1997). *Basic Construction Materials,* 5th ed., Prentice-Hall, Upper Saddle River, NJ.

11. Lai, James, S. (1999). *Materials of Construction,* 2d ed., Kendall/Hunt, Dubuque, IA.

12. Weaver, Martin E. (1997). *Conserving Buildings Guide to Techniques and Materials,* rev. ed., Wiley, New York.

13. ASTM (1999). ASTM Designation E2018-99, "Standard Guide for Property Condition Assessments: Baseline Property Condition Assessment Process," in *Annual Book of ASTM Standards,* ASTM International, West Conshohocken, PA.

14. White, K. H. (1992). "The Client's View: The Public Sector," in G. Somerville, *The Design Life of Structures,* Blackie and Sons, Bishopbriggs, Glasgow, UK, pp. 30–39.

15. Schmalz, T. C., and S. F. Stiemer (1995). "Consideration of Design Life of Structures," *Journal of Performance of Constructed Facilities,* American Society of Civil Engineering, 9(3):206–219.

16. Stillman, J. (1992). "Design Life and the New Code," in G. Sommerville, *The Design Life of Structures,* Blackie and Sons, Glasglow, UK, pp. 3–8.

17. Brick Industry Association Technical Notes on Brick Construction Number 7F. "Moisture Resistance of Brick Masonry—Maintenance," BIA, Reston, VA.

18. British Standards Institute (1989). Document 90/13347, "Draft: Construction Standardisation, Guidance on Drafting and Presentation, Durability Requirements for Products," BSI, London, UK.

19. ASTM (2001). ASTM Designation E2136, "Standard Guide for Specifying and Evaluating Performance of Single Family Attached and Detached Dwellings—Durability," in *Annual Book of ASTM Standards*, ASTM International, West Conshohocken, PA.

20. ASTM (2000). ASTM Designation E2094, "Standard Practice for Evaluating the Service Life of Chromogenic Glazings," in *Annual Book of ASTM Standards*, ASTM International, West Conshohocken, PA.

21. ASTM (1997). ASTM Designation E774, "Standard Specification for the Classification of the Durability of Insulating Glass Units," in *Annual Book of ASTM Standards*, ASTM International, West Conshohocken, PA.

22. ASTM (1998). ASTM Designation F793, "Standard Classification of Wallcovering by Durability Characteristics," in *Annual Book of ASTM Standards*, ASTM International, West Conshohocken, PA.

23. U.S. Bureau of the Census (1992). *Manufacturing and Construction Division—Census of Construction Industries;* U.S. Census Web site: *http://www.census.gov/ftp/pub/const/www /cci/fintro.html.*

24. Wright, R. N., A. H. Rosenfeld, and A. J. Fowell (1995). *NISTIR 5759: National Planning for Construction and Building R&D*, United States Department of Commerce, National Institute of Standards and Technology.

25. ASTM (2003). ASTM Designation WK327, "Standard Guide for Design and Construction of Low-Rise Frame Building Wall Systems to Resist Damage Caused by Intrusion of Water Originating as Precipitation," draft standard under development, ASTM International, West Conshohocken, PA.

CHAPTER 2

Reliability-Based
Condition Assessment

ALLEN C. ESTES, Ph.D., P.E., and DAN M. FRANGOPOL, D.Sc., Dr.h.c.

INTRODUCTION

Reliability-based methods of design and analysis have gained increasing acceptance in academic circles and are beginning to be acknowledged and used by engineer practitioners. Reliability methods take a probabilistic approach to designing a structure in which the result is a reliability index or a probability of failure rather than a factor of safety. In structural design, critical factors such as loads, resistances, deterioration models, and human errors are highly random, and the associated uncertainties must be quantified to ensure the safety of the public.

The factor of safety, which is commonly found in methods such as allowable stress design (ASD), is an empirical attempt, usually based on past experience, to quantify the global effect of all sources of uncertainty associated with a particular situation.

Such an approach is computationally less difficult than a reliability-based approach but makes it impossible to ensure a uniform level of safety across a variety of structures. In general, the requisite safety in ASD usually is achieved through overdesign. However, dangerous underdesign situations also are encountered using the ASD approach.

The load factor resistance design (LRFD) is the beginning of a probability-based design code. For a given failure mode, a structure survives (i.e., continues to perform as it was originally intended) as long as the capacity is greater than or equal to the demand. This requirement is expressed as

$$\phi_n R_n \geq \sum \gamma_i L_i$$

where the left-hand side represents the strength, resistance, or capacity of the structure with respect to the occurrence of a given failure mode, and the right-hand side is the load or demand associated with this failure mode. The nominal resistance is R_n, and the various load effects (caused by live load, dead load, wind load, etc.) are L_i. The factors associated with the loads, γ_i, and with the ability of the structure to resist those loads ϕ_n, are quantified separately. The load and resistance factors are calibrated to achieve a given level of safety. In developing the American Concrete Institute (ACI) code for example, the factors were based on a statistical model that assumed a $1/100$ chance of the material being under strength occurring at the same time as an overload characterized by a $1/1000$ chance of occurring.[1] This method of design allows more consistency of safety for various components and systems.

Reliability-based methods quantify the design uncertainties to a greater detail. The limit-state equation associated with any failure mode is similar to that used in LRFD but without the load or resistance factors. The variables associated with the capacity and demand are random. The yield stress for steel in ASD might be a deterministic quantity of 36 ksi (24.82 kN/cm²).[2] In the reliability-based analysis, the yield stress for steel is more likely defined as a normally, or log-normally, distributed random variable with a mean of 40.3 ksi (27.79 kN/cm²) and a standard deviation of 3.9 ksi (2.69 kN/cm²). When the parameters of all the random variables are considered, the result is a probability of failure (i.e., the probability that the demand will exceed the capacity of the structure). Structures are then designed such that the failure probability presents an acceptable level of risk that is typically a tradeoff between the cost of building and maintaining the structure versus the expected cost of a failure. The most common way of expressing structural reliability is in terms of a reliability index β, where a higher β value indicates a lower probability of failure and, consequently, a safer structure.

Reliability methods are computationally more difficult and complex than traditional deterministic methods. Such methods have only become practical as a result of the huge progress in computer methods and technology over the past two decades. In their complete form, reliability methods often involve complex convolution integrals that have no closed-form solution. Simplified methods that make first- and second-order approximations have been highly successful at reducing the complexity of computation while still producing accurate results. Although it often requires a large number of simulations to obtain good solutions, Monte Carlo methods have produced excellent results.

A structure can be analyzed based on the reliability of a single component or as a system of components. In both cases, the consideration of correlation between random variables is critical to obtaining accurate results, and in the system analysis, the correlation between failure modes is also very important. Reliability methods are often used to optimize the life-cycle cost of a structure and to make future maintenance and repair decisions. This involves time-dependent reliability considerations and the prediction of how a structure will behave over time. Models are needed to predict how loads and resistances will change over time. Deterioration models predict how the structure will weaken owing to such mechanisms as corrosion, scour, fatigue, etc.

These models need to be described in probabilistic terms in order to compute reliability over time and often predict behavior over several decades. The only way to responsibly manage these structures is to periodically inspect the structures, update the various models, and reassess the reliabilities.The inspections are typically either nondestructive evaluations that provide information on a specific defect or periodic visual inspections in which the structural condition is inferred from the data that are available.

The most important drawback to reliability methods is the amount of input data needed to perform a valid analysis. The most rigorous option is to conduct tests to obtain all the input data needed for a specific project, such as strength tests of concrete, traffic surveys on a bridge, corrosion rate tests on steel, storm data analysis at the project site, etc. This is usually prohibitively expensive in terms of cost and time. Past experience and previous studies in the literature are a less costly source of data, but the results may not be applicable to the project at hand. Sensitivity analyses on the respective variables often will help to identify which variables merit the most scrutiny. Unfortunately, reliability results are only as good (or bad) as the input data that support them.

Nevertheless, reliability methods will only increase in usage and prominence as they become more standardized and accepted in the engineering community. Assessing structural condition at a point in time and using that information to forecast future performance in probabilistic terms will only continue to increase in importance.

STRUCTURAL RELIABILITY

No discussion of reliability-based condition assessment would be complete without a description of the underlying theory of structural reliability. For the engineer practitioner, this theory often will be embedded in the software used to solve a particular problem. While it may appear overwhelming, this section is intended to provide an explanation as to why condition assessment needs to be expressed in probabilistic terms and to provide context and background for the case studies that follow.

A reliability analysis begins with a limit-state equation or series of limit-state equations that govern the behavior of the structure. A structure is considered safe or reliable if its capacity R exceeds the demand L placed on it:

$$R > L \quad \text{or} \quad R - L > 0 \quad \text{or} \quad \frac{R}{L} > 1$$

The reliability of a structure p_s is the probability that the structure survives or performs safely. The computation of p_s can be quite complex depending on the number and type of uncertainties and the number of variables that comprise R and L.

Random Variables

The resistance and the load may consist of many variables and many sources of uncertainty associated with them. The elastic moment capacity M_R of a simple beam can be expressed as

$$M_R = \frac{\sigma_y I}{c}$$

where σ_Y = yield stress
I = moment of inertia
c = distance from the neutral axis to the outer beam fiber

The yield stress has uncertainty based on the quality of the beam material. Both c and I depend on dimensions of the cross section and may be random or deterministic variables. There is also uncertainty associated with the ability of the elastic stress equation to approximate real behavior. In addition, there may even be uncertainty associated with human error that occurs during construction of the beam, although this is much more difficult to estimate.

Often the uncertainty can be quantified by data. If 50 or 100 material samples were tested, the results could be plotted in terms of a histogram or a frequency distribution. Statistical parameters can be estimated directly from the data. Quite often such data can be approximated by known theoretical distributions such as the normal, lognormal, and exponential distributions. Statistical tests such as the chi-square test or the Kolmogorov-Smirnov test are available to evaluate how well the data fit a particular theoretical distribution. Random variables can be either continuous or discrete depending on the physical nature of the phenomenon being described. Quantifying random variables can be quite costly, and using the results from similar studies often is desirable. A sensitivity analysis with respect to changes in the main descriptors (e.g., mean values, standard deviations) of random variables will help to identify the relative importance of various descriptors to the structure's reliability and help to evaluate whether the time and effort are justified.

When more than one random variable appears in a limit-state equation, all uncertainties are considered jointly, and the covariance between pairs of random variables is important. In general, the covariance is a measure of the degree to which a linear interrelationship exists between two random variables. The normalized covariance or correlation coefficient between these random variables ρ ranges from a value of $+1$ to -1. At these extreme values, the two random variables are linearly related. Two variables that are statistically independent have no correlation, $\rho = 0$. The magnitude of the correlation coefficient can be computed from individual data points when they exist, but more often it is estimated based on past experience and engineering judgment.

Component Reliability (Single Limit State)

The problem considered in this section involves the analysis of a component with respect to a single failure mode, defined by a single limit state or failure mode.

General. If the capacity R and the demand L are random, and the uncertainties can be quantified, then the reliability or probability of safe performance p_s can be expressed as

$$p_s = P(R - L > 0) = \iint\limits_{R>L} f_{R,L}(r,l)\,drdl$$

where $f_R(r)$ and $f_L(l)$ are the probability density functions of R and L, respectively, and $f_{R,L}(r,l)$ is their joint probability density function.

The capacity R and the demand L may be functions of many other random variables. The generalized structural reliability problem is formulated in terms of a vector of basic random variables $\mathbf{X} = \{X_1, X_2, \ldots, X_n\}^T$, where $X_1, X_2, \ldots, X_n$ are basic random variables that may describe loads, structural component dimensions, material characteristics, and model uncertainties. A limit-state function $g(\mathbf{X}) = 0$ describes the performance of the component in terms of the basic random variables $\mathbf{X}$ and defines the failure surface that separates the survival region from the failure region. If the joint probability distribution of the design variables $X_1, X_2, \ldots, X_n$ is $f_{X_1,X_2,\ldots,X_n}(x_1, x_2, \ldots, x_n)$, the probability of the safe state is[3]

$$p_s = \int \cdots \int_{g(\mathbf{X})>0} f_{X_1,X_2, \cdots, x_n(x_1,x_2, \cdots, x_n)} dx_1 \cdots dx_n$$

which also may be written as

$$p_s = \int_{g(\mathbf{X})>0} f_x(x)dx$$

which represents the volume integral of $f_X(x)$ over the safe region $g(\mathbf{X}) > 0$. The integral describing the joint distribution in the region $g(\mathbf{X}) > 0$ easily can become too complex to solve in either closed form or using numerical methods. Fortunately, effective approximate techniques have been developed.

FORM/SORM. Even if the complex integrals involving joint probability density distributions could be solved easily, there is often not enough information available about the individual distributions or joint distributions to make the effort worthwhile. There are several second-moment formulations, meaning that they rely only on the means and standard deviations of distributions to obtain a solution, that produce good results in most cases. The analytical and approximate probability integration methods such as first- and second-order reliability methods typically involve converting all distributions to equivalent standard normal distributions at a given point, obtaining the required set of uncorrelated (transformed) variables, and defining the reliability index as the shortest distance from the origin to the failure surface in the reduced standard normal space. In first-order reliability methods (FORM), the limit-state function at the design point is linearized. Initial values are chosen for the random variables (e.g., their mean values), and iterative gradient search techniques are used to find this shortest distance in n-dimensional space, where n represents the number of random variables considered in the analysis.

FORM solutions are quite accurate for many problems. For problems which have a high degree of nonlinearity, it is often necessary to approximate limit-state equations using quadratic formulations, which are the basis for second-order reliability methods (SORM). In both types of approximate methods, a computer is required to develop a solution, and there are numerous software programs available such as CALREL,[4] COSSAN,[5] PROBAN,[6] RELSYS,[7] STRUREL,[8] and NESSUS,[9] to name just a few. In most cases the user simply needs to define the limit-state equation and the parameters associated with the random variables.

Monte Carlo Simulation. Monte Carlo simulation is an effective brute-force or intelligent way to avoid the complex mathematical calculations discussed previously. Monte Carlo simulation generates random numbers according to the assumed probability distributions of the variables that appear in the limit-state equation and evaluates the limit-state equation a sufficient number of times to generate an approximate answer. For a given limit-state equation $g(\mathbf{X}) = 0$, the probability of failure p_f is calculated as

$$p_f = \frac{n_f}{n}$$

where n is the number of times a simulation is run, and n_f is the number of times that $g(\mathbf{X}) < 0$. The biggest drawback to Monte Carlo simulation is the large number of simulations required to obtain a valid result. The amount of error is reduced as the number of simulations increases. The percent error is computed as follows[3]:

$$\% \text{ error} = 200\sqrt{\frac{(1 - p_f)}{n(p_f)}}$$

Structures are deliberately designed so that failure is a rare event, and consequently, p_f is very small. It is therefore common that millions and sometimes billions of simulations are required for each reliability calculation. Monte Carlo simulation remains a very effective option for computing reliability for complex situations or even determining the distribution of a random variable that is a function of many other random variables. Of course, there are different suggested possibilities of intelligent Monte Carlo simulation using faster procedures.[10–13]

Reliability Index. The most common means of communicating reliability is through a reliability index β, which is defined as the shortest distance from the origin to the limit-state surface $g(\mathbf{X}) = 0$ in standard normal space.

$$\beta = \frac{\mu_R - \mu_L}{\sqrt{\sigma_R^2 + \sigma_L^2}}$$

where μ is the mean value and σ is the standard deviation of the variables R and L. In the case where R and L are independent, normally distributed variables, the reliability index can be equated to the probability of failure p_f as follows:

$$p_f = \Phi(-\beta)$$

where Φ is the distribution function of the standard normal variate. In this case, Table 2.1 shows the relationship between reliability index and probability of failure. When the variables are not normally or log-normally distributed or the limit-state function is not linear, the reliability index cannot be related directly to the probability of failure, but it remains a highly useful means of communicating the notional level of reliability of a design.

Time-Dependent Reliability. The methods just described apply to computing the reliability with respect to a single failure mode at a specific point in time. When attempting to make decisions about a structure over its useful life, time becomes an important variable. If the load and resistance of the structure can be projected for the future, the simplest approach is a point-in-time method in which the reliability is computed at various specific times in the future. A trend is established, and the structure is planned for a repair when the reliability falls below an acceptable target reliability level. Loads tend to increase over time, and the resistance tends to decrease as

TABLE 2.1 **Relationship Between Reliability Index and Probability of Failure for Normally Distributed Variables and Linear Limit-State Functions**

Reliability Index (β)	Probability of Failure (p_f)
0.0	0.5000
1.0	0.1587
2.0	0.02275
3.0	0.00135
4.0	0.0000316
5.0	0.000000286

the structure deteriorates, so the overall reliability generally can be expected to decrease over time. The weakness of this approach is that it fails to account for previous structural performance.

A more accepted, albeit a much more complex, approach to time-dependent reliability is to compute the probability that a structure will perform satisfactorily for a specified period of time. Whereas reliability is defined as the probability that an element is safe at one particular time, the survivor function $S(t)$ defines the probability that an element is safe at any time t:

$$S(t) = P(T \geq t) = p_S(t)$$

where the random variable T represents time, and $t \geq 0$. The probability that a failure $p_f(t)$ takes place over a time interval Δt is expressed as

$$f(t)\Delta t = P\{t_1 < t \leq t_1 + \Delta t\}$$

where the probability density function is $f(t) = -S'(t)$. It is assumed that the derivative $S'(t)$ exists.

The reliability is often expressed in terms of a hazard function $H(t)$, also called the *conditional failure rate*. The hazard function expresses the likelihood of failure in the time interval t_1 to $t_1 + dt$ given that the failure has not already occurred prior to t_1 and can be expressed as

$$H(t) = \frac{f(t)}{p_s(t)} = -\frac{S'(t)}{S(t)}$$

Thus the hazard function is the ratio of $f(t)$ to the survivor function $S(t)$. All hazard functions must satisfy the nonnegativity requirement. Their units typically are given in failures per unit time. Large and small values of $H(t)$ indicate great and small risks, respectively.[14]

System Reliability (Multiple Limit States)

A structural system may have multiple components and/or failure modes. Many advantages are gained by quantifying the interrelationship between these components and analyzing a structure as an entire system. For example, a system analysis can reveal that some repairs are more important than others. It also may indicate that while each individual component of a structure may have adequate safety, the structure as a whole still may be unsafe.

Series Systems. If the failure of any single component will lead to failure of the entire structure, the system is considered a series, or weakest link, system. If a structural system is treated as a series system of z elements, the probability of failure of the system $p_{f,\text{series}}$ can be written as the probability of a union of events

$$p_{f,\text{series}} = P\left(\bigcup_{a=1}^{z} \{g_a(\mathbf{X}) \leq 0\}\right)$$

where $g_a(\mathbf{X})$ is the performance (also called *state*) function of element a. Therefore, the limit state of element a is defined as $g_a(\mathbf{X}) = 0$, and $g_a(\mathbf{X}) < 0$ is the failure state.[15]

The correlation between failure modes must be taken into account.[15] Consider a series system consisting of two components in which the probability of failure of each individual component is $p_f = 0.01$. If the two failure modes are independent, i.e., the correlation is $\rho = 0.0$; the failure probability of the system is $p_{f,\text{series}} = 1.0 - (1.0 - 0.01)(1.0 - 0.01) = 0.0199$. If the two events are perfectly correlated (i.e., the correlation is $\rho = 1.0$), the failure probability of the system is $p_{f,\text{series}} = 0.01$.

Parallel Systems. A structure is considered a parallel system if the failure of the structure requires failures of all the components. For a parallel system, the probability of failure of the system $p_{f,\text{parallel}}$ can be written as the probability of an intersection of events

$$p_{f,\text{parallel}} = P\left(\bigcap_{a=1}^{z}\{g_a(\mathbf{X}) \leq 0\}\right)$$

For a parallel system consisting of two components whose individual probabilities of failure are $p_f = 0.01$, the system failure probability is upper bounded (first-order) by $p_{f,\text{parallel}} = 0.01$ if the two failure modes are perfectly correlated and lower bounded (first-order) by $p_{f,\text{parallel}} = (0.01)(0.01) = 0.0001$ if the two failure modes are independent. As indicated in this simplified example, there can be huge errors if correlation is neglected.

General Systems. A general system can be modeled as a combination of series and parallel systems. Consider a series system consisting of y parallel systems. Each parallel system a has z_a components. The probability of failure $p_{f,\text{series-parallel}}$, is given as

$$p_{f,\text{series=parallel}} = P\left(\bigcup_{a=1}^{y}\bigcap_{b=1}^{z_a}\{g_a(\mathbf{X}) \leq 0\}\right)$$

The solution methods for evaluating the preceding system probability can be very complex. Upper and lower bounds often are employed to obtain approximate answers. Monte Carlo simulation can provide accurate results, but the computational time can be excessive. FORM and SORM methods still can be used where the reliability of a complex system is solved by sequentially breaking the system into simpler equivalent subsystems,[16] as shown in Fig. 2.1.

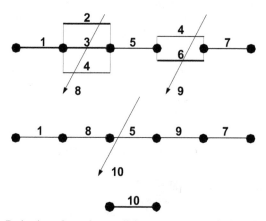

FIGURE 2.1 Reduction of a series-parallel system to an equivalent single component.

RELIABILITY-BASED CONDITION ASSESSMENT

Reliability methods are appropriate for maintenance and repair planning throughout the useful life of a structure. The life-cycle cost includes the costs of initial construction, preventive maintenance, repair, inspection, and expected cost of failure, among others. Life-cycle optimization must balance lifetime cost against acceptable risk. Reliability methods are best for quantifying that acceptable-risk and reliability-based condition assessment is needed to develop and update the life-cycle strategy.

Deterioration and Load Modeling

If the lifetime safety of a structure has to be evaluated and planned, it is necessary to forecast how the structure's capacity and demand will change over time. The loads on the structure may include wind, snow, earthquake, soil pressure, hydrostatic, vehicle, self-weight, or construction loads, among others. Many of these will change over time. Corrosion, rotting, creep, scour, and spalling may cause large changes over time in the capacity of the structure. These changes must be predicted by using adequate probabilistic models.

Loads that occur due to nature, such as wind, waves, and earthquakes, are modeled based on historical data in a specific area. The data, which are measured over a long period of time, are converted to a load or load effect on the structure. One example of this is wind pressure W on a building that has been modeled based on local wind speed V to be

$$W = kC_p V^2$$

where C_p is a pressure coefficient, and k is a factor that accounts for building exposure and wind turbulence. The uncertainties in k, C_p, and V are determined separately, and the distribution of W is computed either in closed form or through Monte Carlo simulation. The annual wind speeds can be used for a point-in-time approach for reliability, whereas a lifetime maximum wind speed is computed for a time-integrated approach.

For man-made loads at specific locations, less data usually are available to develop a probabilistic model. Mathematical models and experience from previous similar structures are used frequently. One example involves modeling the routine loads and lifetime maximum loads over a specific bridge.[2] The live-load model predicts maximum moments and shears for different bridge spans. The study assembled data from 9250 trucks and included number of axles, axle spacing, axle loads, and gross weight of vehicles. The bending moments and shears were calculated for each truck in the survey for a wide range of spans. The result was a series of graphs based on the statistics of extreme values in which the probability of encountering a large truck at the extreme tail of the distribution increases as the number of trucks passing over the bridge increased. As the number of occurrences of a truck passing over the bridge becomes larger, the maximum moment approaches a type I extreme distribution. As a result, it is possible to predict the mean value μ_{M_n} and the standard deviation σ_{M_n} of maximum moment at any time t as[3]

$$\mu_{M_n} = \sigma u_n + \mu + \left(\frac{0.577\sigma}{\alpha_n} \right)$$

$$\sigma_{M_n} = \frac{\pi\sigma}{\sqrt{6}\alpha_n}$$

where $\alpha_n = \sqrt{2 \ln(n)}$

$$u_n = \alpha_n - \frac{\ln[\ln(n)] + \ln(4\pi)}{2\alpha_n}$$

n = the number of times that a truck has passed over the bridge
μ and σ = the mean and standard deviation of the original distribution of trucks

This model is effective if the average daily truck traffic is known and if it is believed that the trucks in the database are representative of the trucks going over the bridge.

Other loading occurrences, such as a major earthquake or a catastrophic collision, are modeled as discrete random variables. Although the exact timing of occurrence of such an event is impossible to determine, past data can provide an average number of occurrences in a given unit of time, also called the *mean occurrence rate* ν. Under specific assumptions, the probability of a specific number of occurrences N_t within a given time interval t is given by the Poisson distribution

$$P(N_t = x) = \frac{(\nu t)^x}{x!} e^{-\nu t} \qquad x = 0,1,2, \ldots$$

There are numerous examples of how any of the existing loads and load combinations can be modeled over time using existing data and probabilistic models. Similarly, the loss of capacity over time can be predicted using deterioration models. Steel corrosion is a common deterioration mechanism that is difficult to predict because it depends on the type of steel, the local environment, the presence of moisture, and even the location of the steel member in the structure. Ideally, one would allow samples of the same type of steel that is used in the project to corrode over time under the same conditions as the structure and use those data to develop a deterioration model. This is costly and time-consuming. Often the more realistic solution is to find results from a similar situation. Albrecht and Naeemi[16] made a comprehensive study to attempt to predict corrosion in different environments for both carbon and weathering steel based on field studies in 46 locations worldwide. Through regression analysis of the field results, they developed a corrosion propagation model that predicts the average corrosion penetration $C(t)$ in micrometers at any time t (years) as follows:

$$C(t) = At^B$$

where A and B are regression parameters based on the environment and type of steel. For carbon steel in an urban environment, for example, A has a mean value of $\mu_A = 80.2$ and a standard deviation of $\sigma_A = 33.68$, whereas $\mu_B = 0.593$ and $\sigma_B = 0.24$. The correlation coefficient between A and B is $\rho_{A,B} = 0.68$. A and B take on different values for different conditions. Since the location of the corrosion is inferred from past experience, the capacity loss in steel beams can be predicted based on section loss due to corrosion. Shear capacity diminishes in proportion to section loss in the web, whereas the reduction in plastic section modulus dictates the loss of moment capacity.

A common deterioration mechanism in concrete bridge decks is the result of penetration of chlorides into the concrete caused by deicing salts. When the penetrated chloride concentration reaches a critical level at the reinforcing steel, corrosion ensues. The section loss in the reinforcing steel causes a loss of moment capacity in the slab, and the increase in volume caused by the corrosion causes the concrete to spall. Thoft-Christensen et al.[17] developed a deterioration model that follows Fick's second law of diffusion and predicts the initiation time to corrosion T_I as

$$T_I = \frac{(d_I - D_I/2)^2}{4D_c} \left(erf^{-1} \left(\frac{C_{cr} - C_o}{C_i - C_o} \right) \right)^{-2}$$

where d_I = concrete cover
    D_I = initial diameter of the reinforcing bar
    D_c = chloride diffusion coefficient
    C_{cr} = critical chloride concentration that will initiate corrosion
    C_i = initial chloride concentration
    C_o = equilibrium chloride concentration on the concrete surface
    erf = the error function.

Once corrosion has started, the diameter of the reinforcement bar $D_I(t)$ as a function of time is modeled as

$$D_I(t) = D_I - C_{corr} i_{corr}(t - T_I)$$

where C_{corr} is a corrosion coefficient, and i_{corr} is a corrosion rate parameter. Using this model, the remaining area of the reinforcing steel at any time is used to compute the moment capacity of the slab. Statistical parameters are provided for each random variable in these equations based on data from the United Kingdom, where the study was prepared. These parameters are a reasonable starting point for computing loss of strength of similar structures in many other locations.

Performance Prediction

Once the time-dependent variables have been modeled, the performance of the structure can be predicted, and tentative life-cycle maintenance forecasts can be made based on that prediction. Figure 2.2[18] shows the deterioration of the plastic section modulus Z of a WF 33 × 132 ft bridge beam based on the corrosion of carbon steel in an urban environment. The mean value of Z drops over time, and the increased standard deviation indicates a greater dispersion of values, which will result in a higher degree of uncertainty over time. The reduced plastic section modulus affects the moment capacity, which, in turn, reduces the reliability of the beam. In conjunction with how the loads on the beam are changing, the reliability can be computed over time. As the acceptable reliability level is assessed, reliability thresholds for repair or conditional failure rates can be established, and the reliability over time then dictates when repairs and maintenance are planned.

Inspection

The reliability of a structure that is forecasted over several decades is only as valid as the input data and models that support it. In some cases, the most desirable data are not available, and analysts must use whatever data exist. A systematic inspection program is needed to verify that the structure is behaving as predicted and to provide new data to modify the prediction models if it is not. Inspections can be expensive, so it becomes important to optimize their schedule in order to maximize the benefit of the information.

NDE Inspection. A nondestructive evaluation (NDE) inspection that is targeted for specific information and does not damage the structure is usually the preferred type of inspection. The information obtained from these inspections is not readily available and often correlates with a specific defect such as fatigue cracks in members, corrosion of girders, or deterioration of concrete. These techniques[19] can include radiography, acoustics travel tomography,[20,21] ultrasonics, thermography, and electrical potential, among others. In some cases, such as the thickness of corrosion on exposed steel, a direct measurement is made. In other cases where a defect is hidden, such as scaling of concrete, acoustics travel tomography, thermography, and radar are useful to infer

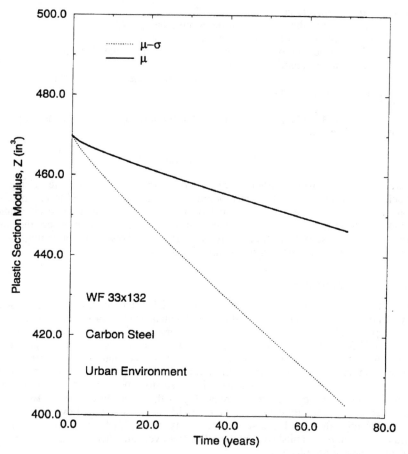

FIGURE 2.2   Reduction in plastic section modulus of a WF 33 × 132 ft girders due to corrosion.[18] (*Reprinted with permission from the American Society of Civil Engineers.*)

damage. Special methods are used when there is a high degree of correlation between a test result such as electrical potential or an acoustical sounding and the defect being sought, such as the presence of corrosion in the steel reinforcement of a concrete structure. Sometimes several techniques may be used in combination to improve confidence in the structural condition. For example, magnetic particle tests, dye penetrants, and ultrasonics are all useful methods to detect fatigue cracks, and if the same conclusion is reached by all three methods, the credibility of the results is enhanced. A probabilistic analysis must account for the ability of the equipment to detect a specific level of damage. Similarly, false-positive results, where the equipment indicates damage where none exists, must be considered.

   There are three main sources of uncertainty that need to be quantified with an NDE inspection: (1) the quality of the inspection equipment and its ability to accurately read whatever measurement is being taken, (2) the correlation between whatever the equipment is measuring and the presence of an actual defect, and (3) the ability to assess the condition of an entire structure based on a finite number of readings. For a probabilistic analysis and an update of a structure's reliability, all these uncertainties should be quantified and considered. Such data are not readily available for most techniques, often are difficult to obtain, and remain a ripe area for continued research.

Marshall[22] quantified the correlation between half-cell potential readings and the presence of active corrosion in concrete. The ASTM guideline[23] states that a half-cell reading more positive than −0.20 V indicates a greater than 90% probability of no active corrosion, and values more negative than −0.35 V indicate 90% probability of active corrosion. The data from 89 different bridges were assembled to determine the probability density functions (PDF) of the half-cell potentials for both sound and damaged deck areas. As shown in Fig. 2.3,[32] the PDF of half-cell potentials in areas where the deck was known to be damaged was a normal distribution with a mean value μ = −207 mV and a standard deviation of σ = 80.4 mV. Similarly, the PDF for half-cell readings in damaged areas of deck was a normal distribution where μ = −354 mV and σ = 69.7 mV. For any half-cell potential reading, the uncertainty associated with whether or not the deck is damaged is quantified. Such data would be useful on every NDE inspection but are difficult to obtain.

Visual Inspection: Condition Ratings. Many structures undergo periodic formalized visual inspections that assess their overall conditions. The Federal Highway Admin-

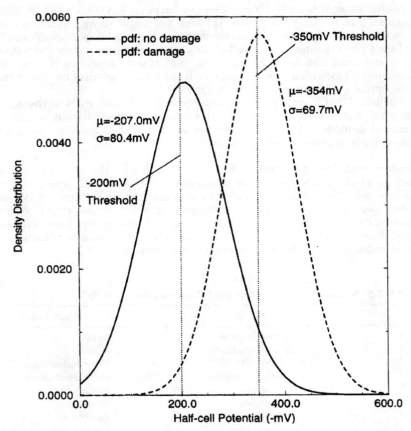

FIGURE 2.3   Probability density functions (PDF) of half-cell potentials in areas where the deck is known to be damaged and undamaged.[32]

istration,[24] for example, requires that every bridge in the National Bridge Inventory be inspected every 2 years. A qualified bridge inspector observes the general condition of the bridge deck, superstructure, and substructure and gives the bridge a condition rating from 1 to 9, as shown in Table 2.2.[25]

The inspection results come from inspector observations rather than a series of specific tests. The advantages of visual inspections are that they are easy to learn, require minimal equipment, are not overly time consuming, and are economical. The primary disadvantage is that a visual inspection only finds surface effects, and the minimum detectable flaws are often a function of lighting, visual acuity, viewing angle, and inspector accessibility to the defect.[26]

As bridge management systems have progressed, many states have developed programs to include much more information than the minimum required by the federal government. Attempts to study how different bridge components behave over time have been made for railings, joints, bearings, and all types of decks, girders, and substructures. The inspection follows a prescribed format and results in a condition assessment or rating for the major elements of the bridge. The PONTIS bridge management system,[27] for example, assigns specialized condition ratings to bridge components such as (307) modular expansion joint, (220) concrete—submerged pile cap/footing, (32) timber deck—w/AC overlay, and (121) steel—bottom chord through truss—painted superstructure.

Another example is the U.S. Army Corps of Engineers (USACE), which maintains a National Dam Inventory Database of 512 dams that require varying degrees of maintenance and repair. A detailed inspection and rating program has been established for steel sheet pile structures, miter gate lock structures, sector gates, tainter and butterfly valves, and tainter dam and lock gates.[28] The result of these inspections is a condition index rating for individual structures between 0 and 100 that describes the current state of the operating equipment, as shown in Table 2.3.[29]

Similarly, USACE requires that a complete reliability analysis be performed to justify any major rehabilitation projects. This includes a current reliability assessment, forecasted demands and deterioration of the structure, hazard functions, and a cost analysis using an event tree of disruptions.[30]

Condition State Transition Models. The difficulty with using the results of visual inspections to update reliability is one of quantifying, in probabilistic terms, what the inspector sees. Most of the condition states are qualitative descriptions that lack numerical quantification. For example, the bridge element that evaluates the amount of corrosion on steel bridge girders is (107) painted open steel girders. Estes and Frangopol[31] demonstrated how this information could be used if the location of the damage was specified and if the damage was quantified as shown in Table 2.4.

TABLE 2.2 National Bridge Inventory (NBI) Condition Ratings (CR)[25]

Condition	Description	Repair Action
9	Excellent Condition	None
8	Very good condition	None
7	Good condition	Minor maintenance
6	Satisfactory condition	Major maintenance
5	Fair condition	Minor repair
4	Poor condition	Major repair
3	Serious condition	Rehabilitate
2	Critical condition	Replace
1	Imminent failure condition	Close bridge and evacuate
0	Failed condition	Beyond corrective action

TABLE 2.3 Condition Index Rating Scale for Inspected Structures[29]

CI Value	Condition Description	Zone	Action
85–100	Excellent: No noticeable defects, some aging or wear visible		Immediate action not required
70–84	Very good: Only minior deterioration or defects evident	1	
55–69	Good: Some deterioration or defects evident, function not impaired		Economic analysis of repair alternatives recommended to determine appropriate maintenance action
40–54	Fair: Moderate deterioriton; funciton is still adequate	2	
25–39	Poor: Serious deterioration in at least some portions of the structure, function inadequate		Detailed evaluation required to determine the need for repair, rehabilitation or reconstruction, safety evaluation required
10–24	Very poor: Extensive deterioration, barely functional	3	
0–10	Failed: General failure or failure of a major coponent; no longer functional		

TABLE 2.4 CDOT Suggested Condition State (CS) Ratings for Element 107: Painted Open Steel Girders[33] (*Reprinted with permission from ASCE.*)

CS	Description	Rust Code	% Section Loss*	Distribution*
1	No evidence of active corrosion; paint system is sound and protecting the girder	—	0–2%	Log normal
2	Slight peeling of the paint, pitting, or surface rust, etc.	Light R1	0–5%	Normal
3	Peeling of the paint, pitting, surface rust, etc.	R1	0–10%	Normal
4	Flaking, minor section loss (<10% of original thickness)	R2		
4	Flaking, swelling, moderate section loss (>10% but <30% of the original thickness); structural analysis not warranted	R3	10–30%	Normal
5	Flaking, swelling, moderate section loss (>10% but <30% of the original thickness); structural analysis not warranted due to location of corrosion on member	R3		
5	Heavy section loss (>30% of original thickness); may have holes thorugh base metal.	R4	>30%	Log normal

*Not part of the PONTIS definition—created to quantify the observed corrosion.

The study assumed three possible inspection programs in which the inspectors were considered very experienced, experienced, and inexperienced and the correct condition state (CS) ratings could be expected 95%, 85%, and 75% of the time, respectively.[84] The quality of the inspection program was determined based on seven criteria[32] in an attempt to quantify human error. For the very experienced inspectors, the density distributions associated with CS1 through CS5 for girder corrosion (element 107) are shown in Fig. 2.4.[33] With each condition state quantified in conservative yet probabilistic terms, the section loss due to corrosion can be related to the decrease in moment capacity, and this allows the computation of structural reliability over time. There are many condition states where such quantification is not feasible.

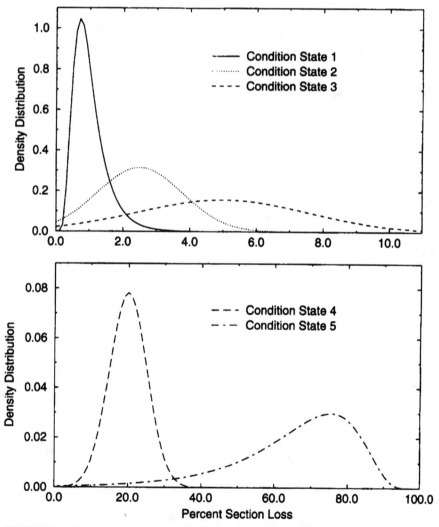

FIGURE 2.4   Density distributions for CS1 through CS5 for girder corrosion using very experienced inspectors.[33] (*Reprinted with permission from ASCE.*)

Condition State as Random Variable. Although the condition state may not always be numerically quantifiable, it has been useful to make the condition state the random variable to be analyzed. If a repair threshold is set at a particular condition state (even though it is nebulously defined by an inspector's observation), one can predict the likelihood of a structure being in that condition state at a point in time. Since thousands of bridge inspections have been conducted and recorded, the results provide a valuable database of information that can be used to forecast structural behavior. Numerous models have been developed that describe the transition of a variety of condition states over time.

Many of the models are based on a linear deterioration of condition states in which the condition rating (CR) at any time t (see Table 2.2) can be computed, and the deterioration rate can be expressed in terms of condition rating loss per year (CR/year). Hearn et al.[34] compiled an extensive list of these models. Some are based on data, whereas others rely on expert opinion. Some of the studies became very specific regarding traffic volume and location. For example, Chen and Johnston[35] indicated 39 years for reaching condition rating 4 for the reinforced concrete (RC) decks where the average daily traffic (ADT) was greater than 4000 rather than the 41 years for all RC decks. Similarly, James et al.[36] found that the condition rating deterioration rate for RC decks on state highways in the western region of the United States was 0.176 CR/year rather than 0.210 CR/year for all RC decks, which equates to 28 years to reach condition rating 4 and 34 years to reach condition rating 3. Jiang and Sinha[37] developed the following polynomial model for a concrete bridge superstructure:

$$CR(t) = 9.0 - 0.28877329t + 0.0093685t^2 - 0.000008877t^3$$

where $CR(t)$ is the condition rating of the bridge at time t, and t is the age of the bridge in years, which translates to 71 years to condition rating 4. Condition rating transition models often conflict, and the analyst must use judgment to choose a model that best applies to the structure at hand.

Two studies by Mlaker[38] and Ayyub et al.[39] used USACE condition indices (CIs) as a direct measure of satisfactory and unsatisfactory performance. Ayyub et al.[39] assembled the actual performance records of 785 hydropower generators to predict how the CIs will change over time in probabilistic terms. The generators were divided into 12 groups based on when they were placed in service and their power capacity. The projected deterioration of the CI was used to develop reliability indices and hazard functions based on predicted time to failure.

Markov Chains. Markov chains can be used to model National Bridge Inventory condition ratings based on the data from large numbers of bridges using transitional probabilities. Jiang and Sinha[37] used Markov chains to model the condition of bridge substructures in Indiana. Table 2.5[37] shows the transitional probabilities for concrete bridge substructures. In this case, the transitional probabilities change as the bridge ages.

The value p_9 indicates the probability that a bridge that is currently in condition rating 9 will remain in this condition state 9 for the next year. For a new bridge that is only 0 to 6 years old, this probability is $p_9 = 0.705$. Assuming that a bridge can only change one condition rating in a given year, the probability that the bridge will fall to condition rating 8 is $1 - p_9$, which for a new bridge is $1 - 0.705 = 0.295$. Once this new bridge (i.e., 0 to 6 years old) has transitioned to condition rating 8, the probability that it will remain in condition rating 8 is $p_8 = 0.818$, and so forth. Using Table 2.5, the time-dependent bridge condition can be modeled easily.

Similarly, Cesare et al.[40] used Markov chains to model many bridge elements in New York State using a database of 850 bridges and 2000 individual spans. The New York condition ratings range from 7 (high) to 1 (low). Based on these New York

TABLE 2.5   Transition Probabilities for Concrete Bridge Substructures Using Markov Chains[37]

Bridge Age (years)	Transitional Probabilities					
	p_9	p_8	p_7	p_6	p_5	p_4
0–6	0.705	0.818	0.810	0.802	0.801	0.800
7–12	0.980	0.709	0.711	0.980	0.980	0.856
13–18	0.638	0.639	0.748	0.980	0.980	0.980
19–24	0.798	0.791	0.788	0.980	0.870	0.824
25–30	0.794	0.810	0.773	0.980	0.980	0.980
31–36	0.815	0.794	0.787	0.980	0.980	0.737
37–42	0.800	0.798	0.815	0.980	0.850	0.980
43–48	0.800	0.800	0.309	0.938	0.980	0.050
49–54	0.800	0.800	0.800	0.711	0.707	0.768
55–60	0.800	0.800	0.800	0.050	0.050	0.505

condition ratings, Cesare et al.[40] developed stationary transition probabilities for numerous bridge elements. For a structural cast-in-place bridge deck with uncoated bars, the transitional probabilities are $p_7 = 0.937$, $p_6 = 0.940$, $p_5 = 0.971$, $p_4 = 0.974$, $p_3 = 0.977$, and $p_2 = 0.961$. Using these probabilities and a simulation of 10,000 bridges, Fig. 2.5 shows the probability distributions of the various condition ratings (called condition states) and the expected number of bridges being in any given condition rating at any time. The reliability using the condition rating as the random variable could be computed and updated.

Reliability-Based Condition Ratings. While the idea is still in its infancy, Frangopol et al.[41] have proposed reliability-based condition ratings for bridges that range from state 1 (unacceptable, $\beta < 4.6$) to state 5 (excellent, $\beta > 9.0$). Bridges are grouped by common type: steel-concrete composite, reinforced concrete, prestressed concrete, and posttensioned concrete. The prediction of the rehabilitation time (i.e., time to failure, defined as the time from construction to when a minimum target reliability level is reached) of similar bridge groups is used to optimize the lifetime maintenance planning. Preventative and essential maintenance actions are planned to extend the service lives of these bridges efficiently. Using steel-concrete bridges as an example, Fig. 2.6[95] shows the expected distribution of reliability-based condition states from a group of 144 bridges built from 1985 to 1994 on which no preventative maintenance has been performed (Fig. 2.6a) or has been performed (Fig. 2.6b). The similarity to Fig. 2.5 illustrates that the same analysis is possible with reliability-based condition states as with the existing condition states. The key difference is that maintenance plans developed from reliability-based data such as shown in Fig. 2.7[95] are directly related to the safety of the structure.

Updating

Once an inspection has been completed (whether it is NDE or visual), the information should be used to update the assessment of structural condition, the load and deterioration models, and the future repair and maintenance strategy. Bayesian updating allows the use of both the prior information and new inspection information in a manner that accounts for the relative uncertainty associated with each.

Assume that prior to an inspection, a random variable Θ was believed to have a density function $f'(\Theta)$, where Θ is the parameter of that distribution. During an inspection, a set of values $x_1, x_2, \ldots, x_n$ representing a random sample from a population

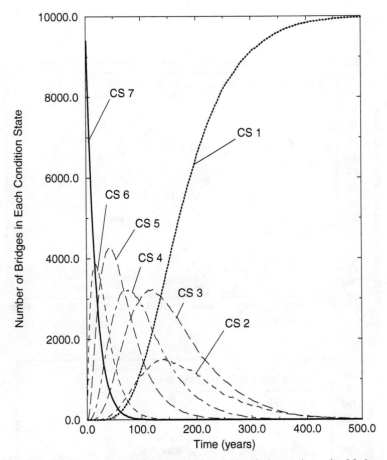

FIGURE 2.5 Condition states (CS) for cast-in-place bridge deck over time using Markov chains and New York State condition ratings.[31] (*Reprinted with permission from ASCE.*)

X with underlying density function $f_x(x)$ is observed and fit to a new density function $f(x_i)$. The updated (also called *posterior*) density function $f''(\Theta)$ that uses both sets of information and provides the best use of both can be expressed as[42]

$$f''(\Theta) = kL(\Theta)f'(\Theta)$$

where $L(\Theta)$ is a likelihood function expressed, for example, as

$$L(\Theta) = \prod_{i=1}^{n} f(x_i|\Theta)$$

and k is a normalizing constant whose value is

$$k = \frac{1}{\displaystyle\int_{-\infty}^{\infty} \left(\prod_{i=1}^{n} f(x_i|\Theta) \right) f'(\Theta)d\Theta}$$

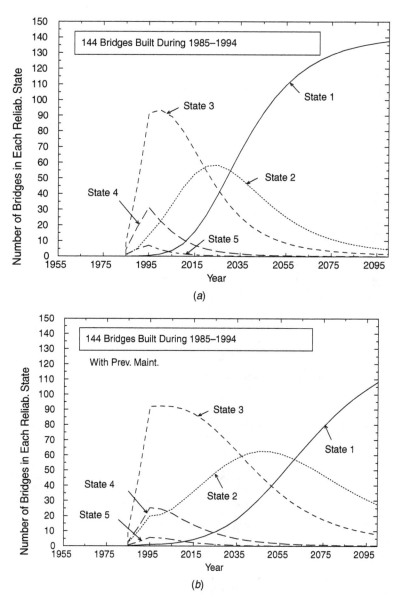

FIGURE 2.6 Expected number of steel-concrete composite bridges without (*a*) and with (*b*) maintenance in reliability states 1 through 5.[95] (*Reprinted with permission from the International Association of Bridge and Structural Engineering.*)

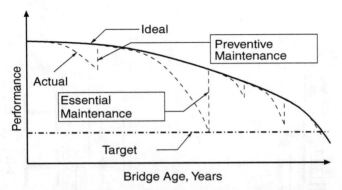

FIGURE 2.7   Different maintenance options involving both preventative and essential mainte-nance.[95] (*Reprinted with permission from IABSE.*)

If the results of the inspection are highly scattered and offer little new information, the effect of the inspection on the new probability distribution is minor. On the other hand, if the inspection results show only minor amounts of dispersion, these results will have a large effect on the revised probability distribution.[43]

CASE STUDIES

Reliability-based methods are more complex than traditional deterministic methods. Such methods are appropriate for important structures where the total life-cycle cost and the consequences of failure are high enough to justify the additional effort and expense. Structures such as offshore platforms, nuclear power plants, bridges, sky-scrapers, tunnels, and dams are common examples where it pays to quantify the risk. This section reports the results from various case studies that use the techniques cited previously to compute the reliability of a structure, project its performance over time, conduct inspections to update the data, and revise the maintenance strategy as a result.

Bridge E-17-AH

Estes and Frangopol[18,33,37,84] outlined a methodology for using system reliability to optimize the lifetime inspection and repair of a deteriorating structure. The approach was illustrated on Bridge E-17-AH, a specific highway bridge (Fig. 2.8)[18] located in the metro-Denver area of Colorado. The bridge system reliability model consid-ered 16 failure modes that included moment failure of the concrete deck, shear and moment failures of the girders, and various failure modes of the substructure. The analysis considered 24 random variables, which included material strengths, mod-eling uncertainties, loads, and load effects. Simplifying assumptions,[18,37,84] such as perfect correlation between spans, symmetry within a span, and the failure of three adjacent girders is required for the superstructure to fail, produced the system model of the bridge shown in Fig. 2.9.

Probabilistic deterioration models predicted the corrosion of girders and the pen-etration of chlorides through the concrete that ultimately corroded the reinforcing bars. Five possible repair options shown in Table 2.6 were examined, and their associated costs were computed.[18,37,84] Based on a 2% discount rate of money and the requirement to make a repair anytime the system reliability index β_{system} fell below 2.0, all possible combinations of the available repair options, as shown in

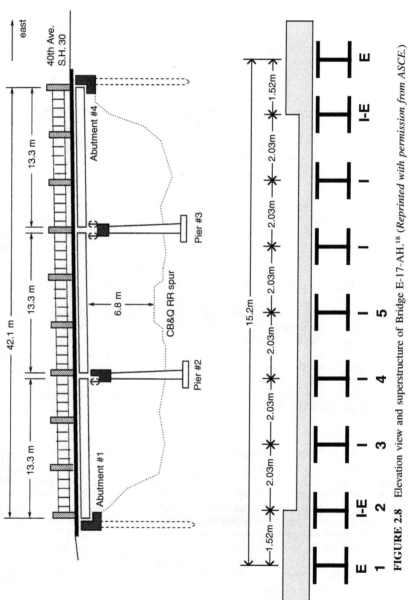

FIGURE 2.8 Elevation view and superstructure of Bridge E-17-AH.[18] (*Reprinted with permission from ASCE.*)

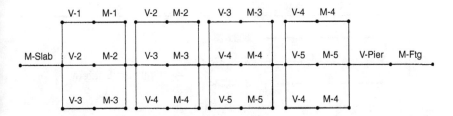

V-1: Failure Due to Shear in Girder 1
M-3: Failure Due to Moment in Girder 3

FIGURE 2.9   Simplified series-parallel model for the Bridge E-17-AH.[18] *(Reprinted with permission from ASCE.)*

Figure 2.10, were considered. Figure 2.11 shows the reliabilities of the components and the system over time using the repair option that the slab gets replaced.

Table 2.7 shows that the optimal reliability-based repair strategy depends on the desired useful life of the structure. System reliability analysis revealed that the component with the lowest reliability was not always the component that most needed to be repaired. It was possible for the reliability index of a component in the parallel portion of the system model to fall below 2.0 without causing the reliability of the system to fall below its threshold value of $\beta_{min} = 2.0$. As time passed and different components deteriorated at different rates, the most critical component early in the life of the structure is not necessarily the most critical component later on. The optimal strategy was updated over time based on the results from NDE inspections (thickness measurements, half-cell potential reading, and three-electrode linear polarization tests) and condition state visual inspections.

Lock and Dam 12

The USACE maintains 238 chambers at 198 lock sites. Miter gates are an important operating component of a lock and dam facility. In many situations, only one lock is available at a site, and if the miter gate does not function, navigation along the entire river is delayed. The Rock Island District of the USACE prepared a time-dependent reliability analysis on Lock and Dam 12 (Fig. 2.12) on the Mississippi River near Dubuque, Iowa, to predict how this structure will perform over its useful life.[44] Estes et al.[45] used the visual inspection data gained from condition index ratings to update the reliability of the miter gate.

The reliability analysis on Lock and Dam 12 miter gates was based on the moment failure limit state on a typical vertical girder, which was determined to be the most critical component. The random variables considered in the analysis were

TABLE 2.6   Repair Options and Associated Repair Costs for Bridge E-17-AH[18]
(Reprinted with permission from ASCE.)

Option	Repair Option	Repair Cost ($)
0	Do nothing	0
1	Replace deck	225,600
2	Replace exterior girders	229,200
3	Replace exterior girders and deck	341,800
4	Replace superstructure	487,100
5	Replace bridge	659,900

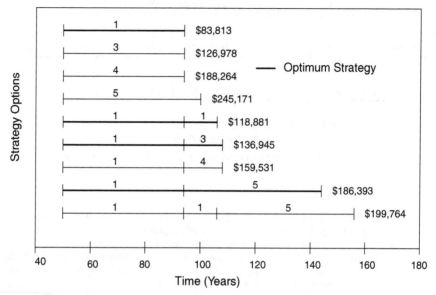

FIGURE 2.10 All feasible repair options for Bridge E-17-AH using a series-parallel model requiring the failure of three adjacent girders.[18] (*Reprinted with permission from ASCE.*)

the yield stress of the steel, the amount of material loss due to corrosion, a modeling uncertainty factor, and the head differential that constituted the load on the structure. Corrosion deterioration was considered in both the splash and atmospheric zones of the miter gate. Based on results from corrosion tests of bare steel under simulated splash zone conditions conducted in Memphis, the following deterioration models were derived for the respective zones[46]:

Atmospheric zone: $\log C = \log 23.4 + 0.65 \log t + \varepsilon_c$

Splash zone: $\log C = \log 148.5 + 0.903 \log t + \varepsilon_c$

where C is the thickness loss due to corrosion in micrometers, t is time in years, and ε_c is an uncertainty factor with a mean of 0.0 and a standard deviation of 0.219 and 0.099 for the atmospheric and splash zones, respectively. With this information, the future performance of the miter gate was estimated using time-dependent reliability, hazard functions, and an event tree to compute the expected costs of disruptions.

The reliability analysis on Lock and Dam 12 extended from 1938 when the miter gate was placed in service to the year 2030. Condition index (CI) visual inspection results were modified to allow them to be used in probabilistic terms. The six levels of corrosion are conservatively quantified within a range of values as shown in Table 2.8. The distributions are defined such that the mean value is based on how far the member is assumed to have transitioned through its condition state, and the standard deviation is based on the capability and credibility of the inspector, as shown in Estes.[32]

The deterioration model was updated based on a hypothetical segment-based inspection in the year 2008.[47] Figure 2.13[124] shows the distribution of the thickness loss for condition state 2 (from Table 2.8) based on the results of the segment-

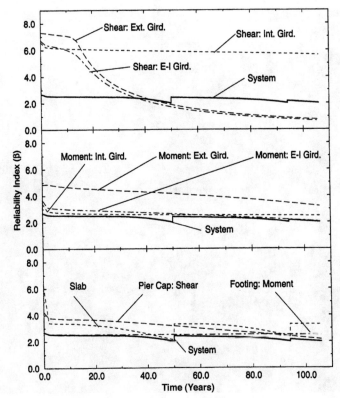

FIGURE 2.11   Lifetime reliability of Bridge E-17-AH using repair option 1 (replace slab).[18] (*Reprinted with permission from ASCE.*)

based inspections. The graphs show the computed distribution when the structure first enters condition state 2 (it is conservatively assumed that the condition state is entered at the halfway point), at the year 1998, and at the year 2008. After the inspections in 1998 and 2008, the deterioration model was updated using bayesian methods and accounted for both the prior and new information. The new models, as shown in Fig. 2.14,[124] are

TABLE 2.7   Optimal Lifetime Repair Strategy for Bridge E-17-AH Based on Strength-Based Series-Parallel System Model[18] (*Reprinted with permission from ASCE.*)

Expected Life (years)	Optimal Repair Strategy	Cost ($)
0–50	Do nothing	0
50–94	1@50	83,813
94–106	1@50, 1@94	118,881
106–108	1@50, 3@94	136,945
>108	1@50, 5@94	186,393

Note: 1@50 indicates option 1 (replace deck) at year 50.

FIGURE 2.12 Photograph of Lock and Dam Number 12.[123] (*Reprinted with permission from Elsevier Publications.*)

TABLE 2.8 Assigned Corrosion Levels from a Condition Index Inspection[124]

Levels of Corrosion			
		Thickness Loss per Side*	
Level	Description	mils	μm
0	New condition	0	0
1	Minor surface scale or widely scattered small pits	0–8	0–200
2	Considerable surface scale and/or moderate pitting	0–20	0–500
3	Severe pitting in dense pattern, thickness reduction in local areas	0–40	0–1000
4	Obvious uniform thickness reduction	40–120	1000–3000
5	Holes due to thickness reduction and general thickness reduction	>120	>3000

* Not currently in CI manual—created to quantify corrosion distress.

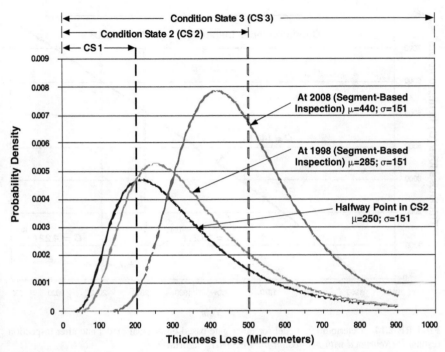

FIGURE 2.13   Distributions for condition state 2 at the halfway point and after inspections in 1998 and 2008.[124] (*Reprinted with permission from Elsevier Publications.*)

After 1998 inspection: $C = 10.23t^{0.903}$
After 2008 inspection: $C = 4.22t^{1.13}$

With the parameters of the thickness loss due to corrosion estimated at any point in time, it is a straightforward process to update the reliability, the hazard functions, and the economic analysis for the miter gate. The original model greatly overestimated the effect of the corrosion.

Innoshima Bridge

In order to develop cost-effective design and maintenance strategies on an existing structure that was designed using conventional methods, it may be necessary to analyze the structure in probabilistic terms. Imai and Frangopol[48] quantified the reliabilities of both the bridge components and the entire bridge system for the Innoshima Bridge, a 1339-m suspension bridge on the Onomichi–Imabari route that connects the islands of Shikoku to Honshu in Japan. A photograph and a schematic view of the Innoshima Bridge are shown in Figs. 2.15 and 2.16, respectively. Since the bridge was designed using the allowable stress design method, the loads, material strengths, and modeling uncertainties all needed to be quantified in probabilistic terms. The wind load, for example, was a function of the air density, drag coefficient, projection area, and correction factors to account for gust effects, bridge elevation, and spatial wind variation. The wind velocity at a 10-m height (V_{10}) was determined using a type I extreme distribution for various return periods, as shown in Fig. 2.17. The suspension bridge computations combined geometrically nonlinear

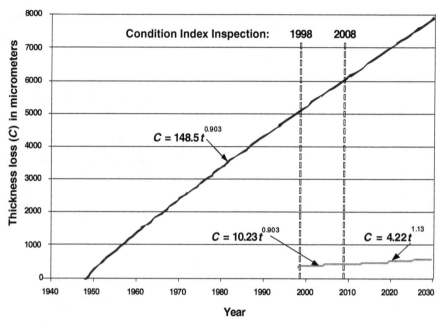

FIGURE 2.14   Deterioration model for miter gate based on original model and after inspection results.[124] (*Reprinted with permission from Elsevier Publications.*)

FIGURE 2.15   Photograph of the Innoshima bridge.

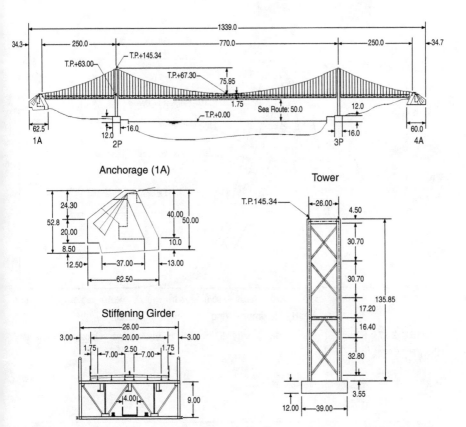

FIGURE 2.16 General view of the Innoshima bridge.[48] (*Reprinted with permission from Elsevier Publications.*)

elastic analysis using the finite-element method with a first-order reliability method (FORM) to obtain results for various damage scenarios. When the bridge was modeled as a series system consisting of the main suspension cables, the hangar ropes, and the stiffening girder under the condition of maximum live load, the system reliability index was $\beta = 5.92$ for the case where the live load was on the entire bridge and $\beta = 3.72$ when the live load was only on one side. Figure 2.18 shows the reliability indices of the three components at points along the bridge when the live load was only on one side. The reliability of the stiffening girders dominated the reliability of the bridge system. Similarly, under the maximum wind load condition, the system reliability index was $\beta = 2.56$.

The bridge was subjected to various failure scenarios in which one member failed and the load was redistributed to other structural members, which required the bridge to be modeled as a series of parallel subsystems. As expected, the reliabilities of the parallel subsystems were higher than the reliability of any individual component. The main cable was designed to a much higher reliability than the stiffening girders because the main cable will exhibit more brittle behavior and has no redundancy. A sensitivity analysis reveals that the material resistances and wind load had the greatest impact on bridge reliability. Reducing the uncertainties with respect to those random variables would most improve the reliability of the bridge. This type of probabilistic condition assessment provides relevant and useful infor-

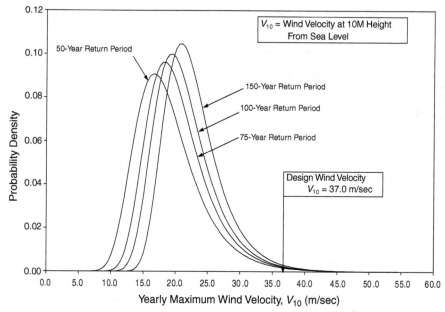

FIGURE 2.17   Effects of time on the density distribution of basic wind velocity.[48] (*Reprinted with permission from ASCE.*)

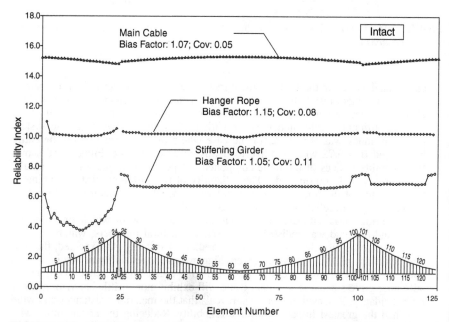

FIGURE 2.18   Reliability indices of main cable, hanger rope, and stiffening girder: load case 2 (D, L, T, SD)–2D model, live load over lateral span.[48] (*Reprinted with permission from ASCE.*)

mation to a planner who needs to spend critical resources to repair and maintain this bridge most efficiently.

Offshore Platform System Design

When a structural system is designed, components must be considered in the context of the entire structure rather than in isolation. Designing the most critical members to the highest standard and designing less critical members to a lesser standard that is compatible with consistent reliability of the system can result in huge cost savings and uniform safety throughout the structure. Gharaibeh et al.[49] presented a criticality ranking of members on an offshore oil platform to use as a basis for decision making and prioritization of various maintenance actions. The reliability-based condition assessment was based on postfailure performance under the 100-year expected loads on the platform. The finite element model and deflected shape of the intact platform are shown in Gharaibeh et al.[49] Under this loading condition, the most critical members were identified and successively removed as the redistribution capacity of the structure was assessed. The residual resistance factor, which is the ratio of the damaged structure ultimate strength over the intact structure ultimate strength, was an important measure for assessing the relative importance of different members of the structural system. The reduction in system reliability was computed as various structural braces were removed.

Rather than choose a target reliability for every member individually, the members were divided into three main groups, denoted as A, B, and C, as shown in Gharaibeh et al.[49] Typically, main leg members were taken as mostly critical (A) because failure of any of these would result in failure of the structure. Main elevation braces and plan members (B) were in the next level of criticality, and the leg pile braces (C) were below that. The results are used to determine target reliabilities for maintenance actions. Target reliabilities are computed to maintain the same probability of occurrence for each failure event of interest. Results obtained for each group (A, B, and C) reflect the level of importance of the members within each group and the postfailure behavior built into the structure. Table 2.9[49] shows the original values of the reliability indices for these three groups and how the target reliabilities of members were able to be reduced without detracting from the reliability of the system. The system analysis identified those members that were less critical and were overdesigned.

Bayesian Updating for a Reinforced Concrete Highway Bridge

When assessing the condition of a structure, the designer has information from prior experience with similar structures. Such studies produce deterioration models and predictions based on expert opinion. At some point, inspection data are used to verify structural performance. The inspection data may be flawed based on the quality of the equipment and the ability of the inspection method to detect the actual flaw. The question becomes, how much credibility is given to the prior de-

TABLE 2.9   Target Reliability Indices for Offshore Oil Platform[49] (*Reprinted with permission from Elsevier Publications.*)

Group	Consequences of Failure	Conservative (Original Values) β_t	Post-Failure Behavior-Based (Revised Values) β_t
A	Very serious	4.25	4.25
B	Serious	3.7	2.95
C	Not serious	3.1	2.25

terioration model versus the new inspection data, and how are the two effectively combined to update and revise the predictive model.

Enright and Frangopol[50] addressed this issue with respect to a reinforced concrete T-beam bridge near Pueblo, Colorado, shown in Fig. 2.19. This 27.3-m. highway bridge contains three equal simply supported spans. Each span has five supporting girders equally spaced at 2.6 m. and labeled girders A through E in Fig. 2.20. Bridge deterioration is caused by salt spray from traffic passing underneath the bridge. The chloride ions from the spray penetrate the concrete and cause the steel reinforcement to corrode once a critical concentration is reached. The section loss in the reinforcement reduces the shear and moment capacity of the bridge. Because the stirrups corrode before the longitudinal reinforcement, shear is the dominant failure mode and occurs near the ends of the girders where the shear forces are highest.

The values for corrosion random variables such as diffusion coefficient, surface chloride concentration, critical chloride concentration, and corrosion rate were taken from the literature and applied to the theoretical model to estimate the deterioration over time. Monte Carlo simulation predicted that the corrosion initiation time was log normally distributed with a mean value $\mu = 2.45$ years and a standard deviation $\sigma = 0.42$ years. Similarly, the corrosion rate is also log-normal with $\mu = 0.15$ mm/year and $\sigma = 0.05$ mm/year. An inspection takes place 25 years after the bridge has been placed in service. The inspection results show the corrosion rate to have a log-normally distribution where $\mu = 0.10$ mm/year and $\mu = 0.04$ mm/year. The prior information and the new inspection results were combined using Bayesian updating to create a new posterior model in which the corrosion rate had a mean value $\mu = 0.12$ mm/year and a standard deviation $\sigma = 0.0288$ mm/year, as shown in Fig. 2.21.

The new model falls between the prior and inspection results and has less dispersion than either of them. The reliability of the bridge (modeled as a system of

FIGURE 2.19 Photograph of reinforced-concrete T-beam bridge near Pueblo, Colorado.[50] (*Reprinted with permission from ASCE.*)

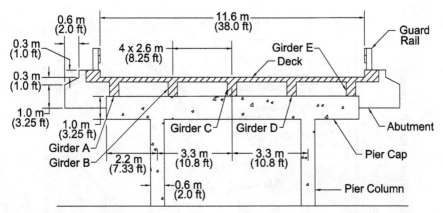

FIGURE 2.20 Cross section of reinforced-concrete T-beam bridge near Pueblo, Colorado.[50] *(Reprinted with permission from ASCE.)*

the girders A through E) is shown in Fig. 2.22[125] based on the prior model, the inspection results, and the updated posterior model. The posterior model computations combined Monte Carlo simulation, regression analysis, numerical integration, and adaptive importance sampling. Because the dispersion associated with the inspection and prior models was approximately the same, both were given approximately equal credibility in developing the posterior model. The study employed several different inspection results for corrosion initiation time and observed how the dispersion of the inspection results has tremendous impact on the posterior

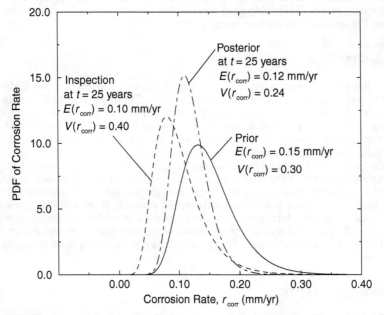

FIGURE 2.21 Bayesian update of corrosion rate on reinforced-concrete T-beam bridge.[50] *(Reprinted with permission from ASCE.)*

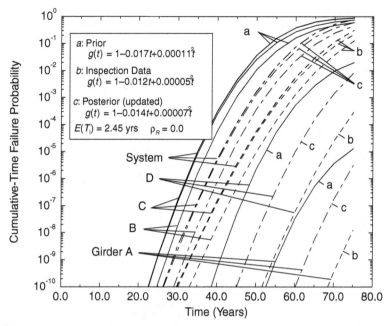

FIGURE 2.22 Updating corrosion rate: prior-, inspection-, and posterior-based time-variant system reliability.[125]

model. Bayesian updating provides a rational method for combining disparate data from different sources into a new model that accounts for the relative uncertainty of those sources. This study used many of the same techniques employed by Mori and Ellingwood[51] on nuclear power plants.

SUMMARY

Reliability-based condition assessment methods are most appropriate for highly important, high-cost structures for which the cost and effort of such methods are justified. Such methods also could be justified for a large number of similar but less important structures that would realize a cost savings due to economy of scale. Probabilistic methods require accurate input data that often are not readily available and difficult to obtain. The input data provide projections that extend over decades. Timely inspection is needed to update the projection models in probabilistic terms. Ideally, the inspections should be targeted for specific information, but efforts should be made to use all available information. As reliability methods gain greater acceptance and wider use, there will be more studies and information available from which to draw input data. None of the structural principles change when a reliability method is employed; the only difference is a result in probabilistic terms that allows an engineer to better quantify what is acceptable risk. Further insights on the topics treated in this chapter are available.[52–127]

ACKNOWLEDGMENTS

The partial financial support that was the basis of the work for this chapter from the U.S. National Science Foundation through grants CMS-9506435, CMS-9522166,

CMS-9912525, and CMS-0217290 and from the U.K. Highways Agency is gratefully acknowledged. The opinions and conclusions presented is this chapter are those of the writers and do not necessarily reflect the views of the sponsoring organizations.

NOTATION

A = regression parameter

a = element in a series or parallel system

B = regression parameter

C_{corr} = corrosion coefficient

C_{cr} = critical chloride concentration

C_i = initial chloride concentration

CI = condition index

C_o = equilibrium chloride concentration

C_p = wind pressure coefficient

CR = condition rating

CS = condition state

$C(t)$ = corrosion penetration

c = distance from neutral axis to outermost fiber on cross section

D_c = chloride diffusion coefficient

D_I = initial diameter of rebar

d_I = concrete cover

$f_L(l)$ = probability density function of load

$f_R(r)$ = probability density function of resistance

$f_{R,L}(r,l)$ = joint probability density function

$f_X(x)$ = observed density function (Bayesian updating)

$f'(\theta)$ = prior density function (Bayesian updating)

$f''(\theta)$ = posterior density function (Bayesian updating)

$g(x)$ = performance or state function

$H(t)$ = hazard function

I = moment of inertia of cross-section

i_{corr} = corrosion rate parameter

k = building exposure and wind turbulence factor

k = normalizing constant (Bayesian updating)

L = load (random variable)

$L(\theta)$ = likelihood function

M_n = maximum truck moment

M_r = elastic moment capacity

N_t = number of specific occurrences

n = number of random variables

n_f = number of failures in a simulation

p_f = probability of failure

p_n = probability that a bridge in condition state n will remain in state n

p_s = probability of survival

R = resistance (random variable)

$S(t)$ = survivor function

T = time (random variable)
T_I = corrosion initiation time
t = time
t_1 = specific point in time
V = local wind speed
W = wind pressure
$\mathbf{X}$ = vector of basic random variables
Z = plastic section modulus
α_n = factor in extreme value distribution
β = reliability index
β_{min} = threshold reliability index for repair
β_{system} = system reliability index
ε_c = uncertainty factor for corrosion
Φ = distribution function of standard normal variate
ϕ_n = resistance factor
γ_i = load factor
μ = mean value
ν = mean occurrence rate
ρ = correlation coefficient
σ = standard deviation
σ_y = yield stress

REFERENCES

1. McGregor, J. G. (1997). *Reinforced Concrete Mechanics and Design,* 3d ed. Prentice-Hall, Upper Saddle River, NJ.
2. Nowak, A. S. (1995). "Calibration of LRFD Bridge Code," *Journal of Structural Engineering,* ASCE, **121**(8):1245–1252.
3. Ang, A. H-S., and W. H. Tang (1984). *Probability Concepts in Engineering Planning and Design,* Wiley, New York.
4. Liu, P-L., H-Z., Lin, and A. Der Kiureghian (1989). "CALREL User Manual," Report No. UCB/SEMM-89/18, Structural Engineering, Mechanics, and Materials, Department of Civil Engineering, University of California, Berkeley.
5. Bucher, C. G. and G. I. Schüeller (1994). "Software for Reliability-Based Analysis," *Structural Safety,* **16**(1 and 2):13–22.
6. Bjerager, P. (1996). "The Program System PROBAN," Appendix 4, in O. Ditlevsen and O. H. Madsen, eds., *Structural Reliability Methods,* Wiley, Chichester, UK, pp. 347–360.
7. Estes, A. C., and D. M. Frangopol (1998). "RELSYS: A Computer Program for Structural System Reliability Analysis," *Structural Engineering and Mechanics,* **6**(8):901–919.
8. Rackwitz, R. (1996). "The Program System STRUREL," in O. Ditlevsen and H. O. Madsen, eds., *Structural Reliability Methods,* Wiley, Chichester, UK.
9. SWRI (1998). Southwest Research Institute, NESSUS (Numerical Evaluation of Stochastic Structures Under Stress), Version 2.4, San Antonio, Texas.
10. Bucher, C. G. (1988). "Adaptive Sampling: An Iterative Fast Monte Carlo Procedure," *Structural Safety,* **5**(2):119–126.
11. Bucher, C. G. and U. Bourgund (1990). "A Fast and Efficient Response Surface Approach for Structural Reliability Problems," *Structural Safety,* **7**:57–66.
12. Bjerager, P. (1990). "On Computational Methods for Structural Reliability Analysis," *Structural Safety,* **9**(2):79–96.

13. Ditlevsen, O., and H. O. Madsen (1996). *Structural Reliability Methods*, Wiley, Chichester, UK.

14. Leemis, L. M. (1995). *Reliability: Probabilistic Models and Statistical Methods*, Prentice-Hall, Englewood Cliffs, NJ.

15. Cornell, C. A. (1967). "Bounds on the Reliability of Structural Systems," *Journal of the Structural Division*, ASCE, **93**(ST1):171–200.

16. Albrecht, P., and A. Naeemi (1984). *Performance of Weathering Steel in Bridges*, NCHRP Report 272, Washington, DC.

17. Thoft-Christensen, P., F. M. Jensen, C. R. Middleton, and A. Blackmore (1997). "Assessment of the Reliability of Concrete Slab Bridges," in D. M. Frangopol, R. B. Corotis, and R. Rackwitz, eds., *Reliability and Optimization of Structural Systems*, Elsevier, New York, pp. 321–328.

18. Estes, A. C., and D. M. Frangopol (1999). "Repair Optimization of Highway Bridges Using System Reliability Approach," *Journal of Structural Engineering*, ASCE, **125**(7):766–775.

19. AASHTO (1994). *AASHTO Manual for Condition Evaluation of Bridges*, American Association of State Highway and Transportation Officials, Washington, DC.

20. Bond, L. J., W. F. Kepler, and D. M. Frangopol (2000). "Improved Assessment of Mass Concrete Dams Using Acoustic Travel Tomography: I. Theory," *Construction and Building Materials*, **14**(3):133–146.

21. Kepler, W. F., L. J. Bond, and D. M. Frangopol (2000). "Improved Assessment of Mass Concrete Dams Using Acoustic Travel Tomography: II. Application," *Construction and Building Materials*, **14**(3):147–156.

22. Marshall, S. J. (1996). "Evaluation of Instrument-Based, Nondestructive Inspection Methods for Bridges," M.Sc. thesis, Department of Civil, Environmental, and Architectural Engineering, University of Colorado, Boulder.

23. ASTM (1987). *Standard Test Method for Half-Cell Potentials of Uncoated Reinforcing Steel in Concrete*, C876–87, American Society for Testing and Materials, Philadelphia.

24. FHWA (2002). Federal Highway Administration, National Bridge Inspection Standards, Internet site: *http://www.fhwa.dot.gov/tndiv/brinsp.htm*, U.S. Department of Transportation, Washington, DC.

25. FHWA (1988). *Recording and Coding Guide for the Structure Inventory and Appraisal of the Nation's Bridges*, FHWA-ED-89-044, Federal Highway Administration, U.S. Department of Transportation, Washington, DC.

26. FHWA (1986). *Nondestructive Testing Methods for Steel Bridges*, Participants Training Manual, Federal Highway Administration, U.S. Department of Transportation, Washington, DC.

27. PONTIS (1995). Release 3.0, *User's Manual*, Cambridge Systematics, Inc., Cambridge, MA.

28. Stecker, J. H., L. F. Greimann, M. Mellema, K. Rens, and S. D. Foltz (1997). *REMR Management Systems—Navigation and Flood Control Structures, Condition Rating Procedures for Lock and Dam Operating Equipment*, Technical Report REMR-OM-19, U.S. Army Corps of Engineers, Washington, DC.

29. Greimann, L. F., J. H. Stecker, and K. Rens (1990). *Management System for Miter Gate Locks*, Technical Report REMR-OM-08, U.S. Army Corps of Engineers, Washington, DC.

30. USACE (1996). U.S. Army Corps of Engineers, Project Operations: Partners and Support, *Work Management Guidance and Procedures*, EP 1130-2-500, Washington, DC.

31. Estes, A. C., and D. M. Frangopol (2001). "Bridge Lifetime System Reliability under Multiple Limit States," *Journal of Bridge Engineering*, ASCE, **6**(6):523–528.

32. Estes, A. C. (1997). "A System Reliability Approach to the Lifetime Optimization of Inspection and Repair of Highway Bridges," Ph.D. thesis, Department of Civil, Environmental, and Architectural Engineering, University of Colorado, Boulder.

33. Estes, A. C., and D. M. Frangopol (2003). "Updating Bridge Reliability Based on BMS Visual Inspection Results," *Journal of Bridge Engineering*, ASCE, **8**(6):374–382.

34. Hearn, G., D. M. Frangopol, and T. Szanyi (1995). *Report on Bridge Management Practices in the United States*, University of Colorado, Boulder.

35. Chen, C. J. and D. W. Johnston (1987). *Bridge Management under a Level of Service Concept Providing Optimum Improvement Action, Time, and Budget Prediction*, FHWA/NC/88-004, North Carolina State University, Raleigh.

36. James, R. P., G. Stukart, W. F. McFarland, A. Garcia-Diaz, R. P. Bligh, S. Baweja, and J. Sobanjo (1993). *A Proposed Bridge Management System Implementation Plan for Texas,* FHWA/TX-92/1259-IF, Texas Transportation Institute, College Station, TX.

37. Jiang, M., and K. C. Sinha (1989). "Bridge Service Life Prediction Model Using the Markov Chain," *Transportation Research Record 1223,* Transportation Research Board, National Research Council, Washington, DC.

38. Mlaker, P. F. (1994). *Reliability of Hydropower Equipment,* COE Waterways Experiment Station under contract DACW39-93-0073, U.S. Army Corps of Engineers, Washington, DC.

39. Ayyub, B. M., M. P. Kaminskiy, and D. A. Moser (1996). *Reliability Analysis and Assessment of Hydropower Equipment,* COE Institute for Water Resources Technical Report under contract USDA-CSRS-95-COOP-2-1792, U.S. Army Corps of Engineers, Washington, DC.

40. Cesare, M. A., C. Santamaria, C. Turkstra, and E. H. Vanmarcke (1992). "Modeling Bridge Deterioration with Markov Chains," *Journal of Transportation Engineering,* ASCE, **118**(6): 820–833.

41. Frangopol, D. M., J. S. Kong, and E. S. Gharaibeh (2001). "Reliability-Based Life-Cycle Management of Highway Bridges," *Journal of Computing in Civil Engineering,* ASCE, **15**(1):27–34.

42. Ang, A. H-S., and W. H. Tang (1975). *Probability Concepts in Engineering Planning and Design,* Vol. 1: *Basic Principles,* Wiley, New York.

43. Melchers, R. E. (1999). *Structural Reliability Analysis and Prediction,* 2d ed., Wiley, Chicester, UK.

44. USACE (1997). *Reliability Analysis of Miter Gates Lock and Dam 12,* U.S. Army Corps of Engineers, Rock Island District, Rock Island, IL.

45. Estes, A. C., D. M. Frangopol, and S. Foltz (2001). "Using Condition Index Inspection Results to Update the Reliability of Miter Gates on Dams," in *Structures 2001: Proceedings of the 2001 Structures Congress and Exposition,* Session 16(4), ASCE/SEI, Washington, DC.

46. Padula, J., C. Chasten, R. Mosher, P. Mlaker, J. Brokaw, and W. Stough (1994). *Reliability Analysis of Hydraulic Steel Structures with Fatigue and Corrosion Degradation,* Technical Report ITL-94-3, U.S. Army Corps of Engineers, Washington, DC.

47. Estes, A. C., D. M. Frangopol, and S. D. Foltz (2003). "Updating the Reliability of Engineering Structures Using Visual Inspection Results," in *Proceedings of the Ninth International Conference on Applications of Statistics and Probability in Civil Engineering,* ICASP9, San Francisco, California, July 6–9; in A. Der Kiureghian, S. Madanat, and J. M. Pestana, eds., *Applications of Statistics and Probability in Civil Engineering,* Millpress, Rotterdam, pp. 1087–1092 and on associated CD-ROM.

48. Imai, K., and D. M. Frangopol (2001). "Reliability-Based Assessment of Suspension Bridges: Application to the Innoshima Bridge," *Journal of Bridge Engineering,* ASCE, **6**(6): 398–411.

49. Gharaibeh, E. S., D. M. Frangopol, and T. Onoufriou (2002). "Reliability-Based Importance Assessment of Structural Members with Applications to Complex Structures," *Computers and Structures,* **80**(12):1111–1131.

50. Enright, M. P., and D. M. Frangopol (1999). "Condition Prediction of Deteriorating Concrete Bridges Using Bayesian Updating," *Journal of Structural Engineering,* ASCE, **125**(10): 1118–1124.

51. Mori, Y., and B. Ellingwood (1993). "Methodology for Reliability-Based Condition Assessment: Application to Concrete Structures in Nuclear Plants," *NUREG/CR-6052,* U.S. Nuclear Regulatory Commission, Washington, DC.

52. AASHTO (1994). *AASHTO LRFD Bridge Design Specifications,* 1st ed., American Association of State Highway and Transportation Officials, Washington, DC.

53. AISC (1994). *AISC Manual of Steel Construction: Load and Resistance Factor Design,* 2d ed., American Institute of Steel Construction, Chicago, IL.

54. Aktan, A. E., D. N. Farhey, D. L. Brown, V. Dalal, A. J. Helmicki, V. J. Hunt, and S. J. Shelly (1996). "Condition Assessment for Bridge Management," *Journal of Infrastructure Systems,* ASCE, **2**(3):108–117.

55. Al-Harthy, A. S., and D. M. Frangopol (1994). "Reliability Assessment of Prestressed Concrete Beams," *Journal of Structural Engineering,* ASCE, **120**(1):180–199.

56. Ang, A. H.-S., and C. A. Cornell (1974). "Reliability Bases of Structural Safety and Design," *Journal of Structural Engineering,* ASCE, **100**(9):1755–1769.

57, Ang, A. H.-S., and D. De Leon (1997). "Determination of Optimal Target Reliabilities for Design and Upgrading of Structures," *Structural Safety,* **19**:91–103.

58. Ang, A. H.-S., D. M. Frangopol, M. Ciampoli, P. C. Das, and J. Kanda (1998). "Life-Cycle Cost Evaluation and Target Reliability for Design," in N. Shiraishi, M. Shinozuka, and Y. K. Wen, eds., *Structural Safety and Reliability,* Balkema, Rotterdam, pp. 77–78.

59. ASCE (2000). *Minimum Design Loads for Buildings and Other Structures,* American Society of Civil Engineers, ASCE Standard, ANSI/ASCE 7-98, ASCE, New York.

60. ASCE (1998). "ASCE 1998 Report Card for America's Infrastructure," *ASCE News,* **23**:24.

61. ASCE (1998). "ASCE Gives Poor Grades to Nation's Infrastructure," *ASCE News,* **23**(4): 1–3.

62. Augusti, G., M. Ciampoli, and D. M. Frangopol (1998). "Optimal Planning of Retrofitting Interventions on Bridges in a Highway Network," *Engineering Structures,* **20**(11):933–939.

63. Chase, S. B., and G. Washer (1997). "Nondestructive Evaluation for Bridge Management in the Next Century," *Public Roads,* July–August:16–25.

64. Clemena, G., D. R. Jackson, and G. C. Crawford (1992). "Inclusion of Rebar Corrosion Rate Measurements in Condition Surveys of Concrete Bridge Decks," *Transportation Research Record 1347,* TRB, National Research Council, Washington, DC, pp. 37–45.

65. Das, P. C. (1999). "Prioritization of Bridge Maintenance Needs," in D. M. Frangopol, ed., *Case Studies in Optimal Design and Maintenance Planning of Civil Infrastructure Systems,* ASCE, Reston, VA, pp. 26–44.

66. Das, P. C., D. M. Frangopol, and A. S. Nowak, eds. (1999). *Current and Future Trends in Bridge Design, Construction, and Maintenance,* Institution of Civil Engineers, Thomas Telford, London.

67. Das, P. C. (2000). "Reliability-Based Bridge Management Procedures," in M. J. Ryall, G. A. R. Parke, and J. E. Harding, eds., *Bridge Management 4,* Thomas Telford, London, pp. 1–11.

68. Das, P. C., D. M. Frangopol, and A. S. Nowak, eds. (2001). *Current and Future Trends in Bridge Design, Construction, and Maintenance,* Vol. 2, Institution of Civil Engineers, Thomas Telford, London.

69. De, S. R. (1990). *Offshore Structural System Reliability: Wave-Load Modeling, System Behavior, and Analysis,* Report No. RMS-6, John A. Blume Earthquake Engineering Center, Department of Civil Engineering, Stanford University, Stanford, CA.

70. Ellingwood, B., T. V. Galambos, J. G. MacGregor, and C. A. Cornell (1980). "Development of a Probability-Based Load Criterion for American National Standard A58," *NBS Special Publication 577,* U.S. Dept of Commerce, Washington, DC.

71. Ellingwood, B. R. (1995). "Engineering Reliability and Risk Analysis for Water Resources Investments: Role of Structural Degradation in Time-Dependent Reliability Analysis," Contract Report ITL-95-3, U.S. Army Corps of Engineers, Washington, DC.

72. Ellingwood, B. R. (1996). "Reliability-Based condition assessment and LRFD for Existing Structures," *Structural Safety,* **18**(2–3):67–80.

73. Ellis, J. H., M. Jiang, and R. B. Corotis (1995). "Inspection, Maintenance and Repair with Partial Observability," *Journal of Infrastructure Systems,* ASCE, **1**(2):92–99.

74. Enright, M. P., and D. M. Frangopol (1998). "Service-Life Prediction of Deteriorating Concrete Bridges," *Journal of Structural Engineering,* ASCE, **124**(3):309–317.

75. Enright, M. P., and D. M. Frangopol (1998). "Failure Time Prediction of Deteriorating Fail-Safe Structures," *Journal of Structural Engineering,* ASCE, **124**(12):1448–1457.

76. Enright, M. P., and D. M. Frangopol (1998). "Probabilistic Analysis of Resistance Degradation of Reinforced Concrete Bridge Beams Under Corrosion," *Engineering Structures,* **20**(11):960–971.

77. Enright, M. P., and D. M. Frangopol (1999). "Maintenance Planning for Deteriorating Concrete Bridges," *Journal of Structural Engineering,* ASCE, **125**(12):1407–1414.

78. Enright, M. P., and D. M. Frangopol (1999). "Reliability-Based Condition Assessment of Deteriorating Concrete Bridges Considering Load Redistribution," *Structural Safety,* **21**(2): 159–195.

79. Enright, M. P., and D. M. Frangopol (2000). "RELTSYS: A Computer Program for Life Prediction of Deteriorating Systems," *Structural Engineering and Mechanics,* **9**(6):557–568.

80. Enright, M. P., and D. M. Frangopol (2000). "Survey and Evaluation of Damaged Concrete Bridges," *Journal of Bridge Engineering,* ASCE, **5**(1):31–38.

81. Estes, A. C., and D. M. Frangopol (2001). "Using System Reliability to Evaluate and Maintain Structural Systems," *Computational Structural Engineering,* **1**(1):71–80.

82. Estes, A. C., and D. M. Frangopol (2001). "Minimum Expected Cost-Oriented Optimal Maintenance Planning for Deteriorating Structures: Application to Concrete Bridge Decks," *Reliability Engineering and System Safety,* **73**(3):281–291.

83. FHWA (2000). "Asset Management: Preserving a $1 Trillion Investment," *Focus,* May:1–2.

84. Frangopol, D. M., and A. C. Estes (1997). "Lifetime Bridge Maintenance Strategies Based on System Reliability," *Structural Engineering International,* IABSE, **7**(3):193–198.

85. Frangopol, D. M., K.-Y. Lin, and A. C. Estes (1997). "Life-Cycle Cost Design of Deteriorating Structures," *Journal of Structural Engineering,* ASCE, **123**(10):1390–1401.

86. Frangopol, D. M., K.-Y. Lin, and A. C. Estes (1997). "Reliability of Reinforced Concrete Girders under Corrosion Attack," *Journal of Structural Engineering,* ASCE, **123**(3):286–297.

87. Frangopol, D. M., ed. (1998). *Optimal Performance of Civil Infrastructure Systems,* ASCE, Reston, VA.

88. Frangopol, D. M. (1999). "Life-Cycle Cost Analysis for Bridges," in D. M. Frangopol, ed., *Bridge Safety and Reliability,* ASCE, Reston, VA, pp. 210–236.

89. Frangopol, D. M., ed. (1999). *Bridge Safety and Reliability,* ASCE, Reston, VA.

90. Frangopol, D. M., ed. (1999). *Case Studies in Optimal Design and Maintenance Planning of Civil Infrastructure Systems,* ASCE, Reston, VA.

91. Frangopol, D. M., and A. C. Estes (1999). "Optimum Design of Bridge Inspection/Repair Programmes Based on Lifetime Reliability and Life-Cycle Cost," in P. C. Das, ed., *Management of Highway Structures,* Institution of Civil Engineers, Thomas Telford, London, pp. 205–230.

92. Frangopol, D. M., and A. C. Estes (1999). "Optimum Lifetime Planning of Bridge Inspection and Repair Programs," *Structural Engineering International,* IABSE, **9**(3):219–223.

93. Frangopol, D. M. (2000). "Advances in Life-Cycle Reliability-Based Technology for Design and Maintenance of Structural Systems," in B. H. V. Topping, ed., *Computational Mechanics for the Twenty-First Century,* Saxe-Coburg, Edinburgh, pp. 311–328.

94. Frangopol, D. M., and H. Furuta, eds. (2001). *Life-Cycle Cost Analysis and Design of Civil Infrastructure Systems,* ASCE, Reston, VA.

95. Frangopol, D. M., E. S. Gharaibeh, J. S. Kong, and M. Miyake (2001). "Reliability-Based Evaluation of Rehabilitation Rates of Bridge Groups," in *Proceedings of the International Conference on Safety, Risk and Reliability—Trends in Engineering,* IABSE, Malta, March 21–23; *Safety, Risk and Reliability—Trends in Engineering,* Conference Report, IABSE-CIB-ECCS-*fib*-RILEM, Malta, pp. 267–272.

96. Ghosn, M., and D. M. Frangopol (1999). "Bridge Reliability: Components and Systems," in D. M. Frangopol, ed., *Bridge Safety and Reliability,* ASCE, Reston, VA, pp. 83–112.

97. Hawk, H., and E. P. Small (1998). "The BRIDGIT Bridge Management System," *Structural Engineering International,* IABSE, **8**(4):309–314.

98. Hearn, G. (1998). "Condition Data and Bridge Management," *Structural Engineering International,* IABSE, **8**(3):221–225.

99. Hendawi, S., and D. M. Frangopol (1994). "System Reliability and Redundancy in Structural Design and Evaluation," *Structural Safety,* **16**(1–2):47–71.

100. Imbsen, R. A., D. H. Liu, R. A. Schamber, and R. V. Nutt (1987). "Strength Evaluation of Existing Reinforced Concrete Bridges," NCHRP Report 292, Washington, DC.

101. Jones, N. P., and B. Ellingwood (1992). "NDE of Concrete Bridges: Opportunities and Research Needs," FHWA-RD-93-040A, 2:1–52.

102. Katsuki, S., and D. M. Frangopol (1994). "Reliability Analysis of Sediment Control Steel Dams," *Structural Safety,* **15**(1–2):131–148.

103. Lipkus, S. E. (1994). "BRIDGIT Bridge Management System Software," *Characteristics of Bridge Management Systems,* Transportation Research Circular Number 423, National Academy Press, Washington, DC, pp. 43–54.

104. Markow, M. J., S. M. Madanat, and D. I. Gurenich (1993). "Optimal Rehabilitation Times for Concrete Bridge Decks," *Transportation Research Record Number 1392,* Transportation Research Board, National Research Council, Washington, DC, pp. 62–71.

105. Mori, Y., and B. R. Ellingwood (1994). "Maintaining Reliability of Concrete Structures I. Role of Inspection/Repair," *Journal of Structural Engineering,* ASCE, **120**(3):824–845.

106. Mori, Y., and B. Ellingwood (1993). "Reliability-Based Service-Life Assessment of Aging Concrete Structures," *Journal of Structural Engineering,* ASCE, **119**(5):1600–1621.

107. Nordall, H., C. A. Cornell, and A. Karamchandani (1987). "A Structural System Reliability Case Study of an Eight-Leg Steel Jacket Offshore Production Platform," in *Proceedings of the Marine Structural Reliability Symposium,* SNAME.

108. Renn, D. P. (1995). "Segment-Based Inspection for Load Rating within Bridge Management Systems," Masters thesis, Department of Civil, Environmental, and Architectural Engineering, University of Colorado, Boulder.

109. Small, E. P., and J. Cooper (1998). "Condition of the Nation's Highway Bridges. A Look at the Past, Present, and Future," *TR News,* **194**(January–February):3–8.

110. Small, E. P., T. Philbin, M. Fraher, and G. Romack (2000). "Current Studies of Bridge Management System Implementation in the United States," *Transportation Research Circular,* **498**(I):A-1/1-16.

111. Sørensen, J. D., and S. Engelund (1998). "Optimal Planning of Maintenance of Concrete Structures," in D. M. Frangopol, ed., *Optimal Performance of Civil Infrastructure Systems,* ASCE, Reston, VA, 169–180.

112. Tao, Z., R. B. Corotis, and J. H. Ellis (1995). "Reliability-Based Structural Design with Markov Decision Processes," *Journal of Structural Engineering,* ASCE, **121**(6):971–980.

113. Thoft-Christensen, P. (1998). "Assessment of the Reliability Profiles for Concrete Bridges," *Engineering Structures,* **20**:1004–1009.

114. Thoft-Christensen, P., and Y. Murotsu (1986). *Application of Structural Systems to Reliability Theory,* Springer-Verlag, Berlin.

115. Thompson, P. D. (1994). "Pontis," *Characteristics of Bridge Management Systems,* Transportation Research Circular, **423**:35–42.

116. Thompson, P. D., E. P. Small, M. Johnson, and A. R. Marshall (1998). "The Pontis Bridge Management System," *Structural Engineering International,* IABSE, **8**(4):303–308.

117. Vassie, P. R. (1997). "A Whole Life Cost Model for the Economic Evaluation of Durability Options for Concrete Bridges," in P. C. Das, ed., *Safety of Bridges,* Thomas Telford, London, pp. 145–150.

118. Yao, J. T. P. (1985). *Safety and Reliability of Existing Structures,* Pitman Advanced Publishing, Boston.

119. Zheng, R., and B. R. Ellingwood (1998). "Role of Nondestructive Evaluation in Time-Dependent Reliability Analysis." *Structural Safety,* **20**(4):325–339.

120. Zorapapel, G. T., G. C. Hart, and D. M. Frangopol (1997). "Performance Assessment of Concrete Masonry Wall Buildings Using Monte Carlo Simulation," *The Masonry Society Journal,* **15**(1):59–66.

121. Frangopol, D. M., and K. Maute (2004). "Reliability-Based Optimization of Civil and Aerospace Structural Systems," in D. M. Ghiocel and E. Nikolaidis, eds., *Engineering Design Reliability Handbook,* CRC Press, Boca Raton, FL (in press).

122. Estes, A. C., and D. M. Frangopol (2004). "Life-Cycle Evaluation and Condition Assessment of Structures," in W.-F. Chen and E. M. Lui, eds., *Structural Engineering Handbook,* 2d ed., CRC Press, Boca Raton, FL (in press).

123. USACE, U.S. Army Corps of Engineers (2003). United States Army Corps of Engineers Web site, *http://www.mvr.usace.army.mil/navdata.*

124. Estes, A. C., D. M. Frangopol, and S. D. Foltz (2004). "Updating Reliability of Steel Miter Gates on Locks and Dams Using Visual Inspection Results," *Engineering Structures,* **26**(3): 319–333.

125. Enright, M. P. (1998). "Time-Variant Reliability of Reinforced Concrete Bridges under Environmental Attack," Ph.D. thesis, Department of Civil, Environmental, and Architectural Engineering, University of Colorado, Boulder.

126. Ellingwood, B. R. (1998). "Optimum Policies for Reliability Assurance of Aging Concrete Structures," in D. M. Frangopol, ed., *Optimal Performance of Civil Infrastructure Systems,* ASCE, Reston, VA, pp. 88–97.

127. Frangopol, D. M., E. Brühwiler, M. H. Faber, and B. Adey, eds. (2004). *Life-Cycle Performance of Deteriorating Structures: Assessment, Design and Management.* ASCE, Reston, VA.

PART II

BUSINESS, CODES, AND LEGAL ASPECTS OF STRUCTURAL CONDITION ASSESSMENT

CHAPTER 3

The Business of Condition Assessment

ANTRANIG M. OUZOONIAN, P.E.

REASONS FOR CONDITION ASSESSMENT

The business and practice of a structural condition assessment must be performed by engineers who are qualified to inspect and report on the conditions of the subject matter. The ability to assess the condition and behavior of structures matures with the structural engineer over the years as he or she becomes more familiar and experienced with the signs of distress and their probable causes in structures. A structural condition assessment survey becomes an "art" with the engineer's knowledge of physics, behavior of materials, and the construction process. Therefore, experienced engineers should take on the task of assessment surveys with the assistance from junior engineers so that the younger engineers will gain experience and knowledge that are required for conducting a structural condition survey for a client.

The reasons for a structural condition assessment survey can be one or more of the following conditions:

- Deterioration due to use and/or exposure to the elements
- Transfer of ownership
- Change of use/occupancy
- Renovation, rehabilitation, and restoration
- Strengthening or hardening
- Damage from wind, earthquake, fire, or impact
- Signs of structural distress

These reasons are discussed briefly below.

Deterioration Due to Use and/or Exposure to the Elements

If you build it, will you maintain it? Prospective owners, whether in the private or public sector, generally are focused on the initial construction costs of a structure and not the cost of maintaining it. More than a decade ago, we all became familiar with the term *deferred maintenance,* most likely generated by greedy owners and accountants on the theory that "if it is not broken, don't fix it." However, having made this statement, the question to ask is: Should it not be maintained continuously so that it does not break? Distress in our bridges, parking garages failures, building facade collapses, and the like have been attributed to lack of maintenance and proper protection from the elements and just normal use. And then someone (probably a corporate manager) came with another term, *reengineering,* and most likely than not applied it to an existing maintenance/repair program, which extended the time period to reduce these program costs. Thus the need for a structural condition assessment survey becomes apparent in order to identify the areas of distress and to establish a repair/maintenance program for extending the life of the structure.

Transfer of Ownership

Prospective buyers of buildings and other structures may require a structural condition assessment prior to purchase in order to use the results of the survey as a negotiating tool for establishing the reasonable purchase price. A process known as a *preacquisition survey,* most likely initiated by realtors and/or attorneys, is used commonly prior to this type of transaction. The purpose of this survey is to assess the condition of the structural components and the exterior envelope of an existing structure in order to formulate an opinion on its safety, stability, and requirement for repairs, if any, for potential future use. Also, there are occasions when a financial institution and/or an insurance company may require a condition assessment of a structure prior to a loan or insurance coverage for a client.

Change of Use/Occupancy

A structure designated for change of use or occupancy generally requires a structural condition assessment for the owner's intended use. This survey will establish the past performance of the structure, and with an analytical engineering evaluation, a basis can be formulated for establishing criteria for the proposed use. Within this investigation, the building attributes, such as code compliance, fire safety, and seismic resistance, would be reported to the client.

Renovation, Rehabilitation, and Restoration

The terms *renovation* and *rehabilitation* have been used interchangeably in the building industry and design profession. For this reason, an attempt is made here to clarify the meaning/interpretation of these two terms. In structural engineering jargon, *renovation* pertains to the rearrangement and/or removal and replacement of building components, both structural and nonstructural. *Rehabilitation* is the term generally used for structures that are in need of repair owing to deterioration from the elements or from use. In common terms, to *rehab* a structure is to give it limited extended life. *Restoration* is definitive, to restore or bring back the condition of the structure to its original or nearly original condition.

Strengthening or Hardening

Structures that require *strengthening* or *hardening* for seismic resistance and/or protection of its occupants and the general public from a blast or vehicle damage from external sources require a structural condition assessment survey. This survey is paramount for gathering information regarding the condition, composition, size, and location of the building's structural components prior to assessing the need for strengthening or hardening the structure.

Damage from Wind, Earthquake, Fire, or Impact

The purpose of a structural inspection following a wind storm, fire, earthquake, or other abnormal event, such as an explosion, is to assess the resulting damage to the structural components. This generally requires a "due diligence" on short notice with little or no preparation regarding inquiry of the type and age of the structure. Further, engineering judgments may be required for the initiation of temporary safety measures for stabilizing the structure until more definitive solutions can be implemented.

Signs of Structural Distress

An inspection of a failure during or after construction would be, in part, an assessment survey and then most likely move into the realm of forensic engineering. This type of structural assessment would encompass the condition or state of the area surrounding the distressed location and would not necessarily formulate an engineering opinion of the probable cause of the distress or failure.

PROTOCOLS

A number of protocols should be reviewed and agreed on between the client and the engineer before beginning a structural condition assessment so as to obviate any misunderstanding and later dispute.

Questions to Ask

When an engineer receives a call from a prospective client, certain pertinent items for the proposed services should be discussed and questions asked to ascertain the owner's needs. The following list of inquires may serve as a general guide and may be tailored to the engineer's specific office practice and/or experience.

What Is the Reason and Purpose of the Survey? It is necessary to inquire about the reason and the purpose of the survey. This will put the engineer in a position where he or she will be able to ask pertinent questions concerning the type of survey to be performed and whether any of the structural assessment survey will become part of litigation. The engineer should inform the client that a multilevel approach for a structural assessment survey is available for consideration. The engineer should then explain briefly that the survey would entail a *preliminary assessment* followed by a *detailed assessment,* if required.

The *preliminary assessment* would include the following activities:

Review of available documents

Performance of a walk-through site inspection, identifying visible distressed areas

Taking field measurements of the distressed area and pertinent measurements of structural components surrounding the area

Performance of a preliminary engineering analysis of a typical or damaged area

Preparation of a report of the preliminary findings with recommendations

The *detailed assessment* is based on the preliminary assessment but is detailed more accurately to assess the structural adequacy of the structure in order to determine if the structure satisfies the required code performance criteria, to identify the building's structural deficiencies, to recommend repairs or rehabilitation, and to prepare a written report of the findings

Who Does the Client Represent? It is important to ascertain your prospective client's position. Is he or she a sole owner or does he or she represent a company/ corporation or public agency and is responsible to a board of directors? The reason: Should the client be other than an individual or a partnership, the engineer then must consider that he or she most likely will have to respond to a number of departments or persons requiring more technical time than a single-person client.

What Does the Client Know of Our Firm? Are you familiar with our firm, have you been referred to us from someone who is familiar with our services, or was our name taken from the Yellow Pages in the phone book? The response will be helpful in understanding the client's expectations and as an opportunity to present your qualifications and experience to the client.

What Is the Type, Age, Location, and History of the Structure? Although this is a basic question, this would allow an experienced engineer, on receiving a verbal description of the project and probable conditions, to visualize the type of structure that will be under investigation and then be in a position ask pertinent questions regarding the type of assessment survey to be performed. Further, if the project is in the vicinity of your office, it would to beneficial to visit the site prior to presenting a proposal to the client. This site visit will enable you to formulate your own opinion of the proposed scope of work for the structural assessment survey.

Are Contract or As-Built Drawings Available? Reference to existing documentation, particularly the original structural, architectural, or as-built drawings, will greatly reduce the engineer's proposed tasks of the survey. On the other hand, should documentation not be available, the engineer's tasks would be more difficult, requiring additional technical investigative time and necessitating probes and possible materials testing to ascertain the various components and materials of the structure.

Will a Written Report Be Required? A written report contains more detailed information than an oral presentation and therefore would be scrutinized more, particularly if attorneys represent opposing parties. Written reports generally require more precise information, which will add to the engineering time in completing the designated tasks.

What Is the Time Frame? The client, in most instances, will request an immediate survey or one as soon as possible with a designated time frame. Scheduling, availability of personnel, and compensation are important factors to be considered in accepting the structural condition assessment survey assignment.

Agreement

The Council of American Structural Engineers (CASE), an affiliate of the American Council of Engineering Companies (ACEC),[1] has prepared an agreement form to assist structural engineers and clients when undertaking a condition assessment survey. At this writing, CASE Document 16, "An Agreement Between Client and Structural Engineer for a Structural Condition Assessment,"[2] is the only formal document form within the engineering profession that states the scope of work to be performed, compensation, and responsibilities between client and engineer. This document is reproduced at the end of this chapter (Exhibit 3.1). In lieu of this formal agreement, a simple short letter of agreement may be proposed to the client stating the tasks the engineer will perform for a fee within a designated time period. A sample letter of agreement is attached at the end of this chapter (Exhibit 3.2) and may be tailored to the engineer's specific mode of practice.

Compensation

Often the prospective client is on a fast-track schedule or has a limited window of opportunity to retain an engineer for the structural assessment of the facility in question. Usually, the client would prefer a fixed fee from the engineer, particularly if it is a preliminary assessment survey. The engineer should be in the position to quote a fixed fee for a preliminary assessment because the scope of services and technical time required for the work could be determined with the information at hand. A detailed assessment survey will require more research and field and office technical time, especially if material tests results need to be analyzed to ascertain the strength and quality of the structural components.

Fee schedules presented to the client also should state compensation for project reimbursables and possible additional services, if any. The form of compensation for engineering services varies with each office. It is not the job for the engineer performing the structural condition assessment survey also to prepare construction bid documents for the areas to be repaired. Should the client desire this service from the inspecting engineer, it would constitute an "additional service." Similarly, should the client request the inspecting engineer to assist in a legal dispute, this also would constitute an additional service. In this case, it would beneficial for the engineer to use CASE Document 12, "An Agreement Between Client and Structural Engineer for Forensic (Expert) Service."[3]

TYPES OF STRUCTURAL CONDITION ASSESSMENTS

The extent, focus, and depth of the assessment are based largely on the client's needs. It may be just a preliminary survey, as outlined earlier in this chapter and discussed below, or it may be a preliminary survey followed by a detailed assessment that includes structural analyses as well as field surveys, as discussed below. Reference is made to the American Society of Civil Engineers, *Guideline for Structural Condition Assessment of Existing Buildings*, SEI/ASCE 11-99.[4]

Preliminary Assessment Survey

The field survey investigation of existing structures should be conducted by a senior engineer and an assistant (the team). The latter may be a junior engineer or a technician. Depending on the size of the structure, there may be occasions when more than one team is required to perform the survey. The following will serve as a guide for performing the necessary procedures for a field condition assessment survey.

It is helpful if the facilities manager or superintendent of the building—if the structure is a building—is available to serve as a guide for the initial tour of the facility. The engineer should interview the guide regarding his or her familiarity with the history of the structure. In the case of a building, a few pertinent questions may include: How long have you been acquainted with the building? Are you aware of any renovations? What is the age of the roofing, or how recently was it repaired or replaced? Have you seen or encountered any distress in the building, such as floor settlement or vibrations, cracks in slabs, water infiltration, visible deterioration, facade displacement, etc.?

After gathering as much information as possible, the team should perform a visual observation of the facades of the building and a walk-through of the interior, noting the condition of the exposed areas of the structure. During this review, one might make observations and cursory notes of typical nonstructural elements such as stairs, handrails, walls, windows, etc. Also, if a portion of the structure is not exposed to view, e.g., hidden by a hung acoustical ceiling or other architectural finishes, requesting your guide to remove a few panels or other finishes for your observation may prove valuable.

In the absence of adequate or any documentation, the team should select one or more typical bays and take measurements of the structural elements for a preliminary structural analysis.

Notes, sketches, and pictures taken in the field are extremely valuable for your report. Therefore, be accurate and clear, this will save you time and enable you to formulate a good report. The team should be equipped with binoculars, a strong flashlight, digital camera, tape recorder, 50-ft tape, note pad for sketches, and even a hammer for sounding concrete, walls, floors, etc.

Detailed Assessment

The preliminary assessment survey is basically a prelude to a detailed field assessment survey. This survey not only reviews the gravity load-carrying systems but also the lateral resisting load system and load paths to the foundations. Suggested guidance is offered in the discussion that follows.

If documents are not available, the primary structural components must be measured and recorded for subsequent analytical review. If steel members are fireproofed, the protective fireproofing material must be removed, in part, for proper measurements.

Reinforcing steel in critical concrete members should be measured and the placement pattern ascertained. This could be performed by using a commercially available R-meter for bar size and location or by intrusive probes in the bottom at midspan and over supports for more complete information for an analytical study. Reinforcing in concrete columns could be investigated in the same manner.

Depth of concrete slab thickness can be measured at floor openings and where piping, duct work, or stairs are located. If none are available for this measurement, a probe can be made by drilling through the slab in order to measure the fill, if any, and slab thickness.

Testing may be required to ascertain material strength. For concrete, cores may be drilled and recovered for testing in accordance with the accepted American Concrete Institute (ACI) requirements. For reinforcing steel, a piece of nonessential steel may be cut and removed for metallurgical and strength tests in accordance with the American Society for Testing and Materials Standards (ASTM). To ascertain the type and strength of structural steel, coupons may be cut from members (at nonessential areas) and tested in accordance with accepted ASTM standards. Representative connection details of main members should be surveyed and reviewed for structural adequacy. Bolted connections should be inspected to ascertain if they are tight. Column bay spacing and filler beam locations should be measured and recorded. Floor-to-floor height measurements also are necessary.

For timber construction, a representative sample of wood may be cut and removed for testing to determine the species and engineering properties of the wood in accordance with the procedures and tests presented in American Institute of Timber Construction standards. It is also possible that foundation probes may be necessary to ascertain the soil carrying capacity and size of footings or pile caps.

There may be cases where the integrity of the structure still may be in question after all the foregoing information is gathered and analyzed. In such cases, field load testing of a portion of the structure may be advisable to satisfy not only the engineer but also the owner and local building department officials as well.

CONTRACT, AS-BUILT, AND AS-CONDITION DRAWINGS

It is important to understand the distinction between the type of drawings that may be used in the structural condition assessment. *Contract* or *construction bid drawings* are those that were used by the contractor when bidding the project to be constructed. These may have been altered during the construction to reflect changes as the work progressed.

As-built drawings pertain to the original construction at completion. As-built drawings are the construction bid drawings that have been revised to reflect approved changes and substitutions by either the design professional, or the owner, or the contractor because of job conditions in the project (structure) during construction. These drawings are also known as a *record set of drawings* that are submitted to the owner for future reference.

As-condition drawings represent the state of the structure at the time of the condition assessment survey. For example, as-condition drawings would be prepared when the construction bid drawings or as-built drawings of a structure are not available for review, and the client requests and commissions the inspecting engineer to perform a complete structural dimensional survey of the size, type, and location of the structural components of the building. In modern practice, the information gathered in this field survey would be transposed to a computer-aided design (CAD) file and become the as-condition drawings for the project.

Field sketches, known as as-condition sketches, have a different format than as-condition drawings. Field sketches of pertinent areas under investigation in existing structures are common and are prepared on site and become an integral part of the investigative report.

REPORT

Preparing and submitting a written report are usually part of a condition assessment survey. Although many engineers may not be trained in report writing, the report writer must be able to express his or her findings in a clear, precise, and factual manner and in language that can be understood clearly by the client. The report should be organized, substantive, but not lengthy and should address the engineer's observations, findings, conclusions and recommendations. The engineer should not philosophize on the architectural design nor establish blame or responsibility regarding the condition of the structure or the areas that were inspected.

Although technical jargon is customary to engineers, the client may not fully comprehend the meaning of some statements in the report. Therefore, it would be beneficial to explain some terms. such as *tension* as pulling apart, *compression* as pressing together, *fatigue* as bending a paper clip back and forth with you hands until it breaks, etc. Further, the engineer should be aware that if the written report is reviewed by adversaries, which is more likely than not, words and statements will be scrutinized for their meanings and implications.

A typical table of contents for the report is as follows:

Executive summary
Project description
Information received and reviewed
Scope of the investigation and assessment
Findings
Summary and conclusions
Recommendations
Disclaimer
Exhibits

These sections should include the information discussed below.

Executive Summary. It is a good idea to begin with an executive summary, particularly if many parties will be reading the report. The summary should state (1) purpose of the structural condition assessment survey, (2) short project description, (3) inspection time period, (4) an overall brief statement of findings, (5) pertinent concerns and the engineer's conclusions, and (6) recommendations.

Project Description. Write a detailed description of the physical project. Refer to photographs and drawings, if any, included in the report

Information Received and Reviewed. List all documentation received and reviewed with the latest issued dates. There may be occasions where the engineer has conducted interviews or was guided through the facilities by persons familiar with the project. List the names, times, and dates of these events.

Scope of the Investigation and Assessment. State the purpose and type of the assessment. Explain the means and techniques employed for the survey. List the various tasks performed and the degree and depth of analyses. Note the areas that were field surveyed for material types, sizes, and thickness. List the locations and types of probes made and samples taken. List the types and numbers of tests performed.

Findings. Identify problem areas. Make field sketches as required, and include them in the report. Remember, paper is cheap; the more field information gathered, written, or sketched, the less "head scratching" and revisit time to the site will be required. Include a summary of test results. Provide pertinent results of your analytical analyses and evaluations of the structure. Indicate the building code requirements at the time of the original design/construction and the differences, if any, from the present accepted building code that are pertinent to the project. Indicate the effects of input from other disciplines, i.e., architect, mechanical engineer, if applicable. Include photographs of inspected areas and general views of the project.

Summary and Conclusions. Provide a summary of the investigation. State and formalize the conclusions based on the information received, results of the site inspection, measurements, results of tests, and analytical analysis.

Recommendations. List recommendations derived from this survey, and if applicable, indicate the consequences if the recommendations are not followed. It is important to state in the recommendations to the client that your professional engineering services are available to him or her for preparing engineering documents for the repairs and/or reinforcement recommendations noted in the report as an additional service.

Disclaimer. The use of an appropriate and carefully worded disclaimer is important to limit the engineer's liability to the specific intent of the report. Refer to the Council of American Structural Engineers (CASE) publication, "National Practice Guidelines for the Preparation of Structural Engineering Reports for Buildings."[5]

Exhibits. Include pertinent photographs taken during the field visit(s). Include drawings or sketches produced for the survey. Provide copies of test data and reports.

REFERENCES

1. Council of American Structural Engineers (CASE), an affiliate of the American Council of Engineering Companies (ACEC).
2. Council of American Structural Engineers (2004). CASE Document 16, "An Agreement Between Client and Structural Engineer for a Structural Condition Assessment," CASE, Washington, DC.
3. Council of American Structural Engineers (1997). CASE Document 12, "An Agreement Between Client and Structural Engineer for Forensic (Expert) Services," CASE, Washington, DC.
4. American Society of Civil Engineers (1999). *Guideline for Structural Condition Assessment of Existing Buildings,* SEI/ASCE 11-99, ASCE, New York.
5. Council of American Structural Engineers (1995). *National Practice Guidelines for the Preparation of Structural Engineering Reports for Buildings.* CASE, Washington, DC.

EXHIBIT 3.1 Sample contract between client and engineer.

CASE Document 16

An Agreement Between Client and Structural Engineer for A Structural Condition Assessment©

Prepared by the Council of American Structural Engineers

Introduction

The purpose of this Document is to provide a sample Agreement for structural engineers to use when providing this service directly to a client. This sample Agreement features:

- ☐ A Letter of Agreement
- ☐ An Exhibit that defines the Summary of Services
- ☐ An Exhibit defining the Terms and Conditions of the Agreement.

There are many occasions where Owners of various types of structures require a detailed structural condition assessment of their existing structure(s). For example: This may be required for up-grading the structure for increase in imposed loads; for the condition assessment of structures due to damage from fire, wind, earthquake; for seismic retrofitting; for historic preservation or change in occupancy; for adding new structure upon or adjacent to an existing structure.

Prospective buyers may require a structural condition assessment survey prior to purchase, other-wise known as a "pre-acquisition survey."

Financial lending institutions and/or insurance companies, at times, require a structural condition assessment survey prior to an acquisition or for insurance purposes.

The Structural Engineering Institute and The American Society of Civil Engineers publication—Guideline for Strutural Condition Assessment of Existing Buildings—SEI/ASCE 11-99 recommends two approaches to a structural condition assessment:

A) Preliminary Assessment—A visual walk-thru, preliminary analysis and estimating the structural adequacy of the existing structure and if needed, establishing the need for a Detailed Assessment.

B) Detailed Assessment—A definitive inspection of the structure (ascertain sizes dimensions and the quality of the members) and an in-depth analysis of the structure in order to increase the reliability of the investigative report and recommendations.

The Summary of Services herein in Exhibit A is a guide for the performance of a condition assessment survey. There services could vary depending upon the specific needs of an individual Project. However, the format used should be similar. Exhibit B is the Terms and Conditions are for both parties to review.

Once the method of compensation for services has been agreed upon, it is important to cross out (or delete in your own firm's version) fee options and other information which are not applicable.

As with all documents that are intended to formalize contractual relationships, the guidance and advice of an attorney is necessary to assure proper usage for specific applications and jurisdictions. We strongly recommend that you have your legal advisor, professional liability carrier, and your accountant review this document. No warranty of any kind is made with respect to this document or other contractual or consequential damages in connection with, or arising out of, the furnishing, performance, or use of this document.

EXHIBIT 3.1   *(Continued).*

CASE Document 16

*An Agreement Between Client and Structural Engineer for
A Structural Condition Assessment©*

Prepared by the Council of American Structural Engineers

Date: _____

Name of Client: _____

Address of Client: _____

Attention: _____

Project: _____
(Name and address/location of structure(s))

Dear: _____

We are pleased to propose the following Agreement for providing an independent structural condition assessment of the above referenced project(s). This proposal will remain open for acceptance for _____ day(s) from the date above.

OBJECTIVE

The objective of this condition assessment is to determine the status of the structure(s) being inspected for the specific requirements of the Client. This survey is limited in scope and will be conducted to the extent necessary to render an opinion regarding the needs, stability and integrity of the structure(s).

DESCRIPTION OF PROJECT

The Project consists of _____

SUMMARY OF SERVICES

The services to be provided are described in the Summary of Services Exhibit A.

TERMS AND CONDITIONS

Terms and Conditions of this Agreement are addressed in Exhibit B.

ENGINEERING CHARGES

Compensation for our services shall be: [*choose one of the following two*]

1. A lump sum fee of _____ dollars ($ _____).

EXHIBIT 3.1 (*Continued*).

2. A fee calculated on an hourly rate basis per our standard rate schedule. At this time we estimate the total fee to be _____ dollars ($_____).
   This total fee is an estimate. If the estimate is exceeded by more than 10% (ten percent), you shall be so advised in advance.

Our current standard hourly rate schedule is:

Principal:	$_____	Field Engineer:	$_____
Project Manager:	$_____	CAD Operator:	$_____
Engineer:	$_____	Technical Assistant:	$_____

Additional Services shall be charged at our then current standard hourly rates

A retainer in the amount of [_____% of the total fee] or [$_____]
shall be paid upon execution of this Agreement. The retainer shall be applied against the final invoice.

REIMBURSABLE EXPENSES [*choose one of the following two*]

1. Reimbursable expenses as described in the Terms and Conditions shall be billed as a multiple of _____ times the cost incurred.

2. In lieu of reimbursable expenses, an administrative fee shall be paid as a lump sum of _____ dollars ($_____)
   which will be invoiced on a pro rata basis throughout the course of the Project.

ADDITIONAL PROVISIONS

This Letter of Agreement, and Exhibits A & B hereto, constitute the entire Agreement between the parties. Two copies of this Letter of Agreement have been provided to you. Please examine these documents and if acceptable, sign one copy of this letter and return it to us along with the retainer. Retain a copy for your records.

We are looking forward to working with you on this Project.

Sincerely,

Offered by: Accepted by and Agreed to:

_____   _____
Structural Engineer   Owner

_____   _____
Print Name and Title   Print Name and Title
Date   Date

EXHIBIT 3.1 *(Continued)*.

CASE Document 16

An Agreement Between Client and Structural Engineer for
A Structural Condition Assessment©

Prepared by the Council of American Structural Engineers

EXHIBIT A—Summary of Services

This is an exhibit attached to and made part of the Letter of Agreement dated _____,
between _____ and _____.
Structural Engineer (SE)   Client

The services of the Structural Engineer (SE) for this proposal include those summarized below.
See Exhibit B—Terms and Conditions, for further details.

Basic Services	Included	Not Included	Remarks
Review available documents and reports.			
Perform a site inspection to ascertain existing structural systems.			
Perform visual observations for: deterioration of materials, damages, modifications, weakness in members or connections, settlement or other structural deficiencies of the Primary Structural System.			
Examine the roof and below grade areas for water infiltration.			
Examine facade by binoculars. Non-Strutural System.			
Prepare a preliminary structural analysis estimating the load carrying capacity of representative members for code compliance.			
Present a verbal report of our findings and recommendations.			
Submit a written report of our findings and recommendations with appropriate photographs.			
Attending meetings			No. of _____

EXHIBIT 3.1 (*Continued*).

Basic Services	Included	Not Included	Remarks
DETAILED ASSESSMENT			
Perform services listed in the **Preliminary Assessment** section			
Perform a field survey of representative members sizes in the absence of existing structural drawings.			
Review load path for the basis gravity load system from origin to foundation. Primary Structural System			
Review the lateral load resisting system.			
Analyze a representative number of structural elements (columns, beams, slabs, bracing, etc.) for gravity and lateral loads.			
Establish a materials testing program for structural members, if required.			
Review architectural and other engineering documents for potential special load or framing requirements. Secondary Structural Elements.			
Prepare a written report of findings and recommendations with appropriate photographs			
Attend meetings			No. of

ADDITIONAL SERVICES

1._____

2._____

3._____

This Exhibit should be expanded to list all the work anticipated on this Project.

EXHIBIT 3.1   *(Continued)*.

CASE Document 16

An Agreement Between Client and Structural Engineer for
A Structural Condition Assessment©

Prepared by the Council of American Structural Engineers

EXHIBIT B—Terms and Conditions

This is an exhibit attached to and made part of the latter of Agreement dated _____,
between _____ and _____.
                Structural Engineer (SE)   Client

Section 1—Geneal

1.1.1   These Terms and Conditions, along with the Letter Agreement, and Exhibit A—
        Summary of Services, form the Agreement as if they were part of one and the same
        document.

1.1.2   The Letter Agreement and Exhibit A may limit or negate the applicabiity of these
        Terms and Conditions.
        Such limitation shall take precedence over provisions of this Exhibit.

1.2   General Obligations of the Structural Engineer and the Client

1.2.1   The Structural Engineer (SE), shall perform those condition assessment services as
        specified in Exhibit A and detailed in these Terms and Conditions. In rendering these
        services, the SE shall apply the skill and care ordinarily exercised by structural engi-
        neers experienced in evaluating existing structures.

1.2.2   The Client shall identify the applicable building code or codes, and provide the SE, if
        available, with copies of each of the following: complete drawings and specifications
        from all disciplines, structural design calculations, and the geotechnical report.

1.2.3   The Client shall provide all criteria and full information with regard to his or her
        requirements for the Project.

1.2.4   The Client and SER shall designate a person to act with authority on his or her behalf
        with respect to all aspects of the Project.

1.2.4   The Client shall arrange for the SE to have access to the site and provide a guided
        tour as may be required.

1.2.5   The Client shall retain a contractor to uncover/expose areas of the structure where
        there is restricted accessibility for the SE to examine the structure and/or take appro-
        priate measurements in order to ascertain member sizes.

1.2.6   The Client shall remove/contain any hazardous materials prior to the SE inspection.

1.2.7   The Client shall pay for all necessary tests required by the SE for the evaluation of
        the structure.

1.2.8   The SE shall submit to the Client a schedule for the performance of the SE's services.

EXHIBIT 3.1 *(Continued)*.

1.3   Definitions

    1.3.1   **Structural Engineer (SE)** is the professional engineer who possesses technical qualifications, practical experience and professional judgment and opinion for the Project.

    1.3.2   **Primary Structural System** is the completed combination of elements which serve to support the building's self weight, the applicable live load (which is based upon the occupancy and use of the spaces), the environmental loads such as wind and thermal, plus seismic loading (if applicable), Curtain wall members, non-loadbearing walls or exterior facade are examples of items that are not part of the Primary Structural System.

    1.3.3   **Secondary Structural Elements** are elements that are structurally significant for the function they serve but do not contribute to the strength or stability of the primary structure. Examples may include but not be limited to: support beams above the primary roof structure which carry a chiller, elevator support rails and beams, retaining walls independent of the primary building, and flagpole or light pole foundations.

    1.3.4   **Non-Structural Elements** are elements of a structure that are not Primary or Secondary Structural Elements. Items in this category could be exterior curtain walls and cladding, non-bearing partitions, stairs and railings.

    1.3.5   **Reimbursable Expenses** are expenses incurred directly or indirectly in connection with the project such as, but not limited to, transportation, meals and lodging for travel, long distance telephone calls and facsimile transmissions, overnight deliveries, courier services, outside consulting services and materials testing laboratories, professional sale taxes and the cost of reproductions beyond those normally required for coordination and information purposes.

Section 2—Basic Services

2.1 General

    2.11   The Basic Services of the SE shall include condition assessment of the Project, as designated in Exhibit A.

    2.12   Preliminary and Detailed Assessment may; stand alone or be combined.

Secton 3—Exclusions

3.1 General

3.1.1   This Agreement is not a check of general requirements, such as Use Group or Type of Construction. It is not a check of life safety or fire protection systems. It is not a check of any code provisions, other than those concerning the stability and integrity of the Primary Structural System.

3.1.2   No attempt will be made to coordinate the structural components of the building with Secondary and Non-Structural Elements shown or specified on the documents of other design disciplines. No attempt will be made to verify dimensions, except to the extent necessary to review the adequacy of a particular structural component.

Section 4—Additional Services

4.1 General

    4.1.1   Services beyond those outlined under Basic Services may be requested. The SE under terms mutualy agreed upon by the Client and the SE may provide these.

EXHIBIT 3.1 *(Continued)*.

Section 5—Fees and Payments

5.1 Fees and Other Compensation

5.11 Fees for Basic Services, Additional Services, and Compensation for Reimbursable Express are set forth in the Letter of Agreement.

5.2 Payments on Account

5.2.1 Invoices for SE's services shall be submitted, at the SE's option, either upon completion of any phase of the service or on a monthly basis. Invoices are payable when rendered and shal be considered past due if not paid within 30 days of the invoice date.

5.2.2 Retainers, if applicable to this Project, shall be credited to the final invoice.

5.2.3 Any inquiry or questions concerning the substance or content of any invoice shall be made to the SE in writing within 10 days of receipt of the invoice. A failure to notify the SE within this period shall constitute an acknowledgement that the service has been provided.

5.3 Late Payments

5.3.1 A service charge will be charged at the rate of 1.5% (18% true annual rate) per month or the maximum allowable by law on the then outstanding balance of past due accounts. In the event any portion of the account remains unpaid 90 days after billing, the Client's pay of collection, including reasonable attorney's fees.

Section 6—Insurance, Indemnification & Risk Allocation

6.1 Insurance

6.1.1 The SE shall secure and maintain professional liability insurance, commercial general liability insurance, and automobile liability insurance to protect the SE from claims for negligence, bodily injury, death, or property damage which may arise out of the performance of the SE's services under this Agreement, and from claims under the Worker's Compensation Acts. The SE shall, if requested in writing, issue a certificate confirming such insurance to the Client.

6.2 Indemnifications

The Client shall indemnify and hold harmless the SE and all of its personnel, from and against any and all claims, damages, loses and expenses (including reasonable attorney's fees) arising out of or resulting from the performance of the services, provided that any such claims, damage, loss or expense are caused in whole or in part by the negligent act or omission and/or strict liability of the Client, anyone directly or indirectly employed by the Client (except the SE) or anyone for whose acts any of them may be liable.

The SER shall indemnify and hold harmless the Client and its personnel from and against any and all claims, damages, losses, and expenses (including reasonable attorney's fees) to the extent they are caused by the negligent act, error, or omission by the SER in performance of its services under this Agreement, subject to the provisions in the paragraph below on Risk Allocation.

EXHIBIT 3.1   (*Continued*).

6.3   Risk Allocation

6.3.1   In recognition of the relative risks, rewards and benefits of the Project to both the Client and the SE, the risks have been allocated such that the owner agrees that, to the fullest extent permitted by law, the SE's total liability to the Client for any and all injuries, claims, losses, expenses, damages or claim expenses arising out of this Agreement, from any cause or causes, shall not exceed the amount of $_____, the amount of the SE's fee (whichever is greater), or other amount agreed upon, $_____. Such causes include, but are not limited to, the SE's negligence, errors, omissions, strict liability, breach of contract or breach of warranty.

Section 7—Miscellaneous Provisions

7.1   Reuse of Documents

7.1.1   All documents including calculations, computer files, drawings and sketches prepared by the SE pursuant to this Agreement are instruments of professional service intended for the one-time use in connection with this Project. They are and shall remain the property of the SE. Any reuse without written approval or adaptation by the SE is prohibited.

7.2   Termination, Successors and Assigns

7.2.1   This Agreement may be terminated upon 10 days written notice by either party should the other fail to perform its obligations hereunder. In the event of termination, the Client shall pay the SE for all services rendered to the date of termination, all reimbursable expenses, and reasonable termination expenses.

7.2.2   The Client and SE each binds himself or herself, partners, successors, executors, administrators, assigns and legal representatie to the other party of this Agreement and to the partners, successors, executors, administrators, assigns, and legal representative of such other party in respect to all covenants, agreements, and obligations of this Agreement.

7.3   Disputes Resolution

7.3.1   All claims, counterclaims, disputes and other matters in question between the parties hereto arising out of or relating to this Agreement or the breach thereof will be presented to non-binding mediation, subject to the parties agreeing to a mediator(s).

7.4   Governing Laws

7.4.1   This Agreement shall be governed by the laws of the principal place of business of the Structural Engineer.

Section 8—Supplement Conditions

[Insert descriptions of any modifications to the Terms and Conditions included in this Agreement.]

EXHIBIT 3.2   Sample letter of agreement form.

An Agreement For the Provision of Limited Professional Services

Introduction

This SAMPLE short form, two-page agreement is recommended for use on small projects, with limited scope of service, where the work will be performed within a relatively short time frame. It contains the essentials of a good contract including scope of service, fee arrangements, and terms and conditions, and is tailored to the small project.

SAMPLE
CASE Document 16

An Agreement for the Provision of Limited Professional Services
Prepared by the Council of American Structural Engineers

Structural Engineer (SE): _____ Client:_____

Project No. _____ Date: _____

Project Name: _____

Location: _____

Scope of Services: _____

Fee Arrangement: _____

Principals	$_____ /Hr.		CAD Operator $_____ /Hr.	
Engineers	$_____ /Hr.		Admin. Assistant $_____ /Hr.	

Retainer Amount: _____

Special Conditions: _____

EXHIBIT 3.2 (*Continued*).

Offered by (SE):   Agreed and Accepted by (Client):

      (signature)   (signature)   (date)

      (printed name/title)   (printed name/title)

The terms and conditions form are part of this agreement.

Terms and Conditions

Structural Engineer (SE) shall perform the services outlined in this agreement for the stated fee arrangement.

Fee

The total fee, except stated lump sum, shall be understood to be an estimate, based upon Scope of Services, and shall not be exceeded by mre than ten percent, without written approval of the Client. Where the fee arrangement is to be on an hourly basis. The rates shall be those that prevail at the time services are rendered.

Billings/Payments

Invoices will be submitted monthly for services and reimbursable expenses and are due when rendered. Invoice shall be considered PAST DUE if not paid within 30 days after the invoice date and the SE may, without waiving any claim or right against Client, and without liability what-soever to the Client, terminate the performance of the service. Retainers shall be credited on the final invoice. A service charge will be charged at 1.5% (or the legal rate) per month on the unpaid balance. In the event any portion of an account remains unpaid 90 days after billing, the Client shall pay cost of collection, including reasonable attorneys' fees.

Access To Site

Unless otherwise stated, the SE will have access to the site for activities necessary for the per-formance of the services. The SE will take precautions to minimize damage due to these activities, but has not included in the fee the cost of restoration of any resulting damage.

Hidden Conditions and Hazardous Materials

A structural condition is hidden if concealed by existing finishes or if it cannot be investigated by reasonable visual observation. If the SE has reason to believe that such a condition may exist, the SE shall notify the Client who shall authorize and pay for all costs associated with the investigation of such a condition and, if necessary, all costs necessary to correct said condition. If (1) the Client fails to authorize such investigation or correction after due notification, or (2) the SE has no reason to believe that such a condition exists, the Client is responsible for all risks associated with this condition, and the SE shall not be responsible for the existing condition nor any resulting damages to persons or property. SE shall have no responsibility for the discovery, presence, handling, removal, disposal or exposure of persons to hazardous materials of any form.

EXHIBIT 3.2 *(Continued)*.

Indemnifications

The Client shall indemnify and hold harmless the SE and all of its personnel from and against any and all claims, damages, losses and expenses (including reasonable attorneys fees) arising out of or resulting from the performance of the services, provided that any such claims, damage, loss or expense is caused in whole or in part by the negligent act or omission and/or strict liability of the Client, anyone directly or indirectly employed by the Client (except the SE) or anyone for whose acts any of them may be liable. This indemnification shall include any claim, damage or losses due to the presence of hazardous materials.

Risk Allocation

In recognition of the relative risks, rewards and benefits of the project to both the Client and the SE, the risks have been allocated so that the Client agrees that, to the fullest extent permitted by law, the SE's total liability to the Client, for any and all injuries, claims, losses, expenses, damages or claim expenses arising out of this agreement, from any cause or causes shalll not exceed the total amount of $50,000, the amount of the SE's fee (whichever is greater) or other amount agreed upon when added under Special Conditions. Such causes include, but are not limited to, the SE's negligence, errors, omissions, strict liability, breach of contract or breach of warranty.

Termination of Services

This agreement may be terminated upon 10 days written notice by either party should the other fail to perform his obligations hereunder. In the event of termination, the Client shall pay the SE for all services rendered to the date of termination, all reimbursable expenses, and reasonable termination expenses.

Ownership Documents

All documents produced by the SE under this agrement shall remain the property of the SE and may not be used by the Client for any other endeavor without the written consent of the SE.

Dispute Resolution

Any claim or dispute between the Client and the SE shall be submitted to non-binding mediation, subjet to the parties agreeing to a mediator(s). This agreement shall be governed by the laws of the principal place of business of the SE.

As with all documents that are intended to formalize contractual relationships, the guidance and advice of an attorney is necessary to assure proper usage for specific appliations and jurisdictions. We strongly recommend that you have your legal advisor, professional liability carrier, and your accountant review this document. No warranty of any kind is made with respect to this document or other contractual or consequential damages in connection with, or arising out of, the furnishing, performance, or use of this document.

CHAPTER 4

Past and Current Structural Codes and Standards

CYNTHIA L. CHABOT, P.E.

INTRODUCTION

Structural condition assessment of facilities will require familiarity with current code requirements as well as the original code requirements under which the structure was designed and built. It is imperative that the engineer performing a condition assessment has familiarity with the provisions as well as the intent, and in some cases the history, of the governing code or standard. Although, the adherence or nonadherence of the original design and construction to building code requirements may or may not have a significant bearing on assessment of the existing condition of the structure, it provides a basis to begin to think about the expected performance of the structure in its current state. Current codes, standards, and guidelines account for existing buildings in their consideration of repair, alteration, addition, and change of use. These codes should be the first point of reference because they may provide strict requirements or allowances depending on particular situations or the jurisdiction's requirements. The building codes, material codes, and specifications, as well as standards used to design and construct the original facility, will provide some understanding or expectation of the actual conditions.

CODES, STANDARDS, AND GUIDELINES FOR EXISTING BUILDINGS

Professionals and the public have recognized the need to assess existing buildings with an understanding toward preservation without sacrificing safety. Codes and standards have been developed for existing structures that have made special considerations toward this view. The facilities that undergo repair, alteration, addition, or change of use may trigger a code for existing buildings depending on the jurisdiction's requirements. These codes and standards try to provide a level of *predictability* to the owner to highlight early what may be expected, as well as to set a level of work required by society through regulations that are *proportional* to the amount of alteration that facility will undergo.

International Existing Building Code

The first edition of the *International Existing Building Code (IEBC)*[1] was published in 2003. The original intent of the code was to become the base code for all work in all existing buildings and that provisions regulating repair, alterations, additions, and change of occupancy in the *International Building Code (IBC;* see section to follow) would be deleted. Although the IEBC is still undergoing changes from code developers, such as plumbing and fire safety, the structural portion is well defined and virtually unopposed. At the time of this writing, the IEBC has been adopted by some states within the United States, such as Georgia, Missouri, Florida, Maryland, Rhode Island, and parts of New York State. With opposition to having the IEBC to account for all existing structures, a compromise was reached at the code hearing in September 2003 that would include the IEBC as a reference in the IBC, replacing Chap. 34, "Existing Structures," and placing this chapter at the end of the IBC as an appendix. This compromise would allow jurisdictions adopting the IBC an opportunity to use either the new IEBC as referenced or maintain the old chapter requirements as provided in the appendix until all parties are in agreement.

SEI/ASCE 11, *Guideline for Structural Condition Assessment of Existing Buildings*[2]

The need to evaluate and more fully use the existing building inventory has come about from changing economic conditions, concern for historic preservation, emphasis on fully using conveniently located structures, space shortages, and the increasing cost of materials and products used in the construction of new buildings. SEI/ASCE 11 provides the design community with tools for assessing the structural condition of existing buildings constructed of combinations of materials, including concrete, masonry, metals, and wood. It consists of an overview of preliminary and detailed assessment procedures, of materials properties and test methods, and of evaluation procedures for various physical conditions of the structure.

SEI/ASCE 30, *Guideline for Condition Assessment of the Building Envelope*[3]

This guideline is a compilation of basic information, procedures, and references for assessing the condition and performance of an existing building's envelope systems and components and identifying problematic and dysfunctional elements.

MODEL CODES AND STANDARDS IN THE UNITED STATES

The development of laws, rules, and regulations to provide for the safety and service-ability of buildings and structures in the United States is somewhat unique. In most countries, the national government oversees the regulatory development and enforcement process, which results in a single national code. In the United States, however, the development of building codes and standards has become a private-sector enterprise involving federal, state, and local government participation but with only minimal influence or control from these groups except as "users" in local enforcement.

Building Codes

Local building codes in the United States are patterned after model building codes. Recently developed model building codes include the *International Building Code* (*IBC*)[4] and the *NFPA 2000 Building Construction and Safety Code*.[5] Previous building codes that are being phased out gradually and replaced with the *IBC* include the *Uniform Building Code*[6] by the International Conference of Building Officials (ICBO), the *National Building Code*[7] by the Building Officials and Code Administrators (BOCA), and the *Standard Building Code*[8] by the Southern Building Code Congress International (SBCCI). The geographic areas of the United States where the various codes had been used are shown in Fig. 4.1. In 1994, International Code Council (ICC) was established which contained the three founding members of BOCA, ICBO, and SBCCI, and its goal was to create a single set of model codes. The provisions set forth in the model building codes are representations of possible regulations and do not become law until enacted by the authority having jurisdiction (state, county, city, etc.). Thus these documents usually are adapted or adopted to satisfy local laws and ordinances and to reflect local building practices.

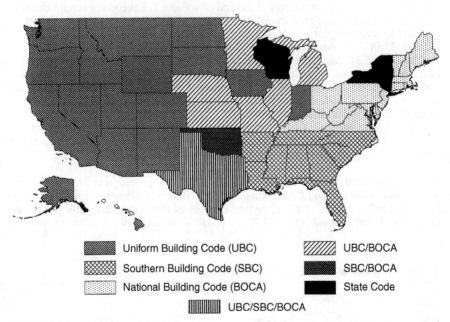

Uniform Building Code (UBC)   UBC/BOCA

Southern Building Code (SBC)   SBC/BOCA

National Building Code (BOCA)   State Code

UBC/SBC/BOCA

FIGURE 4.1   Areas of model building code usage prior to the promulgation of the International Building Code (IBC)[2]

The model codes are consensus documents that have been reviewed by governing bodies and anyone in the general public. Any individual or industry organization may participate in the development of these codes and related deliberations. Industry tends to be heavily involved because code provisions have an obvious impact on the marketplace. The often-stated challenge is to develop provisions that provide an acceptable level of risk with respect to potential hazards and at the same time safeguard the economy. The model codes reference or copy other material codes, manuals, standards, and specifications. By reference or copy, they become part of the model code and thus the requirements of the jurisdiction adopting the model code. Owing to the lengthy code creation process, however, references in the model codes often lag behind the most current material code, manual, standard, or specification. A brief history of the previous model codes follows.

Uniform Building Code.[6] The *Uniform Building Code* was first developed by the International Conference of Building Officials in 1927. Revised editions of the code have been published since that time at approximate 3-year intervals (Table 4.1). New editions incorporate changes approved since the previous edition. The *Uniform Building Code* is designed to be compatible with related publications to provide a complete set of documents for regulatory use.

The provisions of the 1994 edition of the *Uniform Building Code* were reformatted into the common code format established by the Council of American Building Officials (CABO). The new format established a common format of chapter designations for the three model building codes published in the United States at the time. Apart from those changes approved by the conference membership, this reformatting has not changed the technical content of the code.

Provisions of the *Uniform Building Code* and the *UBC Standards* have been divided into a three-volume set. Volume 1 contains administrative, fire safety and life safety, and field inspection provisions. Chapters 1 through 15 and Chapters 24 through 35 are contained in volume 1 in their entirety. Appendices associated with these chapters also are contained in their entirety at the end of volume 1. Excerpts of certain chapters from volume 2 are reprinted in volume 1 to provide for easier usability. Volume 2 contains structural engineering design provisions and specifically contains Chapters 16 through 23 in their entirety. Included in this volume are design standards previously published in *UBC Standards*. Design standards have been added to their respective chapters as divisions of the chapters. Appendices associated with these chapters are contained in their entirety at the end of volume 2. Excerpts of certain chapters from volume 1 are reprinted in volume 2 to provide for easier usability. Volume 3 contains material, testing, and installation standards.

The *Uniform Building Code* was metricated in 1994. The metric conversions are provided in parentheses following the English units. Where industry has made metric conversions available, the conversions conform to current industry standards.

National Building Code.[7] The *BOCA National Building Code* was first adopted by the Building Officials and Code Administrators (BOCA) International, Inc., in 1950. Like the *Uniform Building Code,* revised editions of this code were published at approximate 3-year intervals (Table 4.2). Proposals for changes were either accepted or

TABLE 4.1   *Uniform Building Code* **Edition Dates**

1927	1943	1958	1970	1982	1994
1935	1949	1961	1973	1985	1997
1937	1952	1964	1976	1988	
1940	1955	1967	1979	1991	

TABLE 4.2 *BOCA National Building Code* Edition Dates

1950	1958	1967	1978	1987	1996
1952	1961	1970	1981	1990	1999
1955	1964	1975	1984	1993	

rejected by vote of the organization's active members, who were practicing regulatory code officials.

The 1993 edition of the *BOCA National Building Code* was the first model code to implement the common building code format that was developed cooperatively by the three model code groups together with the American Institute of Architects and the Society of Fire Protection Engineers under the auspices of the Council of American Building Officials. The new format consists of 11 basic subject matter groups: administration and terms, building planning, fire protection, occupant needs, building envelope, structural systems, structural materials, nonstructural materials, building services, special devices and conditions, and reference standards. Structural systems are subdivided into three chapters corresponding to structural loads, structural test and inspections, and foundations and retaining walls. Structural materials include chapters on concrete, lightweight metals, masonry, steel, and wood. *Building Code Requirements for Reinforced Concrete* (ACI 318), *Building Code Requirements for Masonry Structure* (ACI 530), the *AISC Specification for Structural Steel Building, Allowable Stress Design and Plastic Design,* the AISC *Load and Resistance Factor Design Specification for Structural Steel Buildings,* the *National Design Specification for Wood Construction,* and related standards are adopted by reference. Chapter 33 includes requirements for site work, demolition, and construction.

Standard Building Code.[8] The first edition of the *Southern Standard Building Code* appeared in 1946. Like the other model codes, the *Southern Standard Building Code* was revised periodically and finally published at approximate 3-year intervals with annual revisions (Table 4.3). In 1974, the Southern Building Code Congress became the Southern Building Code Congress International, Inc. (SBCCI), and in early 1975, the word *Southern* was dropped from the titles of all *Standard Codes.* In 1994, the SBCCI adopted the common building code format, where the contents were organized in a manner similar to that of *UBC* and *BOCA.* The *Standard Building Code* incorporates, by reference, nationally recognized consensus standards for use in judging the performance of materials and systems.

International Building Code.[4] The *International Building Code* was developed by the International Code Council (ICC). The first edition of the *International Building Code* was prepared in 2000, with amendments in 2001 and 2002 and a second edition in 2003. The intent of the ICC was to draft a comprehensive set of regulations for

TABLE 4.3 *Standard Building Code* Edition Dates

1946	**1965**	1972R	**1979**	1986R	1993R
1948	1966R	**1973**	1980R	1987R	**1994**
1949R	1967R	1974R	1981R	**1988**	1995R
1950	1968R	1975R	**1982**	1989R	1996R
1953–54	**1969**	**1976**	1983R	1990R	**1997**
1957–58	1970R	1977R	1984R	**1991**	1998R
1960–61	1971R	1978R	**1985**	1992R	1999

Note: Bold indicates a complete edition. R indicates revisions to the code versus a complete new edition.

building systems consistent with and inclusive of the scope of the ICBO, BOCA, and SBCCI. The ICC sought to develop an efficient regulatory system for the built environment through the joint and cooperative promulgation of a comprehensive and compatible package of model codes suitable for adoption by governmental entities. The IBC was developed through a governmental consensus system. The ICC is the merger of the three model code organizations—the International Conference of Building Officials (ICBO), the Building Officials and Code Administrators (BOCA), and the Southern Building Code Congress International (SBCCI), which no longer exist independently.

NFPA 5000 Building Construction and Safety Code. The National Fire Protection Association (NFPA) developed the *NFPA 5000 Building Construction and Safety Code,* whose first edition was published in 2002 as the 2003 edition. The code was developed based on a full and open consensus-based procedure accredited by the American National Standards Institute (ANSI). Accreditation by ANSI demonstrates a code development organization's commitment to balanced input from all interested parties. ANSI accredits code developers that adhere to the guiding principles of consensus, due process, and openness.

IBC versus NFPA 5000. The development of *NFPA 5000* created a competing model building code with the *International Building Code.* Many jurisdictions in the United States are currently undergoing reviews of their building codes to decide how to proceed with their own codes. The *IBC* and *NFPA 5000* have many differences, two of which include the code development process and the methods the codes use to reference standards and specifications. Both codes are developed through consensus, however. *NFPA 5000* uses an ANSI-based process for code development, and *IBC* uses a governmental consensus system. From the structural engineer's point of view, the code development process is probably not a major concern, provided that all interested parties have the opportunity for code change proposals. The more important difference to the structural engineer is the method by which reference standards are incorporated into the code. NFPA 5000 references standards without reproducing information in the model code itself. The IBC, however, has reproduced some of the information from material specifications and standards in the tradition of the three former model codes. In some cases, the IBC has made minor modifications to standards to refine the code as it required.

The structural engineering community has recognized advantages to referencing a standard without reproducing or changing the original document in the model code. The standards writing committees have specialized expertise and thus are the proper forum to make highly technical changes. The National Council of Structural Engineers Association (NCSEA) is in the process of reviewing both model building codes to make the codes be, as much as practicable, the same. What this will mean to structural engineers is that either code will have the same requirements for structural engineering. NCSEA will attempt to see that standards that are copied in the model codes are removed, if practical, and adopted by reference. Changes that should be made to a standard will be proposed to the original standards developing organization. This system of reference standards will ensure more uniformity between the two codes. The goal to make both structural portions of the codes the same may take many years of submitting code change proposals to the code development committees and several cycles of code editions. The resulting model codes hopefully will result in consensus structural engineering requirements based on sound engineering and state-of-the-art research. In the meantime, each jurisdiction will choose between the two model codes, and structural engineers will be responsible for using the correct code in their jurisdictions.

International Residential Code. The *International Residential Code (IRC)*[9] was developed by the International Code Council. An alliance was formed between the National Association of Home Builders (NAHB), ICBO, BOCA, SBCCI, and CABO in the 1970s to develop a residential code. Its task was to develop a prescriptive code. CABO produced the first code, *One and Two Family Dwelling Code*[10] (Table 4.4). In 1994, the International Code Council began to develop one code. In 1998, it became the *International One and Two Family Dwelling Code*. In 2000, it became the *International Residential Code (IRC)*. The latest edition of the *IRC* was in 2003, which places the *IRC* editions in line with the *IBC*.

Bridge Code

AASHTO Standard Specifications for Highway Bridges.[11] Compilation of the *AASHTO Standard Specifications for Highway Bridges* began in 1921 with organization of the Committee on Bridges and Structures of the American Association of State Highway Officials (AASHTO). During the period from 1921 until its first printing in 1931, the specifications were developed gradually and were made available in mimeographed form for use by state highway departments and other organizations. A complete specification was available in 1926, and it was revised in 1928. The first edition of the *Standard Specifications* was published in 1931, and it quickly became the de facto national standard for bridges. The *Standard Specifications for Highway Bridges* has been reissued in consecutive editions at approximate 4-year intervals ever since (Table 4.5).

The *AASHTO Standard Specifications for Highway Bridges* serves as a standard or guide for the preparation of state specifications and for reference by bridge engineers. The specifications set forth minimum requirements that are consistent with current practice, with certain modifications necessary to suit local condition. They apply to ordinary highway bridges, and supplemental specifications may be required for unusual types and for bridges with spans longer than 500 ft. Specifications of the American Society for Testing and Materials, the American Welding Society, the American Wood Preservers Association, and the National Forest Products Association are referenced or recognized. Interim specifications usually are published in the middle of the calendar year.[11]

In 1986, the Subcommittee on Bridges and Structures submitted a request to the AASHTO Standing Committee on Research to undertake an assessment of U.S. bridge

TABLE 4.4   *CABO One and Two Family Dwelling Code* to IRC Edition Dates

1971	**1985** Amend	**1995**	
1973 Supp	**1986**	1996–97 Amend	
1975	1987 Amend	**1998**	International One and Two Family Dwelling Code
1976 Supp	1988 Amend	**2000**	International Residential Code
1978 Supp	**1989**	**2003**	International Residential Code
1979	1990 Amend		
1980 Amend	1991 Amend		
1981 Amend	**1992**		
1983	1993 Amend		
1984 Amend	1993–94 Amend		

Note: Bold indicates a complete edition. Supp (supplement) and Amend (amendment) indicate modifications but not a complete edition.

TABLE 4.5 *AASHTO Standard Specifications for Highway Bridges* **Edition Dates**

Date	Edition Number	Date	Edition Number
1921–26	Mimeographed sheets available to state highway departments and other organizations	1961	8th edition
1926	Complete specification available	1965	9th edition
1928	Revised	1969	10th edition
1931	1st edition	1973	11th edition
1935	2d edition	1977	12th edition
1941	3d edition	1983	13th edition
1944	4th edition	1989	14th edition
1949	5th edition	1991	15th edition
1953	6th edition	1996	16th edition
1957	7th edition	1998	17th edition

design specifications, review foreign design specifications and codes, consider design philosophies alternative to those underlying the *Standard Specifications,* and develop recommendations based on these investigations. The principal recommendation of the subcommittee was the development of an entirely new load and resistance factor design (LRFD) bridge design standard. The Federal Highway Administration and the states have established a goal that the LRFD standards be used on all new bridge designs after 2007; only edits related to technical errors in the 17th edition will be made hereafter. The *AASHTO Standard Specifications for Highway Bridges* are applicable to new structure designs prior to 2007 and for the maintenance and rehabilitation of existing structures. New bridges constructed after 2007 will rely on the *AASHTO LRFD Bridge Design Specifications*[12] and its companion, *AASHTO LRFD Bridge Construction Specifications.*[13] The design specifications have been produced in a first edition in 1994 and a second edition in 1998, both of which are produced in English and International System (SI) units. The construction specifications have been produced with a first edition in 1998 only in metric.

DESIGN CODES, STANDARDS, AND SPECIFICATIONS

The model building codes in the United States adopt many of the national design standards developed by organizations involved with building materials, such as concrete, masonry, steel, and wood. The model codes also adopt by reference many of the ASTM standards as the recognized test procedures to ensure construction quality. ASTM *Standards in Building Codes*[15] is a compilation of these standards. Like the model building codes, many of these standards are developed and written in a form that allows them to be adopted by reference in a general building code.

Concrete

ACI 318, Building Code Requirements for Structural Concrete.[17] From the early 1900s until the early 1960s, the principal method of design for reinforced concrete was working stress design (WSD). Since publication of the 1963 edition of the ACI code, there has been a rapid transition to ultimate strength design (USD). The 1963 ACI code (ACI 318-63) treated the WSD and USD methods on an equal basis. However, a major portion of the WSD was modified to reflect USD behavior. The WSD provisions of the 1963 code relating to bond, shear and diagonal tension, and combined axial compression and bending had their basis in USD (Table 4.6).

TABLE 4.6 *ACI Building Code Requirements for Structural Concrete Edition Dates*

Year	ACI Code	Title of Publication	No. of Pages
1910	Standard no. 4	Standard Building Regulations for the Use of Reinforced Concrete	14
1920	Standard no. 23	Standard Building Regulations for the Use of Reinforced Concrete	21
1927	E-1A-27T	Tentative Building Regulations for Use of Reinforced Concrete	36
1928	E-1A-28T	Tentative Building Regulations for Use of Reinforced Concrete	38
1936	ACI 501-36T	Building Regulations for Reinforced Concrete	40
1941	ACI 318-41	Building Regulations for Reinforced Concrete	63
1947	ACI 318-47	Building Code Requirements for Reinforced Concrete	64
1951	ACI 318-51	Building Code Requirements for Reinforced Concrete	64
1956	ACI 318-56	Building Code Requirements for Reinforced Concrete	144
1963	ACI 318-63	Building Code Requirements for Reinforced Concrete	144
1963	ACI SP-10	Commentary on Building Code Requirements for Reinforced Concrete	91
1971	ACI 318-71	Building Code Requirements for Reinforced Concrete	78
1971	Supplement	Commentary on Building Code Requirements for Reinforced Concrete	96
1977	ACI 318-77	Building Code Requirements for Reinforced Concrete	102
1977	ACI Committee Report	Commentary on Building Code Requirements for Reinforced Concrete	132
1983	ACI 318-83	Building Code Requirements for Reinforced Concrete	111
1983	ACI 318R-83	Commentary on Building Code Requirements for Reinforced Concrete	155
1989	ACI 318-89/318R-89	Building Code Requirements for Reinforced Concrete and Commentary	353
1992	ACI 318-89/318R-89	Building Code Requirements for Reinforced Concrete and Commentary	347
1992	ACI 318M-89/318RM-89	Building Code Requirements for Reinforced Concrete and Commentary (in metric units)	347
1995	ACI 318-95/318R-95	Building Code Requirements for Reinforced Concrete and Commentary	369
1995	ACI 318M-95/318RM-95	Building Code Requirements for Reinforced Concrete and Commentary (in metric units)	371
1999	ACI 318-99/318R-99	Building Code Requirements for Reinforced Concrete and Commentary	391
1999	ACI 318-99M/318RM-99	Building Code Requirements for Reinforced Concrete and Commentary (in metric units)	
2002	ACI 318-99/318R-99	Building Code Requirements for Reinforced Concrete and Commentary (retained 1989 designation)	443
2002	ACI 318-02M/318RM-02	Building Code Requirements for Reinforced Concrete and Commentary (metric edition of 1999 was given 2002 year designation)	

In 1971, ACI 318-71 was based entirely on the USD approach for proportioning reinforced concrete members, except for a small section devoted to what was called an "alternate design method," which was WSD. Even in that section, the service load capacities (except for flexure) were prescribed as various percentages of the ultimate strength capacities of other parts of the code. The transition to ultimate strength theories for reinforced concrete design was essentially complete in the 1971 ACI code.

In ACI 318-77, the alternate design method was moved to an appendix. The appendix location served to separate and clarify the two methods of design, with the main body of the code devoted exclusively to the USD method. The alternate design method has been retained in the appendix of ACI 318 since the 1983 edition. Since an appendix location is sometimes not considered to be an official part of a legal document (unless specifically adopted), specific reference is made in the main body of the code (Sec. 8.1.2) to make the alternate design method in the appendix a legal part of the code. Regardless of whether the USD method or the alternate design method is used in proportioning for strength, the general serviceability requirements of the code, such as the provisions for deflection control and crack control, always must be satisfied.

Masonry

ACI 530/ASCE 5/TMS 402, Building Code Requirements for Masonry Structures.[18] In the early 1960s, masonry industry associations began development of a technological database of masonry materials and performance through research and testing programs. The result of this effort culminated in design standards such as the Brick Institute of America's (BIA) *Recommended Practice for Engineered Brick Masonry* in 1966 and the National Concrete Masonry Association's (NCMA) *Specifications for Loadbearing Concrete Masonry* in 1970. Each document addressed only selected masonry materials. In 1970, American Concrete Institute Committee 531 published a report entitled, "Concrete Masonry Structures—Design and Construction," and in 1976 it published *Specifications for Concrete Masonry Construction* (ACI 531.1-76). Both documents served as the basis for *Building Code Requirements for Concrete Masonry Structures* (ACI 531-79), which addressed only concrete masonry.

The American Society of Civil Engineers (ASCE), the American Concrete Institute, and The Masonry Society (TMS) undertook the development of a national design code in the late 1970s. An agreement resulted in the ACI/ASCE/TMS 530 Masonry Standards Joint Committee (MSJC), formed in 1978, to develop a consensus standard for masonry design. A design code and construction specifications were drafted for committee ballot by 1984. Final adoption of a code, specifications, and companion commentaries by ACI, ASCE, and TMS occurred in October 1988.

ACI 530/ASCE 5/TMS 402, *Building Code Requirements for Masonry Structures,* is directed primarily to the designer and code enforcement officials, whereas ACI 530.1/ASCE 6/TMS 602, *Specifications for Masonry Structures,*[19] is directed primarily to the contractors and inspectors. Significant aspects related to these documents are that brick, block, and combinations of brick and block are covered in a single document, and design is based on the premise that all work will be inspected.

Steel

AISC Manual of Steel Construction.[20,21] The AISC *Manual of Steel Construction* has evolved through numerous versions from the first edition, published in 1927. Table 4.7 lists the revisions of the manual and provides brief description of the changes to the code and specification printed in the manual. The first edition of the manual included four parts, and the current edition includes eight parts (Table 4.8). The specifications and codes, Part 5 of both the ASD and LRFD current codes, includes the

TABLE 4.7 *AISC Manual of Steel Construction* **Edition Dates**

Edition and Printing		Comments
		ASD
1st	1927–33	10 printings: Adopted specification June 1, 1923 and revisions Nov. 1, 1928. Adopted code Oct. 1, 1924 and revisions in 1927 and 1928.
2d	1934–36	4 printings: Editorial revisions to specification and code Jan. 1934. Adoption of revised specification June 1936.
3d	1937–41	4 printings: Revised specification and code Jan 1937. Adoption of revised code June 1937.
4th	1941–45	5 printings: Revision to specification and editorial revisions to code July 1941.
5th	1946–62	30 printings: Adoption of revised specification Feb. 1946 and June 23, 1949, Nov. 30, 1961. Revisions to code Nov. 1, 1945, Dec. 1, 1946, and June 26, 1952.
6th	1963–69	4 printings and 4 revised printings: Revisions to specification Apr. 17, 1963 and Feb. 12, 1969. Minor revisions to code Feb. 20, 1963 and Sept. 14, 1966
7th	1970–79	1 printing and 1 revised printing: Adopted suppl. no. 1 and revisions to specification Nov. 1, 1970, adopted suppl. no. 2 and revisions to specifications Dec. 8, 1971, suppl. no. 3 added and revisions to specification June 12, 1974. Revisions to code July 1, 1970 and Oct 1, 1972, where commentary was included for the first time
8th	1980–88	11 impressions, 3 revised printings: Revisions to specification Nov. 1, 1978. Revisions to code Sept. 1, 1976.
9th	1989–present	1 printing, 2 impressions, 2 revisions
		LRFD
1st	1986–93	1 printing
2d	1994–2001	
3d	2002–present	2 printings to date

Note: "Edition" indicates general update for new specification. "Printing" indicates changes made to a supplement issued to the specification or to update or correct material in the manual. "Impression" indicates reprinting with no or minor changes.

"Specification for Structural Steel Buildings" and the "Code of Standard Practice for Steel Buildings and Bridges." These two sections have been developed since the first edition of the code in 1927. Over the years, specifications have been added to the manual to account for new technologies. For example, the "Specification for Structural Joints Using ASTM A325 or A490 Bolts" was first included in the ninth edition.

In 1986, AISC produced the first edition of a new LRFD manual entitled, *Load and Resistance Factor Design (LRFD) Manual of Steel Construction*. The load factors and load combinations prescribed by AISC were developed to be used with the recommended minimum loads given in ANSI A58.1, *Minimum Design Loads for Building and Other Structures* (now SEI/ASCE 7). Building codes either incorporate both the ASD and LRFD approach or adopt them by reference.

AWS D1.1, Structural Welding Code—Steel [25] ***and AASHTO/AWS D1.5, Bridge Welding Code.*** [26] The American Welding Society published the first edition of the *Code for Fusion Welding and Gas Cutting in Building Construction* in 1928. In 1934, a committee was appointed to prepare specifications for the design, construction, al-

TABLE 4.8 *AISC Manual of Steel Construction Section Guide*

1st Edition (1927)	9th Edition ASD (1989)	3d Edition LRFD (2002)

Part I   Standard Specifications
       AISC Code of Standard Practice
       History of Steel and Iron
	Part 1 Dimensions and Properties
	Part 2 Beam and Girder Design
	Part 3 Column Design
		Part 1 Dimensions and Properties
		Part 2 Column Design
		Part 3 Beam and Girder Design

Part II   Properties of Sections
       Formulae for Beam Loadings
       General Mathematical Tables
	Part 4 Connections
	Part 5 Specifications and Codes
	       Miscellaneous Data and
		Part 4 Composite Design
		Part 5 Connections

	Part 6 Mathematical Tables	Part 6 Specifications and Codes Miscellaneous Data and
Part III Strength of Materials General Information	Part 7 Symbols and Index	Part 7 Mathematical Tables Part 8 Index and Nomenclature
Part IV Explanation of AISC Specification	Part 5	Part 5

Rolled Structural Shapes
       Built-up Sections
       Dimensions, Functions
	Specification for Structural Steel
	BuildingsAllowable Stress
	AISC Code of Standard Practice for Buildings and
	Bridges
		Specification for Structural Steel
		Buildings—Load and Resistance
		AISC Code of Standard Practice for Buildings and
		Bridges

Allowable Load Tables Rivet	Specification for Structural Joints Using ASTM A325 or 490 Bolts Specification for Allowable Stress Design of Single-Angle Members AISC Quality Certification Program	Specification for Structural Joints using ASTM A325 or 490 Bolts LRFD Specification of Single-Angle Members Specification for Steel Hollow Structural Sections

teration, and repair of highway and railway bridges. The first bridge welding specification was published in 1936. Until 1963, there were separate AWS committees for bridges and buildings. These two committees joined in 1963 to form the Structural Welding Committee of the American Welding Society. The two documents were consolidated in 1972 into the D1.1 document. However, they were separated once again in 1988 when the joint AASHTO/AWS D1.5, *Bridge Welding Code* was published to address the specific requirement of state and federal transportation departments.

Preparation of the *Bridge Welding Code* was undertaken in response to a need for a common welding code for the fabrication of steel highway bridges by welding. The departments of highways and transportation in the 50 states, the District of Columbia, and Puerto Rico that make up the American Association of State Highway and Transportation Officials (AASHTO) have used the code of the AWS Structural Welding Committee routinely, with appropriate modifications, to produce contract documents suitable for the construction of bridges using federal highway funds. The proliferation of requirements by the constituents of AASHTO resulted in recognition of the need for a single document that could produce greater economies in bridge fabrication while at the same time addressing the issues of structural integrity and public safety.

The Federal Highway Administration of the U.S. Department of Transportation required states using federal funds for the construction of welded highway bridges to conform to the specified standards for design and construction. Conformance to the AWS *Specification for Welded Highway and Railway Bridges* was first specified in the third edition of the AASHTO *Standard Specifications for Highway Bridges* in 1941. In 1962, the Bureau of Public Roads, now the Federal Highway Administration (FHWA), required conformance to the "Circular Memorandum," dated November 13, 1962. This document transmitted additional provisions for welding A36 steel pending publication of the AWS specification that would contain certain essential provisions for welding A36 not then in the code. Another "Circular Memorandum," dated August 19, 1966, modified provisions of the 1966 edition of AWS D2.0-66, *Specification for Welded Highway and Railway Bridges.* An FHWA notice dated July 7, 1971, recommended that ultrasonic inspection not be used for final acceptance of welds made by electrogas or electroslag procedures because of concern that the acceptance levels of AWS D2.0-69, Appendix C, were not suitable to detect or reject piping porosity of major dimensions.

In 1974, AASHTO published the first edition of *Standard Specification for Welding of Structural Steel Highway Bridges.* The eleventh edition of AASHTO *Standard Specifications for Highway Bridges,* dated 1977, directed, "Welding shall conform to the requirements of the AASHTO *Standard Specifications for Welding of Structural Steel Highway Bridges 1974* and subsequent interim specifications." AASHTO published the second and third editions of *Standard Specifications for Welding of Structural Steel Highway Bridges* in 1977 and 1981. All the AASHTO specifications were required to be part of the contract documents as modifications or additions to the AWS *Structural Code—Steel.* This was a cumbersome procedure.[22]

In 1982, a subcommittee was formed jointly by AASHTO and AWS, with equal representation from both organizations, to seek accommodation between the separate and distinct requirement of bridge owners and existing provisions of AWS D1.1. The *Bridge Welding Code* is the result of an agreement between AASHTO and AWS to produce a joint AASHTO/AWS *Structural Welding Code* for steel highway bridges that addresses essential AASHTO needs and makes AASHTO revisions mandatory.

The 1988 versions of the *Bridge Welding Code* provided for the qualification of welding procedures by test to ensure that welds had the strength, ductility, and toughness necessary for use in redundant structures. Nonredundant fracture critical bridge members were not provided for in the first edition of the code. While qualification of welding procedures is required, a major effort has been made to specify the minimum number of tests and the simplest tests that give reasonable assurance of required mechanical properties. Efforts have been made to discourage individual states from requiring duplication of weld testing unless the testing is specified in the bid documents.

Wood

National Design Specification for Wood Construction.[33] In the early part of the twentieth century, structural design with wood was based on general engineering principles using working stresses or design values published in engineering handbooks and in local building codes. These design values often were not in agreement, even for the same species of wood. Further, in most cases, the assigned values were not related to lumber grade or quality level.[27]

The Forest Products Laboratory, an agency of the Forest Service, U.S. Department of Agriculture, in cooperation with the National Lumber Manufacturers Association (NLMA), now the American Forest and Paper Association (AF&PA), prepared a guide for grading and determining working stresses for structural grades of timber that was published subsequently by the U.S. Department of Agriculture, with the title, *Miscellaneous Publication 185—Guide to the Grading of Structural Timbers and the Determination of Working Stresses.*[28] In 1934, the NLMA assembled the information given in *Miscellaneous Publication 185* together with engineering design equations and other technical information on wood in the publication, *Wood Structural Design Data* (*WSDD*).[29] The publication contained extensive span and load tables for various sizes of timber beams and columns. A second edition of *WSDD* was issued in 1938 and a revised second edition in 1941. The first edition of *WSDD,* together with information developed at the Forest Products Laboratory, was published subsequently under one cover as the first edition of the *Wood Handbook* in 1935.[30,31]

When World War II began, it created an urgent need for a comprehensive national design standard for timber structures, including wood connections. After a 3-year effort by the Technical Advisory Committee of the NLMA in close consultation with the Forest Products Laboratory, the *National Emergency Specification for the Design, Fabrication and Erection of Stress Grade Lumber and Its Fastenings in Buildings— Directive 29*[32] was issued in 1943. The directive was prescribed for all federal departments and agencies involved in war construction. The first edition of the *National Design Specification for Wood Construction* (*NDS*)[33] was published by NLMA in 1944 under the title, *National Design Specification for Stress-Grade Lumber and Its Fastenings,* and was primarily the same content as *Directive 29.* This first edition included allowable unit stresses for stress-graded lumber; design formulas; design loads; provisions for timber connectors; bolted, lag-screw, nail, and wood screw joints; and guidelines for the design of glued laminated structural members.

In 1977, the title was changed to the *National Design Specification for Wood Construction* to reflect the new nature of the specification. The specification included allowable unit stresses for stress-graded lumber, design formulas, and design loads and provisions for timber connector, including bolted, lag screw, nail, and wood screw joints. Also included were guidelines for the design of glued laminated structural members (Table 4.9).

LRFD Specification for Engineered Wood Construction was developed by the joint NFPA/ASCE Design of Engineered Wood Construction Standards Committee and adopted by the American Wood Council in 1996. The LRFD specification provided an alternative design methodology to allowable stress design procedures specified in prior editions of the *National Design Specification for Wood Construction* (*NDS*). There are two primary differences between ASD and LRFD procedures for the design of wood structures. In ASD, safety adjustments are applied to strength properties only, and in LRFD, safety adjustments are applied to both loads and strength properties.

The model building codes adopt the *NDS* by reference or by copying sections of the specification directly into the code.

TABLE 4.9 National Design Specification for Wood Construction Edition Dates

Date	Name of Publication	Name of Organization	Comments	Date	Name of Publication	Name of Organization	Comments
1944	NDS SGL&F	NLMA	First edition	1968	NDS SGL&F	NFPA	New edition
1948	NDS SGL&F	NLMA	Revision	1971	NDS SGL&F	NFPA	Machine stress-rated lumber and timber piles introduced
1950	NDS SGL&F	NLMA	Revision	1973	NDS SGL&F	NFPA	New edition
1951	NDS SGL&F	NLMA	Revision	1977	NDS	NFPA	New edition
1952	NDS SGL&F	NLMA	Revision	1982	NDS	NFPA	New edition
1953	NDS SGL&F	NLMA	Revision	1986	NDS	NFPA	New edition
1957	NDS SGL&F	NLMA	New edition	1991	NDS	NFPA	New edition
1960	NDS SGL&F	NLMA	New edition	1997	NDS	AF&PA	New edition
1962	NDS SGL&F	NLMA	New edition	2001	NDS	AF&PA	New edition

Note: NDS SGL&F, *National Design Specification for Stress-Grade Lumber and Its Fastenings;* NDS, *National Design Specification for Wood Construction;* NLMA, National Lumber Manufacturer's Association (now known as AF&PA); NFPA, National Forest Products Association (now known as AF&PA); AF&PA, American Forest and Paper Association.

PS 20 American Softwood Lumber Standard.[34] National standardization of the sizes, grades, and inspection of lumber began in 1924 with publication Simplified Practice Recommendation R16, *Lumber—American Lumber Standards for Softwood Lumber.*[35] R16 was revised subsequently in 1924, 1925, 1926, 1929, 1939, and 1953. In 1969, R16–53 was revised and superseded by DOC Voluntary Product Standard PS 20–70. Separate size standards for dry and green lumber under nominal 5-in. thickness were established to achieve greater uniformity in the dimensions of seasoned and unseasoned lumber at the point of use. *PS 20* was revised subsequently in 1994 and currently in 1999.

Load Standards

SEI/ASCE 7, Minimum Design Loads for Buildings and Other Structures.[36] The National Bureau of Standards published a report of the Department of Commerce Building Code Committee entitled, "Minimum Live Loads Allowable for Use in Design of Buildings," in 1924. The recommendations contained in that document were used widely in revisions of local building codes. These recommendations, based on the engineering data available at that time, represented the collective experience and judgment of the committee members responsible for drafting this document.

The American Standards Association (ASA) Committee on Building Code Requirements for Minimum Design Loads in Buildings subsequently issued a report in 1945 that represented a continuation of work in this field. The committee took into consideration the work of the previous committee and expanded on it to reflect current knowledge and experience. The end result was the *American Standard Building Code Requirements for Minimum Design Loads in Buildings and Other Structures,* ANSI A58.1-1945.[37]

The ANSI A58.1 standard has been revised eight times since 1945 (Table 4.10); the latest revision is referred to as SEI/ASCE 7-02, *Minimum Design Loads for Buildings and Other Structures.* Subsequent to the 1982 edition of ANSI A58.1, the American National Standards Institute (ANSI) and ASCE Board of Direction approved "ASCE Rules for Standards Committee" to govern the writing and maintenance of the ANSI A58.1 standard. The standard prescribes design dead loads, live loads, soil and hydrostatic pressures, wind loads, snow loads, rain loads, and earthquake loads, as well as load combinations and load factors. Like earlier editions of the ANSI standard, ASCE 7 has influenced the development and revision of the other building codes significantly. Some of the provisions of SEI/ASCE 7 are discussed below.

Wind Loads. Because of the complexity involved in defining both the dynamic wind load and the behavior of an indeterminate structure when subjected to wind loads, the design criteria adopted by ASCE 7 are based on the application of an equivalent static wind pressure.

The 1982 ANSI standard was a major revision of the 1972 version. A new wind speed map for annual extreme fastest-mile wind based on an annual probability of exceedance of 0.02 (50-year mean recurrence interval) was introduced, with 70 mi/h as the minimum design basic wind speed. This one map replaced maps for 25-, 50-, or 100-year mean recurrence intervals that were used as a measure of the importance of the facility (anticipated use, life, hazard to personnel, acceptable risk, and other judgment factors). The term *importance factor* was introduced in 1982 as a more consistent approach than selecting one of the three maps.

The most significant change in the ASCE 7-95 was the use of a 3-sec gust wind speed instead of the fastest-mile wind multiplied by a gust factor. This change necessitated revisions of terrain and height factors, gust effect factors, and pressure coefficients for components and cladding.

TABLE 4.10 ASCE 7 Minimum Design Loads for Buildings and Other Structures Edition Dates

Year	Code Designation and Standard Name	Publishing Organization
1924	Minimum Live Loads Allowable for Use in Design of Buildings	National Bureau of Standards
1945	American Standard Building Code Requirements for Minimum Design Loads in Buildings and Other Structures in Buildings	ASA Committee on Building Code Requirements for Minimum Design Loads
1955	American Standard Building Code Requirements for Minimum Design Loads in Buildings and Other Structures in Buildings	ASA Committee on Building Code Requirements for Minimum Design Loads
1972	American Standard Building Code Requirements for Minimum Design Loads in Buildings and Other Structures in Buildings	ASA Committee on Building Code Requirements for Minimum Design Loads
1982	American Standard Building Code Requirements for Minimum Design Loads in Buildings and Other Structures in Buildings	ASA Committee on Building Code Requirements for Minimum Design Loads
1988	Minimum Design Loads for Buildings and Other Structures	ASCE 7 Committee on Minimum Design Loads
1993	Minimum Design Loads for Buildings and Other Structures	ASCE 7 Committee on Minimum Design Loads
1993	Minimum Design Loads for Buildings and Other Structures	ASCE 7 Committee on Minimum Design Loads
1995	Minimum Design Loads for Buildings and Other Structures	ASCE 7 Committee on Minimum Design Loads
1998	Minimum Design Loads for Buildings and Other Structures	ASCE 7 Committee on Minimum Design Loads
2002	Minimum Design Loads for Buildings and Other Structures	SEI/ASCE 7 Committee on Design Loads On Structures During Construction

*With establishment of the Structural Engineering Insitutue (SEI) of ASCE in October 1, 1996, the ASCE 7 designation was changed to SEI/ASCE 7.

Snow Loads. In 1965, the *National Building Code of Canada*[38] first published comprehensive provisions for loads due to snow drifting, together with a commentary, based on prior research. These provisions were reaffirmed in the 1970 edition of the *National Building Code of Canada* and were adopted by ANSI in 1972 (ANSI A58.1-1972) and by the *BOCA Code* in 1975. The snowdrift provisions were revised again in ANSI A58.1-1982 and the 1987 *BOCA Code*.

Earthquake Loads. The history of seismic design codes in the United States is fairly brief but very dynamic. During the 1960s, research was conducted to demonstrate that ductile moment-resisting concrete frames could be designed and constructed in seismic zones. This effort was in response to the Structural Engineers Association of California (SEAOC) requirements that buildings greater than 170 ft in height have a complete moment-resisting space frame. *Design of Multi-Story Reinforced Concrete Buildings for Earthquake Motions* was published in 1961 and provided initial material for the SEAOC Seismology Committee to develop ductile reinforced-concrete provisions.[39] The SEAOC committee subsequently published requirements for reinforced-concrete ductile moment-resisting frames and reinforced-concrete shear walls in the 1966 revision to the SEAOC Bluebook.[40]

In 1974, a cooperative effort was undertaken by the Applied Technology Council (ATC) in California to develop nationally applicable seismic design provisions for buildings. The project, known as ATC-3, was part of the Cooperative Federal Program in Building Practices for Disaster Mitigation initiated in 1972. The primary basis of the provisions was to protect life safety and to ensure continued functioning of the essential facilities needed during and after a catastrophe. Primary consideration was given to the main structural framing systems(s) and to energy-absorbing/dissipating effects of interior partitions, exterior cladding, different types of materials, damping, and drift.

After their publication in 1978, ATC-3 provisions were subjected to an extensive several-year evaluation process by the Building Seismic Safety Council (BSSC) under contract to the Federal Emergency Management Agency (FEMA).[41] The ATC-3 provisions were then modified to better apply to all regions of the United States. This process culminated in 1985 with publication by BSSC of *NEHRP Recommended Provisions for the Development of Seismic Regulations for Buildings, 1985 edition.*[42] NEHRP stands for the National Earthquake Hazard Reduction Program, which is financed by the US government and managed by FEMA.

ANSI A58.1-1982 and ASCE 7-88 contained seismic provisions based on those in the 1985 *Uniform Building Code*. The *UBC* provisions for seismic safety, in turn, were based on recommendations of SEAOC and predecessor organizations. The 1955 and 1972 editions of A58.1 contained seismic provisions based on much earlier versions of SEAOC and *UBC* recommendations.

In 1993, the ASCE 7 seismic provisions were changed substantially from prior editions. The 1993 provisions of ASCE 7 were adopted from *NEHRP Recommended Provisions for the Development of Seismic Regulation for New Buildings, 1991 edition.* These provisions are a direct descendant of *Tentative Provisions for the Development of Seismic Regulations for Buildings,* developed by the ATC in 1978 under sponsorship of the National Science Foundation (NSF) and the National Bureau of Standards (NBS, now the National Institute for Standards and Technology, NIST).

The two most significant differences between the 1993 edition and prior editions are that (1) the 1993 edition is based on a strength-level limit state rather than an equivalent loading for use with allowable stress design and (2) the 1993 edition contains a much larger set of provisions that are not direct statements of loading. The intent is to provide a more reliable and consistent level of seismic safety in new building construction. Further background and commentary on these provisions can be found in the following publications:

Part 2, *Commentary of the NEHRP Recommended Provisions for the Development of Seismic Regulations for New Buildings,* Building Seismic Safety Council, Federal Emergency Management Agency, 1991 edition.

Recommended Lateral Force Requirements and Commentary, Seismology Committee, Structural Engineers Association of California, 1990.

The NEHRP provisions have been selected by the federal government as a benchmark for the design of federal buildings. The 1991 NEHRP provisions were modified and adopted as a supplement to the *National Building Code* by the Building Officials Congress of America and by the Southern Building Code Conference International as an amendment to the *Standard Building Code* in the fall of 1991. The 1991 *UBC* was judged by the federal government to be equivalent in seismic safety to the 1991 NEHRP provisions. These three codes, although used in separate regions of the country, currently serve as the principal seismic design codes in the United States.

SEI/ASCE 37, Design Loads on Structures during Construction.[43] The design standard SEI/ASCE 37, *Design Loads on Structures during Construction,* addresses partially completed structures and temporary structures used during construction. The first edition, which is also the current edition, of the standard was published in 2002. The code development began in 1988, and the code was drafted for review in 1996, 1997, 1998, and twice in 1999. A final committee and public balloting took place in 2001.

The construction loads, load combinations, and load factors were developed to account for the relatively short duration of load, variability of loading, variation in material strength, and recognition that many elements of the completed structure that are relied on implicitly to provide strength, stiffness, stability, or continuity are not present during construction. The load factors are based on a combination of probabilistic analysis and expert opinion. The concept of using maximum and arbitrary point-in-time (APT) loads and corresponding load factors was adopted to be consistent with SEI/ASCE 7.

The basic reference for the computation of environmental loads is also SEI/ASCE 7; however, modification factors have been adopted to account for a reduced exposure period. Furthermore, certain loads may be disregarded owing to the relatively short reference period associated with typical construction projects, and certain loads in combinations effectively may be ignored because of the practice of shutting down job sites during these events, e.g., excessive snow and wind.

CONCLUDING REMARKS

Building codes, standards, and specifications were intended to be standards of practice based on solid engineering principles of the time and to provide *minimum* guidelines within the built environment for the safety of the occupants and the general public. They may provide a guide when performing a structure condition assessment as a starting point to understand the standards of practice at the time a structure was built.

ACKNOWLEDGMENT

Many parts of this chapter rely heavily on Chap. 2, "Design Codes Standards, and Manuals," by John F. Duntemann, in RT Ratay. ed., *Forensic Structural Engineering Handbook,* McGraw-Hill, New York, 2000. In order not to distort the meaning of that chapter, certain passages were reproduced here.

ABBREVIATIONS

AASHTO   American Association of State Highway and Transportation Officials (*http://www.aashto.org*)

ACI   American Concrete Institute (*http://www.aci-int.org*)

AF&PA   American Forest and Paper Association (formerly the National Forest Products Association) (*http://www.afandpa.org/*)

AISC   American Institute of Steel Construction, Inc. (*http://www.aisc.org*)

AISI   American Iron and Steel Institute (*http://www.steel.org/*)

AITC   American Institute of Timber Construction (*http://www.aitc-glulam.org/*)

ANSI   American National Standards Institute (*http://www.ansi.org/*)

ASA   American Standards Association (now known as ANSI)

ASCE   American Society of Civil Engineers (*http://www.asce.org*)

ASD   Allowable Stress Design

ASTM   American Society for Testing and Materials (*http://www.astm.org*)

ATC   Applied Technology Council (*http://www.atcouncil.org*)

AWC   American Wood Council (*http://www.awc.org/*)

AWS   American Welding Society (*www.aws.org*)

BIA   Brick Institute of America

BOCA   Building Officials and Code Administrators International, Inc (now known as ICC)

BSSC   Building Seismic Safety Council (*http://www.bssconline.org*)

CABO   Council of American Building Officials

FEMA   Federal Emergency Management Agency (*http://www.fema.gov*)

FPL   Forest Products Laboratory (*http://www.fpl.fs.fed.us/*)

IBC   *International Building Code*

ICBO   International Conference of Building Officials (now known as ICC)

ICC   International Code Council (*http://www.iccsafe.org*)

IEBC   *International Existing Building Code*

IRC   *International Residential Code*

LRFD   load resistance factor design

MSJC   Masonry Standards Joint Committee (*http://www.masonrystandards.org/*), a collaboration between ACI, ASCE, and TMS)

NAHB   National Association of Home Builders (*http://www.nahb.org*)

NBS   National Bureau of Standards (now known as NIST)

NCMA   National Concrete Masonry Association (*http://www.ncma.org/*)

NCSEA   National Council of Structural Engineers Association (*http://www.ncsea.com*)

NDS   National Design Specification for Wood Construction

NEHRP   National Earthquake Hazard Reduction Program (managed by FEMA)

NFPA   National Forest Products Association (now known as AF&PA)

NFPA   National Fire Protection Association (*http://www.nfpa.org*)

NIST   National Institute for Standards and Technology (formerly NBS; *http://www.nist.gov*)

NLMA   National Lumber Manufacturer's Association (now known as AF&PA)

NSF   National Science Foundation (*http://www.nsf.gov*)

RCSC   Research Council on Steel Connections

SBC   Standard Building Code

SBCCI   Southern Building Code Congress International, Inc. (now known as ICC)

SEAOC   Structural Engineers Association of California (*http://www.seaoc. org/*)

SEI   Structural Engineering Institute (*http://www.seinstitute.org*)

SI   International System of Units (Le Système International d'Unités)

SJI   Steel Joist Institute (*http://www.steeljoist.org/*)

TMS   The Masonry Society (*http://www.masonrysociety.org*)

UBC   Uniform Building Code

USD   ultimate strength design

WSD   working stress design

WSDD   Wood Structural Design Data

REFERENCES

Note that general references to codes cited in this chapter list only the most recent document issued.

1. *International Existing Building Code,* International Code Council, Falls Church, VA, 2003.
2. The Standard Committee Structural Condition Assessment and Rehabilitation of Buildings Standards of the Committee Codes and Standards Activities Division (1999): SEI/ASCE 11-99, *Guideline for Structural Condition Assessment of Existing Buildings,* American Society of Civil Engineers, Reston, VA.
3. The Standard Committee Structural Condition Assessment and Rehabilitation of Buildings Standards of the Committee Codes and Standards Activities Division (2000). *Guideline for Condition Assessment of the Building Envelope,* SEI/ASCE 30, American Society of Civil Engineers, Reston, VA.
4. *2003 International Building Code,* International Code Council, Falls Church, VA, December 2002.
5. *NFPA 5000 Building Construction and Safety Code,* National Fire Protection Association, Quincy, MA, 2002.
6. *Uniform Building Code,* 1997 ed., International Conference of Building Officials, 1997.
7. *BOCA National Building Code,* 1999 ed., Building Officials and Code Administrators International, July 1999.
8. *Standard Building Code,* Southern Building Code Congress International, Inc., 1999.
9. *2003 International Residential Code,* International Code Council, Falls Church, VA, 2003.
10. *1995 CABO One and Two Family Dwelling Code,* International Conference of Building Officials, March 1995.
11. *AASHTO Standard Specifications for Highway Bridges,* 17th ed., American Association of State Highway and Transportation Officials, Washington, DC, 1998.
12. *AASHTO LRFD Bridge Design Specifications (U.S. Customary Units or SI Units),* 2d ed., American Association of State Highway and Transportation Officials, Washington, DC, 1998.
13. *AASHTO LRFD Bridge Construction Specifications (SI Units),* American Association of State Highway and Transportation Officials, Washington, DC, 1998.
14. *Guide to the Chronological Development of the AISC Specification and Code of Standard Practice,* Frank W. Stockwell, Northeast Regional Manager, AISC, New York (date unknown).

15. *ASTM Standards in Building Codes,* 40th ed., American Society for Testing and Materials, Philadelphia, 2003.

16. Traw, J., and B. Tubbs (1996). "Future Perspective of U.S. Model Building Codes," paper prepared for the International Conference on Performance-Based Codes and Fire Safety Design Methods, September 24–26, 1996, Ottawa, Ontario, Canada.

17. ACI Committee 318, ACI 318, *Building Code Requirements for Structural Concrete, (ACI 318-02M) and Commentary (ACI 318RM-02),* American Concrete Institute, Detroit, MI, 2002.

18. Masonry Standards Joint Committee, *Building Code Requirements for Masonry Structures* (ACI 530/ASCE 5/TMS 402), American Concrete Institute, Detroit, MI, 2002.

19. Masonry Standards Joint Committee, *Specifications for Masonry Structures* (ACI 530.1/ ASCE 6/TMS 602), American Concrete Institute, Detroit, MI, 2002.

20. *AISC Allowable Stress Design (ASD) Manual of Steel Construction,* 9th ed., American Institute of Steel Construction, New York, 1989.

21. *AISC Load and Resistance Factor Design (LRFD) Manual of Steel Construction,* 3d ed., American Institute of Steel Construction, New York, 2002.

22. American Institute of Steel Construction (1954). *Iron and Steel Beams 1873 to 1952,* 2d ed., AISC, New York.

23. *AISC Rehabilitation and Retrofit Guide,* Roger L. Brockenbrough, American Institute of Steel Construction, New York, February 2002.

24. *AISC Steel Construction,* 1st ed., American Institute of Steel Construction, New York, March 1928.

25. *Structural Welding Code AWS D1.1/D1.1M:2004—Steel,* American Welding Society, Miami, 2004.

26. AASHTO/AWS D1.5M/D1.5:2002, *Bridge Welding Code,* American Welding Society, Miami, 2002.

27. Wangaard, F. F. (1950). *The Mechanical Properties of Wood,* Wiley, New York.

28. Wilson, T. R. C. (1934). *Guide to the Grading of Structural Timbers and the Determination of Working Stresses,* U.S. Department of Agriculture Miscellaneous Publication 185, Washington, DC, U.S. Department of Agriculture.

29. *Wood Structural Design Data,* National Lumber Manufacturers Association (now National Forest Products Association), Washington, DC, 1944.

30. *Wood Handbook,* Forest Products Laboratory, U.S. Department of Agriculture, Forest Service, Washington, DC, 1935.

31. *Wood Handbook,* Forest Products Laboratory, U.S. Department of Agriculture, Forest Service, Washington, DC, 1999, http://www.treesearch.fs.fed.us/pubs/viewpub.jsp?index=5734.

32. *National Emergency Specification for the Design, Fabrication and Erection of Stress Grade Lumber and Its Fastenings for Buildings, Directive No. 29,* War Production Board, Conservation Division, Washington, DC, 1943.

33. *National Design Specification (NDS) for Wood Construction,* American Forest and Paper Association and the American Wood Council, Washington, DC, 2001.

34. DOC PS 20-99, *American Softwood Lumber Standard,* U.S. Department of Commerce Technology Administration, National Institute of Standards and Technology (NIST), Washington, DC, 1999.

35. *Simplified Practice Recommendation R16, Lumber—American Lumber Standards for Softwood Lumber,* U.S. Department of Commerce, National Bureau of Standards, Washington, DC, 1924.

36. SEI/ASCE 7-02, *Minimum Design Loads for Buildings and Other Structures,* American Society of Civil Engineers and Structural Engineering Institute, Reston, VA, 2002

37. *American Standard Building Code Requirements for Minimum Design Loads in Buildings and Other Structures,* ASA A58.1-1945, American Standard Association, New York, 1945.

38. *National Building Code of Canada,* National Research Council, Ottawa, 1965.

39. Blume, J. A., N. M. Newmark, and L. Corning (1961). *Design of Multistory Reinforced Concrete Buildings for Earthquake Motions,* Portland Cement Association, Chicago.

40. Seismology Committee (1966). *Recommended Lateral Force Requirements and Commentary,* Structural Engineers Association of California, San Francisco.

41. *Tentative Provisions for the Development of Seismic Regulations for Buildings* (ATC-3-06), report for the National Bureau of Standards and National Science Foundation by ATC, Palo Alto, CA, 1978.

42. Building Seismic Safety Council (1992). *NEHRP Recommended Provisions for the Development of Seismic Regulations for New Buildings,* 1991 ed., Federal Emergency Management Agency, Washington, DC.

43. SEI/ASCE 37-02, *Design Loads on Structures during Construction,* American Society of Civil Engineers, Reston, VA, 2002.

■ **CHAPTER 5**

Legal Aspects of Condition Assessment and Reporting

HOWARD W. ASHCRAFT, JR., ESQ.

INTRODUCTION

Structural condition assessments (SCAs) present many challenges to the professionals who perform them. Although the structures being assessed already exist, SCAs create new legal risks for those who perform them. Successful consultants recognize that risks arising out of SCAs often exceed the anticipated profit. Recognizing this, they control the risks at each phase of the assignment. Every element of the assignment— from proposal to contract to report—is planned and executed carefully. This chapter examines the unique characteristics of SCAs and discusses strategies for managing the legal risks.

NATURE OF THE RISK

Why Are Structural Condition Assessments "Different"?

SCAs are not like traditional engineering work. Traditionally, engineers design—they perform calculations, draft drawings, and prepare specifications. Traditionally, engineers also administer the construction of their designs—they review submittals, respond to information requests, review proposed changes, and observe construction. These activities are associated with the construction of new structures, the creation of new improvements. Indeed, even on retrofit projects, the traditional role of the engineer is to design the new elements of the building taking into account the demands of the existing site. In his or her role as designer, the engineer is able to specify the strengths of materials, the size of elements, and the configuration of connections. As designer, the engineer also can request mill certifications, material strength tests, and continuous inspection to ensure that the materials provided and construction performed conform to the intent of the design. Traditionally, engineers are compensated for designing a structure based on a percentage of construction costs—the higher the costs of construction, the higher is the engineer's fee.

Structural condition assessments, however, are quite different. SCAs are performed on existing structures. Engineers who perform SCAs cannot control the design, the materials used, or the quality of construction of the existing structure. Rather, engineers who perform SCAs have to take the structure as is, often without an understanding of the original designer's assumptions, oversight, or concerns. Moreover, by their nature, SCAs require engineers to assess buildings in which all or most of the structural elements and connections are concealed by architectural finishes. Quite often budget constraints prohibit the assessment of every element and force engineers to base their opinions on extrapolations from a limited number of locations. Also, because SCAs are not related to new construction, fees charged for SCAs typically are set on a time and materials basis, not as a function of the risk assumed.

In many jurisdictions, structural condition assessments are not afforded the same statutory protections as traditional design work. Many states have statutes of limitations or statutes of repose[1] that act to limit liability for design. For example, in California, designers may not be sued in connection with the planning or construction of improvements to real property more than 10 years after the project is complete.[2] However, the language of the statute appears to limit this protection to projects where actual construction is performed, arguably excluding SCAs.

Types of Structural Condition Assessment Risks—Consultant's Perspective

Unfortunately, SCA risk is proportionate to the value of the underlying economic transaction, whereas SCA compensation most often is not. For instance, in a preacquisition building assessment, the engineer's liability is measured as the difference between the value of the property free of unidentified deficiencies and its value with the overlooked deficiencies identified. These damages are unrelated to the fee and often are exponentially greater. Additionally, other risk factors, such as fatigue from cyclic loading, geologic conditions, and preexisting deficiencies, cannot be controlled by the consultant. These factors exist independent of the consultant's efforts and typically are identified only after the consultant has performed a significant portion of his or her services. They cannot be reflected in the fee because they are not known until after the contract has been negotiated. Finally, SCA clients—and relying third parties—often are investors or developers. In many instances, they owe fiduciary duties to their partners and investors that will lead them to sue if the investment return or their performance ratings are threatened.

Two scenarios demonstrate types of consequences that arise from SCAs and lead to claims:

Scenario 1: The SCA Overstates the Structural Deficiencies

Seller	• Loses the sale. • Sells the property below its value. • Excessive withholds are placed on the sale. • Closing of transaction is delayed. • Incurs attorney and consultant expenses.
Buyer	• Closing of transaction is delayed. • Incurs attorney and consultant expenses.
Investor/lender	• Closing of transaction is delayed. • Incurs attorney and consultant expenses.

Scenario 2: The SCA Understates the Structural Deficiencies

Seller	• Incurs liability to the buyer for preexisting deficiencies. • Loses the opportunity to sell the property to another or for other uses. • Incurs attorney and consultant expenses.
Buyer	• Pays too much for the property. • Purchases the property without adequate withholds to pay for renovation. • Incurs substantial renovation costs. • Loses the opportunity to invest in more profitable ventures. • Incurs attorney and consultant expenses.
Investor/lender	• Security for the investment or loan is lost or impaired. • Incurs attorney and consultant expenses.

These damages may lead to claims against the consulting firm. Given the potential for these claims, consultants must be aware of and manage their SCA risks.

Types of Structural Condition Assessment Risks—Owner's Perspective

When an owner receives an SCA, he or she also may receive unanticipated potential liabilities. If the SCA gives the structure a clean bill of health, the associated liabilities the owner potentially assumes are minimal. However, when the SCA documents structural deficiencies, owners are placed in a potentially precarious position.

Depending on the seriousness of the deficiencies, owners may be under a duty to alert tenants or other occupants. Failure to alert tenants may result in liability in the event the deficiency results in property damage or personal injury. Additionally, tenants confronted with serious structural deficiencies may feel uncomfortable occupying the structure and may seek to end the tenancy or seek a reduction in rent.

Owner's receiving notice of serious deficiencies also may be faced with an obligation to warn or protect third parties with reason to enter the structure or lenders holding the structure as collateral. Accordingly, prudent owners will work with the engineers performing the SCA to ensure that the scope is well defined and that the budget exists to perform an accurate evaluation.

RISK MANAGEMENT STRATEGIES

Risk can be managed in three ways: It can be avoided, minimized, or transferred. A combination of these strategies can be used to balance profitability and risk.

Risk Avoidance

The ultimate risk management tool is the ability to say, "No." In *True Professionalism*,[3] David Maister observes that successful professional firms do not seek more business; they seek better business. They analyze each engagement on a fully costed basis to determine if the engagement is profitable. Before agreeing to perform an SCA, the consulting firm should calculate the entire cost, including the allocable overhead and risk.

Risk is expensive. Even if the risk is covered by professional liability insurance, most policies have substantial deductibles. In addition, claims drain a firm's internal resources. Rather than concentrate on managing and marketing the firm, senior professionals are diverted to working with the firm's attorneys and attending meetings, depositions, and hearings regarding the claim. Honestly evaluated, these risk costs may exceed the anticipated profit on the project. Blindly working on marginal projects exemplifies the rationalization, "We lose a little on every project, but we make it up with volume."

Client Risk Factors. The first step in risk avoidance is careful screening of clients. Issues to investigate include

How stable is the client? Will the client disappear after the transaction? Is the client a joint venture, syndicate, or other temporary entity? Is the client a limited liability entity, such as a limited-liability partnership or corporation? Does the client have significant assets? Is it a single-asset entity? Can the client withstand a large structural loss?

How sophisticated is the client? Is the client an experienced building owner? Does the client understand the risks involved? What are the client's risks during and after the SCA? Can the client weigh the risks and rewards?

Who will rely on the report? Is the report intended solely for the client? How widely will it be distributed? What third parties, such as investors, lenders, or buyers, will rely on the report? Would you willingly perform the assessment for each of these third parties?

Does the client regularly sue? Does the client have a record of suing consultants? Has the client refused to pay other consultants their fees? Is the client a speculator or developer?

Project Risk Factors. The consulting firm also should scrutinize the type of project and the structure of the economic transaction. Issues to consider include

What type of economic transaction is involved? Is the assessment being performed to secure financing? Is the assessment being performed to obtain hazard insurance? Or is the assessment to be used as a factor in the negotiation of the purchase price or renovation withholds?

What structural deficiencies might exist? Is the structure likely to contain undersized members? What is the age of the structure? What were the detailing practices at the time the structure was built? What were the construction practices at the time the building was built? Is the structure subject to cyclic loading? Has the structure experienced a major seismic event?

What geologic factors may exist? What is the structure's proximity to fault lines? Are the underlying soils stable? Is liquefaction a concern? Is the existing foundation system compatible with the soil profile?

What populations might be affected? Will the property be used for industry, commerce, or habitation? What is the owner's or prospective owner's loss tolerance? Is the structure a necessary facility?

How good is available information? Is detailed and accurate information available? Does the client have good records concerning the original construction? Are people available who know how the structure was built? Do accurate as-built drawings exist? Are there photographs of the building during construction? How accessible are the critical structural elements?

Few projects scores well on all these criteria. However, if a project scores poorly on many of these issues, the professional should determine carefully whether the risk exceeds the possible profit. Consider whether you are being paid adequately for the increased risk inherent in this project. If you are not, is the client willing to minimize your risk by limiting your liability or indemnifying you from loss? Are any other protections being offered to you? If you are being asked to undertake a risky project at normal profit margins, without liability concessions, then the project should be declined.

Risk Minimization

Real-world projects do not have perfect risk profiles. How, then, do you minimize risks in the projects you perform?

Quality Creation. Many consulting firms have quality assurance/quality control programs. Too often these programs attempt to "inspect in" quality and only focus on the technical work being performed. Most claims, however, do not result from pure technical errors but are caused by failures in process, communication, or expectations.[4] Rather than rooting out errors through inspection, firms should focus on training personnel and instituting procedures that create quality. The quality program should include the full scope of the firm's activity from initial marketing to final report. Additionally, consultants should recognize that when performing SCA's, they will not be afforded the opportunity to "inspect in" quality because the structure has been long since constructed.

Risk is minimized if personnel are trained and supervised properly. Checklists, procedure manuals, and similar standardized approaches reduce the risk of error.

Proposals, Contracts, and Reports. The proposal phase establishes the client's expectations. Firms routinely tout their quality and efficiency when trying to obtain work. Clients perceive these representations as promising specific results, *even if the contract states otherwise.* And in some instances, the representations may create a basis for liability. Even when they do not, if the client's perception of what is being done—and can be done—differs from the contract or actual practice, the seeds of dispute have been sown.

Marketing personnel should receive the same training given to project managers and should understand the limits of what may be promised. They also should understand the risk concessions that are necessary for challenging projects. Senior personnel should review written proposals for reasonableness and terms, as well as cost. The firm's standard terms and conditions—or more stringent terms and conditions for a risky project—should be included with the proposal. If the firm knows of third parties that will be relying on the report, these parties should be advised of any limitations on their reliance that will be conditional to the firm performing the assessment.

The written contract is critical to minimizing liability. Oral agreements lead to disputes over scope of work, payment, and liability. They are, in the words of Samuel Goldwyn, "not worth the paper they are printed on."

Few consulting firms are capable of drafting effective agreements without legal assistance. It is tempting to copy a form used by another firm and assume that it is sufficient. However, liability issues and contract provisions are subject to local variation and change with time. The "borrowed" form may be effective—but maybe not.

A firm providing structural assessments should invest in legal advice to develop good contracts. The attorney should specialize in representing structural firms. Not only will the quality of the contracts be better, but the attorney also can explain what the contract language means and what other contracting options are available. Experienced attorneys also will know what types of contract language are being used, and accepted, in your locality. To reduce costs, you may want to provide the attorney with copies of contracts similar to the type of contract you would like to develop.

When drafting contracts for SCAs, it is important to acknowledge that the reliability and accuracy of the SCA are directly dependent on the visibility and accessibility of the elements being evaluated. It is also important to acknowledge that the quality of the SCA is a direct function of the amount of inspection, sampling, and testing performed. These variables are a function the client's budget. Professional services agreements for the performance of SCAs should alert clients unambiguously that the recommendations of the report are only as good as the information available and should advise clients that the information available is limited by the SCA's budget.

Your report is the culmination of the process and must be consistent with your proposal and your contract. This can be achieved by using a report skeleton that is prepared in conjunction with your standard contract and has been reviewed by your counsel. Project personnel should not be able to modify the standard language without senior authorization.

Each report should receive senior-level review before issuance. The review should check for consistency between the proposal, contract, and report and should ensure that each contract task has been completed and the results discussed as required in the report. The review should check for reasonableness and compliance with industry standards or standards provided by the client that have been incorporated into the contract.

Contractual Clauses.

Scope of Work. There are two aspects to your scope of work: what tasks you are going to accomplish and the standard against which your work will be judged. Both need to be defined clearly. Your work scope provisions should state the specific tasks you will accomplish, the information you will be provided by your client or others, and the tasks that are outside your scope of work. The listing of omitted tasks is critical if you are not performing an SCA based on a published standard. Otherwise the client—or relying third parties—may argue that you breached the standard of care by omitting critical tasks.

The standard of care should be defined explicitly. The standard of care for design professionals typically is defined as that care exercised by like professionals practicing in the same local and under similar circumstances. Accordingly, structural engineers are required to use the training, care, and skill used by similar engineers under similar circumstances. Clients often try to elevate this standard of care by requiring work to the "highest standard for engineering professionals" or similar language. Certification requirements and warranties also are used to elevate the standard of care. These increase the consultant's risk significantly and often create insurance coverage problems. Professional liability policies do not cover liability assumed by contract—precisely what you have done if you agree to a heightened standard of care.

Regional standards of care are being overtaken by published standards. For instance, ASTM E2026-99 describes a process for the estimation of building damageability in earthquakes.[5] You should review carefully the ASTM—or any other referenced standard—to determine if it matches the normal practices for your organization or community. If it does not, you need to modify your standard practices, or you need to state explicitly that your services are not being provided in accordance with ASTM or other published standards. Otherwise, in any dispute you will have difficulty justifying why you did not comply with all the requirements of the published standard.

Reliance on Client-Supplied Data. SCAs are based, in large part, on information obtained from the client or other persons who have controlled the property. Rarely does your budget allow you to verify the completeness or accuracy of all supplied data. Your contract should require the client to provide you with all material information and permit you to rely on this information without independent verification.

Third-Party Reliance. SCAs often are provided to persons other than your client. For instance, building owners often will provide an existing SCA to potential buyers or lenders. This greatly expands the scope of people who may rely on your work—and sue you if the report is inaccurate or misleading. This may be acceptable if you know who the third party is and have appropriate concessions in fee and contract terms. However, it is not acceptable without these controls.

Third-party reliance may undermine protections you have built into your contract. Because the third party has not signed your contract, it can argue that it is not bound by protective terms, such as limitations on liability or consequential damage waivers. By allowing third party reliance, you may have thrown away hard-won contract protections.

Your contract should state that your services are solely for the benefit of your client and that you do not intend third parties to benefit from or rely on your work. It is also advisable to restrict the distribution of your report. Your client should be prohibited contractually from distributing your report without your consent. This prohibition can be strengthened by requiring your client to indemnify you from third party actions arising from their distribution of your report.

Limitation of Liability. Limitation of liability is an important risk allocation tool. Properly negotiated and drafted, a limitation of liability clause caps your liability to your client. In general, it is not a defense against claims brought by third parties. Other provisions, such as indemnity agreements or reliance letters, must be combined with limitation of liability to create a complete risk allocation plan. Professional associations such as ASFE support using limitation of liability clauses.[6]

Enforceability of limitation of liability clauses varies among jurisdictions. In a few states, limitation of liability clauses are not effective. In almost all states, the clause must be drafted carefully. Before using a limitation of liability clause, you should ask your lawyer whether limitation clauses are effective in your state, how they should be drafted, and what steps—such a notice and negotiation—are important to enforcing the clauses. Although not a panacea, limitation of liability clauses are an important element in a risk management plan.

Consequential Damages Waiver. A consequential damage waiver releases you from liability for consequential damages. Many clients will agree to a consequential damage waiver even if they reject a limitation of liability clause.

Consequential damages are losses that are not inherent in the breach of an agreement and were not contemplated by the parties when the contract was made. For example, in a contract to develop and print film, loss of the film is inherent in faulty processing of film and is a direct damage. If you sue the processing laboratory, it will replace the film. If your roll of film contained a rare and important photograph, the anticipated profits you lost when the film was ruined are consequential damages. Since your agreement with the laboratory waived consequential damages, you cannot recover the profits you could have made by selling the photograph. In many instances, the consequential damages are the critical loss suffered, and their waiver can be a substantial benefit.

Consequential damage waivers are becoming more common in construction documents. For instance, the American Institute of Architects *General Conditions for the Contract for Construction* now contains a clause waiving consequential damages between the owner and the contractor.[7]

Transfer of Liability

Some portion of liability cannot be avoided or minimized. If possible, this liability should be transferred through indemnification or insurance.

Indemnification. An indemnity provision transfers responsibility for a loss to another party. For example, if you indemnify your client, you will be responsible for paying any judgment or settlement—and possibly the attorneys fees—related to a covered claim. Your client's risk—being sued—had been transferred to you. Similarly, you can have your client indemnify you against risks, such as preexisting contamination.

However, indemnification has three weaknesses:

1. Indemnification is useless unless the indemnitor exists and is solvent at the time of the claim.
2. It is difficult to allocate risk precisely through indemnification. Indemnification generally is an all-or-nothing proposition.[8] It is too strong or too weak. Also, court interpretations have reduced the predictability of the result.
3. Many states have anti-indemnification statutes that limit the scope and nature of indemnification. These usually apply to construction contracts, but *construction* often is defined broadly enough to include SCAs. It is better to be indemnified than not, but the value of indemnification must be evaluated critically before you rely on the provision.

Drafting indemnification provisions is best left to professionals. Interpretation of the clauses is highly technical and varies significantly among states. Small language differences can cause greatly differing results. Unless you have significant experience with indemnification agreements, you should not draft or evaluate indemnity agreements without legal assistance.

Many clients require that you indemnify them against certain risks. Agreeing to do so can greatly increase your potential liability, though. Also, this increased risk may not be insurable. Most professional liability policies exclude coverage for liability assumed by contract. Thus, if your liability is expanded by an indemnification agreement, the difference between your liability with the indemnification clause and your liability if the clause did not exist is not covered. If you are being asked to sign an indemnification agreement, you should have it reviewed by your insurers or your insurance broker to ensure that your insurance coverage will not be affected.

Who should indemnify whom? Many owners unthinkingly demand that their consultants indemnify them from a broad range of risks. However, if you assume that *indemnification* is designed to balance allocation of risk, then *indemnification* should follow these basic principles. The person who indemnifies is (1) the person who already has the risk, (2) the person best able to control the risk, or (3) the person who is best able to insure the risk. Two examples illustrate this approach.

Consider the risk of your employee being injured while performing the assessment. This is a risk that the owner does not have until your employee comes on the site. By supervising your employees and providing safety equipment and training, you are in the best position to control the risk. Through workers compensation and general liability coverage, you can insure against this risk. Under these circumstances, you should be willing to indemnify the owner against the risk your employees will be injured on their site.

However, the risk of preexisting structural deficiencies is quite different. The owner, and not the consultant, has this risk before the assessment is done. This is also a risk the consultant cannot control. The owner, however, can purchase insurance to cover this risk or may have profited from the deficiencies by purchasing the building at a discount. Not only should the consultant not indemnify the owner from this risk, but the owner also should indemnify the consultant.

As a general rule, your risk should not increase just because you signed a contract. If it does, then you are providing insurance, not services.

Insurance. Insurance is an effective tool for transferring risk. Although insurance generally will protect you from catastrophic losses, you should not assume that all risk issues are resolved.

Most professional liability policies have substantial deductibles. Because engineering malpractice cases are complex, even a meritless claim can exhaust the entire deductible. For example, if you have one claim per hundred assessments, a $25,000 deductible, and a 10% profit, and your SCAs average $2500, you will just break even—almost. The time spent in hearings, depositions, and assisting your attorneys will further reduce your profitability. Thus you need a sound risk management strategy, and you must follow it.

SPECIAL PROBLEMS

Third-Party Reliance

The degree to which third parties can rely on your reports is a complicated and thorny issue. Choosing your client is a basic risk management technique. By letting third parties rely on your report, however, you have effectively taken on a multitude of unknown "clients." Ironically, because these third parties have not signed your contract, they may not be bound by the carefully crafted risk allocation provisions, such as limitations of liability.

The scope of third-party reliance varies from state to state. In some jurisdictions, parties cannot rely on the findings of the SCAs unless they are parties to your contract or the contract expressly states that your services are intended to benefit the third party (actual privity). Others allow third parties to rely on your report when there is a relationship that approaches privity (privity of relationship). Most states allow reliance where the work is intended to benefit a specific party (intended beneficiary, *Restatement of Torts,* 2nd, Sec. 552). A few states permit reliance if it is foreseeable that a class of plaintiffs will rely on the report (foreseeability). Because of the variability between jurisdictions, you should get legal advice concerning what rules will apply to your work and craft your agreements to match the applicable rules.

Your contract should state that your report is prepared solely for your client and that third parties cannot rely on the report. If you must allow third parties to rely, they should sign a reliance letter that makes the terms, conditions, and limitations of your contract applicable to them. A good third-party reliance letter binds the third party to the contract terms, requires that risk allocation provisions apply to the third party, and ensures that a single aggregate limitation applies to all claimants.

Certificates and Warranties

Some third parties require consultants to certify the structure's soundness or to certify that the SCA has been performed pursuant to the applicable published standard. Although you should perform SCAs to published standards, certifications should be avoided. In many jurisdictions, certifications are not harmless technical descriptions of the work performed. Rather, they are guarantees of the work's legal effect and expand liability well beyond the "standard of care."

The standard of care is a consultant's basic liability standard. It is a negligence standard that is met if work is performed in the manner of competent consultants performing similar work under similar circumstances at that time in the same geographic area. In contrast, a certificate or warranty is a strict liability standard that is breached if the representation is untrue, regardless of negligence.

Certificates and warranties greatly increase your risk. They also create insurance problems. Liability for certificates and warranties is implicitly limited by the contracted-for-liability exclusion in your professional liability policy and is expressly limited under some insurance policies.

There are three strategies for dealing with certificates and warranties. Refusing to sign is the first option. Second, you can require that all certificates and warranties are acceptable to you prior to performing your work. Third, you can water down the certificate from a positive statement of fact to an expression of professional opinion.

Disclosure of Hazardous Conditions

What do you do if you become aware of hazardous conditions that the owner has no intention of curing? Unfortunately, there is no clear answer. Some argue that the consultant's sole duty is to its client and that there is no obligation to inform the public of an unsafe structure or a hazardous condition. Others argue that engineers exist to promote the general welfare of society and therefore have a duty to inform the public of potentially hazardous structures. Indeed, many registration statutes and codes of ethics contain statements that the professional should act in the best interests of the public.[9]

To compound the problem, contracts often contain confidentiality provisions that conflict with a right, or duty, to disclose Thus the consultant may have responsibilities to the public that would require him or her to breach the contract. To reduce this tension, any confidentiality provision should permit the consultant to make any disclosures required by law.

OTHER SOURCES OF INFORMATION

Several good sources of information concerning risk management and contract drafting are

American Consulting Engineers Council. ACEC focuses on the business aspects of professional practice. Their Web page is located at *http://www.acec.org.*

ASFE, Professional Firms Practicing in the Geosciences. ASFE maintains a Web site listing available publications. The ASFE *Contract Reference Guide* contains detailed discussions of contract language and has suggested contract language. Contact ASFE at *http://www.asfe.org.*

DPIC Companies. This insurer publishes a version of the ASFE *Contract Reference Guide* and has online risk management materials. The site can be found at *http://www.dpic.com.*

Terra Insurance Company, A Risk Retention Group. Terra is a risk retention group owned by environmental and geotechnical firms. Terra publishes a Multimedia Contract Seminar on a CD-ROM that is available from Terra or HWAC. Terra can be reached via e-mail at *terrarrg@worldnet.att.net* or by telephone at (415) 927-2901.

Professional Practice Insurance Brokers. This insurance broker maintains an online library of risk management papers and advisories. Information can be found at *http://www.ppip.com.*

RA&MCO Insurance Services. This insurance broker/managing agent maintains an online library of risk management papers and advisories. Access this on the Internet at *http://www.ramco-ins.com.*

DOS AND DON'TS

	Yes	No
1. Assess your risk avoidance strategy.		
Do you carefully choose your clients?	○	○
Do you carefully choose your projects?	○	○
Have you included risk in the cost of performing the services?	○	○
2. Evaluate your risk minimization management.		
Have you had your proposal, contract, and report forms reviewed by a lawyer who represents structural firms?	○	○
Do you have a quality assurance/quality control program that includes client selection, contracting, and documentation?	○	○
Do you have procedures in place to ensure that work is performed in accordance with specific contractual requirements or published standards?	○	○
Have you refrained from signing contracts with heightened standards of care?	○	○
3. Consider your risk transfer options.		
Do you use risk allocation tools, such as limitation of liability and indemnification provisions?	○	○
Have you refrained from signing indemnity agreements unless your insurers have approved them?	○	○
4. Assess your preparation against special problems.		
Do you use reliance letters if you must permit a third party to rely on your report?	○	○
Have you included language in your legal documents to prevent third parties from relying on your report unless you use a reliance letter?	○	○
Have you avoided signing warranties or certifications unless reviewed by your lawyer?	○	○

REFERENCES AND NOTES

1. A *statute of limitation* defines the amount of time within which a suit must be filed after injury occurs. A *statute of repose* defines that amount of time from completion of the project after which no suits may be filed. For example a 4-year statue of limitation for breach of contract means that suit must be filed within 4 years from the date the breach occurs, whereas a 4-year statute of repose for patent defects means that no suit for patent defects can be brought more than 4 years after building completion.
2. California Code of Civil Procedure, Sec. 337.15, provides, in part: "No action may be brought to recover damages from any person, or the surety of a person, who develops real property or performs or furnishes the design, specifications, surveying, planning, supervision, testing, or observation of construction or construction of an improvement to real property more than 10 years after the substantial completion of the development or improvement. . . ."
3. Maister, D. (1997). *True Professionalism,* Free Press, New York, Chap. 20.
4. The Construction Industry Institute has analyzed construction projects to determine the factors statistically relevant to predicting success or failure. They discovered that people factors and process factors had far greater influence than project factors. *Disputes Potential Index,* CII documents VC-405, SP23-3.
5. ASTM E 2026-99, *Standard Guide for the Estimation of Building Damageability in Earthquakes,* ASTM, New York, 1999.

6. Sample clauses are found in ASFE's *Contract Reference Guide*, 3d ed., pp. 103–107; and *Limitation of Liability: A Handbook for Consulting Engineers, Environmental Consultants, Architects, Landscape Architects, and Other Design and Technical Consultants*, ASFE, New York, 1992.

7. *General Conditions for the Contract for Construction*, American Institute of Architects Document A201-1997, Paragraph 4.3.10, AIA, Washington, DC, 1997.

8. Indemnification provisions can use a comparative negligence standard. This is a better approach, where you are required to indemnify your client. But if you are being indemnified, it may not provide you with the practical result you seek. Since the obligations are on a comparative negligence basis, they cannot be determined without fully litigating the issues. This means that you will be responsible for your full defense costs until—and if—there is a final comparative determination. Since most cases settle, this may never occur.

9. Relying, in part, on the general obligations of a professional engineer, the California Attorney General published an opinion that a structural engineer observing conditions of "imminent danger" and knowing that its client did not intend to remedy the problem had a duty to warn the persons at risk of the dangerous condition. Attorney General opinions need not be followed by the courts but are often persuasive.

PART III

SURVEY AND ASSESSMENT OF STRUCTURAL CONDITIONS

CHAPTER 6

Buildings

DAVID B. PERAZA, P.E., and ERIC C. STOVNER, P.E., S.E.

INTRODUCTION

There are many different types of structural condition assessments of buildings, depending on the circumstances and the objective of the client. The assessments can vary widely in terms of their scope, the level of detail, and their comprehensiveness. However, some types of inspections are recurring. The sale and purchase of any significant facility triggers the need for a due diligence inspection; the threat of a seismic event prompts a vulnerability study; natural hazards, accidents, and terrorist attacks may require rapid assessments of large numbers of buildings; and the need for long-term planning prompts owners to perform "checkups" on their buildings.

This chapter discusses the most common types of structural condition assessments performed on buildings or on portions of buildings. It includes the purposes of the assessments, common problems, methodologies, and strategies for performing a high-quality assessment. Guidance for preparing the condition assessment report is given elsewhere in the book.

DUE DILIGENCE EVALUATION

This type of evaluation is required commonly by a potential buyer of a property. The objective is generally to identify whether there are any readily apparent structural issues associated with the property. The results of the study typically will be used to help the client determine whether or not to purchase the property, to assist in budget planning in the event the sale goes ahead, and possibly to negotiate the final purchase price.

The client generally has a short amount of time to perform its investigation and often has a limited budget that it can allocate in this early stage. Consequently, this type of investigation tends to be superficial in nature. As such, although it is useful for detecting gross structural issues, it may not detect every structural issue, especially subtle or hidden conditions.

There is considerable liability associated with due diligence evaluations, as illustrated by the following anecdote. A potential buyer of an industrial building retained a consulting engineering firm to do a full due diligence evaluation, which included evaluation of the structural systems. Structural drawings were made available to the consultant. As part of its work, the consultant performed a simple calculation showing that certain joists would be overloaded during a design snow load owing to snow drifting from an abutting higher building. The consultant clearly identified this serious deficiency in its report. The building was purchased, and the owner hired another engineering firm to design structural modifications related to raising the roof to accommodate large equipment. These modifications were made. But the owner had not provided the due diligence report to the second engineering firm, so the issues that had been raised by that report were never addressed. Several weeks later there was a record snowstorm, and the portion of the building that had been questioned collapsed. The firm that had performed the due diligence evaluation, identified the problem, and reported it properly ended up embroiled in a lawsuit initiated by the owner.

Planning

Proper planning of an evaluation can help to ensure that the client's objectives are satisfied and that the work is performed in an effective manner.

Client Objectives and Concerns. The client's objectives and concerns will determine the scope of the project. This type of study usually is commissioned by a client who is considering purchasing a property. On a typical project, the client will be interested in knowing whether there are any unusual structural conditions that will require remediation.

Sometimes the client will want to know whether the building meets certain structural criteria that are important to the client for his or her purpose. If the client has any special criteria, he or she is responsible for providing that information to the consultant who will be performing the assessment. For example, a client might require that floors have certain "flatness" so that forklift vehicles can operate properly, or the client may need to know whether the floors are capable of supporting heavy file loads. The consultant should advise the client regarding the feasibility of being able to determine whether the building meets the special criteria and the range of costs associated with the effort.

In most cases, the client will have no suspicions of any structural problems. In this case, usually it will be the responsibility of the consultant to propose a scope of work that includes the observation of components that are representative of the building structure, that are readily accessible, or that in the consultant's experience may be problematic.

In other cases, the client may have some specific concerns as a result of his or her inspection of the property and will want to know whether the problem is structural in nature. For example, a brick masonry building may exhibit cracking. Depending on the nature of the cracking, it could be indicative of foundation settlement, corrosion of the underlying steel, or simply thermal expansion and contraction. Each of these conditions will have a different remedial scope of work associated with correcting the problem, and the costs may vary widely.

If structural work is found to be necessary, the client in most cases will need a budget cost estimate. Who will prepare the cost estimate should be discussed specifically in advance and addressed in the consultant's proposal. Likewise, the required level of detail and accuracy of the estimate should be discussed. Generally, budget estimates based on square-foot costs are sufficient, but a client may require additional detail. In some cases, the consultant may have the necessary experience and resources to prepare an estimate in-house. In other cases, the client may elect to hire a cost estimator to work with the consultant, or the client may prefer to perform its own cost estimates.

A due diligence assessment typically is limited to a "walk through" type of inspection of elements that are readily accessible and visible. For many building types, this means that most of the structure will not be observed because it is concealed by finishes. If ceilings can be readily removed, such as a lay-in acoustical ceiling, the consultant may make observations at selected locations. Probes and tests normally are not included in this type of assessment. Calculations usually are not performed unless the client has raised a specific concern in advance that requires calculation in order to be addressed properly.

Document Gathering and Review. A more meaningful evaluation can be performed if the original structural and architectural drawings are available. It is usually the responsibility of the client to obtain or provide access to the drawings, and this is an item that should be addressed in the agreement. Occasionally, the client can obtain the drawings in advance from the building's owner. Usually, however, partly because of time pressures, the professional will not be given the opportunity to review the drawings until he or she arrives at the site. In some cases, drawings will not be available at all.

The drawings will yield basic information, such as the vintage of the building, the bulk of the building, the design criteria, the type of structural systems, and other important information. Coupled with the investigator's previous experience, a review of the drawings may assist in identifying areas or representative components that may be problematic and that therefore should be inspected.

Basic information that can be gleaned from a review of the documents and which would be appropriate to report on includes the names of the design professionals, when the plans were prepared, governing building codes, key design criteria, type of construction, overall plan dimensions, number of stories, usage and occupancy, location of expansion joints, and building orientation.

It is valuable to identify in advance the general type of construction. The type of construction may affect the scope of the work, the sequence of inspection, and which elements should be inspected. It also will allow the consultant the opportunity to seek out relevant resources and may influence staffing for the project. The type of construction is readily obtainable from the original drawings. If drawings are not available, then a "reconnaissance" visit may be useful. Some basic structure types include conventional steel building, preengineered building, cast-in-place concrete, posttensioned, precast concrete, wood frame, and bearing-wall construction. Older structures also may include archaic structural systems, such as cast-iron columns or cinder concrete floor slabs.

Review of the drawings also will assist the consultant in identifying the main structural systems, such as typical vertical load-carrying elements and spacing, lateral load-resisting systems, foundation system, typical floor system, nontypical floor systems, roof system, location and nature of special load-transfer elements, and construction of below-grade areas.

If there are special features, such as large canopies, retaining walls, pedestrian bridges, or atrium skylight structures, the level of inspection of these features should be addressed.

Identification of the applicable codes and standards often will be useful and sometimes may be necessary for a proper evaluation. The design documents normally will identify the principal codes and standards that governed the design and construction of the building. Key parameters include the design live loads, the design snow loads, wind or seismic design criteria, and allowable stresses or limit-design capacities. Comparison against current standards and codes can be useful, particularly for vintage buildings. For example, modern standards generally allow a higher allowable stress for steel members than was permitted in the early 1900s, even though modern steel is virtually identical metallurgically. This "reserve" often may be used justifiably to support additional load. On the other hand, the allowable stresses for some types of wood members have decreased considerably and may create a potential vulnerability. Likewise, design loads may have changed considerably—for better or for worse. Chapter 4 includes valuable information on how key codes and standards have changed through the years.

Inspection

In most commercial or residential buildings, the structural components are typically concealed by finishes, and therefore, the areas where the structure can be observed directly often will be limited. Much of the inspection therefore will focus on searching for distress of nonstructural components that may be indicative of an underlying structural problem. Some symptoms of possible structural problems are the following:

Water stains and damage to ceilings. Water penetration may be causing damage to structural members.

Difficulty in operating doors or windows. This may be an indication of overall building settlement or of a local sagging of the floor system.

Cracking of plaster or gypsum wall board. This may be an indication of foundation settlement or of a localized problem with the floor system. Cracking in floor finishes may be indicative of excessive deflection of the floor system or improper reinforcement of the slab

Excessive deflection of ceilings. This may be indicative of either a problem with the framing from which the ceiling is hung, or a problem with the ceiling suspension system itself.

Distress or displacement of facade elements. This may be symptomatic of improper support of the facade, inadequate connections, or corrosion of the underlying structural elements

The following subsections discuss inspection of some of the components that most often are readily observable.

Roofs. Roof structures, particularly in older buildings, are vulnerable to damage from water penetration. Steel structures may exhibit corrosion, and wood structures may suffer from decay.

Although roofing per se usually is not included within the scope of a structural condition survey, it is often valuable to perform a casual inspection of the roofing and to inquire about its age and any history of leakage. The casual inspection would assist in determining the general condition of the roofing, general drainage patterns, and whether there are any obvious points of water entry. The conditions observed may lead the consultant to investigate the underlying structure further at certain locations. Conditions that are of interest include blisters, alligatoring, and open seams or tears in the membrane, ponding, and blocked drains or scuppers (Fig. 6.1).

Some codes require that parapets have scuppers as a secondary means of drainage or else that the roof structure be designed to support the depth of water that might accumulate, assuming that the main drains are clogged. In these localities, it therefore would be important to confirm that scuppers are present.

It is important to try to determine whether there are any areas where there has been a history of leakage. An effective technique is to walk the space with a knowledgeable maintenance person. Seeing the spaces may serve to refresh the person's memory, and it would allow the consultant to inquire immediately about any unusual conditions he or she is observing (Fig. 6.2).

If a ceiling conceals the underside of the roof slab, an effort should be made to locate any signs of water stains or damage on the ceiling itself. If such areas are found, the location should be noted, preferably on a scale drawing, and if possible, the structure above should be inspected. If inspection above the ceiling is not possible, the report should recommend that access be gained to allow inspection.

It is helpful to realize that the smaller or thinner a steel structural member is, the more likely that corrosion will be structurally significant. Therefore, the metal deck and slab reinforcement are the items most commonly in need of repair. Corrosion of slab reinforcement typically would be accompanied by spalling of the surrounding

FIGURE 6.1 A casual inspection of the roofing at this low-rise residential project revealed numerous points of water penetration, including open seams in the newly installed membrane. Water penetration was causing deterioration of the wood roof deck and even causing the facade stucco to separate from the building. The roofing contractor for this project had been terminated before the installation was complete.

FIGURE 6.2   This case illustrates the importance of interviewing personnel who are knowledgeable about the building's maintenance problems and following up on potential issues that they raise. An area of the roof slab of this hotel collapsed into a guest room without warning, killing the occupant. It was found that the slab reinforcement and much of the steel framing were completely corroded away. The ceiling, which originally was the underside of the structural slab, had been a constant source of leakage over the years and had required constant maintenance. Well-intentioned maintenance personnel had replastered spalled areas and eventually installed a gypsum board ceiling to conceal the water damage, never realizing the structural consequences. The source of moisture penetration ultimately was traced back to a missing length of flashing.

concrete. Corrosion is rarely allowed to progress undetected to the point that it damages large structural members (Fig. 6.3).

If the ceiling is of a type that is readily removable, such as a lay-in system, or if the ceiling has access hatches, the consultant should take advantage of them to inspect the area above the ceiling at random locations, even if there are no signs of water penetration. This would allow the consultant to observe the general type of structural system, its overall condition, the condition of fireproofing, and other items.

The consultant should record carefully the extent of any areas inspected, as well as the extent of areas not inspected. This information can prove to be useful if a question arises later about what the consultant could have or should have seen.

Basement Areas.   Unfinished basement areas can provide an opportunity to observe directly some foundation elements. These may include perimeter reinforced concrete walls, columns, slabs on grade, and shearwalls.

One common problem is water penetration through cracks in perimeter concrete walls. This can cause corrosion of reinforcement and consequent spalling. Slabs on grade often display cracking either as a result of subgrade settlement or due to an insufficient number of control joints. The crack patterns provide additional information.

FIGURE 6.3   Water penetration through this slab over a long period of time corroded the slab reinforcement. This caused the concrete cover to spall away and fall into the apartment, along with the direct-applied plaster, narrowly missing an occupant.

Framed Slabs. Framed slabs normally have a finish material that conceals the condition of the surface. If the finish is flexible or can move independently of the slab, it will conceal underlying conditions such as cracks. This is the case with carpeting or "floating" wood floors. Semiresilient materials, such as vinyl composition tile, may or may not reflect cracks. On the other hand, a brittle finish, such as ceramic tile, that is bonded to the slab often will reflect underlying cracks and movement (Fig. 6.4).

If a slab does not have a finish over it, its condition can be inspected readily. This is often the case for below-grade parking areas or other "back of the house" areas. Items of slab inspection include cracking, spalling, delamination (may not be possible visually), exposed reinforcement, scaling, impact damage, ponding, and the presence of expansion joints,

The underside of a framed slab also can be inspected for cracking, spalling, exposed reinforcement, and signs of water penetration. Correlating cracks or leakage that appears on the bottom surface with corresponding conditions on the top surface can be useful in diagnosing the cause of the damage (Fig. 6.5).

Ceilings. Ceilings usually are not included in the structural consultant's scope of work in a due diligence evaluation. However, heavy ceilings, such as those made of gypsum board or plaster, have the potential to cause serious injury or even death if there is a failure (Fig. 6.6).

The ceilings that are most prone to having problems are those which are located outdoors, such as under canopies and soffits, and those over natatoriums. Such environments can quickly cause severe corrosion of the relatively thin members that make up the suspension system. This is especially true if unprotected steel, which is common for dry environments, is used. There is even a case study of a ceiling collapse over a

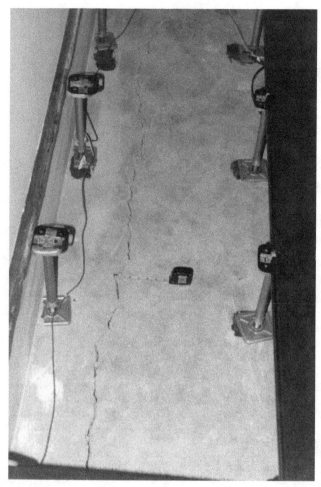

FIGURE 6.4   These wide cracks in a reinforced slab-on-metal deck were noticed when one of the raised floor panels was removed. This triggered an intensive investigation, which found that the reinforcement had been installed in the wrong direction throughout this 250,000-ft^2 facility. A battery of full-scale laboratory load tests was conducted, from which it was concluded that, fortunately, the form deck was acting compositely with the concrete. The integrity of the slab was confirmed with in situ load tests, and no repairs were required.

natatorium where the "stainless steel" hangers failed owing to stress corrosion, causing 12 fatalities.[1]

In some cases, the ceiling is part of the system that provides the floor construction with its fire rating. Penetrations through such a ceiling must be sealed properly to protect the fire rating of the floor.

If inspection of ceilings is not in the scope of work, but a concern presents itself to the consultant during the course of his or her work, it would be appropriate to raise the concern in the report and to recommend further study. Ceiling inspections are

FIGURE 6.5   This below-grade parking area had been surfaced with an asphalt overlay. Localized "heaving" of the asphalt was noticed, which led to its removal. This revealed large areas of spalled concrete and of severely corroded reinforcement in the structural slab and joists. Extensive structural repairs had to be designed and implemented.

covered in more detail later in this chapter under "General Structural Integrity Inspections."

Facades.   Facade inspection often is treated as a specialty and may or may not be included in the structural engineer's scope of work. Chapter 8 is dedicated to assessing the condition of facades. However, the condition of a facade, particularly one made of masonry, often is symptomatic of underlying conditions with the structure and therefore deserves some mention here.

Foundation settlement often manifests in the form of cracking of masonry facades. The classic settlement crack occurs at approximately 45 degrees on the diagonal in stepwise fashion along brick mortar joints. Typically, a series of these diagonal cracks appears between windows that are "stacked" vertically. The direction of the cracking indicates which side the settlement occurred on (Figs. 6.7 and 6.8).

Corrosion of steel spandrel beams and perimeter columns often causes bulging and cracking of the masonry in front of the affected member. This is particularly true when the exterior masonry is relatively tight against the steel. At beams, this usually manifests as a horizontal band of bulged brick, possibly accompanied by a horizontal separation along one of the beam flanges. This condition eventually can lead to the band of brick being pushed off the building entirely. Parapets sometimes can be heaved upward by the expansive corrosion forces. Column corrosion typically manifests as a pair of vertical cracks. This condition often occurs at building corners. Initially, the crack may occur owing to other causes, such as thermal expansion and contraction of the masonry. Left unrepaired, this starts a vicious cycle where water enters the fine crack, causing a small amount of corrosion, and then expansion from the corrosion

FIGURE 6.6   This plaster ceiling in this commuter rail terminal collapsed one morning during rush hour, killing two persons and injuring dozens others. The ceiling had been noticeably sagging, which had prompted an inspection within the space above the ceiling by maintenance personnel. It was found that the hangers had been hooked through tabs punched from the metal deck and that the tabs had broken. The weight of the maintenance worker triggered the collapse.

widens the crack slightly, which admits more water, and eventually, the cracks become so wide that the masonry "column" becomes unstable and separates from the remainder of the wall (Figs. 6.9 through 6.11).

Steel lintels are used frequently to support the weight of masonry over window openings. A common ailment is corrosion, which causes downward deflection owing to the expansive nature of the corrosion process (Fig. 6.12).

Warehouse and Industrial Buildings. In warehouse and industrial buildings, the roof structure typically is exposed and can be inspected directly. Often, however, owing to the height of the roof, it is only possible to observe these members from a distance. Binoculars may be useful for better inspecting selected locations (Fig. 6.13).

Preengineered structures are used often for these buildings. They typically use cold-formed Z- or C-shaped purlins to support a standing-seam roof. For longer spans, sometimes open-web joists are used. The rafters typically consist of built-up tapered members spaced 20 to 30 ft apart that run over the tops of intermediate columns.

Items of inspection may include:

Rotation and sag of purlins. Overloaded Z purlins typically fail by first rotating about their longitudinal axis and then by sagging excessively. If purlins have

FIGURE 6.7   The patterns of diagonal cracking in this brick facade between the windows are clear indications of foundation settlement.

been overloaded, perhaps by a heavy snowfall, they may not rebound elastically, and the twist and sag will be permanent. Such a condition, if observed, is serious and should be reported immediately.

Added hung loads. The roof systems of preengineered buildings typically are designed very efficiently for specific loadings that are identified in the design phase. Therefore, it is not advisable to hang any additional loads without prior approval from the manufacturer. Owners and tenants may not be aware of this common limitation and so may hang additional items, such as ceilings, winches, or other items. In addition, these items may be attached in an inappropriate manner, such as from the tip of a Z purlin or by drilling a hole in the bottom flange of a purlin. The consultant should attempt to identify items that appear to have been added or that appear to be attached improperly and recommend that the owner review them with the manufacturer.

Bridging. Bridging often is required for the top chords of open-web joists and top flanges of cold-formed purlins, where the standing-seam roof is not capable of providing sufficient lateral bracing. The bottom chord and flanges of these members also may require bridging for lateral bracing for wind uplift load cases. To be effective, the bridging must be anchored either into the adjacent construction (masonry wall, for example) or to a self-braced system. Bridging normally is considered to be a tension-only member, so anchorage is required at both ends.

Bottom-flange bracing of rafters. Where rafters are continuous over intermediate columns, the bottom flange is in compression, and it is necessary therefore to brace it against lateral buckling. A similar condition occurs near the haunches where the rafter is moment-connected to the sidewall column. The bracing often is achieved by diagonal braces attached to the flange and to a purlin. The presence of these braces is critical to the proper performance of the rafter, and the

FIGURE 6.8 The step cracking throughout this building indicated that there had been foundation settlement. It was suspected that the wood piles on a certain side had deteriorated owing to the water table having been drawn down many years earlier because of adjacent construction. Excavation of test pits confirmed the severe deterioration of the wood piles and wood cap and, in one area, the total absence of the pile tops.

(a)   (b)

FIGURE 6.9 *a.* Corrosion of the steel corner columns manifested as damage to the limestone cladding. *b.* Limestone had to be removed and replaced in order to repair the steel.

FIGURE 6.10 Bulging of the masonry along horizontal bands was found to be due to severe corrosion of the underlying spandrel beams.

(a)

(b)

FIGURE 6.11   *a.* This pair of vertical wide cracks created a column of masonry that was not secure and that could very conceivably have fallen onto the adjacent building. The local building department was notified of the potential hazard, and the masonry was removed immediately. *b.* The underlying steel column was found to be severely corroded such that extensive reinforcement had to be designed and installed.

FIGURE 6.12   These steel lintels from a brick masonry facade were almost completely corroded and had to be replaced.

braces are items that conceivably could be omitted inadvertently by the erector. Therefore, the presence of these braces should be confirmed. In open-web joist construction, it is common to extend the joist's bottom chord to the bottom flange of the rafters in negative-moment zones for the same reason (Fig. 6.14).

Bracing column tops. In cases where bottom flange bracing in not needed because the rafters or girders are not continuous over the columns, it still may be necessary to brace the column top. This usually is achieved with either fitted stiffeners in the girder end or by extending the joist's bottom chord.

GENERAL STRUCTURAL INTEGRITY ASSESSMENTS

Building owners sometimes commission a more detailed assessment of a building's structural integrity. There are several potential reasons for such a study:

FIGURE 6.13   The roof structure of this industrial building is completely exposed, facilitating inspection.

FIGURE 6.14   Bracing of the column tops and of the bottom flanges of the continuous girders was absent throughout this industrial building. This was not noted during a due diligence inspection. Portions of the building collapsed shortly afterward, and one of the issues debated was whether the professional had exercised due care.

- The owner wishes to identify any imminent hazards. This sometimes will include nonstructural items, such as ceilings or facades, that may present falling hazards.
- The owner wishes to identify maintenance or repair items that will prolong the life of the structure.
- The owner wishes to prepare a building-specific inspection program.
- The owner wishes to prepare a capital expenditure program.
- The owner has experienced a specific problem and wishes to investigate the extent of similar problems.

In most cases the building will have been in service for a considerable length of time and may be showing signs of wear and tear or possibly damage. In many cases these studies are performed for public facilities, such as for transportation terminals, museums, convention centers, and other places of assembly.

This type of investigation typically is more in-depth and detailed than a due diligence inspection. This is partly due to the fact that the owner is committed to the property and therefore can allocate funding for a more complete investigation. The field inspection may take place over several days or even weeks and may include physical probes and testing.

Planning

Proper advance planning of the evaluation is important so that the client's concerns and objectives can be addressed in a way that makes efficient use of the consultant's time and resources.

Client Concerns and Objectives. Some owners, particularly government agencies who manage portfolios of buildings, will have criteria regarding the scope of the inspection and reporting requirements.[2] Obviously, these need to be provided to the consultant in advance, and the consultant needs to review and incorporate these criteria. Often it will be necessary, through discussion with the client, to make adjustments to the general criteria to adapt them to a specific project.

Establishing priorities for the work that needs to be done is often important for the client. A prioritized list will assist the client in short- and long-term planning and budgeting. Categories should be discussed with the client to arrive at a system that will be most useful to the client. One possible priority system would consist of three simple categories:

Immediate. These are items that present an immediate structural or falling hazard. These items normally would be reported to the owner immediately, even before the report is issued, so that the necessary steps can be taken. These steps may include cordoning off an area, installing shoring, or removing hazards.

Priority. These are items that do not present an immediate hazard but that require action in an expedited manner. This may include interim repairs, further investigations, probes, and similar items

Routine. These are items that require work that can be performed as part of a regularly scheduled maintenance program.

Another possible system is to prepare a prioritized work plan that is tied to a timeline:

- Immediate (performed prior to issuing the report
- Short term (within 1 year)

- Medium term (within 5 years)
- Long term (within 10 years)

This approach will be attractive to organizations that need very structured guidance as to what work should be done when or to organizations that need to make long-term budget decisions

A critical factor will be how often the client intends to repeat the inspection cycle. If a facility is inspected frequently, then it may be possible to revisit the condition of a component at the next inspection, thereby deferring the decision to repair it. If the next scheduled inspection is too distant to allow monitoring of a component, then the consultant will have to recommend remedial action of items that otherwise may not perform well in the interim. The consultant should consider recommending an inspection cycle for questionable or critical items.

The cost of recommended repairs will be an important factor in the client's planning. In some cases, the consultant may prepare estimates. If the consultant does not have this capability, then the client may wish to hire a cost estimator to work with the consultant. Some clients prefer to prepare their own cost estimates. The responsibility for preparing cost estimates must be discussed with the client in advance.

Document Gathering. The documentation available for review is often extensive because the owner frequently has a library of building documents. The documentation may include a complete set of design drawings, tenant improvement drawings, shop drawings of some components, and even calculations.

Typically, the client will supply or make available the drawings for the structure to be inspected. The consultant normally relies on being provided with the latest drawings. If it is obvious to the consultant that the drawings are not the latest or that key drawings are missing, then the consultant needs to alert the owner.

Documents filed with local building officials may not be the latest because it is common to continue developing the drawings subsequent to the initial filing. But these may be useful if the drawings are not available from a better source.

At a minimum, the consultant will need the original structural drawings and possibly portions of the specifications. If there have been structural alterations or additions, those drawings will be needed also. Depending on the project, it also may be necessary to obtain architectural floor plans.

If previous inspections of the facility have been conducted, these reports should be reviewed carefully. The reports may contain baseline information regarding the condition of certain elements or may include recommendations for follow-up inspection.

Staffing. It is usually desirable for the inspection to be performed by or under the supervision of a qualified registered professional engineer. To be qualified, an engineer should be experienced in building design, construction, and inspection, usually with at least 10 years' experience. For large or complex projects, the field inspection often is led by a degreed engineer with several years' experience, possibly with assistance from other technical personnel. The qualifications of the proposed team should be discussed with the client.

Level of Inspection. The consultant must plan the level of inspection that will be afforded to each component, taking into account ease of access, likelihood of problems being present, the criticality of the component, and the nature of the problems that may be present. The levels of inspection typically considered are

  Visual inspection. Inspection from a reasonable distance, such that an initial assessment can be made. Appropriate for typical components where no problem is suspected.

Hands-on. Close-up inspection at arm's length. Appropriate for critical or one-of-a-kind items, especially if they are prone to having problems. Examples of items that may warrant this type of inspection are truss bearings, cable anchorages, and sliding connections. Probes or removal of finishes may be necessary to expose the item.

Testing. Appropriate where the nature of a potential problem is such that its presence may not be ascertained from a hands-on inspection. Types of tests are discussed later in this chapter.

The level of inspection to be used for various systems and components is a topic that should be discussed with the client.

Checklists. For large projects, especially when several inspectors will be needed, it is useful to prepare inspection checklists. The checklists can serve as reminders to help ensure that components are not overlooked, can help standardize reporting and terminology, and can define what types of conditions are being looked for. It may be appropriate to share these checklists with the client in advance so that the client can provide input and suggest modifications. The following subsections include suggestions that may be incorporated into checklists.

Field Work

In most buildings, the structural elements are concealed behind architectural finishes, such as gypsum wallboard and ceilings. Therefore, it is normally not cost-effective to observe all the structural elements. The consultant therefore will have to select elements carefully that he or she wants to observe, taking into consideration the likelihood of an underlying structural problem and balancing it against the difficulty of gaining access.

Building maintenance personnel often are a valuable source of information regarding the condition and history of the building. They can provide information such as the location of leaks, repairs that have been made, maintenance history, and other information that may help the consultant in identifying areas of concern.

Field Notes. The value of properly documenting first-hand field observations cannot be overstated, yet it is something that often is overlooked until it is too late. Poor documentation can lead to expensive reinspection, ambiguity about what was or was not inspected, and possibly loss of the client's confidence in the consultant.

The goal is to record all the relevant information observed by the inspector in such a way that it can be understood readily by other persons, perhaps years later.

Some suggestions for preparing high-quality field notes are

- Sign, date, and title every page.
- Use handwriting that is clear and neat.
- Label sketches so that they are self-explanatory.
- Record observations using either a hardbound book or sequentially numbered pages.
- Use waterproof ink as opposed to pencil or felt-tip pens.
- If photographs were taken, cross-reference them to the notes.
- Consider using a photograph log sheet, which includes a roll identifier, photograph number, and a brief caption.
- Label and organize photographs so that they can be correlated with field notes or logs.

- If observations are noted on a separate drawing, cross-reference the drawing. The marked up drawing also should be signed, dated, and titled.
- Indicate the names of other persons present during the inspection.
- Indicate the weather conditions, if relevant.
- Inspectors should submit field notes and photographs to the project manager on a regular basis, and the project manager should review the notes and provide feedback to the inspector.

Visual Inspection. Following are items of inspection that apply to various common structural systems and materials.

Steel Construction. Deterioration and damage that are visible on steel members and connections includes corrosion, deformation, cracks, and weld defects. Following is a more detailed list of these conditions.

Almost all commonly used steels are susceptible to corrosion if they are in a moist environment and do not have a coating. It is valuable during a visual inspection to make a qualitative assessment of the degree of corrosion. This may help to determine whether more detailed testing or analysis is needed. Following are useful categories to indicate the degree of corrosion:

Minor. Surface corrosion that has resulted in negligible loss of section. It is useful to identify this condition so that preventive measures can be taken to prevent further corrosion.

Moderate. Corrosion that has resulted in a measurable loss of section. Structural repairs may not be needed, but maintenance is definitely needed to arrest further corrosion.

Severe. Corrosion that has resulted in significant loss of section, to a degree that repairs probably will be needed. An example would be a beam whose flange has been reduced to a knife edge.

Signs of distress and other conditions that may be visible on members include

- Excessive deflection
- Severe stress concentrations, as evidenced by cracking paint
- Cracks at notches, web penetrations, or coped areas
- Kinks from thermal strain
- Web crippling at bearings and locations of concentrated loads
- Diagonal web buckling near supports in built-up plate girders
- Bowing
- Misalignment
- Deformation
- Twisting of flexural members, possibly indicating insufficient lateral torsional bracing
- Presence/absence of bridging or other lateral bracing and its anchorage

Conditions that may be visible on connections include

- Cracks in welds
- Weld undercut
- Obviously undersized or missing welds

- Irregular welds
- Loose, broken, or missing bolts or rivets
- Bolt threads not fully engaged
- Bolt holes irregularly spaced or shaped
- Gaps between faying surfaces
- Evidence of slippage
- Inadequate edge distance of bolt holes
- Prying of clip angles
- Deformations
- Evidence of movement at bearing plates and sliding connections

In some applications, steel members are protected with fireproofing. The fireproofing could consist of a sprayed-on or troweled-on cementitious material, gypsum board, or concrete encasement. Areas of missing or damaged fireproofing should be noted.

Concrete Construction. Visible conditions that should be noted include

Cracks. It is valuable to report the crack width either in numerical terms or, if there are many cracks, in categories. A simple and useful categorizing scheme is

Hairline:	Barely visible
Fine:	$\frac{1}{32}$ to $\frac{1}{16}$ in.
Medium:	Between $\frac{1}{16}$ and $\frac{1}{8}$ in.
Wide:	Greater than $\frac{1}{8}$ in.

The location and orientation of cracks, and especially any patterns, are important characteristics in helping to determine whether the cracks are stress-related. For example, cracks that occur regularly in areas of high stress, such as negative-moment regions along column gridlines, are stress-related, whereas cracks that are intermittent, changing direction, and occur at random locations probably are not stress-related.

It is also useful to know whether the crack extends through the entire thickness of a slab because this may help to determine whether the crack was caused by flexure, thermal changes, or something else. If there is a vertical differential or a "step" across the crack, this may be an indication of subgrade settlement. A valuable piece of information that is useful for both diagnosing the cause of a crack and developing an appropriate repair is whether the crack is continuing to move over time. Surface-mounted crack monitors can provide this information in a simple and direct manner. For special applications, more sophisticated systems are available, some of which can report movement wirelessly and over the Internet (Fig. 6.15).

Scaling. Scaling is the flaking or loss of material from a concrete surface. It is caused most often by freeze-thaw damage and/or by a weak cement paste layer at the surface. Scaling is best characterized according to depth. For example:

Light:	Less than $\frac{1}{4}$ in. deep
Medium:	Between $\frac{1}{4}$ and $\frac{1}{2}$ in. deep
Heavy:	Deeper than $\frac{1}{2}$ in.

Spalling. Spalling is the detachment of concrete fragments. Spalling often is caused by expansive forces generated by the corrosion of embedded reinforcement, but

FIGURE 6.15 Surface-mounted crack monitors can determine accurately if a crack continues to move over time.

it also can be caused by other factors. Spalls are best categorized by size. For example:

Small:	Less than 6 in. in diameter
Medium:	Between 6 and 12 in. in diameter
Large:	Larger than 12 in. diameter

Exposed reinforcement. In areas of spalled concrete, the reinforcing bars often will be exposed. The degree of corrosion should be noted and reported. In addition, the depth of concrete cover over the bars may provide useful information regarding the cause of the corrosion and may affect repair recommendations (Fig. 6.16).

Signs of water penetration. Usually manifested as efflorescence, a whitish deposit of minerals, along cracks and joints. If the efflorescence is stained a rust color, it probably indicates corrosion activity.

Delamination. Delaminations are incipient interconnected spalls usually extending over larger areas. They can be difficult to detect visually but are detected readily by sounding. If delamination is suspected, then a systematic sounding survey and mapping may be warranted. This is covered later in the "Testing" section of this chapter. Depending on the visual observations and their evaluation, it may be desirable to perform testing. Basic concrete testing also is covered later in this chapter, and Chapter 15 describes the entire spectrum of concrete testing.

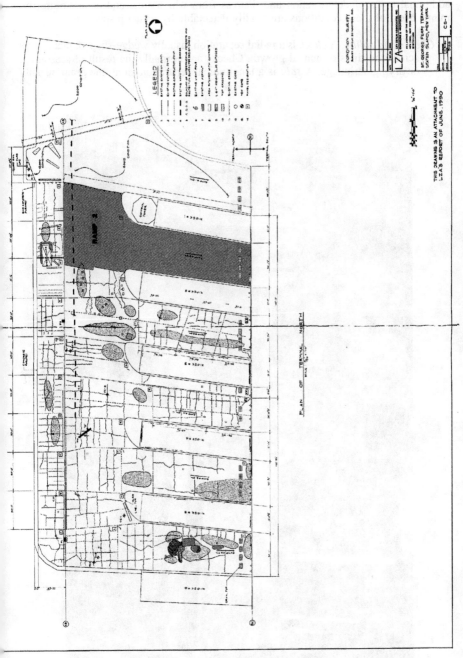

FIGURE 6.16 A detailed mapping of cracks and other conditions can be invaluable in detecting patterns and making an accurate diagnosis.

Wood Construction. Owing to its organic nature, wood construction is subject to additional types of deterioration and distress from which steel and concrete are immune. Most of these conditions are readily detectable by visual inspection.

    Checks and splits. A *check* is a radial separation, usually wedge-shaped, that occurs across the rings of annual growth. Checks are normally the result of seasoning, or drying shrinkage. A *split* is a longitudinal separation along the grain, usually

FIGURE 6.17 This wood column in a turn-of-the-century building has a longitudinal split, which considerably reduces its buckling strength.

through to the opposite surface, and is caused most often by stress or mechanical damage. Depending on their length and location in a member, checks and splits can diminish the strength of a member drastically. For example, a split at the notched end of a floor joist increases the shear stress dramatically, or a check along the height of a wood column can decrease its buckling resistance (Fig. 6.17).

Insect damage. Termites, carpenter ants, and other insects can hollow out wood members. Wood shavings at entry holes are telltale signs of insect activity but may not be indicative of the full extent and severity of internal damage. If this type of damage is suspected, further investigation may be required.

Decay. Fungi in an environment with the proper temperature, oxygen, and moisture can consume and destroy wood. Wood moisture content needs to be above 20% to sustain fungal growth. Therefore, wood in a conditioned space and not subjected to leakage or other unusual moisture would be unaffected by decay (Fig. 6.18).

Impact damage. Impact by vehicles or equipment can cause serious immediate damage to wood structures.

Older wood truss long-span roofs deserve special attention for two reasons. First, the allowable stresses for truss members were lowered dramatically in 1971. For example, the 1962 *National Design Specification* (*NDS*)[3] specified an allowable stress of 1900 lb/in.[2] for tension parallel to the grain for Douglas fir–larch, select structural. This value had been in use for decades. However, the 1971 NDS Supplement cut this allowable stress in half, to 950 lb/in.[2], based on research. A similar but not as dramatic reduction was made in the allowable stress for compression parallel to the grain. Obviously, these reductions in allowable stresses directly affect the safety of wood trusses that were designed using the older values and has led to reinforcement of many of these structures. Another common problem with older trusses is decay of the bearing area, where moisture has a tendency to collect. Since this is such a critical zone, consideration should be given to gaining access to this area to allow hands-on inspection. In one case that we are familiar with, as part of a due diligence project, a structural engineer inspected the condition of long-span wood trusses and recommended repairs to several of the trusses. The repairs were made. Several years later, under a snow load, the roof trusses sagged substantially and eventually collapsed. It appears that the collapse initiated at the trusses on which the engineer had *not* called for repairs. The incident led to litigation, which eventually was settled out of court (Fig. 6.19).

Another common problem in wood frame construction has been the use of plywood with a fire-retardant treatment (FRT). Between 1985 and 1995, the roof sheathing of over 750,000 housing units experienced structural degradation to the point where replacement was necessary, at a cost of over $2 billion dollars.[4,5] In some case the deterioration was so severe that it was unsafe to even walk on the sheathing. The cause of the degradation was the use of certain fire-retardant chemical formulations in combination with exposure of the plywood to temperatures over 100°F, as can occur in roof applications. Plywood manufacturers have since changed their formulations for new product, but the condition is still present in some existing facilities and should be watched for (Fig. 6.20).

Composite wood members, such as glued laminated members, often depend on adhesives. Deterioration of the adhesive can result in delamination and severe weakening of the member.

Ceilings. Ceilings often are not included in a structural consultant's scope of work. However, heavy ceilings, such as those made of gypsum board or plaster, can result

(a)

(b)

(c)

FIGURE 6.18 Statewide inspections of wooden salt domes for a highway department revealed a variety of conditions in these long-span structures, such as (a) checks, (b) impact damage, and (c) decay of sheathing.

FIGURE 6.19 The wood roof trusses of this chapel collapsed under a snow load. Issues investigated included changes in allowable stresses, decay, and juvenile wood.

FIGURE 6.20 Premature deterioration of fire-retardant-treated (FRT) plywood was a widespread problem for frame construction built in the 1980s and 1990s. (*Photo courtesy of the U.S. Department of Agriculture, Forest Products Laboratory.*)

in serious injury or fatality if there is a failure. Ceilings over natatoriums and outdoor areas, such as canopies, are especially vulnerable to having problems. It therefore would be prudent for these ceilings to be inspected from time to time.

There are numerous different types of ceilings, both proprietary systems and conventional systems. System vary in weight, material, and type of attachment (suspended versus direct adhered). Lightweight ceilings include acoustical tile (lay-in or concealed spline, metal pan, egg crate, plastic grids, and spray-on) and perforated metal. These ceilings generally do not pose a significant hazard owing to their minimal weight and the small size of the units.

Of more concern are the medium- and heavy-weight ceiling types, such as plaster, gypsum board, ceramic tile, and wood. Besides their weight, these ceilings often are assembled into continuous sheets, which could result in a progressive failure of a large area (Figs. 6.21 and 6.22).

A typical suspension system for a plaster ceiling consists of the following main elements:

FIGURE 6.21 This plaster ceiling over a classroom collapsed suddenly, narrowly missing students sitting at their desks. The ceiling had been in place for more than 50 years.

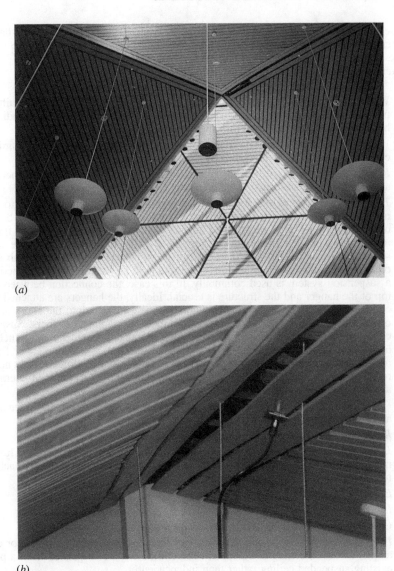

(a)

(b)

FIGURE 6.22   *a.* A wood slat from this new ceiling fell into the sanctuary seating below. This prompted the erection of scaffolding throughout to allow a hands-on inspection of the attachment of every slat. *b.* Inspection from close range revealed visibly distorted slats throughout the ceiling and hundreds of instances of improper attachment.

- Hangers—typically spaced on a 4-ft grid.
- Carrying channels—typically spaced on 4-ft centers
- Furring channels—typically spaced at 1-ft centers
- Metal lath
- Three coats of plaster—the scratch coat (first coat), brown coat, and finish coat

Design weights for this type of system range from 10 lb/ft^2 for a gypsum-based plaster to 15 lb/ft^2 for a cement-based plaster. Exterior ceilings and soffits also need to be able to resist wind pressures and suctions, as well as corrosive action.

Items to note during the inspection include

- *Visible deflection.* This is an indication that there *already* has been a failure, probably of a hanger or one of its connections. It is important that this condition be reported immediately and with high urgency.
- *Water stains.* Water penetration can result in disintegration of gypsum board or can cause corrosion of the ceiling suspension system.
- *Broken, disconnected, or missing hangers.* This would be visible only if there is a catwalk or similar access to the area above the ceiling.
- *Type of attachment to the structure.* Some ceilings are attached directly to the underside of the structure. In residential construction, for example, gypsum board or plaster ceilings often are attached directly to the underside of wood joist framing. In this case, it is important to know whether the attachment was made with screws, as is the modern practice, or with nails. Depending on their length and other factors, nails may not provide reliable support. For commercial construction, a suspension system is used commonly. In this case, the connection between the top of the hanger and the structure is crucial. Ideally, the hangers are attached via embedments, such as T-shaped heavy straps that were cast into the slab during original construction, or via positive attachment to framing members. Suspect connections include expansion anchors, powder-actuated fasteners, and punched tabs in the underside of metal deck. Localized removal of concrete cover from reinforcement in order to loop a hanger over the exposed reinforcement is not a recommended technique because it imposes bending stresses on the reinforcement and removes the protective concrete cover.
- *Connections between the various elements,* such as between carrying channels and furring channels.
- *Deformation or twisting of the suspension grid members.*
- *Local deformation of framing members to which the ceiling is attached.* This is most likely in cases where light-gauge framing is used, and the ceiling is attached improperly to the light-gauge purlins.
- *Evidence of repairs.*
- *Hanger spacing and type.*
- *Cracking or other distress of brittle ceiling finish.*
- *Items hung from components that were not intended to provide support.* For example, an added ceiling or equipment sometimes may be suspended from a pre-existing suspended ceiling rather than independently.
- *Was the ceiling designed to be accessible for maintenance personnel to walk on or for storage?* If not, then are there proper warnings or impediments to prevent such use?

A number of other items besides ceilings often are suspended from a structure. Some of these items are heavy, and their proper attachment is critical. These include heavy mechanical piping and pipe racks, mechanical units and air handlers, signs and scoreboards, and large lighting fixtures. Inspection of the suspension systems for these elements is similar to that for ceilings.

Probes. Probes are made for the purpose of exposing elements that otherwise would not be visible. They generally are expensive but can provide valuable information.

Probes can be used to determine the nature of the construction, to confirm the good performance of critical or vulnerable components, or to determine the underlying causes of observed distress.

The services of a contractor generally are required to open a probe and later to patch the area. Probes require considerable time on the part of the consultant, who must select the locations, coordinate with the contractor, and make observations once the probes are made. Sometimes the client also will require that the consultant inspect the adequacy of the patching (Fig. 6.23).

Probes often are made at representative locations where distress is manifested. In some cases it also may be prudent to make probes at critical components or components that are prone to having problems. They include

- Connections of hangers and cables and of heavy suspended items such as scoreboards
- Support systems for heavy ceilings, especially in outdoor areas or over humid spaces
- Cantilevers
- Long span members and their connections
- Members exposed to the elements or to humid environments

Testing. Depending on the results of the visual inspection or of the probing, there may be a need to perform testing. Many types of tests and examinations can be performed; some can be performed in the field, and others must be performed in the laboratory. Following are the ones that are used most commonly. Additional tests are described in SEI/ASCE 11-99, *Guideline for Structural Condition Assessment of Existing Buildings.*[6]

Steel

Weldability. When renovating vintage steel structures, there is often a concern that the steel may not be weldable or that special welding techniques may be necessary. Thus it is prudent to examine coupons removed from members metallurgically and to determine whether the equivalent carbon content is in the acceptable range.

Strength. It is often valuable to test coupons for yield strength and ultimate strength either to assist in confirming that the steel meets a certain standard or perhaps with the hope that it has a higher-than-specified strength that can be taken advantage of.

Magnetic particle. This test is used for locating discontinuities or cracks, usually at welds. It involves magnetizing the area and then distributing a fine magnetic powder over the area. The patterns displayed by the powder indicate whether there is a discontinuity.

Chapter 17 describes in detail the range of tests available for steel (Fig. 6.24).

Concrete. A number of tests are performed commonly on concrete. Following are some of the more commonly used tests.

Compressive strength. This key concrete property can be determined by drilling cores from the concrete and testing them in the laboratory. The height-to-diameter ratio of the core must meet American Society for Testing and Materials

FIGURE 6.23 The fractures in the welds joining these moment plates to the columns could only be detected by making probes through the concrete encasement.

FIGURE 6.24   Techniques ordinarily used for locating buried pipeline were adapted to locate and map steel roof beams.

(ASTM) requirements, and the strength of the core can be related to the concrete cylinder strength.[7,8] This is the most accurate method of determining strength.

Schmidt rebound hammer. This test estimates the concrete strength by measuring the rebound of a spring-driven mass in a handheld impactor. The test is easy to perform in the field and provides immediate results. However, frequent calibration is needed, and the method is best for determining the relative strength of one area versus another.

Windsor probe. This test estimates concrete strength by measuring the penetration of a probe fired into the concrete. Performed in the field, it has advantages and disadvantages similar to those of the Schmidt rebound hammer.

Petrographic examination. A trained petrographer using high-power magnification can determine visually some important characteristics.

• *Air content.* The size and distribution of air bubbles play an important role in the freeze/thaw resistance of concrete.

• *Proportions of ingredients.* The water-cement ratio is a key factor in the strength and durability of concrete. Petrographic examination can provide a qualitative assessment of this characteristic.

• *Cracks.* Whether cracks formed when the concrete was still plastic or after hardening can be important.

Chemical tests. Laboratory testing is sometimes necessary. Common tests include

• *Chloride content.* The presence of chlorides over the corrosion threshold value[9] greatly accelerates the corrosion process.

- *Ph.* Lowering the Ph depassivates the concrete, allowing corrosion of the steel.
- *Carbonation.* Carbonated concrete no longer provides a protective environment for embedded reinforcement.

Pachometers (R-meters). These are devices for magnetically locating reinforcement. Some devices also can determine the depth of cover or bar diameter.

Half-cell potential. This test measures the electrical potential for the purpose of determining the probability of corrosion activity. Contours of the measured values can be mapped to provide a visual indication of areas with high potentials.[10]

Chapter 16 provides a detailed description of concrete testing (Fig. 6.25).

Wood

Species identification. The structural properties of wood vary widely depending on the species. A trained individual can determine the species based on visual examination under low magnification of a small sample.

Visual grading. Allowable stresses for wood members depend on the size, type, and location of defects. Defects include knots, splits, and checks. Trained individuals can classify a member according to the visual grading rules for the particular type of wood. For example, the Western Wood Products Association sets grading rules for softwoods in 12 western states. These rules result in classifications such as select structural, no. 1, no. 2, etc, each of which has a set of allowable stresses associated with it.

FIGURE 6.25   Testing of the concrete determined that alkali-silica reaction contributed to the cracking experienced by this prestressed concrete girder.

Machine stress rating. This is a nondestructive test performed on an entire member in a laboratory. It accurately determines the modulus of elasticity E of the member in bending. The tests results are used together with visual grading techniques to arrive at a more accurate classification.

Strength tests. If visual grading or machine stress grading does not provide sufficient accuracy, destructive tests can be performed to determine mechanical properties, such as bending strength, density, compressive strength, shear strength, tensile strength, etc.

Moisture testing. The moisture content of wood affects the allowable stress, as well as how much additional shrinkage the member might be expected to experience. Handheld moisture meters can quickly provide reasonably accurate values of moisture content. If more precise values are needed, samples will have to be removed, sealed, and tested in a laboratory.

Drilling and probing. Drilling and probing with an awl can be used to locate pockets of decay or insect damage, as well as to determine the depth of decay or fire damage.

Coring. Small cores can be removed and examined in the laboratory to determine species, moisture content, depth of damage, and type and amount of preservative treatment.

Chapter 19 discusses in detail the many tests that are appropriate for wood materials

SEISMIC EVALUATIONS

Seismic evaluations of existing buildings generally are performed prior to potential seismic events for one of the following reasons: when the building owner is preparing (either voluntarily or by jurisdictional regulation) to retrofit or upgrade the building, as part of a prepurchase or due diligence evaluation of a building to identify potential vulnerabilities of the building or to estimate the potential damage losses of the building for use when considering or obtaining earthquake insurance coverage.

The seismic evaluation of a building is not only a "what *is*" investigation but also a "what *if*" investigation. The evaluator is concerned with identifying current systems and conditions but also is concerned with the ways in which the building likely will perform when subjected to a potential earthquake of uncertain dynamic characteristics, duration, source direction, and distance. The best way to gain knowledge of how buildings perform during earthquakes is to perform investigations of buildings damaged by earthquakes (see following section). Also useful are the review of earthquake reconnaissance reports published by the Earthquake Engineering Research Institute (EERI), EERI Slide Collection CD-ROM,[11] and FEMA/ATC Postearthquake Safety Evaluation of Buildings CD-ROM.[12] The FEMA/ATC CD-ROM shows examples of typical damage caused by different earthquakes to specific building types. The Stratta textbook[13] also has many useful photographic illustrations of typical damage.

Planning

Before performing a seismic evaluation, it is necessary to obtain and determine certain information.

Client Concerns and Objectives. In order to determine the scope of the evaluation, it is critical to understand the client's concerns and objectives. Objectives often are defined clearly when evaluations are mandated by jurisdictional regulation. However, when evaluations are undertaken voluntarily by a client, the objectives may need formulation. Issues to consider include whether the building is of historical merit and what are acceptable modifications to the building, to what performance level should the building be upgraded, to what confidence level should the building be evaluated (determining the appropriate level of testing and analysis), and evaluation of the most cost-effective and performance-enhancing upgrade options.

Scope. The scope of work must define whether the seismic evaluation is limited to structural elements of the building or also includes nonstructural elements. Structural elements are elements that are relied on to carry and transfer to other elements the structure's seismic inertial forces. Nonstructural elements are not relied on to carry seismic loads but are subjected to their own inertial forces and the building structure's deformations and include chimneys, mechanical equipment, and ceilings.

The appropriate level of evaluation needs to be defined. At the start of a project, the appropriate level of evaluation may not be determinable, and a phased approach can be used in which the scope of each successive phase is determined by the results of the preceding evaluation phase.

The evaluation is presented to the client as a report. The scope of work usually includes the development of conceptual (at the minimum) recommendations for seismic rehabilitation measures that would upgrade the building to the objective performance level. Rehabilitation is beyond the scope of this chapter; the reader is referred to ICBO's *Guidelines for Seismic Retrofit of Existing Buildings*[14] and FEMA 356, *Prestandard and Commentary for the Seismic Rehabilitation of Buildings*.[15]

Document Gathering. Original and remodel drawings that contain structural information are very helpful for the evaluation. Generally, there are distinct structural drawings for a project, although for older construction, structural information sometimes was presented in the architectural drawings. Original specifications also can be useful. Possible sources for copies of the original drawings are the building owner, building maintenance department/facility engineer, city building or planning department, original architect or engineer or previous remodel architect or engineer, and historical or archival consultants. When obtaining drawings from a city department, some jurisdictions require the evaluator to seek a letter from the original design firm that prepared the drawings requesting permission for the evaluator to have copies of the drawings.

To determine the levels of seismic design and detailing that the building should have been constructed for originally, it also may be useful to determine and review the building code in effect at the time that the building was submitted originally for permit application.

Identifying Applicable Codes and Standards. If not already familiar with the authority having jurisdiction, the evaluator should determine whether there are any mandated evaluation standards or ordinances pertaining to the potential seismic rehabilitation of the building to be evaluated. For example, from the 1970s to 1990s, most cities in California had enacted mandatory requirements for the seismic rehabilitation of unreinforced masonry structures. Understanding the jurisdiction's specific rehabilitation ordinance is necessary when performing the seismic evaluation of a building. Copies of such regulations should be obtained along with any referenced standards or guidelines that are identified therein.

Premobilization. Before mobilizing for the evaluation, planning is required to determine the following: Are there regulations or codes that may dictate current strength/performance requirements for the building? Is the building of historical merit? What is the required or desired seismic performance level? Does the evaluation include nonstructural elements? Are original drawings available? Will a geotechnical investigation report be required or helpful? Will testing of materials be required? Is there a need for other consultants, including historical, hazardous materials identification, and termite or pest damage evaluation?

Evaluation

When the planning is completed, the engineer is prepared to visit the site and then to evaluate the building.

Field Work. The field work for seismic evaluations is similar to that for due diligence and general structural integrity evaluations described in this chapter. Careful photographic and sketch documentation is required. Probes and/or physical tests may be required, particularly for postscreening phases. The building must be evaluated for symptoms of degradation to structural elements that might affect the capacity of the lateral-load-resisting system (see sections in this chapter).

Seismic Design Philosophy. When evaluating a building for potential seismic performance, it is important to understand the seismic design philosophy used for most building designs: Buildings are expected to suffer damage during significant earthquakes. From the SEAOC "Blue Book," [17] "Structures designed in accordance with recent building codes should, in general, be able to

1. Resist a minor level of earthquake ground motion without damage.
2. Resist a moderate level of earthquake ground motion without structural damage, but possibly experience some nonstructural damage.
3. Resist a major level of earthquake ground motion—of an intensity equal to the strongest earthquake, either experienced or forecast, for the building site—without collapse, but possibly with some structural as well as nonstructural damage."

Good seismic performance requires several characteristics of buildings. They are continuous load path, demonstrated performance of system type, regular structural configuration, ductility, and redundancy.

A continuous load path is essential for good seismic performance; e.g., if a house sill plate is not bolted to the foundation, the house can slide off the foundation during an earthquake. The evaluator must review whether continuous load path is provided in the building. At times, particularly when reliable drawings are not available, a visual evaluation may not be entirely conclusive; finishes may conceal load path connections, and confirming probes may be required.

Ductility, demonstrated structural systems, regular configuration, and redundancy are all ideal but may not occur in existing buildings because of their materials and systems in place and architectural layouts and other limitations. For example, a mid-block retail building functionally may require an open storefront for easy access by shoppers. If the lateral-load-resisting system consists of shearwalls, a shearwall cannot be used at the storefront, but the structure can still be stable with a three-sided lateral system (shearwalls at the rear and on two sides).

Structural System. The evaluator must identify the lateral-load-resisting structural system; i.e., is it a concrete shearwall building or steel moment frame building? All the different systems perform in a different manner. See FEMA 154[16] and the SEAOC "Blue Book" commentary[17] for basic types of lateral-load-resisting systems and their behavior.

It is also necessary to recognize and identify structural and/or mass irregularities. "Irregularities in load paths and in structural configuration are major contributors to structural damage and failure due to strong earthquake ground motion." [13,17] Therefore, when performing a seismic evaluation, it is essential to review for the presence of any structural irregularities. See Figures 6.28 and 6.29 for conceptual illustrations of structural irregularities. A building also must be reviewed as to whether it is immediately adjacent to another building and vulnerable to pounding during a seismic event (Figs. 6.26 and 6.27).

A seismic evaluation also includes review of nonstructural elements that may be detrimental to building performance or safety of the occupants. For example, residential masonry chimneys have collapsed repeatedly during various earthquakes. Other examples of nonstructural elements requiring review include inverted pendulum-type masses, rooftop water tanks, heavy unanchored mechanical equipment, and ceilings and light fixtures immediately above egress routes.

Evaluation References. ASCE/SEI 31-03, *Seismic Evaluation of Existing Buildings,*[18] is a broad-reaching tool for the seismic evaluation of various types of buildings

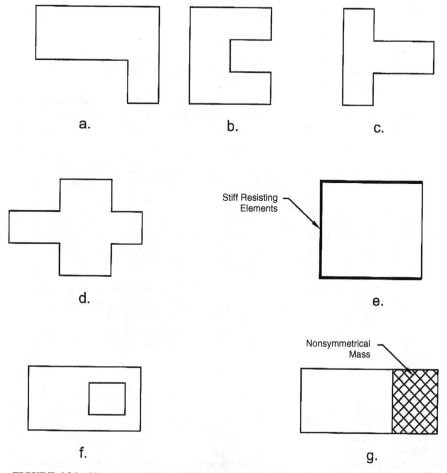

FIGURE 6.26 Plan structural irregularities. (*Adapted with permission from SEAOC, 1999.*[17])

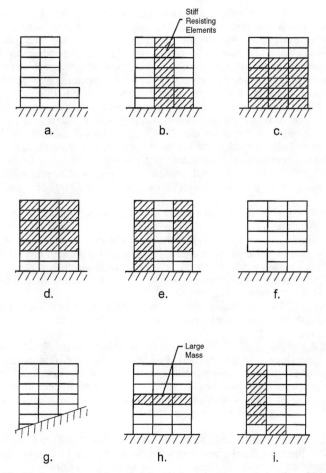

FIGURE 6.27   Vertical structural irregularities. (*Adapted with permission from SEAOC, 1999.*[17])

in any region of seismicity. It is a consensus-based standard that can be adopted by authorities having jurisdiction. Evaluation in accordance with ASCE 31 (or its predecessor FEMA 310, *Handbook for the Evaluation of Buildings*[19]) is currently required for buildings in which a federal agency is considering leasing space or purchasing the building.

The evaluated structure must be categorized as to building type based on the lateral-force-resisting system and the diaphragm type. There are 24 different building types. A building that cannot be categorized as a specific building type can be evaluated as a general building. The evaluation may be made for either a life safety or immediate occupancy performance level.

ASCE 31 uses a three-tiered seismic evaluation process as follows: screening phase, evaluation phase, and detailed evaluation phase, with successive phases performed should the preceding phase so warrant. During the screening phase, the evaluator determines whether the building postdates the edition of a model building code for that given building type, the model building code being deemed to provide sufficient benchmark design requirements. The evaluator must determine the level of seismicity, which is a function of the mapped response accelerations for the site and the site class (soil or rock characteristics). The site class may not be known definitively for the building being evaluated—site-specific geotechnical information may not be available—and therefore, a site class will need to be assumed. Depending on the level of seismicity and the performance level, the evaluation of additional screening phase checklists (and performance of additional structural calculations) may be required. The screening phase also uses a structural checklist for the given building type, with responses to various evaluation statements either compliant or noncompliant. Evaluation checklists are also provided for foundations and geologic site hazards and nonstructural elements. Noncompliant responses identify specific issues that require further investigation. Some rudimentary calculations may be performed during the screening phase. The evaluation phase involves a structural analysis—linear static or linear dynamic—of the building structural system and, at times, an analysis of noncompliant nonstructural elements. The detailed evaluation phase consists of a nonlinear analysis of the building.

The checklist statements are all derived from instances of previous damage or failure caused by earthquakes. While the checklists are a valuable tool, they need to be used with latitude and an eye toward conditions not encompassed in the checklists. For example, there is at least one instance of significant failure of cantilever canopies that affected a life safety performance level, but there is currently no evaluation statement related to review of cantilevers having high gravity demands and subjected to potential additional seismic demands.

While ASCE 31 is a valuable tool for the evaluation of buildings of most types and materials, there are additional resources that can be used alone or in conjunction with ASCE 31.

ATC-50, *Simplified Seismic Assessment of Detached, Single-Family, Wood-Frame Dwellings,*[20] provides a simplified four-sheet evaluation form that evaluates various foundation, structural, and nonstructural elements of a house and identifies vulnerabilities that potentially can be retrofitted to improve the performance of the house. The form was developed initially with structural and regional seismic hazard scoring systems specific to the City of Los Angeles, but the form can be revised to be used in other areas.

ATC-40, *Seismic Evaluation and Retrofit of Concrete Buildings,*[21] provides a recommended methodology and commentary for the evaluation of cast-in-place concrete buildings designed and built in California prior to the late 1970s.

SEAONC, *Guidelines for Seismic Evaluation and Rehabilitation of Tilt-Up Buildings and Other Rigid Wall/Flexible Diaphragm Structures,*[22] provides detailed background or tilt-up buildings and their design and construction and guidelines for the prioritized evaluation, based on the importance of vulnerabilities demonstrated in past earthquakes, of various potential deficiencies.

FEMA 351, *Recommended Seismic Evaluation and Upgrade Criteria for Existing Welded Steel Moment-Frame Buildings,*[23] describes in detail the various potential failure mechanisms for steel moment frame connections. It provides an analysis procedure to determine the probable structural performance of a moment-frame building for a given site seismicity, as well as a more detailed procedure for an evaluation having higher confidence levels.

Loss Estimation / Probable Maximum Loss Analysis. Loss estimation is useful to insurers and lenders because it quantifies a dollar value (usually expressed as a percentage of the building value) that the building may suffer from an earthquake.

Structural engineer Karl Steinbrugge developed the probable maximum loss (PML) methodology and defined it as follows:

> The probable maximum loss for an individual building is that monetary loss expressed in dollars (or as a percentage of insured value) under the following conditions: (1) located

on firm alluvial ground or on equivalent compacted man-made fills in a probable maximum loss zone, and (2) subjected only to the vibratory motion from the maximum probable earthquake, that is, not stride a fault or in a resulting landslide.

The building class probable maximum loss is defined as the expected maximum percentage monetary loss which will not be exceeded for 9 out of 10 buildings in a given earthquake building class under the conditions stated in the previous paragraph. The loss to the tenth building may be quite anomalous due to unknown design or construction peculiarities, or to unusual earthquake motions and building response or geologic hazards, which result in a "poor fit" building classification. This 9 out of 10 definition tends to introduce slight error on the low side for low PML values and slight error on the high side for high PML values.[24]

Loss estimation is best suited to the evaluation of a portfolio of buildings such that the impact of the standard deviation of loss is spread among multiple buildings. However, it is used in the industry for individual buildings with the reasoning that an insurer (or lender) is insuring many buildings, and the risk of deviation/probability is spread among those holdings.

Depending on the method and scope of work used for a loss estimation, it may not be required to visit the building or review a geotechnical report. Peak ground accelerations can be obtained from The National Earthquake Hazards Reduction Program (NEHRP) maps and from the U.S. Geological Survey (USGS) Web site by zip code or latitude/longitude.

ATC-13, *Earthquake Damage Evaluation Data for California*,[25] is the loss estimation method used by most engineers. It provides PMLs based on databases of previous earthquake losses and expert opinions. Since there have been several significant earthquakes and new loss data since the publication of ATC-13 in 1985, some engineers will adjust the loss estimates from ATC-13 qualitatively. Engineers also may modify ATC-13 PML values to account for specific structural features associated with the evaluated facility. Buildings are classified into different building classifications that are further distinguished by three ranges of heights of buildings: low rise (1–3 stories), midrise (4–7 stories), and highrise (8+ stories). To determine the PML, the seismicity of the site must be determined. ATC-13 uses the Modified Mercalli Intensity (MMI) Scale to represent seismicity.

The ATC 13-1 commentary[26] emphasizes that ATC-13 was never intended for loss estimation for individual buildings and that it should not be used in such a manner. The commentary provides additional guidance for the evaluation of above- or below-average construction and provides examples in calculating PML for several building types.

FEMA 351, *Recommended Seismic Evaluation and Upgrade Criteria for Existing Welded Steel Moment-Frame Buildings*,[27] provides methodologies for a rapid loss estimation and a detailed loss estimation for individual buildings. The rapid loss estimation method uses graphs plotting separately the restoration cost of moment frame connections and the restoration cost of nonstructural elements, both expressed as a percentage of the building replacement value. The detailed loss estimation method requires nonlinear analysis of the building under consideration.

Hazards U.S. (HAZUS) is a loss-estimation tool that is used primarily with Geographic Information System–based computer software but also can be used with the methodology in the HAZUS *Technical Manual*.[28,29] It is intended for loss estimation for a large inventory of buildings. Databases of buildings and building types in regions across the country continue to be updated and published. Training and user groups are available.

Because loss estimation is a relatively new technique, different engineers use different terms and methods. ASTM 2026-99, *Standard Guide for the Estimation of Probable Loss to Buildings from Earthquakes*,[30] seeks to standardize the terms and methods.

Proprietary computer software that uses data compiled from in-house resources in addition to published databases of loss damages, is available for loss-estimation analysis.

Custom Evaluations. In addition to performing seismic evaluations based on a guideline or standard method, seismic evaluations also may be commissioned that involve the review of selected identified elements or a conceptual review as to the likely performance of a certain building.

INSPECTION OF DAMAGED BUILDINGS

Inspection of buildings is required following a sudden event, such as an earthquake, hurricane, fire, explosion, or terrorist act. Sometimes the inspection involves one or two buildings, or at the other extreme, it may involve hundreds or thousands of buildings. Such inspections often are conducted under the aegis of a local government agency for purposes of determining whether the building presents a hazard either to its intended occupants or to the general public.

In this section, the procedures used commonly for evaluating buildings following a seismic event will be summarized. In addition, a more detailed description will be given of the methods used to rapidly evaluate the buildings surrounding the World Trade Center following the September 11, 2001, terrorist attacks.

Postearthquake Evaluation

Evaluation of large inventories of buildings following a seismic event is often performed using the procedures set forth in ATC-20[31] and the addendum ATC 20-2.[32] The corresponding field manual, ATC-20-1,[33] is a pocket distillation that is intended to be carried by the inspecting engineer for ready reference.

The ATC 20 procedures contain two levels of inspection: rapid evaluation and detailed evaluation.

> *Rapid evaluation* is the first level of evaluation that is typically performed, and it is intended to take approximately 10 to 20 minutes per structure. It is intended to quickly identify buildings that are obviously unsafe and those which appear to be safe. Questionable structures would be subjected to a detailed evaluation.
>
> *Detailed evaluation* is the second level of examination, which can take anywhere from 1 to 4 hours per structure. It consists of a thorough visual examination, inside and out (Fig. 6.28).

Engineering evaluations, such as those that would be performed for the owner of the building, are not covered in ATC 20.

One of the hallmarks of the ATC 20 system is its posting system. On each inspected building, the inspector posts an official placard indicating the results of the inspection (Fig. 6.29). The placard indicates to owners, tenants, and authorities whether the building can be occupied or entered and any attendant restrictions. The meanings of the placards are described in Table 6.1.

This ATC 20 method also includes standardized reporting forms that the inspector fills out for each building. The system has been used successfully in earthquake-prone regions, such as California. It literally allows thousands of buildings to be inspected quickly by "volunteer" inspectors and engineers while presenting a consistent system that is easy for the public to understand.

Several resources are available for detailed postearthquake evaluations and include the following. FEMA 306[34] and FEMA[35] provide formalized procedures to evaluate

ATC-20 Rapid Evaluation Safety Assessment Form

Inspection

Inspector ID: _____

Affiliation: _____

Inspection date and time:_____ ☐ AM ☐ PM

Areas inspected: ☐ Exterior only   ☐ Exterior and interior

Building Description

Building name: _____

Address: _____

_____

Building contact/phone: _____

Number of stories above ground:____ below ground: __

Approx. "Footprint area" (square feet): _____

Number of residential units: _____

Number of residential units not habitable: _____

Type of Construction

☐ Wood frame   ☐ Concrete shear wall
☐ Steel frame   ☐ Unreinforced masonry
☐ Tilt-up concrete   ☐ Reinforced masonry
☐ Concrete frame   ☐ Other: _____

Primary Occupancy

☐ Dwelling   ☐ Commercial ☐ Government
☐ Other residential   ☐ Offices   ☐ Historic
☐ Public assembly   ☐ Industrial ☐ School
☐ Emergency services   ☐ Other: _____

Evaluation

Investigate the building for the conditions below and check the appropriate column.

Estimated Building Damage (excluding contents)

Observed Conditions:	Minor/None	Moderate	Severe
Collapse, partial collapse, or building off foundation	☐	☐	☐
Building or story leaning	☐	☐	☐
Racking damage to walls, other structural damage	☐	☐	☐
Chimney, parapet, or other falling hazard	☐	☐	☐
Ground slope movement or cracking	☐	☐	☐
Other (specify) _____	☐	☐	☐

☐ None
☐ 0 –1%
☐ 1 –10%
☐ 10 – 30%
☐ 30 – 60%
☐ 60 – 100%
☐ 100%

Comments: _____

Posting

Choose a posting based on the evaluation and team judgment. Severe conditions endangering the overall building are grounds for an Unsafe posting. Localized Severe and overall Moderate conditions may allow a Restricted Use posting. Post INSPECTED placard at main entrance. Post RESTRICTED USE and UNSAFE placards at all entrances.

☐ **INSPECTED** (Green placard)   ☐ **RESTRICTED USE** (Yellow placard) ☐ **UNSAFE** (Red placard)

Record any use and entry restrictions exactly as written on placard: _____

_____

Further Actions   Check the boxes below only if further actions are needed.

☐ Barricades needed in the following areas: _____

_____

☐ Detailed Evaluation recommended: ☐ Structural   ☐ Geotechnical   ☐ Other:_____

☐ Other recommendations: _____

Comments: _____

FIGURE 6.28   These are the standard ATC-20 inspection forms that are used for (*a*) the rapid assessments of buildings following a seismic event and (*b,c*) for detailed assessment of a damaged building.

ATC-20 Detailed Evaluation Safety Assessment Form

Inspection

Inspector ID: _____

Affiliation: _____

Inspection date and time: _____ ☐ AM ☐ PM

Final Posting
from page 2

☐ Inspected
☐ Restricted Use
☐ Unsafe

Building Description

Building name: _____

Address: _____

_____

Building contact/phone: _____

Number of stories above ground: ____ below ground: ___

Approx. "Footprint area" (square feet): _____

Number of residential units: _____

Number of residential units not habitable: _____

Type of Construction

☐ Wood frame   ☐ Concrete shear wall
☐ Steel frame   ☐ Unreinforced masonry
☐ Tilt-up concrete   ☐ Reinforced masonry
☐ Concrete frame   ☐ Other: _____

Primary Occupancy

☐ Dwelling   ☐ Commercial   ☐ Government
☐ Other residential   ☐ Offices   ☐ Historic
☐ Public assembly   ☐ Industrial   ☐ School
☐ Emergency services   ☐ Other: _____

Evaluation

Investigate the building for the conditions below and check the appropriate column. There is room on the second page for a sketch.

	Minor/None	Moderate	Severe	Comments
Overall hazards:				
Collapse or partial collapse	☐	☐	☐	_____
Building or story leaning	☐	☐	☐	_____
Other_____	☐	☐	☐	_____
Structural hazards:				
Foundations	☐	☐	☐	_____
Roofs, floors (vertical loads)	☐	☐	☐	_____
Columns, pilasters, corbels	☐	☐	☐	_____
Diaphragms, horizontal bracing	☐	☐	☐	_____
Walls, vertical bracing	☐	☐	☐	_____
Precast connections	☐	☐	☐	_____
Other_____	☐	☐	☐	_____
Nonstructural hazards:				
Parapets, ornamentation	☐	☐	☐	_____
Cladding, glazing	☐	☐	☐	_____
Ceilings, light fixtures	☐	☐	☐	_____
Interior walls, partitions	☐	☐	☐	_____
Elevators	☐	☐	☐	_____
Stairs, exits	☐	☐	☐	_____
Electric, gas	☐	☐	☐	_____
Other_____	☐	☐	☐	_____
Geotechnical hazards:				
Slope failure, debris	☐	☐	☐	_____
Ground movement, fissures	☐	☐	☐	_____
Other_____	☐	☐	☐	_____

General Comments: _____

_____

Continue on page 2

FIGURE 6.28 (*Continued*).

ATC-20 Detailed Evaluation Safety Assessment Form | **Page 2**

Building name: _____ Inspector ID: _____

Sketch (optional)
Provide a sketch of the building or damaged portions. Indicate damage points.

Estimated Building Damage
If requested by the jurisdiction, estimate building damage (repair cost + replacement cost, excluding contents).
☐ None
☐ 0–1%
☐ 1–10%
☐ 10–30%
☐ 30–60%
☐ 60–100%
☐ 100%

Posting
If there is an existing posting from a previous evaluation, check the appropriate box.

Previous posting: ☐ INSPECTED ☐ RESTRICTED USE ☐ UNSAFE Inspector ID: _____ Date:_____

If necessary, revise the posting based on the new evaluation and team judgment. *Severe* conditions endangering the overall building are grounds for an Unsafe posting. Local *Severe* and overall *Moderate* conditions may allow a Restricted Use posting. Indicate the current posting below and at the top of page one.

☐ **INSPECTED** (Green placard)   ☐ **RESTRICTED USE** (Yellow placard)   ☐ **UNSAFE** (Red placard)

Record any use and entry restrictions exactly as written on placard:_____
_____
_____

Further Actions Check the boxes below only if further actions are needed.
☐ Barricades needed in the following areas:_____
_____
☐ Engineering Evaluation recommended: ☐ Structural   ☐ Geotechnical   ☐ Other: _____
☐ Other recommendations: _____
_____
_____

Comments: _____
_____

FIGURE 6.28 (*Continued*).

INSPECTED

LAWFUL OCCUPANCY PERMITTED

This structure has been inspected (as indicated below) and no apparent structural hazard has been found.

☐ **Inspected Exterior Only**

☐ **Inspected Exterior and Interior**

Report any unsafe condition to local authorities; reinspection may be required.

Inspector Comments:

_____
_____
_____

Facility Name and Address:

_____
_____
_____

Date _____

Time _____

(**Caution:** Aftershocks since inspection may increase damage and risk.)

This facility was inspected under emergency conditions for:

_____
(Jurisdiction)

Inspector ID / Agency

_____
_____

**Do Not Remove, Alter, or Cover this Placard
until Authorized by Governing Authority**

RESTRICTED USE

Caution: This structure has been inspected and found to be damaged as described below:

_____
_____
_____

Entry, occupancy, and lawful use are restricted as indicated below:

☐ Do not enter the following areas: _____

☐ Brief entry allowed for access to contents: _____

☐ Other restrictions: _____

Facility name and address:

_____
_____
_____

Date _____

Time _____

(**Caution:** Aftershocks since inspection may increase damage and risk.)

This facility was inspected under emergency conditions for:

_____
(Jurisdiction)

Inspector ID / Agency

_____
_____

**Do Not Remove, Alter, or Cover this Placard
until Authorized by Governing Authority**

FIGURE 6.29 These placards are posted prominently on buildings after they have been inspected using the ATC-20 procedures.

UNSAFE

DO NOT ENTER OR OCCUPY
(THIS PLACARD IS NOT A DEMOLITION ORDER)

This structure has been inspected, found to be seriously damaged and is unsafe to occupy, as described below:

Date _____

Time _____

This facility was inspected under emergency conditions for:

_____

(Jurisdiction)

Do not enter, except as specifically authorized in writing by jurisdiction. Entry may result in death or injury.

Inspector ID / Agency

Facility Name and Address:

**Do Not Remove, Alter, or Cover this Placard
until Authorized by Governing Authority**

FIGURE 6.29   (*Continued*).

damage to lateral-force-resisting systems of concrete or masonry walls or infilled frames. FEMA 352[36] provides guidance on conducting inspections to detect damage in steel moment-frame buildings following an earthquake and determining their safety in the postearthquake environment.

Finally, new technologies, such as satellite imaging of areas recently stricken by earthquake, can be very useful in quickly prioritizing areas requiring evaluation.

World Trade Center Disaster

The structural engineering response to the September 11, 2001, attacks in New York City is described briefly because it can serve as a model for future postdisaster evaluations of large numbers of buildings. Following the terrorist attacks on the World

TABLE 6.1   ATC-20 Placard Meaning

Placard Color	Placard Title	Usage
Green	Inspected	This is for structures that appear to be safe. Occupancy is permitted.
Yellow	Restricted use	For damaged structures. The placard should identify what the restrictions are, such as areas that may not be entered. It also may indicate whether brief entry is permitted for the purpose of removing contents.
Red	Unsafe	The building cannot be entered or occupied.

Trade Center towers, the New York City Department of Design and Construction (DDC) took the lead in coordinating the engineering and contractor efforts required to support the rescue and recovery operations. DDC immediately enlisted the firm of LZA/Thornton-Tomasetti (LZA/TT) to provide all structural engineering services.

On the afternoon of September 11, DDC provided access to the site to LZA/TT and representatives of contracting firms. This advance team immediately began strategizing and marshalling personnel and equipment. As requested by DDC, 30 LZA/TT structural engineers assembled at the site the morning of September 12. These structural engineers were divided into seven teams and then were merged with personnel from DDC, the Department of Buildings, the Port Authority, the Fire Department, and the Police Department. These multidisciplinary teams performed emergency inspections on September 12 and 13 of the buildings and structures immediately surrounding the World Trade Center. The purpose of these "triage" inspections was to identify buildings with major structural damage and to determine whether nearby trains on subway lines could operate without triggering further collapses (Fig. 6.30).

Two buildings received special attention during these first two days, for completely opposite reasons. 130 Liberty, the Bankers Trust building, had suffered serious structural damage over many floors, and there was a concern that more areas of the building might collapse. Engineers entered the building and, by observing the type of construction at several floors, were able to determine that the probability of further collapse was minimal. In contrast, 1 Liberty Plaza had no structural damage at all. However, rescue workers repeatedly sounded the evacuation horn, falsely believing, owing to an

FIGURE 6.30 On September 12 and 13, 2001, organized teams that included structural engineers, police, firefighters, and others conducted "triage" inspections of the buildings surrounding the World Trade Center.

optical illusion, that the building was about to collapse. Radio stations and newspapers even reported that the building actually had collapsed. A team of engineers was assigned to spend a full day examining the structure from top to bottom, and their findings helped allay the fears of many people.

Almost immediately, the Police Department and the National Guard had established a perimeter boundary around the area to restrict access and protect the public. Initially, the boundary was placed as far north as 14th Street, and the restricted area included a mix of residential, retail, and commercial uses. It therefore became a high priority to identify which buildings were structurally safe for occupancy, to identify overhead debris that presented a potential falling hazard, and to help determine whether the perimeter of the restricted zone could be contracted safely. A systematic inspection of the buildings surrounding the site therefore was needed.

LZA/TT assigned the bulk of this task to engineers who had been organized by the Structural Engineers Association of New York (SEAoNY). The area surrounding the World Trade Center was divided into 15 sectors, and a team of engineers was assigned to each sector.

Three sets of inspections were performed by the teams:

1. Rapid evaluation of about 400 buildings was performed on September 17 and 18.
2. Detailed evaluations of 31 buildings were performed starting on September 21. These were buildings that had suffered either structural damage or substantial collateral damage (such as window breakage).
3. Another rapid evaluation of all the buildings was conducted between October 4 and 10. This was performed to determine whether specified protective measures had been installed or repairs had been made (Fig. 6.31).

The ATC 20 procedure was selected as the basis for the rapid inspections because it is well suited for dealing with large numbers of buildings. Modifications were made to adapt the procedure to the special circumstances and requests of building officials. The modifications consisted of the following:

Placards were not posted on the buildings. The purpose of the placards in ATC 20 is to inform occupants whether a building can be occupied. However, at the World Trade Center, entire neighborhoods had been evacuated and remained closed to the public, so there was no need to post placards on individual buildings. As groups of buildings were deemed to be safe, the boundaries of the restricted zone were adjusted to allow access.

The ATC 20 classification system was modified. Based on discussions with building officials, the classifications evolved to reflect the nature and extent of damage rather than whether a building was occupiable. Table 6.2 shows the classification system.

The interiors of buildings were not inspected. Unlike a seismic event, which can cause damage to interior structural members that may not necessarily be visible from the exterior, the damage to the surrounding buildings in this case was caused by the impact of debris and projectiles and by a blast of air generated by collapse of the towers. Therefore, it was sufficient to inspect the facade and roof of a building to determine if the building had been damaged.

Roofs were inspected for the presence of debris, including powdery debris. There was a concern that the powder deposited by the collapse, which was up to 1½ in. deep, might clog roof drains during heavy rains, which could result in a roof overload. The contractors therefore were directed to remove this debris. Access to the roofs was not always possible, so the visual inspections were supplemented with examination of high-resolution aerial photographs.

The report for each building included recommendations for protection of the public, if appropriate. Typical recommendations included to board up broken windows, to

B9	54	1
SECTOR	BLOCK	LOT

RAPID VISUAL EVALUATION - INITIAL ASSESSMENT

BUILDING DESCRIPTION

ADDRESS 1:	130 Liberty St	TYPE OF CONSTRUCTION:	Steel frame
ADDRESS 2:	133-151 Washington St		

BUILDING NAME:	Bankers Trust Deutsche Bank	NUMBER OF STORIES:	39

INSPECTION

INSPECTOR:	DOMENICO ANTONELLI	INSPECTION DATE:	Monday, September 17, 2001
GROUP NUMBER:	B9	EVALUATION AREA:	Exterior and Interior

EVALUATION

OBSERVED CONDITIONS:		EVALUATION COMMENTS:
Collapse:	Severe	NORTH FAÇADE, IMPACT DAMAGE @4TH TO 20TH FL. COLUMN MISSING ALONG WITH CONNECTED SPANDRELS AND AREAS OF FLOOR SLAB
Building Leaning:	Minor/None	
Racking:	Minor/None	
Falling Hazard:	Severe	
Roof Damage:	Moderate	
Other:	Unknown	

ESTIMATED BUILDING DAMAGE (excluding contents): 　**30-60%**

RATING

POSTED RATING:	**Unsafe - Structural**
RESTRICTIONS:	

FURTHER ACTIONS

BARRICADES REQUIRED:	FOUR SIDES	
DETAILED EVALUATION RECOMMENDED:	Structural	AT NORTH SIDE, EXTERIOR COLUMN MISSING FROM 4TH TO 20TH FLOORS. PLAZA AREA DESTROYED.
OTHER RECOMMENDATIONS:		

GENERAL COMMENTS

Adapted from ATC-20 form copyright Applied Technology Council

FIGURE 6.31 These forms were used for rapid inspections of 400 buildings following the September 11, 2001, terrorist attacks of the World Trade Center. *a.* This form was used for the initial round of inspections. *b.* This form was used for the second, and final, round of inspections.

WTC EMERGENCY: CITY OF NEW YORK
STRUCTURAL ENGINEERS ASSOCIATION OF NEW YORK

B9	**54**	**1**
SECTOR	BLOCK	LOT

RAPID VISUAL EVALUATION - FINAL ASSESSMENT

BUILDING DESCRIPTION

ADDRESS 1:	130 Liberty St	**TYPE OF CONSTRUCTION:** Steel frame
ADDRESS 2:	133-151 Washington St	

BUILDING NAME: Bankers Trust Deutsche Bank   **NUMBER OF STORIES:** 39

INSPECTION

INSPECTOR:	GNA	**INSPECTION DATE:** 10/12
GROUP NUMBER:	B9	**EVALUATION AREA:** Exterior only

EVALUATION

OBSERVED CONDITIONS: **DESCRIPTION:**

Collapse/Leaning:	Severe
Structural Damage:	Severe
Non-Structural/Falling Hazard:	Severe
Roof Damage/Debris/Water:	Moderate
Other (specify):	Unknown
Other (specify):	

ESTIMATED BUILDING DAMAGE (excluding contents):   **30-60%**

COMMENTS:

RATING

POSTED RATING: Major Damage (Blue)

RESTRICTIONS:

FURTHER ACTIONS

TYPE:		**LOCATION:**
Netting:	Yes	N, E, W facades
Window/Facade:	Yes	N, E, W facades
Shoring:	No	exterior only
Sidewalk Bridge:	Yes	E and W façade
Other:	No	

COMMENTS:

BARRICADES RECOMMENDED:	No
DETAILED EVALUATION RECOMMENDED:	
ROOF DAMAGE:	Clean

GENERAL COMMENTS

FIGURE 6.31 (*Continued*).

TABLE 6.2 Building Classifications Used at World Trade Center

Color	Condition
Black	Collapsed
Red	Partially collapsed
Blue	Structural damage
Yellow	Collateral damage
Green	Undamaged

erect a protective sidewalk bridge, to install facade netting to contain glass and small debris, and to install shoring. Follow-up inspections were conducted to determine whether the recommendations had been implemented.

The initial rapid assessment identified about 30 buildings with structural damage or with significant facade damage. These buildings were then subjected to a more detailed multidisciplinary life-safety inspection. These inspections were conducted floor by floor and included structural systems, as well as fire alarms, sprinklers, and other life-safety systems. Individual reports were prepared, and they were provided to building owners to assist them in determining the necessary repairs. These assessments were conducted by engineers and architects from LZA/TT, with assistance on several buildings from other structural engineering firms organized by SEAoNY.

A second round of rapid inspections was conducted. This allowed engineers to observe whether recommendations or repairs had been implemented and provided an opportunity to double-check the initial assessments. This reinspection resulted in two notable corrections. First, the initial inspection map did not show the St. Nicholas church building near the southwest corner of the site. This small frame building had been completely flattened by heavy steel debris from the fall of the towers, leaving virtually no trace that it had ever existed. Second, the reinspection also identified structural damage to a four-story building several blocks to the north, where aircraft landing gear had crashed through the roof. This damage had not been visible from the street in the initial inspection.

The inspection forms for the rapid assessments were collected after each inspection and entered into a Microsoft Access database. The final product of each inspection was a building-by-building tabulation and an accompanying color-coded map that was circulated widely and published. All the procedures and results are documented in a report[37] (Fig. 6.32).

SUMMARY

This chapter has provided some insights gained from the inspection of hundreds of buildings over the years. It addressed both the nontechnical and technical principles because a successful condition assessment depends on both.

The nontechnical principles include that it is vital to understand the client's objectives, to communicate clearly and frequently with the client, to stay within the budget and schedule, and to present the results in a way that is meaningful to the client. These principles apply whether the assessment is performed on an "ordinary" building or on a special type of structure,

The technical principles vary depending on the type of structure that is being evaluated. This chapter dealt with conventional buildings, such as commercial buildings, industrial structures, apartments, and hotels. The following chapters provide guidance on the assessment of a variety of other structures, such as historic monuments, stadiums, parking decks, and others.

WTC EMERGENCY

NYC DDC/DoB Cooperative Building Damage Assessment
07 November 2001

BUILDING RATINGS

SEAoNY
LZA/Thornton-Tomasetti

No Damage
Minor Damage
Major Damage
Partial Collapse
Collapse

FIGURE 6.32   This map summarizes the findings regarding the condition of the buildings in the vicinity of the World Trade Center after September 11, 2001.

REFERENCES

1. Ortega, Ilias, and Søren Bisgaard, (2000). "Quality Improvement in the Construction Industry: Three Systematic Approaches," Report No. 10, Quality Management and Technology, Institute for Technology Management, University of St. Gallen, Switzerland.

2. *Guidelines for Condition Surveys of Buildings.* Port Authority of New York and New Jersey, Engineering Department, New York, April 2000.

3. *National Design Specification for Wood Construction.* National Forest Products Association, Washington, DC, 1962.

4. Winandy, Jerrold E. (2000). *Thermal Degradation of Fire-Retardant-Treated Plywood* U.S. Department of Agriculture, Forest Service, Forest Products Laboratory, Publication VI-19, Madison, WI, http://www.fpl.fs.fed.us/documnts/techline/vi-19.pdf.

5. NAHB National Research Center (1990). *Home Builders Guide to Fire Retardant Treated Plywood: Evaluation, Testing, and Replacement.* NAHB, Upper Marlboro, MD.

6. SEI/ASCE 11-99 (1999), *Guideline for Structural Condition Assessment of Existing Buildings.* American Society of Structural Engineers, Reston, VA.

7. ASTM C42/C42M-03 (2003), *Standard Test Method for Obtaining and Testing Drilled Cores and Sawed Beams of Concrete.* ASTM International, West Conshohocken, PA.

8. ASTM C39/C39M-03 (2003), *Standard Test Method for Compressive Strength of Cylindrical Concrete Specimens.* ASTM International, West Conshohocken, PA.

9. Hausmann, D. A. (1967). Steel Corrosion in Concrete: How Does It Occur, *Materials Performance* **6**(19), pp. 19–23.

10. ASTM C876-91 (1999). *Standard Test Method for Half-Cell Potentials of Uncoated Reinforcing Steel in Concrete.* ASTM International, West Conshohocken, PA.

11. Slide Collection CD-ROM-98-2, Earthquake Engineering Research Institute (EERI), Oakland, 1998.

12. FEMA/ATC (2002). "Postearthquake Safety Evaluation of Buildings," CD-ROM, Applied Technology Council, Redwood City, CA.

13. Stratta, James L. (1987). *Manual of Seismic Design.* Prentice-Hall, Englewood Cliffs, NJ.

14. *Guidelines for Seismic Retrofit of Existing Buildings.* International Conference of Building Officials (ICBO), Whittier, CA, 2001.

15. FEMA 356 (2000). *Prestandard and Commentary for the Seismic Rehabilitation of Buildings.* Federal Emergency Management Agency, Washington.

16. FEMA 154 (1988). *Rapid Visual Screening of Buildings for Potential Seismic Hazards: A Handbook.* Federal Emergency Management Agency, Washington.

17. *Recommended Lateral Force Requirements and Commentary.* Structural Engineers Association of California (SEAOC), Sacramento, CA, 1999.

18. SEI/ASCE 31-03 (2003). *Seismic Evaluation of Existing Buildings.* American Society of Civil Engineers (ASCE), Reston, VA.

19. FEMA 310 (1998). *Handbook for the Seismic Evaluation of Buildings—A Prestandard.* Federal Emergency Management Agency, Washington.

20. ATC-50 (2002). *Simplified Seismic Assessment of Detached, Single-Family, Wood-Frame Dwellings.* Applied Technology Council, Redwood City, CA.

21. ATC-40 (1996). *Seismic Evaluation and Retrofit of Concrete Buildings,* Vol. 1 and 2. Applied Technology Council, Redwood City, CA.

22. *Guidelines for Seismic Evaluation and Rehabilitation of Tilt-Up Buildings and Other Rigid Wall/Flexible Diaphragm Structures.* Structural Engineers Association of Northern California (SEAONC), San Francisco, CA, 2001.

23. FEMA 351 (2000). *Recommended Seismic Evaluation and Upgrade Criteria for Existing Welded Steel Moment-Frame Buildings.* Federal Emergency Management Agency, Washington.

24. Steinbrugge, Karl V. (1982). *Earthquakes, Volcanoes, and Tsunamis: An Anatomy of Hazards.* Skandia America Group, New York.

25. ATC-13 (1985). *Earthquake Damage Evaluation Data for California.* Applied Technology Council, Redwood City, CA.

26. ATC-13-1 (2002). *Commentary on the Use of ATC-13 Earthquake Damage Evaluation Data for Probable Maximum Loss Studies of California Buildings.* Applied Technology Council, Redwood City, CA.

27. FEMA 351 (2000). *Recommended Seismic Evaluation and Upgrade Criteria for Existing Welded Steel Moment-Frame Buildings.* Federal Emergency Management Agency, Washington.

28. HAZUS99-SR2 (1999). *Technical Manual,* Federal Emergency Management Agency, Washington, http://www.fema.gov/hazus/dl_sr2.shtm.

29. HAZUS99-SR2 (1999). *User's Manual.* 1999, Federal Emergency Management Agency, Washington.

30. ASTM 2026-99 (1999). *Standard Guide for the Estimation of Probable Loss to Buildings from Earthquakes,* American Society for Testing and Materials, West Conshohocken, PA.

31. *Procedure for Postearthquake Safety Evaluation of Buildings,* Applied Technology Council, Redwood City, CA, 1989, www.atcouncil.org.

32. *Addendum to the ATC-20 Postearthquake Building Safety Evaluation Procedures,* Applied Technology Council, Redwood City, CA. 1995, www.atcouncil.org.

33. *Field Manual: Procedure for Postearthquake Safety Evaluation of Buildings.* Applied Technology Council, Redwood City, CA, 1989, www.atcouncil.org.

34. FEMA 306 (1998). *Evaluation of Earthquake Damaged Concrete and Masonry Wall Buildings: Basic Procedures Manual.* Federal Emergency Management Agency, Washington.

35. FEMA 307 (1998). *Evaluation of Earthquake-Damaged Concrete and Masonry Wall Buildings: Technical Resources.* Federal Emergency Management Agency, Washington.

36. FEMA 352 (2000). *Recommended Postearthquake Evaluation and Repair Criteria for Welded-Steel Moment-Frame Buildings.* Federal Emergency Management Agency, Washington.

37. Guy Nordenson and Associates and LZA Technology/Thornton-Tomasetti (eds). *World Trade Center Emergency—Damage Assessment of Buildings, Structural Engineers Association of New York Inspections of September and October 2001,* prepared for the New York City Department of Design and Construction, 2003, New York, NY.

CHAPTER 7

Historic Buildings and Monuments

MARK J. TAMARO, P.E., and R. WAYNE STOCKS, P.E.

INTRODUCTION

Structural condition assessment of historic buildings and monuments requires a different level of expertise and experience than that for conventional modern buildings. There are several factors that distinguish historic structures from their modern counterparts. Historic structures were built by design standards and construction methods that were vastly different than those of the modern era. The engineer conducting an evaluation of a historic structure must be fully aware of the parameters under which the structure was designed and constructed originally. The original structural design theory, the intended building use, and construction practices are all important conditions that need to be studied and understood in order to address the current condition of a historic structure appropriately. Unlike in modern buildings, structural material properties often are unknown in historic structures, and drawings defining the structural system may be limited or may not exist at all. The fact that a building or monument is designated as "historic" also means that it is statistically more prone to have developed long-term structural problems than younger buildings, and greater care must be taken to investigate possible deterioration. Lastly, engineers must be more sensitive to the architectural fabric of historic structures as compared with modern buildings when evaluating them because preservation of the original materials and finishes is paramount (Fig. 7.1).

The U.S. National Park Service National Register of Historic Places generally uses 50 years as the minimum age for designating a structure as historic. The examples and case studies used in this chapter generally are taken from projects performed on the East Coast of the United States, with dates of original construction ranging from the late 1700s to just prior to World War II. It is our intent to concentrate on archaic materials and building systems. Pre–World War II is used in this chapter as a convenient, if not precise, benchmark for the time of transition between historic and modern construction. Although the specific examples and tabulated data listed herein most commonly apply to the 1800s to 1940s era, the overall recommended processes and procedures for survey and evaluation listed in this chapter can be used successfully for other time frames and geographic locations as well.

FIGURE 7.1   Scaffold access system for the Washington Monument. This system was designed and constructed without fastening mechanically to the historic fabric of the monument facade.

COMMON REASONS FOR ASSESSMENT OF HISTORIC STRUCTURES

Historic structures may require structural assessments for a number of reasons. Several common reasons are discussed below.

Visible structural deterioration or known failure. The most obvious need for assessment may exist where structural deterioration is exposed to view, such as corrosion of iron or steel roof members. An example of this shown in Fig. 7.2, which shows the architecturally exposed historic wrought iron roof trusses at the Smithsonian Institution's Museum of Arts and Industries, built in the 1880s. In this case, areas of concern regarding the roof structure were readily visible from the floor below. The Smithsonian Institution's facilities personnel could very easily make the determination that a structural assessment was necessary based on seeing the extent of corrosion on the exposed trusses.

Structural failure. Structural assessments also will be called for after an unanticipated structural failure. A failure often will indicate a hidden problem that may have the potential to occur in other locations of a similar construction type. This was evident in the case of the balcony collapse at the University of Virginia, where a wrought iron hanger rod had corroded to the extent that it failed in tension when it was subjected to large live loads. This hanger, designed by Thomas Jefferson, was on the exterior of the building with a portion hidden within the wood balcony floor. Deterioration at this location had gone undetected prior to the collapse. In this instance, prudence dictated that balconies of similar construction on the campus be inspected in order to prevent further failures.

Deterioration of building enclosure. When a building or monument exhibits deterioration of a particular building system, such as its facade or roof, a survey of the underlying structure may be warranted. Water infiltration through the building's skin is usually obvious to the building user, but resulting deterioration to the structure may be concealed. Rust of iron and steel framing and rot of wood framing can progress at a slow pace and go unnoticed for many years until a more dramatic event, such as sagging of roof members, is noticed.

Lack of documentation regarding the structural capacity of the building. Many historic monuments and buildings do not have the design live load capacities

FIGURE 7.2   Exposed wrought-iron roof trusses.

indicated on the original construction documents, and analysis may have to be performed to determine existing capacity. When original construction documents are no longer available, field verification of the existing structure must be done first in order to perform an analysis.

Change of use or increase capacity required. Adaptive reuse of a historic building often changes the live load requirements for the building. For example, a residential building, historic or modern, requires a live load capacity of 40 lb/ft^2 according to most current building codes. If such a building were to be renovated for use as a museum or office, generally 100 lb/ft^2 may be required as occupancy loads. The original gravity-load-resisting system then would need to be assessed to determine if reserve capacity is available to accommodate the higher loads.

Severe events. Condition assessment normally will be required for a historic structure if it is subjected to forces caused by either severe natural or human-made events. Examples of such events include hurricanes, earthquakes, floods, explosions and fires, and undermining or subsidence of foundations.

Executive order and other federal, state, or local requirements. Assessments of existing buildings, including historic structures, often are mandated by government bodies and regulating agencies. These mandates frequently are directed toward assessment for particular types of structural performance as opposed to a general structural review. Examples include federal regulations requiring the investigation of seismic performance, vulnerability to blast loads, and progressive collapse as related to security concerns.

INFORMATION GATHERING PRIOR TO PHYSICAL INSPECTION

Archival Research

It is important to gather as much documentation as possible about a building prior to beginning its physical inspection. While records of construction documents and renovations often are available for modern buildings, it is likely that they are not readily available or no longer exist for historic structures. Typically, copies of building drawings are kept on site by the building management or are retained and stored off site by the building owners. The older a building becomes, the more likely it is that these documents are lost, damaged, or deteriorated. Over the course of time, changes can occur in building management or ownership, and building drawings may be lost in the transition. This is particularly true for structural drawings because they are rarely used by building operations staff, and frequently they go unnoticed if they are removed and not returned.

In the event that structural building drawings are not available from building management, other outside sources can be used to locate copies. Local building departments maintain records of construction projects that are submitted for permits and can be contacted for acquiring duplicates of record documents. Unfortunately, owing to storage limitations, some jurisdictions discard documents past a certain age. Beyond the local municipalities, many federal agencies have archive records of construction documents for the buildings in their inventory. The General Services Administration (GSA) maintains microfilm records for thousands of buildings that it managed previously or is managing currently. Other examples of federal agencies that can be used as resources for buildings and monuments under their jurisdictional control include the National Parks Service and the Architect of the Capitol in Washington, D.C. Professional archival research firms and organizations such as the Historic American Buildings Survey (HABS) also can be used to search for and acquire copies of existing building drawings. When all else fails, there is a possibility that drawings still may be in the possession of the original design firms involved or their predecessor organizations. If some records exist but the original designers are not known, it may be possible to track down the firm responsible for the structural design from the title block on a drawing for a different discipline.

Original Construction Documents

A building's original construction documents are an invaluable asset to an engineer performing an assessment. Structural systems, individual member sizes, and spacing and location of key components all can be determined readily by reviewing a set of existing drawings. When the existing drawings are available and minor selective demolition is used to spot check the information, the process of documenting the structure's existing condition is greatly simplified. By contrast, when no documentation is available, the only alternatives are to rely on a limited knowledge of the structure based on visual observations of exposed elements or to embark on a program of extensive nondestructive and destructive testing or probing. The latter option can be extremely time-consuming, expensive, and destructive to the historic fabric that conceals the structure.

Date of Construction

When original documents are not available, determining the date of construction may be very helpful in establishing the possible structural systems used in a particular building. This piece of information allows the engineer to use information regarding construction practices of a certain era to zero in on particular construction types used in the building. The date of construction can be determined from the building itself based on dedication plaques or corner stones. Other sources include building department records, neighborhood histories, and media archives such as newspapers and periodicals that may have featured the events during construction and may even include construction photographs.

Many sources that list the approximate starting dates for the common uses of particular building systems are available currently, such as Rabun.[8] A brief list of archaic construction systems in chronological order is given in Table 7.1.

Original Building Codes

Determining what codes and standards were in effect during construction of a particular structure can provide an engineer with critical information such as design loads, methods of fireproofing, construction practices, and allowable design stresses for building materials. Local building codes became common in major cities such as New York City, Chicago, and San Francisco in the middle to late 1800s and usually had major revisions in response to catastrophic events such as building collapses, fires, or earthquakes. These early codes established regulations for design, such as minimum bearing-wall thickness and uniform live loads. Knowledge of these codes can provide an engineer with valuable insight into the structural system and can reduce the number of destructive probes required.

Depending on the location and date of original construction, the design of a historic structure may or may not have been governed by a national or local building code. In this case, it is useful to research period design texts, handbooks, or proprietary product literature. Examples of these include *A Treatise on Masonry Construction,* by Ira Baker, and Mahan's *Treatise on Civil Engineering,* which provided design and analysis theories and design stresses for materials. Product manuals such as the Carnegie Steel Company *Pocket Companion* provided precise material properties for steel produced by Carnegie Steel Company, as well as extensive design aids and typical details.

TABLE 7.1 Building Systems Timeline for the Eastern U.S.
Based on Authors' Experience and Observations

Building System	Estimated Initial Date of Use	Estimated Date of Obsolescence
Bearing walls of brick or stone with lime-sand mortar	Colonial era	1850s–1900s
Masonry load-bearing walls with Portland cement mortar	1850s	Still in use
Masonry arches and groin-vaulted floor systems	Colonial era	1860s–1900s
Wrought-iron members	1840s–1860s	1890s–1920s
Cast-iron members	1830s	1920s
Timber frames of sawn lumber,[8] balloon frame (nailed connections)	1830s	1900s
Timber frames of sawn lumber,[8] platform framed	1900s	Still in use
Wrought-iron frames with built-up (riveted) members	1880s	1920s
Wrought-iron or steel beams with tied-brick arch floors	1840s	1880s
Terra cotta floors	1870s	1900s
Steel beams		
With terra cotta floors	1880s	1920s
With concrete floors	1890s	1960s
Plain concrete foundations	1880s	1900s
Concrete ribs with terra cotta infill and concrete-encased steel frames	1910s	1940s
Reinforced-concrete slabs and frames	1920s	Still in use
Welded steel	1930s	Still in use
Bolted-steel connections	1950s	Still in use

CLASSIFICATION OF CONSTRUCTION

In order to assess a building properly, the types of structural systems must be determined. Many historic buildings contain archaic structural systems of which the modern engineer may not be familiar. The following discussion briefly describes many systems common to their respective eras but used rarely today. Early construction types for foundations, floor and gravity-load systems, and lateral-load systems are described below.

Foundations

Prior to the late nineteenth century, building foundations supporting load-bearing masonry walls generally consisted of unreinforced masonry, typically rubble, stone, or brick. The distribution of load to the supporting soil was accomplished through corbeling action of the footings. Although much of modern geotechnical engineering theory was not standardized until the early twentieth century, sophisticated forms of masonry foundations were evident in the United States as far back as the turn of the eighteenth century. This is illustrated in Fig. 7.3, which shows an inverted-arch foundation connecting isolated piers to a continuous masonry strip footing. Inverted arches

FIGURE 7.3   Inverted-arch foundation at the Baltimore Basilica (1801).

were used to distribute the large concentrated loads uniformly from the masonry piers to the supporting soils. Early deep foundations for buildings generally consisted of timber piles or hand-dug caissons. Decay of timber foundations is a significant concern and can be related to variations in the water table, as well as biologic infestation such as marine borers. The use of precast piles is documented as early as the 1890s, and these allowed for deeper foundations that were not susceptible to organic decay. A critical aspect for assessing historic pile capacities is the determination of the pile length. Figure 7.4 shows the use of pulse-echo methods to estimate pile integrity and length of precast piles. New developments in concrete before the 1920s lead the way

FIGURE 7.4   Pulse-echo method used to determine pile length and integrity, National Archives Building, Washington, D.C.

to modern concepts of individual spread footings, which used steel I-beam rails to form grillages, distributing concentrated loads to larger areas through flexural strength and stiffness.

Although many historic structures were supported on foundations that were based on empirical practices and rules of thumb, the concept of calculating adequate soil bearing pressures was not disregarded by all early designers. This is evidenced by the complex underpinning system used at the Washington Monument in 1880, which is shown in Fig. 7.5. In this case, the original foundation design resulted in substantial overstresses in soil-bearing pressure causing settlements measured during construction. After a halt in construction, a concrete underpinning scheme was accomplished successfully.

Floor Systems

Colonial-era construction relied extensively on timber beam and post construction owing to the availability of inexpensive lumber. In addition to wood-framed floors, brick and stone masonry bearing walls were used regularly. Brick barrel and groin vaults also were used in floor construction for larger, more noteworthy buildings. These masonry vaulted systems generally consisted of one- or two-way brick arches with loose fill placed above to create a level surface for finished wood floors. Horizontal thrusts from the vaults normally were resisted by the mass of exterior walls, which were several feet thick. Attempts to deal with the issue of horizontal thrust led to the introduction of wrought-iron ties between the piers of groin-vault construction (Figs. 7.6 and 7.7). Because these ties can be hidden within masonry partitions between piers, it is essential to confirm the existence or absence of ties in a groin-vault system to prevent damaging the ties during subsequent building renovation.

As technological advances of the industrial revolution in Europe reached the United States, wood-joist floors and masonry vaults had begun to be replaced with new structural floor systems. Although the use of structural iron predates the eighteenth century in the Far East, it was its application in European industrial buildings and railroad bridges that led to its widespread use. Wrought-iron roof trusses and floor beams, as well as cast-iron columns, provided significant technological improvements for designers. Iron had high strength-to-weight ratios that could greatly increase span lengths, and its noncombustible properties were a great improvement over timber construction. Numerous floor systems were developed between 1850 and 1900 that incorporated wrought-iron beams with masonry, terra cotta, or concrete floor infill. The earliest beam sections often resembled train rail profiles with a rounded top flange and wider flat bottom flange but eventually evolved into sections resemble modern wide flanges

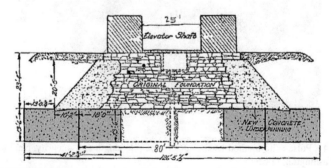

FIGURE 7.5   Sketch of Washington Monument foundation underpinning.

FIGURE 7.6   Masonry groin-vault roof system, Tariff Building, Washington, D.C.

FIGURE 7.7   Masonry groin-vault system with wrought iron tension ties to accommodate thrust, Patent Office Building, Washington, D.C.

by the 1890s. Iron beam depths were commonly in the 8- to 12-in. depth range, and typical spans were roughly 18 to 22 ft. Beams were spaced at 4 to 6 ft on center, with various types of floor construction filling in between the beams. The earliest of these systems (1860s) used shallow brick arches one or two courses deep with the spring line of the arches starting at the beam bottom flanges (Figs. 7.8 and 7.9). This system required heavy walls to resist the unbalanced lateral thrust at the building exterior. Later, wrought-iron ties were used running perpendicular to the beams to resist the thrust of the arches. Lighter systems were developed by using hollow terra cotta blocks in the form of tied arches, as well as flat or "jack" arches. As portland cement concrete became available as a building material, proprietary systems began to emerge in which reinforced-concrete floors could be poured between beams using combinations of small iron tee sections and terra cotta blocks to create forms supporting the wet concrete. Concrete floor systems became common in the early 1900s. One-way reinforced-concrete slabs spanning between concrete-encased steel beams replaced the brick and terra cotta systems described earlier, and concrete joist systems were used extensively during the 1920s through the 1940s. The concrete joist systems typically were 7 to 8 in. deep spaced at roughly 16 in. on center, with the void spaces being formed by lightweight terra cotta blocks (Fig. 7.10). Both one- and two-way (waffle slab) systems were used to span to concrete-encased steel beams at the column grids or to reinforced-concrete beams. These systems typically were used in bays with 18 to 22 ft between columns. Numerous graphic and written descriptions of the floor systems just discussed can be found in Friedman[6] and Rabun.[8]

Both wrought- and cast-iron columns were used extensively in the latter half of the nineteenth century. Cast-iron columns were used with both timber and iron beam floor systems and gave early designers relatively high-strength compression members that could be cast into numerous cross-sectional shapes with bases and bracket connections formed as one integral element. Major drawbacks to cast-iron columns were the fact that their production was labor-intensive and that the material exhibits substantially less tensile capacity than its compressive capacity. Limited flexural resistance and susceptibility to cracking gave way to wrought-iron and eventually steel column sections. Wrought-iron and steel columns typically were fabricated into built-up sections

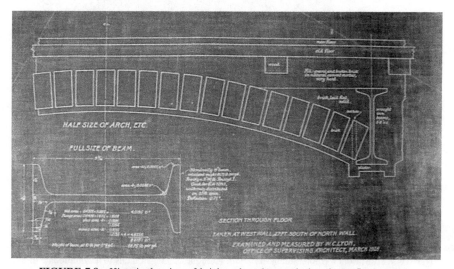

FIGURE 7.8   Historic drawing of brick arch and wrought-iron-beam floor system.

FIGURE 7.9   Cut through brick arch and wrought-iron-beam floor system.

by bolting or riveting together plates or other shapes. It is difficult to distinguish between these two materials, and often a material sample is required to perform a chemical composition test. Certain proprietary systems are more identifiable by shape, such as the wrought iron columns produced by the Phoenix Company (Fig. 7.11).

Wrought iron and steel trusses also were commonly used by the end of the nineteenth century (Figures 7.2 and 7.12). Numerous geometric configurations were built, many of which employed systems that are now very uncommon or obsolete, such as the lattice truss system (Fig. 7.12).

Lateral Load Systems

Prior to introduction of the iron or steel frame, lateral loads on buildings were resisted primarily by masonry exterior walls and interior partitions that served as both gravity-

FIGURE 7.10   Reinforced-concrete joists with terra cotta tile infill floor system.

FIGURE 7.11   Section of a Phoenix column.

load-bearing elements and unreinforced shear walls. Most buildings had closely spaced walls and partitions and generally did not exceed five or six stories in height; therefore, resistance against lateral loads was not a critical concern of early designers. Wood sheathing or siding also provided lateral resistance for wood-framed structures, and knee-braced systems with mortise and tenon connections were common practice in wood buildings with large roof spans and few interior walls. However, by the turn of

FIGURE 7.12   Wrought-iron lattice roof truss.

the 1900s, elevator technology led to a new age of high-rise construction. Both steel-and iron-framed buildings were constructed with heights many times greater than their base widths. Engineers had to design lateral systems based on rational theories of mechanics considering member stresses, lateral drift, and connections. Lateral systems included braced frames, knee braces, and moment frames. Design of statically deter-minate braced frames (analyzed as pin connections) did not differ greatly from modern designs. However, indeterminate moment frame systems required more complex anal-yses. Semirigid moment frames also were used commonly. Riveted girder-to-column connections were designed to resist wind-induced moments but were assumed to ex-hibit nonelastic rotation when subjected to gravity loads. These girders were designed as "simple beams" for the full gravity loads and continuous frames for the lateral wind loads.

DETERMINING PROPERTIES OF BUILDING MATERIALS

Once the structural systems of the structure being investigated have been established and the structure's physical condition has been documented, it is usually necessary to determine the material properties in order to check the existing capacity of members. The strength and stiffness characteristics of materials used in historic structures vary widely based on age and quality of the original material, but with proper identification of material types, an engineer can assess members with confidence. When undocu-mented, the geometry of the structure is commonly determined by exposing members and physically measuring them (Figs. 7.13 and 7.14). Nondestructive techniques such as ground-penetrating radar also can be used (Fig. 7.15).

Material properties generally are established by two techniques. One involves iden-tifying the material from original documentation, date of construction, or manufacturer markings. The other is by testing. Choice of techniques will depend on the level of detail of the assessment, the amount of money available, the relative importance of the structure, and the degree of accuracy needed to confirm whether the structure is adequate.

Documents, Dates, Markings

The first technique, at a minimum, gives the engineer a starting point and an approx-imation of material values. The existing drawings may list early American Society for

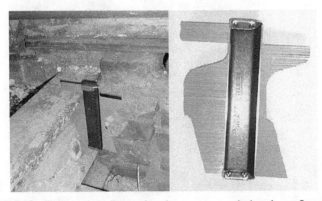

FIGURE 7.13   Carpenter's scribe used to document wrought-iron-beam flange profile.

FIGURE 7.14 Caliper used to measure wrought-iron-beam web thickness.

FIGURE 7.15 Photograph of GPR data collection for mapping original wall flues and chases in load-bearing masonry walls.

Testing and Materials (ASTM) material designations (or predecessors to the ASTM) or simply strength values, such as yield stress for metals or mix proportions for concrete. This information, when available, usually represents conservative values for the actual material strengths. However, the engineer should be wary of material changes through either substitutions during construction or deterioration. Wood, masonry, and concrete all have the potential to loose strength as a result of environmental conditions, and this may not be evident during a visual inspection. Date of construction can be used to hone in on the material properties by identifying the materials available or commonly used during the time period. If accurate values are required, date of construction should be used only as a starting point for further determination. Manufacturer markings stamped or painted on iron, wood, or masonry can be a suitable method for identifying the material and thus its properties, provided that the engineer has access to the appropriate historical manufacturer data. For example, it is common to use manufacturer markings in conjunction with dates of construction to cross-reference wrought-iron members with yield strength information provided in American Institute of Steel Construction (AISC) references.[2]

All these techniques have the benefit of being nondestructive and can be accomplished quickly without the additional costs of laboratory testing. However, when materials cannot be classified confidently as just described, or when more precise values are necessary, the methods described below can be used. Note that these tests are only a few of the multitude of destructive and nondestructive methods currently available and that those listed below represent common and practical methods as related to historical structures. It also should be noted that while testing of existing materials can produce highly accurate values for their properties, use of these acquired data for rating of a structure's capacity depends on the frequency of samples taken and the variability of the material. For example, three coupon samples may be used for determining an average yield strength of wrought-iron beams (of similar size and configuration), whereas much more frequent samples or, alternatively, higher factors of safety would be prudent for a less uniform material such as masonry.

Material Property Tests

The following is a list with brief characterizations of the basic tests used on common construction materials. More complete discussions of tests can be found in the chapters of this book dedicated to specific materials.

Steel, Wrought Iron, and Cast Iron

Chemical composition. Laboratory test of sample to identify type of metal and other properties, such as weldability for repair.

Tensile test. Used to determine yield and ultimate stress, as well as modulus of elasticity. When combined with chemical composition tests, it can be used to determine the closest ASTM designations.

Compressive test. Similar to the aims of a tensile test but particularly suitable for cast iron, which exhibits high compressive capacity but low tensile capacity.

Fatigue test. Used to determine cyclic loading limits for particular stress ranges. May be useful for members subjected to repetitive loading cycles, such as framing used to support machinery.

Concrete

Laboratory compressive test. Cylindrical core samples taken from existing structures, tested to reveal the concrete's ultimate compressive strength. The modulus of elasticity and modulus of rupture can be estimated from the result.

Petrographic tests. Microscopic inspection of concrete samples, which provides useful chemical composition information and can be used to determine water-to-cement ratio, among other properties.

Chloride content. Very commonly used test to determine chloride ion profile in concrete. Provides information relating to deterioration of concrete affecting the original strength characteristics.

In-situ compressive test (Windsor probe or Swiss hammer). Easy to perform in the field and provides a rough approximation of compressive strength. It potentially can have inaccurate results owing to lack of proper calibration or probes being driven into large, hard aggregate.

Masonry

Laboratory compressive test. Can provide compressive strength data for stone, brick, or terra cotta tiles. Note, however, that nineteenth-century mortars (lime-sand in particular) are often significantly weaker than the masonry units themselves.

In-situ compressive test. Measures in-place compressive strength by inserting jacks in slots in the mortar joints or by removing units adjacent to the test specimen to provide space for jacks. This test also can be modified to determine shear strength and bond strength between the masonry units and the mortar.

Chemical analysis of mortar. Samples are analyzed to determine amounts of different mortar components (e.g., whether or not the mortar contains portland cement).

Flexural strength or modulus of rupture strength tests. Used primarily on stone samples and useful for assessing the strength of large stone lintels, beams, or stone floor panels.

Wood

Strength tests. Strength tests, both parallel and perpendicular to grain, are the most reliable way of grading wood, but variations in material may require numerous samples for a reliable range of results.

Coring. Core samples can provide information regarding degree of decay caused by fungi or insects, amount of moisture content (which is related to strength), and depth of fire damage or char.

Hand probing. The easiest but least quantitative method, and it involves penetrating the wood with a sharp metal object such as a screw driver to determine softness or areas of local decay and high levels of moisture.

MONITORING PROGRAMS

Assessment of historic buildings often involves the implementation of monitoring programs to document building movement. Buildings can be monitored to determine whether or not damaging movements are ongoing, or they can be monitored in anticipation of a significant environmental change, such as an adjacent excavation or blasting. Although monitoring of historic buildings differs from that for modern structures in that older structures generally were constructed of brittle materials that are more susceptible to damage caused by building movement, the methods used to monitor historic structures are essentially the same as those used on modern structures. However, higher thresholds for acceptable levels of movement should be established for historic structures based on their relative importance and the rigidity of their facades and framing. Displacement monitoring points often are set up at various elevations and corner points of a building to document relative movements of the entire structure when new construction work is being performed adjacent to the historic structure. These points typically are surveyed on a regularly scheduled and frequent basis to observe displacements before they become destructive. Both horizontal and vertical movements must be watched closely because disturbances to building foundations can result in lateral translation owing to tilting, as well as vertical settlement. Monitoring also may be required in discrete locations such as at existing cracks in masonry walls. Strain gauges can be used to collect large amounts of data at such critical points. While building movements commonly are caused during adjacent excavations, dewatering, or installation of tie-back anchors, another cause of damage is vibration. Driving of piles, jack-hammering, and operating heavy equipment adjacent to or within a historic structure can result in cracking of masonry walls and floor systems, as well as historic finishes. It is critical to record vibrations as part of a comprehensive monitoring program for a historic structure. Accelerometers can be set to collect acceleration values over long periods of time, and peak values can be correlated with events occurring near the structure being monitored. As with displacement, thresholds normally are established to limit activities as soon as the potential for damage occurs. For example, an acceleration threshold of 1 in./sec is used commonly for historic masonry structures.

With all types of monitoring, it is important to note that daily and seasonal changes in temperature and moisture should be anticipated so as to not lead to confusion when interpreting the resulting data.

STRUCTURAL ANALYSIS OF ARCHAIC BUILDING SYSTEMS

Historic structural systems generally are analyzed to establish the true load carrying capacity of the structure when it is unknown owing to lack of information or weakened as a result of deterioration. Discussions regarding statically determinate structures, indeterminate structures, and load tests are presented below as they relate to historic and archaic structures.

Determinate systems. Most historic structures are statically determinate systems that were designed through the use of hand calculations and are relatively straightforward to analyze once the geometry and material properties have been established. The most obvious examples are floors supported by beams and columns constructed of wood, metal, or concrete. Floors with slabs spanning to beams or girders, which, in turn, are supported by columns and individual footings, usually can be analyzed as individual elements (either continuous or simple span) based on tributary loads and direct load paths. Load-bearing masonry walls and arches are also generally straightforward to analyze. Stresses in arches typically can be estimated by assuming boundary conditions for which a determinate analysis can be performed, such as a three hinged arch, or by using closed-form solutions documented for various geometric and support configurations. These generic solutions are documented in numerous texts such as *Roark's Formulas for Stress and Strain,* by Warren C. Young.[9] It should be noted, however, that while this approach can yield quick and fairly accurate results, it does not address conditions such as large concentrated loads. The engineer also should be aware that determinate systems may have nonstructural elements that are contributing to the load carrying system and create a more complex structure. Individual structural members also may behave differently than anticipated owing to mechanisms such as composite action. An example of this is given in the load-test discussion in the next section of this chapter, where wrought-iron beams were found to act compositely with the brick arch floor they supported.

Indeterminate systems. Some historic structures do require sophisticated calculations in order to be analyzed properly. Complex trusses, domes, and groin vaults found in many historical buildings and monuments typically are statically indeterminate. The most effective way to analyze these structural forms is through the use of three-dimensional finite element computer models; see U.S. Capitol dome finite-element model in Fig. 7.16. While this method is well suited for high-profile projects in which time and budget constraints allow for a very labor-intensive analysis, it may not be an option for most assessments. In the latter case, the engineer may rely on approximate or empirical methods to establish approximate stress ranges or use similar structures with known behaviors as a guide. For example, masonry groin vaults were common in buildings prior to the development of iron beams, and their original design methodology is some-

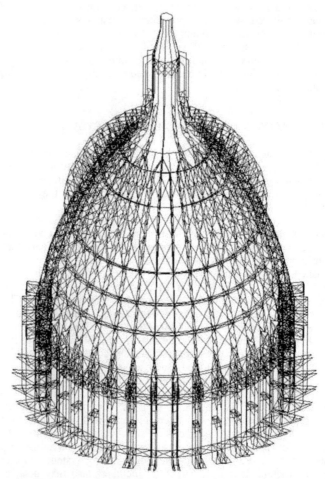

FIGURE 7.16 Three-dimensional finite-element model of U.S. Capitol dome.

what ambiguous. Renovations of these types of structures often require vertical penetrations through the vaults and through masonry walls below the vaults. Determining stresses and load paths in the masonry system is critical prior to locating such penetrations. For determining approximate stress levels, three approaches are described in Williams.[4] We typically use the stress distribution shown in Fig. 7.17, which defines main diagonal and orthogonal arches between corner piers and assigns a theoretical distribution of the total load to each arch, which ultimately is supported by four corner piers.

LOAD TESTING

Renovation or modernization of historic structures typically requires determining the load-carrying capacity of a building's floor or roof system. For a specified use, local and national building codes specify the minimal live-load capacity. In order to evaluate the adequacy of a given structural system in a historic building, the structural engineer may be forced to rely on limited information, as discussed earlier. Documentation of historic buildings often is limited to architectural sections that do not indicate the size and spacing of wrought-iron or steel beams. Similarly, drawings that indicate concrete slab thickness and size and spacing of reinforcing steel often are not available. A program of limited destructive and nondestructive testing that could reveal the necessary information may be considered too disruptive to the tenants or may not be allowed owing to concerns for the historic fabric of the structure. In the absence of sufficient structural information to evaluate the theoretical load-carrying capacity, a full-scale load test of a portion of the floor or roof system is a common procedure for

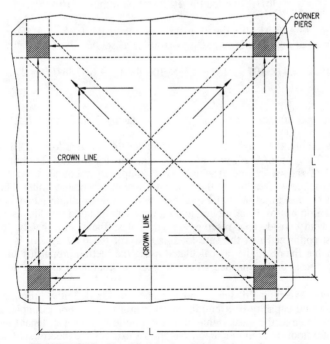

FIGURE 7.17 Simplified load path for groin-vault system.

determining the load-carrying capacity in an existing historic building. Even with adequate material and section properties, the theoretical load capacity may not satisfy the requirements of the desired use. A full-scale in situ load test may prove that the system is in fact capable of supporting a desired load.

Imagine the difficulty in explaining to a building owner or federal agency that the floor system in their historic building has a theoretical live-load capacity of almost nothing, although the building has performed exceptionally well for over 140 years. In a particular federal building in Washington, D.C., a structural assessment report completed in 1893 concluded that the floors could support a total safe load of 112 lb/ft². With the weight of the floor construction averaging 90 lb/ft², the engineer confidently noted that 22 lb/ft² was available for movable loads. The preliminary structural assessment approximately 100 years later yielded similar results (total safe load of 115 lb/ft², dead load of 107 lb/ft², available live load capacity of 8 lb/ft²) after limited destructive testing determined section properties and topping slab thickness, assuming allowable bending stresses consistent with the time period. This is not exactly the answer an owner likes to hear. This case justified the expense of performing a load test to rate the floor capacity accurately and is outlined in the following case study.

Case Study: Federal Office Building, Washington, D.C.

The floor construction of the 1865 federal building used in this case study consists of 9-in. deep wrought-iron beams spaced at 5 ft, 8 in. on center and spanning approximately 20 ft, 6 in. Brick arches supported on the bottom flanges of the beams support a topping slab, wood sleepers, and wood flooring. The estimated dead load of the floor system is 107 lb/ft². For the specified use, the desired service live-load totals are

$$80 \text{ lb/ft}^2 \text{ live load} + 20 \text{ lb/ft}^2 \text{ partitions} = 100 \text{ lb/ft}^2$$

According to ACI-318-99, the ultimate load intensity is $0.85 \times (1.4D + 1.7L)$. The required additional loading to achieve 100 lb/ft² service live-load rating is

$$0.85 \times (1.4 \times 107 \text{ lb/ft}^2 + 1.7 \times 100 \text{ lb/ft}^2) - 107 \text{ lb/ft}^2 = 272 - 107$$
$$= 165 \text{ lb/ft}^2$$

The test load of 165 lb/ft² was applied by placing 2 ft, 8 in. of fresh water over the test area (Fig. 7.18).

Remarkably, the maximum deflection under the full live load of 166 lb/ft² was only 0.22 in., or $L/1120$, and remained constant for the full 24-hour period. On removal of the water, the system almost completely rebounded, with a final remaining deflection reading of 0.02 in., well within acceptable limits. This load test confirmed what intuition and experience with similar buildings had implied. The wrought-iron beams, brick arches, and topping slab function together as a composite system with considerable reserve strength and stiffness. Fortunately, the $10,510 load test confirmed that the safe load capacity of the floor system is at least 100 lb/ft² rather than 8 lb/ft² (as calculated using the section properties and allowable bending stress) and confirmed what 140 years of continued use had demonstrated.

A properly executed full-scale load test can provide definitive information about the behavior and capacity of a building's structural system. For example, a load test may be justifiable as the most viable option if a building owner or tenant requires that a portion of a floor be used to support a load that exceeds the theoretical load carrying capacity of the floor system. Naturally, special care must be taken to ensure that load

FIGURE 7.18   Photograph of floor load test.

test produces the maximum stresses and deflections that the "real" loading would create. In multispan conditions, pattern loading should be used to achieve realistic results. A successful load test may justify increased service loading, eliminate the need for costly strengthening, avoid disturbance to the historic fabric, or in the case of the federal building discussed previously, justify continued use without modification or strengthening.

Convincing a building owner that a load test is the most appropriate solution for determining the load-carrying capacity of a structural system represents a challenge to the structural engineer and the design team. Assuming that conventional methods of analysis have been exhausted, the relatively high cost of conducting a load test and the potential risk of damage must be considered carefully. The potential reward in proving that the tested structural system is adequate for a proposed use may outweigh the cost and risk factors.

REFERENCES

1. American Society of Civil Engineers (1999). *Guideline for Structural Condition Assessment of Existing Buildings*. SEI/ASCE 11-99, Reston, VA, ASCE.
2. American Institute of Steel Construction (1999). *Historical Record Dimensions and Properties Rolled Shapes Steel and Wrought Iron Beams and Columns*, AISC New York.
3. Brown, J., and R. J. M. Sutherland (1997). *Studies in the History of Civil Engineering*, Vol. 9: *Structural Iron, 1750–1850*. Ashgate, Brookfield, VT.
4. Williams, C. C. (1930). *The Design of Masonry Structures and Foundations*, 2d ed. McGraw-Hill, New York.
5. Baker, I. O. (1902). *A Treatise on Masonry Construction*, 9th ed. Wiley, New York.
6. Friedman, D. (1995). *Historical Building Construction*. Norton, New York.
7. Carnegie Steel Company (1917). *Pocket Companion for Engineers, Architects and Builders*, 19th ed. Carnegie Steel Company, Pittsburgh, PA.
8. Rabun, J. S. (2000). *Structural Analysis of Historic Buildings*. Wiley, New York.
9. Young, W. C. (1989). *Roark's Formulas for Stress and Strain*, 6th ed. McGraw-Hill, New York.

CHAPTER 8

Building Facades

KIMBALL J. BEASLEY, P.E., and MARK K. SCHMIDT, S.E.

INTRODUCTION

The design and delivery of building facades, in combination with roofing systems, comprise up to one-fifth of all building-related construction.[1] Remediation of building facade failures also can be very costly—in rare cases approaching the original construction cost of the entire structure. In addition to the financial repercussions of facade failures, some conditions may have an impact on the safety of building occupants and the general public. Therefore, a proper assessment of building facades is paramount to the early detection and remediation of problematic conditions.

The building facade often is exposed to a hostile environment. The exterior walls must endure weather extremes during the life of the building. In recent decades, economic pressures have forced designers to create thinner, less costly wall systems with higher performance expectations. Consequently, the potential for building facade problems and failures has increased.

A specialized practice dedicated to evaluating the cause of facade failures and developing remediation schemes has evolved as an outgrowth of the modern complex built environment. Such facade assessment involves the application of engineering principles in a systematic and logical manner. However, the assessment cannot be reduced to a prescriptive series of steps that must be followed without variation. Such rigid methods discourage the inquisitiveness and flexibility that are essential to the collection and synthesis of data.[2]

COMMON FACADE TYPES

Building facades may be divided into two broad structural categories: load-bearing and non-load-bearing. Non-load-bearing facade systems need only support their own weight and forces (external or internal) acting directly on the facade, whereas load-bearing systems support these forces as well as loads from a portion of the structure itself.

Most facades also can be categorized as either barrier or cavity wall systems, the later also referred to as *drainage wall systems*. Figure 8.1 illustrates three common wall systems schematically. Cavity walls contain an internal mechanism designed to capture any water that penetrates the outer skin and drain it back to the exterior. This type of wall system has essentially two lines of defense against water leakage: the outer seals and an internal drainage system. The integrity of the internal drainage system depends on watertight through-wall (internal) flashings and effective wall drainage.

There are essentially two types of barrier wall systems: skin barriers and traditional barriers. A skin-barrier wall is one that relies totally on exterior surfaces and seals to prevent water leakage to the interior. The success of a skin-barrier wall lies in maintaining sometimes thousands of feet or even miles of exterior joint sealants. Traditional barrier walls, those built prior to or in the early part of the twentieth century, rely on the mass of the wall to absorb water and then disburse it as water vapor to control

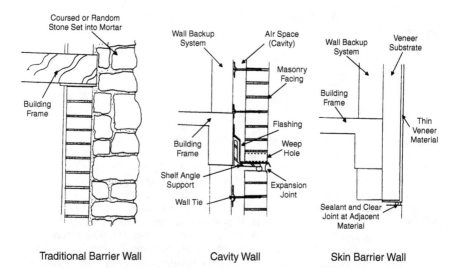

Traditional Barrier Wall   Cavity Wall   Skin Barrier Wall

FIGURE 8.1   Common types of wall systems.

leakage. Joints solidly filled with mortar keep the water from penetrating in great volume or from accumulating within the wall.[3]

There is a great deal of flexibility in the choice of materials, appearance, and configuration of building facades. Industrial facilities, for example, simply may require a utilitarian envelope of light-gauge metal sheets over an economical building frame. First-class commercial office buildings, however, may require a highly stylized envelope combining glass and architectural metal, masonry, or stone veneer panels. In most cases, the successful facade employs features that are not overly complex together with materials, connections, and details that are time-proven. While there are countless materials and methods of construction for building facades, this chapter focuses on the more common types of facades used on significant mid- to high-rise buildings in the United States. Descriptions of other types of wall systems and facade components can be found in other standards.[4]

Facades of older structures often consist of load-bearing masonry or dimension stone; however, both facade types also can be used in a non-load-bearing configuration, where the masonry or stone is supported by steel shelf angles or other connection assemblies attached to the structural frame. Newer structures with skeletal building frame systems typically support non-load-bearing, curtain wall–type facades sometimes consisting of metal and glass, precast concrete, and thin stone veneers. Each of these facade systems is described briefly in the following subsections.

Masonry

Masonry units used in building facades vary from man-made fired clay or formed concrete to natural stone units. Clay masonry, such as brick, terra cotta, and "structural" tile, usually are preformed by extrusion or molds and fired to vitrifying temperatures. Concrete masonry is commonly a low-water-content mixture of portland cement and aggregate tamped or vibrated into forms. Natural stone masonry may be either field stone collected and used without alteration or stone roughly trimmed or shaped to better fit into the wall. A characteristic common to almost all masonry facades is the use of mortar, a mixture of finely ground lime and sand aggregate (and within the last century, portland cement), to bond masonry units together. The masonry usually is interlocked for strength and tied back to the structure via wall ties or bonding units. The exposed facade material and configuration must be able to endure the climate and resist water penetration without deterioration or water leakage to the interior. In most mid- to high-rise masonry facade construction, the self-weight of the masonry is supported at each floor level directly by the floor framing or by steel shelf angles or plates attached to the floor framing. In older structures, the masonry facade may form load-bearing walls that are supported by the foundations. Lateral support is provided by constructing the facade integral with interior masonry backup walls (using masonry headers, for example) or by connecting it to backup walls or the structural frame using metal ties.

Precast Concrete

Architectural precast concrete panels have been used as an exterior cladding material for more than 40 years, with some limited applications dating back to the 1920s and 1930s.[5] Production of these panels in a controlled plant environment allows for quality control and a variety of concrete finishes that would be difficult to achieve with cast-in-place concrete. The controlled manufacturing environment allows the precast concrete panels to meet exacting tolerances for dimensions and anchorage placement; however, the structural frame to which the panels are connected generally is not constructed to such precise tolerances. Typical anchorages of precast concrete panels consist of steel plates or structural sections cast into the panel, which are welded or fastened mechanically to plates or structural sections cast into a structural concrete frame or are attached to a structural steel frame.

Cast stone is another type of architectural precast concrete that is manufactured to simulate natural stone. It is used mostly as ornamentation and architectural trim where natural stone or terra cotta otherwise might be used. Cast stone can be installed as a laid-in-place product or attached to the structure similar to precast architectural panels. Cast stone can be formed using either a wet-cast or a dry-tamp procedure to create a smooth finish without bug holes and with better color consistency. While similar to conventional precast concrete, cast stone differs in its high cement content, small aggregate size, and characteristic lack of air entrainment. However, recent studies have shown that air entrainment can increase the durability of cast stone significantly in a freeze-thaw environment.[6]

Dimension Stone

Dimension stone facades consist of natural stone quarried and saw cut to prismatic shapes. Dimension stone gained popularity as a facade material early in the twentieth century when buildings were soaring to greater heights. Contemporary dimension stone units usually are not set in mortar and most often are much larger than ashlar masonry stone units. With contemporary dimension stone construction, each stone unit is usually supported independently by metal shelf angles at each floor level and tied back to the structure with metal straps or ties. Thinner stone facades sometimes are supported on a framework of structural shapes called a *strongback system* or are affixed to precast concrete panels with thin "hairpin" ties.

Dimension stone finishes may vary from highly polished granite or marble to irregularly surfaced travertine or sandstone. The durability of the dimension stone facade is generally a function of the stone material, the effectiveness of the internal water drainage system, and the method of support, attachment, and movement accommodation.[7]

Metal and Glass Curtain Walls

These curtain-wall systems typically consist of a metal framework infilled with a combination of vision or spandrel panels. The vertical framework members, which generally extend one or two floors in height, are fastened to the structure at each floor level using metal anchorage assemblies. Vision panels—usually windows—can be either fixed or operable, whereas spandrel panels typically are fixed and opaque to visually obscure the interstitial floor spaces. Spandrel panels can be a variety of materials, such as glass, metal, stone, or a composite.

Most curtain walls are aluminum framed; however, isolated "punched" windows used within various building facades may be constructed of wood, steel, bronze, aluminum, or polyvinyl chloride (PVC) framing. Aluminum framing offers several advantages. Aluminum is light in weight and resistant to corrosion. Because the metal has poor thermal characteristics, many aluminum framing systems manufactured today incorporate thermal breaks or some type of thermal improvement.

A number of glass types are used in windows and curtain walls, including annealed, heat-strengthened, and tempered. Nearly all the glass products used in modern building construction are manufactured by the float-glass process. Slowly cooled (annealed) float glass is inherently weak in tension. The most common method of increasing the ultimate tensile stress of glass is heat treating. By heating glass to about 1100°F (590°C) and then cooling it rapidly, the outer one-fifth of the glass thickness goes into compression, and to maintain equilibrium, the inner portion goes into tension. The more rapidly the glass is cooled, the higher is the compression. Currently, glass produced with a surface compression ranging from 3500 to 7500 lb/in.^{2} (24 to 52 MPa) is termed *heat-strengthened,* and glass with a surface compression of over 10,000 lb/in.^{2} (69 MPa) is termed *tempered.*[8]

Glass also may be tinted or coated with reflective or thermal films. Individual glass sheets can be combined into insulating glass units (IGUs) or laminated glass products. Glazing methods (means of securing the glass in the frame) include varieties of sealants and gaskets.

TYPES OF FACADE FAILURES

Failures of modern mid- and high-rise facades and their components can be classified as serviceability (operational or aesthetic) failures or safety-related failures. The engineering community generally is most concerned about the safety of the facade. However, since the satisfaction of tenants and the acquisition of profitable long-term lease agreements largely determine the value of many structures, owners and property managers also have strong concerns about serviceability issues. Failures in either category may have tremendous cost implications for building owners and managers, as well as for design professionals.[9]

Operational Failures

Operational failure of a building facade may be described broadly as a condition that inhibits the facade's ability to perform as expected. The facade need not detach from the structure and collapse to be considered a failure. Operational failures such as water or air leakage, unacceptable cracking, or movements are far more common and in aggregate more costly than complete collapses. Operational failures can become safety concerns if left unattended.

Water-Related Facade Damage. The majority of facade problems reported by building owners and managers involve water leakage. This inordinate focus on facade leaks is understandable because scores of full-time leak inspectors, otherwise known as building occupants, are always on the job. Even small amounts of water can damage expensive interior finishes and render insulating materials within the wall ineffective. Chronic water leakage also can promote mold and mildew growth, creating potential health risks.

Shortcomings or deficiencies in the original facade design, such as a lack of effective flashings, inadequate drainage, and poor-quality materials, can result in significant water infiltration problems.[10] Problems also can develop when the wall does not accommodate construction tolerances (leaving poorly sealed joints or seals), when inadequate materials are substituted, and when poor workmanship is employed. Finally, weathering and aging of exterior building components, most notably sealants and gaskets, can compromise their effectiveness.

Sealant Joints. To some degree, nearly all facades rely on sealants for maintaining watertightness. Water leakage often can be attributed to the failure of these sealants. Although the criteria for proper sealant joints are given in most sealant manufacturers' literature, failures resulting from faulty joint design and construction are common. There are five essential elements of proper sealant joint design and installation: (1) proper sealant dimensions, (2) adequate substrate preparation, (3) appropriate backer rod selection, (4) proper tooling of the sealant, and (5) sealant compatibility with substrates and adjacent materials.

For maximum elasticity, joint sealants should have a width-to-depth ratio of 2:1 and a practical maximum width of 1 to 2 in. (2.5 to 5 cm). This optimal width-to-depth ratio minimizes bond stresses at the sealant-to-substrate interface. Surface preparation may include grinding of the substrate (if required to achieve desired bond), cleaning with an approved solvent, and applying primer if required. To avoid absorbing water that eventually may penetrate the joint, either a closed cell or a proprietary open/ closed cell combination foam backer rod should be used. However, caution should be exercised with predominantly closed cell backer rods that may outgas when punctured during installation, causing bubbles to form in fresh sealant. Backer rods should be sized approximately 25% larger than the joint to allow for adequate tooling. The sealant should be tooled shortly after installation to enhance the bond, improve the elastic performance, and improve the overall aesthetic. Prior to installation, the proposed sealant should be checked for compatibility with adjacent sealants, gaskets, glazing tapes, stone substrates (for staining potential), laminated glass edges, and insulating glass unit seals. Figure 8.2 illustrates some of the essential features of a proper sealant installation.

In addition to design and construction deficiencies, urethane-based sealants may experience a loss of their elastomeric properties and degrade to a "chewing gum" consistency through a process known as *sealant reversion*. Although the cause of this phenomenon is difficult to pinpoint, the necessary elements typically include high temperatures, ultraviolet light exposure, and prolonged exposure to water or water vapor working alone or in combination.[11]

Corrosion of Embedded Steel. Concrete, masonry, or stone facade components with mild steel reinforcement or embedments are prone to spalling as a result of corrosion. The by-products of corrosion (i.e., rust) can occupy up to 10 times the volume of the

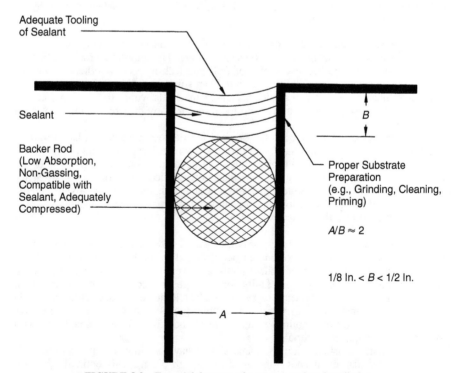

FIGURE 8.2   Essential features of a proper sealant installation.

original steel. This expansion of the steel creates high splitting/bursting stresses in the adjacent concrete, masonry, or stone that eventually can lead to cracking and spalling. Factors that increase the probability for corrosion of embedded steel include inadequate cover (distance from the embedded steel to the exposed surface), cracking or mortar separations (creating avenues for water penetration), carbonation of the concrete or mortar cover, and water-soluble chlorides in the concrete or mortar.

Carbonation is the process by which concrete or mortar slowly loses its alkalinity from a chemical reaction with carbon dioxide in the atmosphere. This process starts at the exposed concrete or mortar surface and progresses inward at a rate of approximately ⅛ to ¼ in. every 10 years in high-quality, normal-weight concrete. However, these rates usually are significantly greater for lightweight concrete and mortar. Factors tending to increase the depth and rate of carbonation include high porosity of the concrete, high relative humidity, cracks within the concrete, and exposure to higher atmospheric carbon dioxide concentrations owing to conditions such as urban or industrial environments or direct-fire heaters. Depending on the presence of these factors, the depth of carbonation may reach the embedded steel reinforcement or steel anchors, fostering corrosion of the steel and subsequent spalling of the concrete or masonry, as shown in Fig. 8.3.

Chlorides sometimes are added intentionally to the concrete or mortar during construction to accelerate setting in cold weather. Chlorides can be present in the concrete along coastal regions from environmental exposure or from brackish water used in the original mix water. Studies[12] have shown that acid-soluble chloride contents above about 0.2% by weight of cement (approximately 0.02% to 0.03% by weight of concrete) typically can promote corrosion of embedded steel, which otherwise is "protected" by the alkalinity of the portland cement paste. Exceptions to this corrosion threshold can occur if the chloride detected is from certain calcareous aggregates containing bound chloride because this chloride is not significantly water soluble.

Restrained Movement. Expansion and contraction of the facade material relative to the structural frame may lead to facade failure. These failures are related mainly to the thermal movements of exposed materials restrained by the sheltered structural

FIGURE 8.3   Concrete spall on architectural precast concrete facade panel.

frame, which is relatively free from thermal fluctuations. Moisture expansion of fired-clay products (brick, terra cotta, etc.) also can cause differential movement and resulting distress. Additionally, the building frame may move diametrically to the expanding facade from load-induced elastic deformation and shrinkage or creep of concrete columns.[13] Depending on the degree of restraint and rigidity of facade connections or supports, localized damage, cracking, bulging, or even collapse of the facade can result from restrained differential movement. The use of hard mortar (with a high concentration of portland cement) in masonry patches or in tuck pointing may make the problem worse by creating stiffer portions of the facade.

Precast concrete is also prone to distress from restrained movement if connections do not provide the degree of movement capability recommended by the Precast/Prestressed Concrete Institute.[14] Generally, separate connections should be employed to resist gravity and wind loads. Some well-intentioned erectors and even fabricators have used fully welded or bolted connections that may not provide sufficient flexibility to accommodate thermal or wind-induced movements, including interstory drift.

Marble Hysteresis. *Marble hysteresis* is a phenomenon that results when the marble calcite crystals fracture along their boundaries. Marble is composed of rhomboid-shaped crystalline calcite grains tightly interlocked with their neighboring grains. This complex crystal morphology leads to intergranular fractures from nonisotropic expansion and contraction. During thermal movements, the crystals tend to dislocate and not return to their original position, often resulting in significant strength loss. The dislocation or loosening of grains near the exposed surface causes "sugaring" (erosion) and potentially irreversible growth of the affected marble. When this occurs, thin marble panels tend to expand unevenly and bow. Thicker panels tend to resist bowing[15] by virtue of the additional restraint afforded by the larger proportion of marble that is not exposed to the elements.

Glass Breakage. Other than breakage resulting from vandalism, accidents, and the relatively rare occurrences of hurricanes, typhoons, tornadoes, and earthquakes, the two main glass failure mechanisms are spontaneous breakage and thermal energy breakage.

Although the term *spontaneous* would seem to apply to most types of glass breakage, it refers specifically to breakage caused by nickel sulfide impurities within glass. It is usually the mode of failure for unexplained breakage of tempered glass and older production heat-strengthened glass with an unusually high surface compression. Current production heat-strengthened glass should have a surface compression of less than 7500 lb/in.^{2} (54 MPa),[8] which generally precludes spontaneous glass breakage. Nickel sulfide may be introduced unintentionally into the batch through the raw materials or manufacturing equipment. Formerly, it also was introduced intentionally into certain tinted glass products.

Nickel sulfide impurities undergo a complete phase change during the production of annealed and heat-strengthened glass. For at least the last 30 years, the glass industry has been aware that in the production of tempered glass, the phase change of certain forms of relatively pure nickel sulfide impurities is incomplete. With subsequent cumulative exposure to heat, the nickel sulfide inclusion, or stone, as it is commonly called, increases in volume. If the stone, which usually ranges from about 4 to 80 mil (0.1 to 2 mm) in diameter, is located within the inner tension zone of the tempered glass, it may lead to spontaneous breakage. The amount of stored energy in tempered glass generally causes it to break into small pieces and evacuate the opening on application of even small pressures. By definition, the cumulative size of the largest 10 pieces of the broken tempered glass should be less than 10 in.^{2} (64.5 cm^2).[16]

Susceptible glass sheets located on sunny elevations of a building will exhibit spontaneous breakage earlier from increased exposure to heat. Most of the spontaneous breakage generally occurs within approximately 7 years following installation on a building facade[17] and seldom exceeds 1% of the tempered-glass sheets on a single building.

By the middle to late 1980s, the glass industry was issuing warnings about spontaneous breakage, with some manufacturers recommending against the use of tempered glass in exterior facades unless required by code. This latter recommendation appears prudent unless special steps are taken during manufacturing to minimize the chances that harmful inclusions are present or the tempered glass is laminated or otherwise restrained from falling (e.g., by anchored film) should breakage occur.

Thermal-energy breakage generally occurs when the central portion of a glass sheet becomes much warmer than its captured edges. The resulting thermal stress gradient in the glass creates high tension stresses along the edge that may exceed the edge strength. Factors that increase the likelihood of thermal breakage include edge defects; larger edge area; thicker glass; tinted, coated, or filmed glass; indoor shades or blinds; outdoor shading patterns; framing in contact with masonry or concrete; and an eastern or southern exposure.

Because of its lower edge strength, typically about 4000 lb/in.^{2} (28 MPa), annealed glass is the most susceptible to thermal breakage. Temperature differentials of 40 to 60°F (4 to 16°C) between the center and edge of annealed glass can cause breakage. Unless some of the factors listed earlier change materially, most of the thermal energy breakage occurs within the first 2 to 3 years of service, although some exceptions exist.[17]

Aesthetic Failures

Facade conditions that affect the overall appearance of the building potentially can have a great impact on the building's marketability. Since appearance and location are two key factors that attract tenants and prospective purchasers, it stands to reason that a condition that has marred the building facade may reduce the value of the structure substantially. Some of the more common aesthetic failures of building facades involve staining or etching of glass products, delaminated or discolored metal finishes, and "bleeding" sealants. Remedies for mitigating glass staining, metal coating deterioration, and sealant staining of adjacent porous materials are laborious and costly. In some cases, early detection and treatment can help to mitigate the damage.

Coating and Finish Degradation. Beginning in about 1980, fluoropolymer coatings applied over epoxy-based primers experienced delamination between the topcoat and primer. Fluoropolymer coatings spray-applied over the epoxy-based primers are particularly susceptible to failure, as shown in Fig. 8.4.

While fluoropolymer coatings in general are exceptionally durable, they will pass ultraviolet (UV) light. Where the topcoat is not sufficiently thick to protect the epoxy primer, chalking of the UV-sensitive epoxy and associated debonding of the topcoat occur. Studies have indicated that where the commonly specified minimum topcoat thickness of 1.0 mil (0.025 mm) was not achieved, the coating failed after approximately 10 years of exposure.[18] Other factors affecting the tendency of the fluoropolymer topcoat to debond include the proximity of the coated surface to reflective glass, which increases coating surface temperatures, and to standing water.

Prior to about 1975, most of the metal and glass curtain walls produced in the United States used an anodized finish on the aluminum framing members. Anodizing is a process of artificially producing a surface oxide film, which is thousands of times thicker than natural film. This thickened film produces a surface that is more resistant to weathering, corrosion, and abrasion. In the 1950s, the industry became aware that an anodic film thickness of 0.5 mil (0.013 mm) was inadequate in preventing aluminum

FIGURE 8.4   Debonding of spray-applied fluoropolymer coating.

corrosion in city environments. Subsequently, the standard minimum anodic film thickness was increased to 1.0 mil (0.025 mm).

A corrosion study performed on a landmark structure built in the 1960s in the midwestern United States revealed that corrosion, as shown in Fig. 8.5, was occurring primarily at locations where the anodic film thickness was measured to be less than 1 mil (0.025 mm).[18] Observations of this structure also followed the established patterns: The areas that received the least amount of rain washing or experienced extend periods of condensation were most susceptible to corrosion.

For anodized aluminum facade components that are not cleaned periodically, the accumulation of soot and grime and the acidic nature of condensation and dew may cause pitting corrosion. While corrosion of aluminum curtain wall components seldom affects structural integrity, the resulting unsightly pitted surface cannot be corrected by cleaning. In addition, the pits will continue to gather dirt and proliferate the corrosion process.

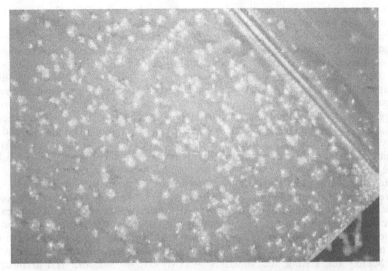

FIGURE 8.5   Pitting corrosion of anodized aluminum column cover.

Glass Etching and Sealant Staining.   A fairly common facade configuration, especially in low- to mid-rise construction, is one using glass in vision areas and precast concrete panels in spandrel areas. When subjected to the washing effect of rain water, alkalis can leach out of the concrete surfaces and come into contact with the glass. Alkaline materials allowed to remain in contact with glass will etch its surface permanently, as shown in Fig. 8.6. Initially, this condition may resemble water spots or

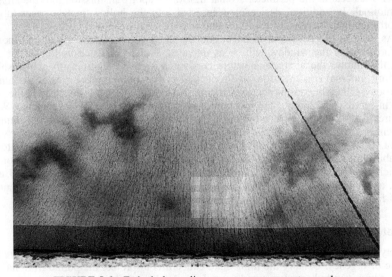

FIGURE 8.6   Etched glass adjacent to precast concrete panels.

stains on the glass. With continued exposure, the damage to the glass surface can become more severe and objectionable to the building occupants.

Facades constructed with the glass recessed inboard of the concrete surface or with drip strips or eyebrows at window head locations to divert surface runoff water usually will not experience this problem. Application of certain clear or opaque coatings will seal concrete surfaces, thus minimizing or eliminating leaching of harmful alkalis.

Light-colored porous stone cladding panels, such as granite, limestone, or marble, are particularly vulnerable to staining due to migration of sealant plasticizers or incompletely polymerized polymers. While staining has been observed with a variety of sealant types, staining from silicone sealants is most common. The amount and viscosity of the fluid migrating from the sealant largely determine the degree of stone staining. Silicones that contain more than about 13% silicone fluid are more likely to cause staining.[19]

Staining of the stone panels adjacent to sealant joints may result in wet or dark-colored bands also termed *oil-penetration stains,* dark streaking, or dry or light-colored bands also termed *halo stains* that occur often under wet conditions. Dark streaking is the result of airborne particles accumulating on areas where fluid migration has developed into lines or strips both above and below the sealant joint. Other than real-time testing to determine the propensity for staining, the most appropriate test is American Society for Testing and Materials (ASTM) C1248, *Standard Test Methods for Staining of Porous Substrates by Joint Sealants.* This standard, however, does not necessarily predict the development of staining over the long term.

Insulating Glass Unit Failures. Sealed insulating glass units (IGUs) are used extensively in all types of contemporary window construction. Typically consisting of two sheets of ¼-in.-thick (6 mm) glass separated by a ½-in.-thick (13 mm) sealed airspace, the units provide a significantly greater thermal resistance (R value) than monolithic sheets of glass.

The separation of the glass sheets is maintained by a hollow metal tube that runs continuously around the IGU perimeter between the glass sheets. The tube is adhered to the glass by sealant or a combination of sealants in an attempt to create a nearly impervious seal at the perimeter so that moisture vapor transmission into the airspace is minimal. In addition, desiccant is inserted into the hollow metal spacer tube to absorb nominal amounts of moisture that may penetrate the seal.

As long as the edge seal is intact and the airspace is dry, the visual and thermal performance of the window will be as intended. Failure or deterioration of the edge seal that allows the airspace moisture content to increase significantly will result in fogging (internal condensation) and reduction of the thermal properties of the unit.

Today, IGU fabrication processes and sealant capabilities have advanced to the point that in-service seal failures are less common than they were 10 to 15 years ago. When failures do occur, they are most often the result of an inadequate internal drainage system or barrier-wall seals that allow prolonged IGU edge exposure to water or water vapor. Ponding of water within the glazing cavity is particularly detrimental to the durability of the IGU edge seal. In addition, incompatibility of the glazing components (which may release solvents, curing by-products, or plasicizers) and the IGU edge seal, as well as perimeter edge wraps or bands that trap water, also can contribute to IGU failures.[20]

Safety-Related Failures

Safety-related failures of building facades are the most dramatic and create the greatest sense of urgency. Usually the immediate concern is to protect the public from hazardous or imminently hazardous conditions. Initially, this often involves restraining facade components from falling to the ground using netting, anchors, and sidewalk protection.

Subsequently, attention is focused on determining the extent of the problem and designing and implementing appropriate long-term repairs.

Structural and Connection Failure. As shown in Fig. 8.7, anchorage failures of facades can be spectacular. In this case, normal cyclic movements of the facade caused several overhead bolts to loosen and eventually dislodge, leaving a portion of the curtain wall with no lateral support to resist wind forces.

More common anchorage failures may be associated with precast concrete wall panels. Lack of consideration of building frame tolerances, unintentional restraint of thermal movements, creep and shrinkage of the panels, elastic shortening of the building frame, and connection eccentricities have resulted in anchorage failures of precast concrete panels.[21] Many of these issues can be addressed in the design phase by providing adjustable connections to accommodate construction tolerances and movements. As-built verification of the building frame coordinates prior to panel fabrication can reduce the number of required field modifications.

Many failures of precast concrete panel anchorages result from construction deficiencies, which typically involve a combination of improperly located or misaligned building-frame connections and improper field modifications. These field modifications may involve an excessive number of erection shims (Fig. 8.8) or inadequately welded connections (Fig. 8.9). Minimizing field modifications is advantageous because most are not engineered and may be performed by workers who do not fully understand the structural implications.

Occasionally, facade components are subject to wind loads and seismic movements that are higher than anticipated in design. An example of the former condition is a prominent high-rise building in the midwestern United States that experienced wind-induced glass failure near the building corners after construction. For structures built more than 15 to 20 years ago in earthquake regions, the facades may not accommodate

FIGURE 8.7 Mechanical anchorage failure of glass and aluminum curtain wall.

FIGURE 8.8 Misaligned shims and partially collapsed shim stack.

in-plane movements induced by earthquake ground motions properly. Even more recently constructed facades could experience localized failures when slip connections do not work as intended. Earthquakes can cause widespread damage to a variety of facade components, including glass, as shown in Fig. 8.10, often causing additional property damage or injury.

Hazardous Materials. Although not of specific concern in the structural assessment of building facades, the presence of hazardous materials should be taken into consideration, especially as it may pertain to sample removal for inspection or testing and

FIGURE 8.9 Inadequately welded precast panel connection.

FIGURE 8.10   Shard of glass that fell from a building during the Northridge earthquake in California.

possible remediation of certain facade conditions. It should be pointed out that toxicology is an inexact science. While acceptable exposure levels have been established for many materials (and are subject to change with additional data), there are other materials and organisms that are potentially hazardous to humans for which no regulations exist. Nevertheless, a brief description of some of the more prominent hazardous materials and their probable locations within a building facade is given below.

Older steel components (e.g., flashings, solder, and shelf angles) may contain or be coated with paints containing lead or other toxic metals that require special abatement, handling, and disposal for either repair or replacement. Vintage sealants (i.e., caulking or putty) used on facades may contain asbestos fibers or polychlorinated biphenyls (PCBs) that require special abatement and disposal. Some forms of cement board, siding, insulation, fireproofing, plaster, stucco, and gypsum board also contain asbestos fibers. These types of facade components should be tested for the presence of hazardous materials prior to implementing a preservation or rehabilitation program to determine proper abatement, handling, and disposal methods.

Products used in the cleaning or restoration of facades, as well as some coatings or adhesives, may contain volatile organic compounds (VOCs). As a result of their history as carcinogens or toxins, several states have enacted legislation to regulate VOCs. In cases where occupants complain of headaches, stomachaches, dizziness, or respiratory problems following recent facade work, products containing VOCs may be the cause.

In addition to water staining and damaged finishes, prolonged water leakage behind facades can lead to biologic hazards. Environments that provide sustained water and nutrient sources encourage colonization of bacteria and fungi such as mold. Some of these biologic particles can be harmful to building occupants, particularly if they are exposed to elevated levels for extended periods of time. Because of the limited data relating mold concentrations to health risks, there are currently no governmental regulations regarding permissible concentrations in indoor environments. Since mold and mold spores are present virtually everywhere as part of the natural environment, the complete eradication of mold in typical structures is not practical or warranted.

Over the past several years, more attention has been paid to the potential health hazards of mold within buildings. At this time, the engineering and medical communities are deliberating over what constitutes appropriate caution and proper counsel. In the future, a consensus response to mold may evolve. However, given the differing environments, variances in the susceptibility of individuals, conflicting medical opinions, and variables in mold growth itself, it is unlikely that an effective set of absolute rules will be developed.

FACADE SURVEY AND EVALUATION METHODS

Facade surveys may range from cursory visual inspection and photographic documentation from afar to close-up detailed examination, which may include testing, monitoring, or analysis of facade components and supports. The scope and depth of the facade survey depend on the impetus for the inspection. If a survey is needed to assess existing facade conditions without specific identified problems or failures (e.g., a prepurchase inspection or a baseline survey for insurance purposes prior to adjacent construction), a cursory visual inspection may be adequate. If the survey is part of an investigation to determine the cause of a complex failure or distressed facade conditions, careful planning and meticulous data- and sample-collection protocols usually are needed. Controlled testing and monitoring also may be part of a comprehensive investigation.

Document Review

Existing documents may help to establish a record of original design and construction, as well as prior maintenance or repairs subsequent to construction. Original construction documents may include project specifications and design drawings, change and field orders, shop drawings, or as-built details. Documents may be furnished by others or acquired by research. Building owners, managers, and maintenance staff are obvious sources of relevant documents. In the absence of written documentation, these parties should be interviewed to determine a history of the facade problems and related repairs. Records kept by the local building department are sometimes the only source of original construction documents.

An assessment of the quality of information contained in the document is required. Who produced the document? What was their interest? What is the completeness or quality of the supporting data? Published works may help to establish the standard of care of design or construction practices.

Condition Survey

The objectives of the condition survey include identifying the type, location, and extent of facade deterioration and distress; determining locations for subsequent close-up inspections, potential field tests, inspection openings, and removing samples; identifying potential causes of distress; and establishing potential scope and locations for repair work.

The initial exterior visual condition survey may involve recording conditions on copies of building elevation drawings and photographing distressed as well as typical conditions. In the absence of existing building elevation drawings, high-resolution photographs along with a detailed log can be used to record conditions. Later, the photographed overall views and details can be annotated manually or electronically. Binoculars and telephoto equipment often are used to facilitate observation and documentation of cracks, bulges, and spalls. An interior visual survey also should be performed if water leakage has been reported or if unfinished inner wall surfaces allow details of the exterior wall construction to be observed. The condition survey drawings (plans and elevations) or annotated photographs offer a valuable tool for visualizing the overall patterns of distress and identifying underlying causes (Fig. 8.11).

Close-up inspections of distressed and nondistressed facade areas will provide information related to the nature of the distress, the as-built construction (which may be contrary to the design drawings), conditions that may have existed prior to occurrence of the distress, and possible repair techniques. Notable conditions, measurements, and dimensions may be recorded on sketches of various facade components. Examination of subsurface wall elements and facade connections can be accomplished by creating inspection openings (removal of surface materials) or by observing conditions via an optical borescope or subminiature camera. Hidden or embedded metal elements can be located and possibly sized using metal detectors.

Safe access for close-up inspections may be from roof setbacks, balconies, pipe scaffolding, aerial lifts, rope descent systems, or swing stages. Contractor assistance usually is required to provide close-up access and to create and patch inspection openings.

Removal of samples for evidence or for laboratory testing also may be performed during the close-up inspection. Samples of visibly deteriorated or corroded elements

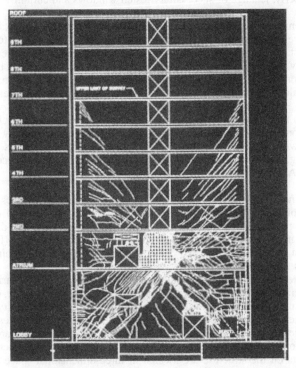

FIGURE 8.11   Annotated elevation drawings help to establish patterns of building facade distress.

and companion samples of apparently intact elements may be secured for comparison. Since the opportunity to collect samples may be limited in the future, taking more samples than needed usually is prudent. The quantity, locations, and types of samples taken will be based on the nature of the failure and on the preliminary hypotheses developed. Since the samples may need to be representative of the entire construction, sample selection needs to be of sufficient size and relatively random. Since purely random sample selection is seldom possible owing to limitations, such as access restrictions, sample bias is almost always present. The sample bias need not affect the test results adversely if the bias is anticipated and its implications fully understood.

Field Tests and Monitoring

A great number of tests and monitoring programs are available or adaptable for in situ assessment of existing building facades. Described below are some of the more common alternatives for field use.

Water Leakage Testing. The American Architectural Manufacturers Association[22] (AAMA) provides recommendations for test pressures (static air pressure) below which water leakage should not occur. These test pressures are based on the design pressure of the window or curtain wall elements and their classification. Windows that meet a minimum water resistance test pressure of 6 to 8 lb/ft^2 in the field typically do not have water leakage problems. However, with aging and deterioration of surface sealants and gaskets, the internal drainage systems of windows and curtain walls will be subjected to greater water exposure. If these internal systems have not been designed and installed properly, leakage of successfully tested windows and curtain walls will eventually occur.

When attempting to replicate and pinpoint the source of water leakage, most systems that have experienced water leakage in service will leak during testing without any applied static air pressure difference. Therefore, water testing often can be performed simply by using a spray rack (Fig. 8.12) conforming to ASTM E1105[23] without the need to construct a costly pressure chamber. When testing in this manner, the test duration should be extended as necessary beyond the 15 minutes prescribed in the standard.

In addition to the standardized water test methods, a 1.5-in.-diameter (40 mm) hose and nozzle connected to a building fire-suppression riser can be a very effective tool when faced with relatively inaccessible curtain wall surfaces such as barrel vaults, sloped glazing, and skylights. Once water leakage develops, the source can be pinpointed by AAMA 501.2[24] nozzle testing (Fig. 8.13).

Nondestructive Testing. The borescope, a fiberoptic device, enables direct visual examination of the hidden structural members or anchorages in wall cavities. This alternative to removing finishes and facade components uses a small-diameter $\frac{1}{4}$ or $\frac{3}{8}$ in. (6 or 8 mm) metal tube inserted through a small drilled opening (Figs. 8.14 and 8.15). Modern microminiature video cameras also can be used to examine underlying elements with minimum damage.

Infrared thermography uses special cameras that detect minute variations in the surface temperature of the object being imaged. Since internal cracking, delamination, trapped moisture, or deterioration often affects the thermal conductivity of the wall, such surface temperature variations can reveal underlying discontinuities.

Photogrametry is a technique in which high-quality photographs are taken from varying positions and digitized to create a stereoscopic view of the structure's surface. Selected points on the structure are located in three-dimensional space (usually with the aid of a reflective device called a *pentaprism*), enabling measurement of surface displacement. By repeating this process periodically, a record of progressing movement

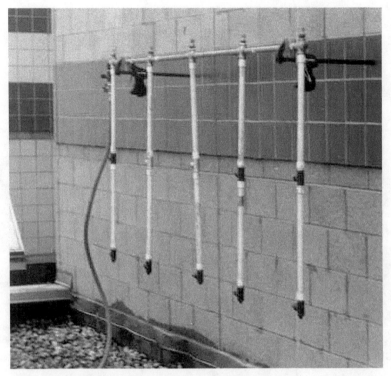

FIGURE 8.12   Water spray testing with grid of nozzles based on ASTM E1105.

FIGURE 8.13   Nozzle spray test based on AAMA 501.2.

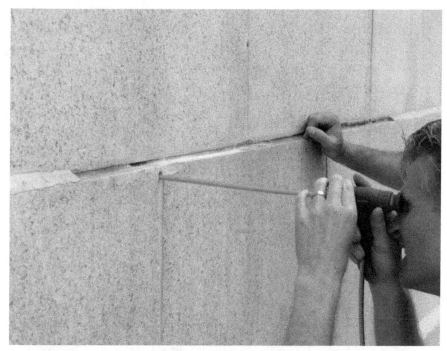

FIGURE 8.14 A fiberoptic borescope used to examine underlying facade support elements.

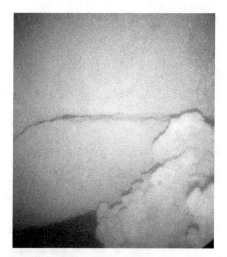

FIGURE 8.15 Fractured terra cotta support viewed through borescope.

can be obtained. The photographs also can serve to document existing conditions and to identify prospective repair locations.

Pulse velocity or impact echo techniques use short-duration, reproducible, low-frequency mechanical energy pulses introduced into the subject material. Flaws within the material disrupt or reflect the waves that are received by a surface transducer and are analyzed for pulse velocity or transient resonant frequencies. Response wave patterns may be examined to determine the soundness of the element and the character of internal flaws (Fig. 8.16).

The location of embedded steel reinforcement and steel support elements within a wall can be determined with a metal detector. This device also can be used to identify the depth of concrete cover and reinforcing bar size. The metal detector measures changes in electrical inductance to indicate the presence and configuration of a ferromagnetic material (Fig. 8.17).

Instrumentation and Monitoring. When assessing building facades, a great deal of information can be gained from measuring minute movements and strains over a period of time. Judiciously placed strain gages or displacement transducers (Figs. 8.18 and 8.19) can help to determine the behavior of the facade system under a variety of loading conditions. Stresses or displacements may be detected and correlated with wind or thermal loading effects. Some complex facade geometries or connection details may defy analysis and require in situ monitoring to determine actual behavior.

Laboratory Tests and Analyses

Numerous laboratory tests and analysis methods are available to help determine the properties of facade materials, the nature of deterioration, and the effectiveness of remedial measures. Following is a description of several common laboratory methods.

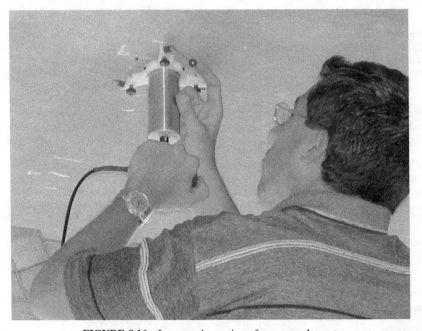

FIGURE 8.16   Impact echo testing of concrete element.

FIGURE 8.17   Metal detector used to locate and measure depth of embedded steel.

Petrographic Microscopy. This standardized microscopic examination of stone, concrete, brick, or mortar is based on the methods outlined in ASTM C856.[25] The objectives of a petrographic examination are to identify potentially deleterious conditions, such as unsound or reactive aggregate, unhydrated cement, microfractures, and carbonation of concrete or mortar. Petrographic examination is also used to determine the cement content, water-cement ratio, percent of entrained air, characteristics of the air void system, and degree of consolidation of concrete or mortar, as well as the overall matrix composition of stone. In general, petrography assesses the overall quality and soundness of the stone, brick, concrete, or mortar.

Chemical and Spectroscopic Analyses. Methods involving wet chemicals, Fourier transform infrared spectroscopy (FTIR), x-ray diffraction (XRD), scanning electron microscopy (SEM)/energy dispersive spectroscopy (EDS), atomic absorption spectroscopy (AA), UV visible spectroscopy (UV-vis), and electrochemical analyses can be used to evaluate a wide variety of facade materials. For example, one test procedure outlined in ASTM C1152[26] and ASTM C1218[27] is employed frequently to determine the chloride ion content and profile (variation in chloride concentration with distance from the surface) in concrete or mortar. The resulting information is used to quantitatively evaluate the potential for corrosion of embedded steel elements.

Physical Property Tests. Properties of materials, such as compressive or tensile strength, modulus of elasticity, freeze-thaw susceptibility, water absorption, coefficients

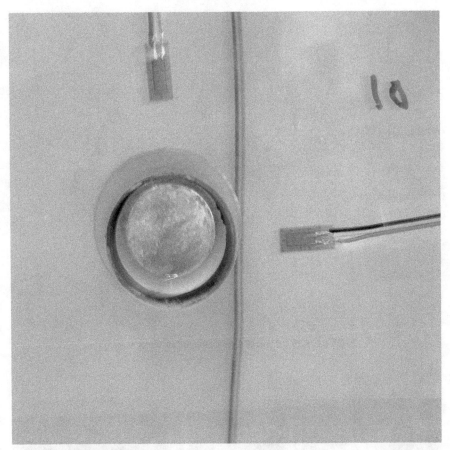

FIGURE 8.18   Strain gages for determining stress levels in glass adjacent to support.

of thermal or moisture expansion, and flexural strength (Fig. 8.20), are used to establish relevant characteristics of facade components. For most types of facades, cores or prism samples are removed from the wall and tested in the laboratory in general accordance with appropriate ASTM and other empirical standards. In the case of glass assessment, practical physical tests are limited; general principles of glass fractography, as outlined in ASTM C1256,[28] are used postmortem to evaluate the cause of failure.

Mock-Ups and Models. Testing of custom mock-ups and models may be very useful to understand the behavior of complex building facade systems. Service loads (whether static, cyclic, or dynamic) can be replicated while monitoring displacements, strains, and overall behavior in a controlled environment. The model can be tested to failure or to a selected proof load, which typically is mandated by codes or standards to be 1.5 to 2.5 times the design loads. Tests may focus on connections or facade components of concern. Actual facade elements preferably should be used, but comparable materials may be substituted if removal is not practical. Duplicating an actual failure

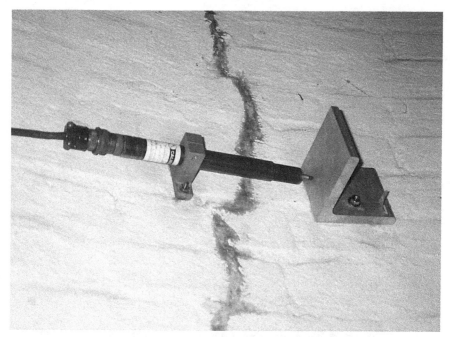

FIGURE 8.19 Typical displacement transducer (LVDT).

FIGURE 8.20 Flexural strength test of marble specimen.

with a mock-up of the as-built elements offers a convincing verification of a failure hypothesis.

Structural Analysis

A mathematical model of a facade component or wall assembly may be beneficial to determine its theoretical capacity to resist loads or its load deflection characteristics. Such analyses frequently are used to determine if the component or wall assembly meets the requirements of codes and standards. The analysis may involve finite-element computer models or hand calculations. Caution must be exercised with interpretation of computer analyses because accuracy of the results typically depends on many assumptions. Other inaccuracies can occur when inelastic deformations or large deflections are modeled with analytical methods that are based on linear elastic behavior and small deflection theory.

Following the initial structural analysis of a portion of the facade, one of three conclusions is drawn with respect to code conformance: (1) the facade has an adequate factor of safety, (2) the facade is grossly overloaded and the calculated factor of safety is approximately 1 or less, or (3) the facade possesses a factor of safety greater than 1 but less than that required by code.[29] If analytical results point to the second conclusion but the facade exhibits no significant distress, the analysis should be examined closely for errors or erroneous assumptions. If results point to the third conclusion, a refined analysis or a physical load test may be performed in an attempt to prove the adequacy of the facade in question.

Frequently, load tests on in situ facade components/assemblies or a laboratory mock-up reveals capacities that exceed those determined by analysis. This phenomenon occurs owing to real-world factors (e.g., mechanical properties in excess of the published minimum values and complex imprecise load paths) and complex redistribution of forces resulting from conditions such as local yielding and nonlinear behavior, which typically are not considered in the analytical process. As a result, load tests and mock-ups sometimes are conducted to prove the adequacy of components and assemblies that are analytically marginally deficient.

STANDARDS AND CODE REQUIREMENTS

The information contained in standards and codes can be important to the proper assessment of building facades, particularly in cases involving litigation. Building codes and standards usually contain the *minimum* requirements (i.e., minimum levels of safety) for the *design and construction* of building facades and related components. These minimum requirements are arbitrary and subject to change. The codes and standards also may contain procedures and loading criteria for proof testing of assemblies. What codes and standards typically do not provide is information related to the cause of facade distress or failures and the means of assessing the residual strength of such components. Many engineers and architects mistakenly have deduced a cause for failure merely on the basis that certain facade elements did not meet the design criteria of a certain building code or standard.

That which is not contained in standards or codes—namely, a keen understanding of the material and structural behavior of facades and facade components—is more critical for determining existing or potential failure mechanisms, as well as the residual strength of distressed facade components. Frequently, such an understanding is developed through a combination of field, laboratory, and testing experience. Without this understanding, erroneous assessments of building facades may result from unreasonable assumptions.

When it is necessary to distinguish between possible design and construction deficiencies, codes and standards governing the design and construction of the subject structure should be examined. These codes and standards usually will be referenced in the project specifications, if available; otherwise, the codes in effect in the period immediately preceding permitting or construction are assumed to apply. When significant deficiencies are determined using the codes and standards governing the design and construction of the subject structure, the required remedial work typically is designed to meet the current codes and standards. Depending on the scope of the remedial work, codes also may require upgrading entire facade assemblies to meet current standards.

Wind loading is frequently the most critical design loading for facades. ASCE 7, *Minimum Design Loads for Buildings and Other Structures,*[30] is a nationally recognized consensus standard and perhaps the most accurate treatment of wind loads available in a standard format. The ASCE 7 standard has been adopted or referenced by many national and local building codes. It should be noted that code bodies often do not adopt the most recent version of standards owing to the slow process of approving code changes. Therefore, governing design criteria for many facades frequently are based on older and sometimes outdated bodies of knowledge.

General standards for the evaluation and testing of a variety of materials, including many facade components, are presented in Part IV, "Evaluation and Testing of Structural Materials and Assemblies." Some specific standards are included as references to this chapter.

REFERENCES

1. Perry, J. (2001). "Unwrapping the Industry—Perspectives on Building Envelope Design and Delivery," in *Proceedings of the International Conference on Building Envelope Systems and Technologies (ICBEST)*. National Research Council Canada and Institute for Research in Construction, Ottawa, Ontario Canada.

2. Beasley, K., H. Greenspan, J. O'Kon, and J. Ward (1989). *Failure Investigation.* American Society of Civil Engineers, New York.

3. Beasley, K. (1998). "Contemporary Skin Barrier Wall Design and Water Leakage," in *The Construction Specifier*. Construction Specifications Institute, Alexandria, VA.

4. SEI/ASCE 30-00 (2000). *Guideline for Condition Assessment of the Building Envelope.* American Society of Civil Engineers, Reston, VA.

5. Jester, T. (ed.) (1995). *Twentieth-Century Building Materials: History and Conservation.* National Park Service, Washington, D.C.

6. Kaskel, B., Wonneberger, and S. Bortz (2001). "Freeze-Thaw Durability of Cast Stone," in *Concrete International*. American Concrete Institute, Farmington Hills, MI.

7. Lewis, M. (1995). *Modern Stone Cladding: Design and Installation of Exterior Dimension Stone Systems.* ASTM Manual 21, American Society for Testing and Materials, West Conshohocken, PA.

8. ASTM C1048-97b (1997). *Standard Specification for Heat-Treated Flat Glass—Kind HS, Kind FT Coated and Uncoated Glass.* American Society for Testing and Materials, West Conshohocken, PA.

9. Nugent, W., and M. Schmidt (2001). "Failures of Modern High-Rise Building Facades and Components," in *Safety, Risk and Reliability—Trends in Engineering Conference Report.* International Association for Bridge and Structural Engineering, Valetta, Malta.

10. Beasley, K. (1997). "Trouble Prone Architectural Features, Masonry Walls Likely to Leak," in *The Construction Specifier*. Construction Specifications Institute, Alexandria, VA.

11. NiCastro, D. (1997). *Failure Mechanisms.* American Society of Civil Engineers, New York.

12. ACI 222RT-01 (2001). *Protection of Metals in Concrete Against Corrosion.* American Concrete Institute, Farmington Hills, MI.

13. Beasley, K. (1987). "Masonry Cladding Stress Failures in Older Buildings," *Journal of Performance of Constucted Facilities* Vol. 1, No. 4, pp. 229–238.

14. *PCI Design Handbook,* 5th ed., Precast/Prestressed Concrete Institute, Chicago, IL, 1999.

15. Tschegg, W., C. Wildham and W. Eppensteiner (1997). "Accoustic Emissions and Anisotropic Expansion When Heating Marble," *Journal of Performance of Constructed Facilities* Vol. 11, pp. 35–41.

16. ANSI Z97.1-1984(R1994) (1994). *American National Standard for Safety Glazing Materials Used in Buildings—Safety Performance Specifications and Methods of Test*, American National Standards Institute, New York.

17. Schmidt, M. (2001). "Glass—Why It Breaks," in *Proceedings of the International Conference on Building Envelope Systems and Technologies (ICBEST)*, National Research Council Canada and Institute for Research in Construction, Ottawa, Ontario, Canada.

18. Nugent. W., and M. Schmidt (2000). "Preventing Failures in Metal and Glass Facades," in *Proceedings of the Second Congress on Forensic Engineering*. American Society of Civil Engineers, San Juan, Puerto Rico.

19. Scheffler, M., and J. Connolly (1996). "History of Building Joint Sealants," in *Science and Technology of Building Seal, Sealants, Glazing and Waterproofing*, Vol. 5. ASTM STP 1271, American Society for Testing and Materials, West Conshohocken, PA.

20. SIGMA TM-3000(97) (1997). *Glazing Guidelines for Sealed Insulation Glass Units*. Sealed Insulating Glass Manufacturers Association, Chicago.

21. Popovic, P., and R. Arnold (2000). "Preventing Failures of Precast Concrete Facade Panels and Their Connections," in *Proceedings of the Second Congress on Forensic Engineering*. American Society of Civil Engineers, San Juan, Puerto Rico.

22. AAMA MCWM-1-89 (1989). *Metal Curtain Wall Manual*. American Architectural Manufacturers Association Technical Information Center, Schaumburg, IL (reissued 2003).

23. ASTM E1105-00 (2000). *Standard Test Method for Field Determination of Water Penetration of Installed Exterior Windows, Curtain Walls, and Doors by Uniform and Cyclic Static Air Pressure Difference*. American Society for Testing and Materials, West Conshohocken, PA.

24. "AAMA 501.2-03 (2003). *Quality Assurance and Diagnostic Water Leakage Field Check of Installed Storefronts, Curtain Walls and Sloped Glazing Systems*. American Architectural Manufacturers Association Technical Information Center, Schaumburg, IL.

25. ASTM C856-95e1 (1995). *Petrographic Examination of Hardened Concrete*. American Society for Testing and Materials, West Conshohocken, PA.

26. ASTM C1152/c1152M-97 (1997). *Acid-Soluble Chloride in Mortar and Concrete*. American Society for Testing and Materials, West Conshohocken, PA.

27. ASTM C1218/C1218M-99 (1999). *Standard Test Method for Water-Soluble Chloride in Mortar and Concrete.* American Society for Testing and Materials, West Conshohocken, PA.

28. ASTM 1256-93 (1998). *Interpreting Glass Fracture Surface Features*. American Society for Testing and Materials, West Conshohocken, PA.

29. Department of Housing and Urban Development (1983). *Rehabilitation Guidelines 1982 9: Guidelines for Structural Assessment*. HUD-0002958, Office of Policy Development and Research, Washington, D.C.

30. ASCE 7-98 (2000). *Minimum Design Loads for Buildings and Other Structures*. American Society of Civil Engineers, Reston, VA.

CHAPTER 9

Parking Structures

PREDRAG L. POPOVIC, P.E., S.E., JAMES P. DONNELLY, P.E., S.E.,
and BRIAN E. PULVER, P.E.

INTRODUCTION

Although they are classified and constructed as buildings, parking structures are unique. When not enclosed, they are subject to ambient weather conditions, which may vary widely based on the geographic location. In cold climates, they are often exposed to snow, ice, and water, as well as to the corrosive action of deicing salts. Unlike a bridge deck, the inside of a parking structure is not rinsed by rain, and its exposure to chlorides may be aggravated by poor drainage.

Parking structures are subjected primarily to loads from moving vehicles, and their roof levels are exposed to weather similar to bridge decks. Since they are frequently very large in plan, they experience greater volume changes (temperature, shrinkage, and creep) than enclosed structures, which usually are smaller and are exposed to more uniform temperatures, humidity, and moisture. Restraint of these volume changes can cause cracking of garage structural members, which allows for ingress of water and chlorides, leading to accelerated deterioration.

Parking structure slabs usually are constructed of concrete and are supported by concrete beams and columns, except in rare instances where steel framing is used. The functional and traffic requirements dictate that the open-floor spans be in a range of approximately 55 to 60 ft (16 to 18 m) to allow concurrently for two traffic lanes and two parking lanes. The access to different floors sets the floor height, ramp slopes, and turning radii, whereas the building code requirements affect the magnitude of loads and fire protection and ventilation criteria.

Because of the longer lengths of parking structures, volume-change movements have to be accommodated and designed for. Properly specified and designed expansion joints will reduce the effect of volume changes. Also, because of the longer spans of the floor framing system and the moving concentrated loads, the magnitude of floor deflection and vibration will vary based on the structural system. A typical parking structure is shown in Fig. 9.1.

All these factors make parking structures different from other types of building structures, and assessment of the condition of parking structures has to take these factors into account.

CODES AND STANDARDS

Although parking structures are a distinct type of structure, there are no codes that specifically address parking structures alone. As such, the design and evaluation of a parking structure must be based on the appropriate national standards and codes and the local building code. Since parking garages almost invariably are constructed with concrete, the general provisions of the American Concrete Institute's (ACI) *Building Code Requirements for Structural Concrete* (ACI 318)[1] apply, with those involving concrete cover and other durability requirements being of particular value. The appropriate code from the American Institute of Steel Construction (AISC) also would apply in a steel-framed parking structure. General provisions regarding the assessment of structures can be found in the *Guidelines for Structural Condition Assessment of Existing Buildings,*[2] by the American Society of Civil Engineers (ASCE).

The general provisions of the appropriate local building code also apply. Although codes will vary from locale to locale, the most relevant provisions are those involving design loadings. In addition to the dead load of the structure, the codes usually specify 50-lb/ft^2 (2.40-kN/m^2) live load, which may be reduced in some instances based on the tributary area of the floor slab or the beam supporting the area. However, a typical garage floor fully packed with cars bumper to bumper will receive only about 30-lb/ft^2 (1.44-kN/m^2) load. The floor slabs also usually are checked for a concentrated 2000-lb (8.9-kN) load to simulate the effect of a car wheel. The roof level also should

FIGURE 9.1   Long, open parking structures experience greater volume changes.

be designed for additional snow load, if applicable. The seismic and wind loads will be the same as for any other building.

Lateral impact of an automobile in a garage is modeled with a single horizontal ultimate load of 10,000 lb (44.5 kN) applied at a height of 18 in. (0.46 m) above the driving surface and distributed over a 1-ft^2 (0.1-m^2) area. This load is used to check the lateral barriers along the perimeters of floors.

In addition to the building codes, several other documents are applicable to the design and evaluation of parking structures. The ACI committee on parking structures has produced two valuable documents, *Guide for Structural Maintenance of Parking Structures* (ACI 362.2R-00)[3] and *Guide for the Design of Durable Parking Structures* (ACI 362.1R-97),[4] which soon will be replaced by the forthcoming *Standard Practice for the Design and Construction of Durable Concrete Parking Structures.* These documents all contain valuable information unique to parking structures. The National Parking Association has produced a *Parking Garage Maintenance Manual*[5] that contains general information about the upkeep and deterioration of these facilities. Technical information directly applicable to parking structures is also available from trade associations such as the Post-Tensioning Institute and the Precast/Prestressed Concrete Institute, each of which has produced guides for the design and construction of a particular type of parking structure.[6,7]

A book entitled, *Parking Structures: Planning, Design, Construction, Maintenance, and Repair,*[8] is another source on parking structures. Many articles and papers about specific parking structure projects also have been published, and several of them are listed in references at the end of this chapter.[9–12]

STRUCTURAL SYSTEMS

As for other concrete structures, parking garages generally are constructed using one of three types of structural systems: cast-in-place conventionally reinforced concrete, cast-in-place posttensioned concrete, or precast concrete. A fourth type of construction involving structural steel framing elements in conjunction with a cast-in-place concrete floor slab is a less common structural system for parking structures. Precast concrete and cast-in-place posttensioned concrete are currently the most common types of construction for new free-standing parking structures. These two structural systems can readily accommodate the 55- to 60-ft (16- to 18-m) spans desired in most parking structures. Conventionally reinforced cast-in-place and steel-framed parking garages usually are found in mixed-use structures where the parking garage is located beneath or adjacent to a building, and its design is thereby constrained by the structural system selected for the remainder of the structure.

Cast-In-Place Conventionally Reinforced Concrete Systems

Conventionally reinforced concrete parking structures typically use steel reinforcing bars with a yield stress of 60,000 lb/in^2 (414 MPa) as the embedded reinforcing, although 40,000 lb/in.^{2} (276 MPa) steel has been used in some older structures. This type of construction offers significant flexibility in terms of the available structural systems. The various structural systems that have been used historically include one-way joist systems, two-way joist (or waffle-slab) systems (Fig. 9.2), two-way flat-slab systems (Fig. 9.3), and one-way slab systems (Fig. 9.4). Two-way flat-slab systems and one-way joist systems are probably the most common, but since conventionally reinforced concrete parking garages often are constructed as a part of multiuse developments, the structural floor system in the parking structure may be based on what structural system is used elsewhere in the project.

One drawback to conventionally reinforced concrete floor systems is that they cannot span as great a distance or may require larger-sized members as compared with

FIGURE 9.2 Overall view of a typical two-way conventionally reinforced waffle slab parking structure.

FIGURE 9.3 Overall view of a typical two-way conventionally reinforced flat-slab parking structure.

FIGURE 9.4 Overall view of a one-way slab parking garage.

those built with precast or posttensioned concrete. As a result, parking structures built with conventionally reinforced concrete floor systems generally have maximum column spacings in the range of 30 ft (9 m), thereby necessitating columns in the area of the parking stalls.

Cast-in-Place Posttensioned Concrete Systems

Posttensioned concrete structures consist of cast-in-place concrete with high-strength steel prestressing tendons providing all or a major portion of the structure's reinforcing. These tendons are stressed and anchored at the ends of the structural elements and at intermediate construction joints. The steel tendons usually are covered with a protective corrosion-resistant greaselike coating and contained inside a protective plastic sheathing.

This type of structure also offers a great degree of flexibility in terms of the structural systems that can be used for garage decks. Many of the structural systems used in conventionally reinforced cast-in-place concrete parking garages also can be used in posttensioned concrete garages. The most common of these systems in newer posttensioned concrete garages is a one-way posttensioned slab spanning between posttensioned beams. These slabs are relatively thin, typically 5 to 7 in. (130 mm to 180 mm) and generally span 15 to 25 ft (5 to 8 m) between floor beams. The floor beams are generally 2 to 3 ft (0.6 to 0.9 m) deep and 12 to 18 in. (0.3 to 0.45 m) wide, enabling them to span 60 ft (18 m) across a typical bay. The beams typically are supported by conventionally reinforced concrete columns located along the edges of the structural bays.

Other structural floor systems that have been used frequently in cast-in-place posttensioned concrete parking garages include two-way flat slab systems and one-way joist systems. As with one-way slab systems, the structural elements in these systems tend to be thinner than in conventionally reinforced concrete structures of the same configuration. One disadvantage of two-way systems is that they cannot accommodate longer spans, so some columns are required in the areas of parking stalls.

Precast/Prestressed Concrete Systems

Precast and prestressed concrete systems are composed of multiple concrete elements manufactured prior to erection of the structure. Most precast concrete parking structures have similar structural systems for the garage levels, typically consisting of multiple 8- to 12-ft-wide (2.4- to 3.6-m) precast, prestressed concrete elements that span across 50- to 60-ft-wide (15- to 18-m) structural bays. For newer garages, these main deck elements almost always are double tees ranging from 18 to 32 in. (0.45- to 0.81 m) deep, but single tees and deeper hollow-core planks have been used in some older garages. To provide the driving surface, these precast elements are covered by a concrete topping, either cast in place after erection or applied at the precasting plant (pretopped).

To support the ends of the main deck elements, inverted-tee or L-shaped ledger beams usually are provided, with the ends of the deck elements supported on bearing pads positioned on the beam ledges. Along interior columns lines where the adjacent bays are at the same elevation, 2- to 3-ft-deep (0.6 to 0.9 m) inverted-tee beams generally are used. L-shaped ledger beams usually are located along exterior column lines or at interior column lines where the elevation of the two adjacent bays differs. As a result, the L-shaped ledger beams almost always are made tall enough to serve as barrier walls, typically 4 to 5 ft (1.2 to 1.5 m). On the perimeter of a garage structure, pocketed spandrel beams sometimes are used in lieu of ledger beams to reduce the amount of rotation that results from the eccentric load applied by the deck elements bearing on the ledge. These supporting elements, like the double-tee deck elements, usually are prestressed, except in some instances where the spans along a column line are short enough that conventional reinforcing can be used. An interior view of a typical precast parking structure is shown in Fig. 9.5.

In a precast parking structure, the vertical load-bearing elements are generally columns with brackets or corbels fabricated into the column section. The brackets and corbels provide a bearing seat to accept the ledger and spandrel beams. Bearing pads consisting of neoprene or another elastomeric material are provided at these bearing locations. One alternative to columns is wall panels, which also can be provided with corbels or bearing pockets to support the beams and/or deck framing elements. Although wall panels usually are found at the stairwells, they also have been used to support deck element ends in lieu of ledger beams at interior column lines.

Systems with Structural Steel Framing

Some parking structures have been built using structural steel elements as the structural framing system. In such a structure, rolled-steel sections are used as the columns and beams to support a conventionally reinforced concrete floor slab. Steel joists also can be incorporated, spanning between beams at a spacing as little as 3 ft (0.9 m) on center, resulting in a thinner slab section. The concrete floor slab usually is designed as a one-way slab spanning between adjacent steel beams or joists, typically with stay-in-place metal decking used to form the underside of the concrete floor. Figure 9.6 shows a steel-framed parking structure with closely spaced steel joists supporting the metal decking.

FIGURE 9.5   Overall view of a typical precast, prestressed concrete parking structure.

FIGURE 9.6   Overall view of a typical steel-framed parking structure.

TYPICAL DETERIORATION AND DISTRESS

Because of their direct exposure to the environment and to vehicular traffic, parking structures can be subject to many different types of deterioration. The main sources of deterioration include corrosion of embedded reinforcing, cracking, degradation of the concrete itself, or surface-related deterioration resulting from the exposure of the concrete to the environment. These deterioration mechanisms are described below.

Corrosion-Related Distress

Perhaps the most common form of deterioration found in concrete parking structures is delamination and spalling of the concrete surface resulting from corrosion of the embedded reinforcing steel. The highly alkaline environment provided by the cement paste in concrete protects the embedded reinforcing steel by forming a passivating layer around the steel. However, if a sufficient quantity of chloride ions is able to enter the concrete and penetrate to the level of the embedded steel, this passivating layer will break down. A chloride ion content of about 0.2% by weight of cement (0.03% by weight of concrete) is accepted commonly as the corrosion threshold for concrete. With a chloride content above this threshold at the level of the reinforcing steel, the corrosion reaction can proceed if sufficient amounts of moisture and oxygen are present. All parking structures in northern climates where deicing salts are used are susceptible to this form of deterioration on account of the presence of chlorides in most deicing chemicals. Chloride-induced corrosion also can occur in areas where the structure is exposed to salt through the air or groundwater, as occurs near large bodies of salt water.

As reinforcing steel, prestressing/posttensioning strand, and other embedded metals begin to corrode, corrosion by-products (rust) are formed on the surface of the corroding metal. These corrosion by-products generally are five to eight times larger in volume than the metal from which they form. As the rust expands, pressures build up within the concrete until the tensile stress of the concrete is exceeded, and a crack forms. For fairly widely spaced reinforcing bars or those located near a corner of a structural element, the crack will tend to emanate diagonally from the reinforcing bar toward the nearest face of the structural element. A diagram of a typical concrete spall is shown in Fig. 9.7.

For closely spaced reinforcing bars, the crack will tend to propagate in the plane of the reinforcing, thereby separating the concrete cover from the concrete beneath. Separation of the concrete cover from the base concrete by such planar cracks is commonly referred to as a *delamination*. As the corrosion process continues and corrosion by-products expand further, the delaminated concrete will become loose and eventually fall off the concrete structure, resulting in spalls or potholes. A diagram of typical delaminations and potholes in a parking structure is provided in Fig. 9.8. Vibrations from vehicular traffic can help to expedite spalling of the concrete in parking garages once delaminations begin to form. An area of spalled and delaminated concrete on the top surface of a parking deck floor slab is shown in Fig. 9.9.

Of particular concern is the loss in structural load-carrying capacity that occurs during the corrosion process as a result of the loss of reinforcing cross-sectional area and a decrease in bond between the reinforcing and the concrete. A related problem in structures built with unbonded posttensioning is that moisture can enter the tendon sheathing and initiate corrosion of the posttensioning steel. Because of the annular space within the sheathing, corrosion of the tendon may not create a delamination or spall, hindering discovery of this deterioration. In an unbonded posttensioned system, the failure of a corroded tendon will render it ineffective for its entire length, unlike conventional reinforcing, which is only affected in the region of section loss because of continual bond to the surrounding concrete.

Corrosion of the embedded reinforcing also can result from carbonation of the concrete. Carbonation is a natural reaction between the cement paste and carbon dioxide in the air that causes calcium carbonate to form in the concrete, reducing the alkalinity of the concrete to a level that allows the passivating layer around the reinforcing steel to break down. Fortunately, this reaction is quite slow and can take many years to advance into the concrete to the level of the reinforcing steel. As a result, carbonation-related corrosion usually is limited to older structures with fairly permeable concrete, such as lightweight concrete, or with minimal cover over the reinforcing. Carbonation should not be a problem for many years for newer parking structures constructed with denser concretes.

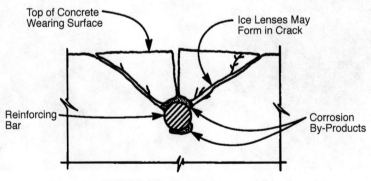

FIGURE 9.7 Typical concrete spall.

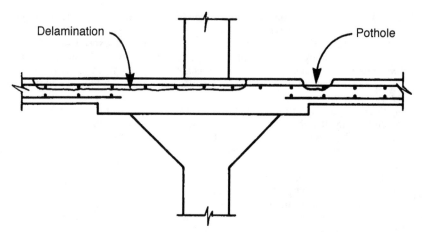

FIGURE 9.8   Typical deck delamination and pothole.

Cracking

Cracking of one form or another occurs in most concrete structures. The formation of cracks can have a significant effect on the durability of a parking structure because it provides moisture on the deck surface with an easy path to enter the concrete and reach the level of the reinforcing steel. In areas where deicing salts are used or that are near the sea, chloride ions can be washed into the cracks with the moisture. This

FIGURE 9.9   Spalling and delamination of a top deck surface.

can result in the premature corrosion of the embedded steel. At cracks, the passivating layer that protects the reinforcing steel is discontinuous, allowing corrosion to occur locally without the presence of chlorides.

There are many potential causes for cracking in a concrete structure, although all generally are associated with tensile stresses that exceed the cracking strength of the concrete. One possible source of such cracking in parking structures is applied load. This is particularly true for slabs and beams in a conventionally reinforced concrete structure because there is no precompression in the concrete as in posttensioned structures or prestressed structural elements. In such members, the tensile strength of the concrete often is exceeded in the tensile zone of the member when loaded, which is necessary to allow the reinforcing steel to participate fully in resisting the applied load. Because of exposure to moisture and vehicular traffic on their top deck surfaces, such cracking creates a problem in negative-moment regions of parking structures, where cracking will occur on the top surface of the concrete deck, thereby allowing moisture and deicing salts to enter the concrete. Loadings that exceed the design loads also may result in cracking. Figure 9.10 shows an example of cracking on the top surface of a parking deck slab.

Another significant potential source of cracking in a concrete parking structure is shrinkage of the concrete, particularly drying shrinkage and plastic shrinkage soon after construction. *Drying shrinkage* occurs as the excess moisture in new concrete evaporates, resulting in a slight decrease in the overall volume of the concrete. Over 90% of such shrinkage occurs in the first 1 to 3 years of a concrete structure's life. Restraint of this contraction can be expected to cause cracking, which often extends through the entire thickness of a structural element. *Plastic shrinkage* is also associated with drying of the concrete but only occurs during the brief period of time that the concrete is in a plastic state immediately after placement. For plastic shrinkage to occur, water must be evaporating from the surface of the concrete faster than it can

FIGURE 9.10   Cracking on a top-slab surface around a column.

be replenished by bleed water, causing the surface to dry out and contract relative to the concrete below the surface, thereby resulting in fairly shallow cracks in the top layer of concrete. Proper mix design, placement, finishing, and curing techniques can prevent plastic shrinkage cracking and help to reduce drying shrinkage cracks.

The restraint of volume changes such as those resulting from drying shrinkage or thermal changes is a potential source of cracking in a concrete structure. This is particularly true in posttensioned concrete structures, which also are subject to volume changes from creep and elastic shortening. Parking structures are particularly susceptible to volume-change issues because they tend to be rather long and wide structures. Furthermore, structural elements with sufficient lateral stiffness to resist movement caused by volume changes, such as walls, stairwells, larger columns, etc., are often located near the corners of parking structures, where such movement is greatest. Rigid connections between the floor system and laterally stiff structural elements can be expected to result in cracking due to restrained volume changes (Fig. 9.11).

Material-Related Degradation

Deterioration due to improper, inadequate, or nondurable components in the concrete also has been observed in many concrete parking structures. In older structures, cyclic freezing is a common source of such deterioration. Other material-related deterioration mechanisms that sometimes are found in parking structures include alkali-silica reaction, sulfate attack, and delayed ettringite formation. Nondurable aggregates also can result in premature concrete deterioration. Although these types of deterioration are not unique to parking structures, their direct exposure to moisture and temperature changes make parking structures particularly susceptible to these mechanisms.

FIGURE 9.11 Full-depth slab crack at corner of structure resulting from restraint of volume changes.

Cyclic Freezing. Damage from repeated freezing and thawing, also known as *freeze-thaw damage,* occurs when concrete without proper air-entraining freezes repeatedly in a saturated condition. When water freezes, it expands. When this phenomenon occurs within a hardened material such as concrete, localized expansive pressures can develop in the concrete around the water particles, which can result in the initiation of microcracks and eventually cause disintegration of the concrete matrix if this occurs repeatedly over the long term (Fig. 9.12).

The presence of air-entrainment can prevent this problem by providing additional voids within the concrete into which the water can expand when it freezes. As a result, older concrete parking structures built prior to the advent of air-entraining agents most likely are susceptible to this form of deterioration.

Alkali-Silica Reaction. Alkali-silica reaction (ASR) can occur when a reactive aggregate such as a chert is used in the concrete. In the presence of moisture, such aggregates can react with the alkali in the cement to form a gel-type product that is larger in volume than the components of which it is made. As a result, expansive pressures build up within the concrete. With continued growth of the alkali-silica gel, microfracturing of the concrete will occur. This type of degradation usually manifests itself as a series of parallel planar cracks within the concrete or areas of disintegration of the concrete matrix.

Sulfate Attack and Delayed Ettringite Formation. Sulfate attack involves the introduction of sulfate from an external source into the hardened concrete, which then reacts with the cement paste to form a compound known as *ettringite.* This compound, like alkali-silica gel, is larger in volume than the components from which it is formed, thereby creating expansive pressures within the concrete as it forms and grows under

FIGURE 9.12   Freeze-thaw damage of a concrete column.

further exposure to moisture. Delayed ettringite formation (DEF) is similar to sulfate attack in the way it works, but in DEF, the sulfate comes from within the concrete, specifically from excess sulfur in the cement. Otherwise, both mechanisms are similar, requiring the presence of sulfur and sufficient moisture to proceed. Since parking garages are exposed directly to moisture, they can be susceptible to these two types of deterioration, although reported cases in parking structures are fairly rare. Sulfate attack most likely would be found in a parking structure built below grade in a soil with a high sulfate content, whereas DEF most likely would be found in a precast structure where a cement with a high sulfur content in the cement clinker was used.

Nondurable Aggregates. Coarse aggregates that are fairly weak or susceptible to cyclic freezing or wear-related deterioration also can result in premature degradation of a concrete. Such deterioration can be prevented during the original construction by careful selection of the aggregate source used in the structural concrete.

Surface Durability

Other deterioration that parking structures face includes surface-related degradation mechanisms such as scaling and abrasion or wear. Scaling involves the disintegration of the cement paste at the surface of the concrete, usually in conjunction with cyclic freezing. This results in a rough, pockmarked top surface of the concrete and has been observed to result in a loss of the surface concrete for a depth of up to 1 in. (Fig. 9.13).

In a number of instances this has been attributed to poor finishing practices that resulted in a weaker, low-air-content, high-water-cement-ratio material at the surface. Premature exposure of new concrete to chloride-based deicing chemicals also can be a factor by creating high osmotic pressures in the uppermost portion of the concrete before the concrete is fully cured.

FIGURE 9.13   Scaled concrete surface.

Abrasion or wear of the concrete surface has been observed in some instances, resulting in a gradual wearing away of the concrete surface. Such deterioration generally will be limited to locations where large amounts of vehicular traffic occur, particularly if the concrete is somewhat weaker. A strong, properly finished and cured concrete with an appropriate mix design would not be readily susceptible to this type of deterioration, except in areas where tire chains may be used.

Deterioration in Steel-Framed Structures

The exposed framing members in a steel-framed parking structure can be very susceptible to corrosion-related deterioration because of their exposure to water and chlorides. This deterioration is similar to reinforcing bars that corrode in cracked concrete because they are exposed directly to moisture and are not surrounded by concrete with its associated passivating layer and alkaline environment. As such, it is important that the steel members in a steel-framed parking structure receive adequate protection, usually in the form of a protective coating. This is similar to the methods used to protect steel bridge structures.

Where stay-in-place steel form decking is used, additional deterioration of the concrete floor slab and decking may occur. This is caused by moisture moving through the floor slab, particularly at cracks, which then becomes trapped and collects on the top surface of the metal decking. This can result in unsightly corrosion of the metal decking, which may not be structurally significant if the concrete slab is reinforced adequately. Trapped moisture and chlorides also can result in increased deterioration of the concrete floor slab by increasing the moisture content in the concrete and providing a source of chlorides to begin the corrosion process for the bottom reinforcing.

Deterioration at Expansion/Construction Joints

Because parking structures tend to be wider and longer than many buildings, they usually are built with several construction joints, and expansion joints commonly are required to minimize the amount of volume-change-related movement occurring at any given location. Many such joints are not detailed properly or sealed adequately to prevent moisture from entering the concrete beneath. As a result, one location where all parking structures are particularly susceptible to deterioration is the area immediately adjacent to or below expansion and construction joints. Long-term regular maintenance of the joint seals and sealants is required. If such seals not maintained properly, sections of the structure underside can be exposed to moisture and chlorides. The resulting deterioration is typically in the form of corrosion-induced cracking, delaminations, and spalling, although other moisture-related deterioration mechanisms also can occur (Fig. 9.14).

CONDITION SURVEY

Several methods can be used to assess the condition of a parking garage structure. The methods selected will vary from structure to structure and project to project depending on the extent and type of deterioration, the information required, and the desired end result of the investigation. The following subsections describe the various elements of a typical condition survey.

Design Drawing and Report Review

The first step of a condition survey should be to review available design or as-built drawings and any maintenance or inspection reports. Design drawings show the intended construction and indicate expected behavior and load paths. Problematic con-

FIGURE 9.14 Deterioration of a concrete structure along the underside of a leaking expansion joint.

struction details can be identified. The drawings can be used during the field survey to compare with the actual construction. Maintenance and inspection reports may indicate maintenance and performance problems, which can help focus the investigation. Both drawings and reports can assist in developing plans and checklists for the field survey.

Visual Survey

The visual survey is one of the most useful methods for an experienced engineer to assess the condition of the structure. The initial information obtained during this survey helps to direct the investigation. The visual survey allows the engineer to get an overall view of the structure's condition quickly and to identify representative areas or locations for more thorough or specific investigative techniques or materials testing. An inspection checklist should be developed prior to beginning the visual survey and should be revised as necessary while the survey progresses. Checklist items should focus on the objectives of the investigation, the information desired from the investigation, and individual tasks that need to be accomplished to obtain this desired information.

Documentation. If design drawings are available, it is very useful to make reproductions of plan views and elevations of the structure for field-survey base sheets prior to performing the visual survey. If no design drawings are available, sketches showing column lines and approximate dimensions should be made before beginning the survey. Notes of the type, location, and quantity of deterioration, along with other documentation such as photograph numbers and inspection opening locations, should be placed on the field sheets. The field sheets with the areas and extent of the observed deterioration shown in approximate scale then can be reviewed to ascertain general conclusions or trends regarding the structure's condition.

Crack Evaluation. Concrete structures almost always crack. Typically, these cracks are due to drying shrinkage of the concrete, thermal-induced expansion and contraction of structural elements that are restrained, or loading of the structure. Often cracks are not an indication of any structural problems but require repair and maintenance to improve durability. However, crack patterns may indicate deficiencies in the structure. These deficiencies could be in the original design or the result of improper construction.

All cracks of concern should be noted on the field sheets in their approximate locations, showing the general orientation and length of the cracks. The size of the cracks can be measured using crack comparators, which vary from clear plastic cards to handheld magnifiers that have lines of varying widths on them against which the cracks can be compared.

Cracks of structural concern include but are not limited to

- Wide or opening cracks in regions of high bending moments or shear forces
- Slab cracks emanating radially around columns, indicating possible punching shear problems
- Cracks in precast, prestressed concrete connection and support elements such as corbels, ledger beams, and double-tee beam stems
- Linear cracks corresponding to prestressing or posttensioning tendons
- X cracks in columns, especially short columns
- Leaking cracks

If cracks are present that could be due to a structural deficiency, then additional testing, investigation, or analysis should be performed to determine the nature of this deficiency so that the structure can be repaired or strengthened properly.

Delamination Detection

The quantity of delaminations can vary significantly in different areas of the structure. For instance, at the entrance to a parking garage, the concentration of deicing salts and chloride ions is usually much higher than on enclosed or upper floors. A delamination survey of the whole parking structure is required to determine the location, distribution, and quantities of delaminations. If only a partial delamination survey can be performed, the areas to be surveyed must be representative of the various exposure and service conditions.

Several methods can be used to detect delaminations in a garage. These range from physically sounding the structure to more high-tech methods such as infrared thermography and impact echo. Each of these methods has its advantages and disadvantages.

Sounding. The most tried and true method for detecting delaminations involves physically sounding the structure. For slabs, this method usually involves dragging a chain across the top surface of the slab (Fig. 9.15). In nondelaminated areas, the chain produces a clear ringing sound. When an area of delamination is encountered, a distinct drumlike hollow sound is produced by the chain because of the presence of a near-surface planar crack within the slab cross section.

For vertical and overhead surfaces, delaminations typically are detected by striking the structure with a hammer or steel rod, as shown in Fig. 9.16. Such overhead surveys usually are focused around areas where corrosion, moisture, and mineral staining are observed.

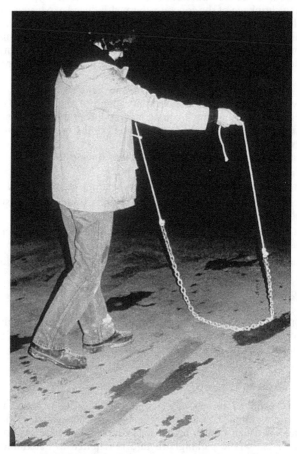

FIGURE 9.15   Chain-drag delamination survey.

FIGURE 9.16   Sounding slab underside using a steel rod near areas of spalled concrete.

Specialized Delamination Detection. If delaminations are present deep below the surface of the structure, a chain-drag or hammer-tap survey may not be effective in detecting them. One method that can be used to detect deep delaminations is a nondestructive sonic wave technique known as *impact-echo testing*. Impact-echo testing uses mechanical energy in the form of a short pulse, or wave, that is introduced into a structural element to detect flaws, discontinuities, or other anomalies within concrete members. A transducer is placed on the surface of the structure, and the surface is impacted mechanically, usually with a small ball bearing. This impact produces surface, shear, and longitudinal waves in the concrete. The longitudinal wave propagates into the concrete until it reaches the back surface of the structure and reflects back toward the impacted surface, where the transducer detects the return wave. If internal discontinuities are present, the longitudinal wave energy is reflected from these imperfections or discontinuities, resulting in multiple or earlier detection of the return wave. A schematic diagram of this testing method is shown in Fig. 9.17.

This method requires determining the propagation velocity of the concrete, which allows the reflected waves to be analyzed with a fast Fourier transform (FFT) analyzer

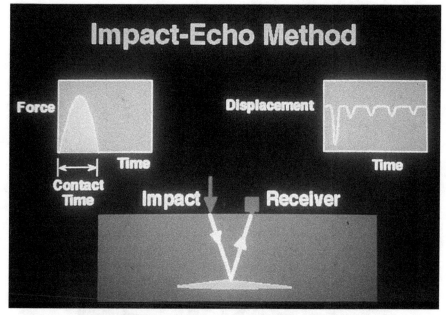

FIGURE 9.17   Schematic diagram of the impact-echo method of delamination detection.

to determine the internal characteristics of the concrete and therefore the approximate thickness of a structural element, or depth and presence of a planar defect (delamination). Actual thickness or depth to an internal discontinuity should be confirmed either by selected direct verification or by coring or drilling through the thickness of the structure. The method provides valuable data at a localized area near the test site and requires a relatively frequent grid of test points in order to evaluate an entire slab or member. This results in the method being somewhat labor-intensive.

Infrared thermography is another specialized method that can be used to detect delaminations in an exposed deck of a parking structure. This technique employs a thermally sensitive camera that is directed at portions of the structure to detect temperature gradients within different areas of the viewed surfaces. If no delaminations are present in the structure, the surface temperature of the structure is fairly uniform. However, during either natural or forced heating or cooling of the structure, the surface temperature of delaminated concrete will change faster than that of the adjacent sound concrete and will be detected by the infrared camera as a higher or lower temperature zone. Typically, infrared measurements of garage decks are performed either on a sunny day with the sun striking directly on the structure to detect how quickly the structure heats up or at night after the sun has been striking the surface to detect how the structure cools down.

A distinct advantage of this method is that large areas can be surveyed fairly quickly. This method can be very useful for parking decks that are covered with an asphalt overlay, where performing a chain-drag survey or impact-echo test would not be possible because the concrete surface is not accessible. Drawbacks to this method include its being somewhat dependent on the weather and the need for the test area to experience some time when the temperature is changing (either by exposure to radiant heating from the sun or nightly cooling cycles). Additionally, the determination of delaminations can be questionable if the concrete is covered by an asphalt overlay, so the location and extent of delaminations need to be verified by coring or drilling into the structure.

Nondestructive Testing

Several nondestructive methods can be used to determine selected conditions in a parking structure. It is advantageous to use nondestructive methods to determine these conditions because these methods are generally noninvasive and do not damage the structure. Several of these methods are described below.

Reinforcing-Bar Locators. Reinforcing-bar locators magnetically find and measure the concrete cover for reinforcing bars embedded in the structure. The use of a reinforcing bar locator is shown in Fig. 9.18. The size of the reinforcing bars needs to be known to measure the concrete cover. This technique is effective when it is necessary to analyze the existing load-carrying strength of the structure because the reinforcing bars at several locations can be surveyed relatively quickly without damaging the

FIGURE 9.18 Magnetically locating and measuring concrete cover for embedded reinforcing bars.

structure. For deeper reinforcing bars, the accuracy of the concrete cover readings is reduced, so the concrete cover readings for this condition should be determined by either electromagnetic (ground-penetrating radar) testing or by drilling.

Radiographic Techniques. In some cases it is critically important to know the locations of the embedded reinforcing. One method that can be used to determine this is radiography. An isotope source is located on one side of the member to be examined, and sensitive film is secured to the backside. During exposure, the steel and concrete will absorb differing amounts of the emitted waves due to differences in their densities, resulting in visible indications of the position of the steel on development of the film. Once an entire area has been surveyed, these images can be put together to show where the reinforcing bars are located. This technique is very useful for locating post-tensioning tendons and reinforcing bars in concrete without excavation. Special site precautions are needed when using this technique.

Ground-Penetrating Radar. This method is an effective nondestructive method that is used principally to locate steel reinforcing bars or posttensioning or prestressing tendons in a structure without core drilling, chipping, or other invasive techniques. In general, ground-penetrating radar (GPR) detects the arrival time and energy level of a reflected electromagnetic pulse. Since electromagnetic wave propagation is affected by changes in dielectric properties, variations in the condition and configuration of a structure will cause changes in the signal. Information is obtained by observing the return time, amplitude, shape, and polarity of the signal. In concrete structures, GPR is used commonly to find voids under on-grade slabs, areas of honeycombing, the locations of embedded conduits, anchors, and reinforcing bars, etc. When calibrated properly, GPR also can be used to measure the thickness of a concrete member.

Half-Cell Potential Corrosion Survey. It can be advantageous to determine the corrosion activity through the electrical potential of mild reinforcing steel within a structure. Electrical potential indicates the probability of corrosion activity and usually is measured using a copper–copper sulfate half-cell. This half-cell is essentially an open-ended tube containing a copper rod and a saturated solution of copper sulfate, with a porous plug at one end of the tube. A schematic diagram of this testing method is shown in Fig. 9.19.

Readings are obtained by placing the porous plug on top of a series of electrical junction devices, typically wet, soapy sponges arranged in a grid over the area in question, to create electrical contact with the concrete. A voltmeter is connected to an existing embedded uncoated reinforcing bar and to the half-cell. The voltage displayed by the voltmeter is the electrical potential difference between the half-cell and the reinforcing bar. A contour plot can be developed showing the different potential readings, which can be used to identify anodic (corroding) areas and cathodic (noncorroding) areas. American Society for Testing and Materials (ASTM) C876-91[13] describes this procedure and states that there is a 90% probability that corrosion is occurring if the $CuCuSO_4$ potential is numerically greater than -0.35 V. If the measured electrical potential is numerically less than -0.20 V, then there is a 90% probability that corrosion is not occurring. For potentials between -0.20 and -0.35 V, the extent of the corrosion activity is highly variable. The steepness of the potential contours is also a good indicator of corrosion activity.

Corrosion Rate Testing. While corrosion potential testing allows rapid determination of areas where corrosion is likely, corrosion rate testing, using the polarization resistance technique, returns a corrosion rate value. Unlike corrosion potential readings, the corrosion rate is a quantitative measurement of the amount of steel corroding at

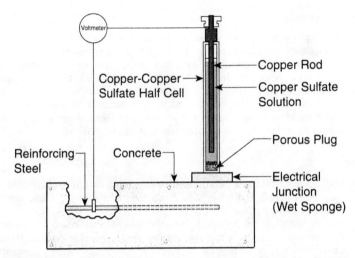

FIGURE 9.19 Schematic diagram of copper–copper sulfate half-cell testing.

the time when the measurement is taken. The measured current even can be converted to the amount of section loss per year using Farley's equation. The equipment used to measure the corrosion rate also can measure corrosion potential and concrete resistivity. Corrosion rate measurement is much slower than corrosion potential measurements, and it can take 2 to 5 minutes to complete one reading.

Rebound Hammers. A rebound hammer, typically called a *Schmidt* or *Swiss hammer*, is a device that measures surface hardness, which can be used as a relative indicator of concrete strength. The end of the hammer's plunger is placed in contact with the surface of the concrete, and as the body of the hammer is pushed toward the concrete, a spring-loaded mass is released, resulting in an impact through the plunger onto the concrete surface. The rebound of the mass is measured by the device and is displayed as a nondimensional number. Several readings need to be taken in a grid pattern at each location and averaged together. This rebound number often is misused by investigators because it is a poor predictor of actual concrete strength. The rebound numbers are proportional to changes in near-surface concrete strength and therefore can serve as a quick method to identify zones of relatively higher- or lower-strength material. Often the hammer is used to locate areas where subsequent core samples are taken and compressive strength is measured directly by testing of those samples.

Concrete Moisture Testing. For corrosion to occur, there must be moisture in the concrete. Moisture testing that measures the humidity in the concrete and the concrete's resistivity can indicate if there is sufficient moisture for corrosion to occur or continue. Two methods typically are used to measure the concrete's humidity. In one, a hole is drilled in the concrete, a humidity probe is inserted, and the humidity is measured directly over a period of time. Another method involves collecting powder samples at various depths in the concrete and measuring the humidity of the powder in the laboratory.

Exploratory Openings

In some cases exploratory openings need to be made in the structure, perhaps to verify the results from nondestructive testing, to inspect posttensioning tendons and other reinforcing, to examine the condition of buried waterproofing membranes and their terminations, or to expose portions of the structure that are otherwise covered with overlays for delamination surveys. An example of an exploratory opening is shown in Fig. 9.20. Information from the visual survey, such as areas of corrosion or mineral staining, should be used to help select the locations of these openings. All inspection openings should be repaired once the inspection is completed.

Sampling

Other methods to determine the condition of a parking structure require that samples be taken from the structure and brought to a laboratory for testing. These samples can be taken in several different ways.

Concrete core samples can be taken from the structure and used to determine the compressive strength of the concrete, to determine the chloride content through the elements' cross sections, or to provide a sample of the concrete for petrographic examination. The cores are removed from the structure using a coring rig, which uses various diameter diamond-tipped core barrels to cut out cylindrical concrete core samples. A core hole at a sampling location is shown in Fig. 9.21. A reinforcing-bar locator or other testing techniques should be used to locate the embedded reinforcing so that the reinforcing is not damaged by the coring process.

One method that can be used to determine the chloride content in a concrete structure is to take powder samples from the structure. Powder samples are taken by drilling

FIGURE 9.20 Exploratory opening to examine condition of concrete deck and waterproofing membrane under an asphalt topping.

FIGURE 9.21   Core hole at a slab delamination.

into the structure at different locations and collecting this powder at the various depths where determination of the chloride content is desired. Care should be taken to not contaminate the samples that are to be tested with powder from other locations. Core samples are preferred for accurate determination of chloride ion profiles.

Laboratory Testing

To determine the strength and condition of a structure, laboratory tests often need to be performed. The most common of these tests for parking structures include compressive strength testing, chloride ion testing, petrographic analysis, and carbonation testing. Which tests are selected depends on the deterioration observed and the information that is needed.

Compressive Strength Testing. When concrete core samples are removed from the structure, they can be taken to a testing laboratory and tested in compression to verify that the concrete has the expected compressive strength. It is ideal to take cores that will have a 2:1 length-to-diameter ratio. If this ratio cannot be achieved, then ASTM C42[14] provides correction factors that can be used to adjust the compressive strength obtained from the compression test.

Chloride Ion Content. Another test that can indicate the susceptibility of a concrete structure to future corrosion is to measure the chloride content through the depth of a concrete member cross section. When the chloride ion content reaches a concentration in the range of 0.2% by weight of cement, corrosion can begin. Samples for testing can be either powder samples or concrete core samples. At a given sampling location, chloride content should be determined for at least three different depths within the structure cross section, with one of the samples being taken deep in the concrete to be used as a baseline chloride ion content. Typically, other test locations should correspond to the level of existing reinforcing bars. When calculating chloride diffusion coefficients, slicing cores at a minimum of four to five depths is recommended.

Petrographic Analysis. A petrographic analysis can help to determine the overall soundness of the concrete and involves examining a core or chunk sample microscopically, as shown in Fig. 9.22. This examination can help to determine approximate air content, water-cementitous ratio, aggregate distribution, depth of carbonation, surface-finish characteristics, and susceptibility to various degradation mechanisms. Additionally, any material problems or incompatibilities can be determined. Sometimes petrography also can indicate the need for additional testing, so it is helpful if it is performed in the beginning of the laboratory testing process. A highly trained and experienced petrographer is needed for this evaluation.

Carbonation Testing. Determining the depth of carbonation in a concrete structure is relatively simple. The testing can be performed in the laboratory or the field, and it

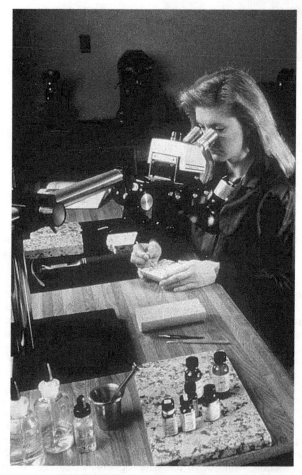

FIGURE 9.22 Petrographic examination.

is performed by applying a pH indicator such as phenolphthalein to a freshly fractured concrete surface. Areas of the concrete where the phenolphthalein was applied that do not turn pink are carbonated. This test generally is performed in the laboratory on core samples. In the field, the depth of carbonation can be determined by drilling a hole in the concrete, cleaning the powder out of the hole, and putting the pH indicator on the sides of the hole.

STRUCTURAL ANALYSIS AND LOAD TESTING

Parking structures are analyzed as any other structure using standard methods of structural analysis. However, the application and analysis of the live loads are different. The application of the live loads and the analysis are of a particular importance for structural systems with thin concrete slabs subjected to concentrated wheel loads.

Structural Analysis

When the original structural design drawings are available in a routine structural assessment, a typical garage bay is checked for the design loading. This analysis will use design data shown on the drawings, including dimensions of structural members, specified compressive strength of the concrete, yield strength of reinforcing steel, and the position of the reinforcing inside the concrete structural members. This analysis will result in determination of the as-designed capacity of structural members.

It is very rare that the actual strength of concrete and the location of reinforcing steel in the as-built structure are the same as shown in design drawings. In most structures, the actual strength of concrete is higher than the specified strength because the compressive strength of concrete generally increases over time. The location of reinforcing bars inside concrete structural members is never the same as shown on the design drawings. Smaller variations in the amount of concrete cover and location of the reinforcing bars can be expected and generally would not affect the structural capacity of concrete members significantly. However, in two-way slab systems and thin one-way slabs, the mislocation of reinforcing bars could result in a significant reduction in the as-built load-carrying capacity of the slabs and their connections to the beams and columns. For example, misplacement of the top reinforcing layer in a 6-in. (150-mm) thick slab where concrete cover increases from 1 to 3 in. (25 to 75 mm) results in reduction of structural capacity of more than 40%. Therefore, in order to assess the actual structural capacity of an existing parking structure, the actual dimensions and concrete strengths of the structural members and the as-built location of the reinforcing steel should be used. These data will be obtained by field investigation and testing.

When different repair options for a parking structure are considered, an additional structural analysis may be required that will take into account the weight and capacity of the repaired structure. For example, if the placement of a concrete overlay over the top surface of an existing garage slab as a protection option is considered, an analysis may show that the structure will be loaded beyond its as-built capacity. This may require the use of a thin waterproofing membrane system as a protection option during consideration of various repair options.

Load Testing

Sometimes it is difficult to apply analytic methods to portions of a parking structure to assess its strength and load-carrying capacity. For example, it is almost impossible to properly model the connection of a posttensioned concrete girder and reinforced concrete column where posttensioned tendon anchorages are embedded between vertical and horizontal steel reinforcing bars in the column. Furthermore, analytic methods cannot fully account for the variation in spacing and location of each individual reinforcing bar embedded in concrete slab, variation in concrete strength, presence of cracks in concrete members, and the possible presence of delaminations in some areas of concrete slabs.

To determine if a parking structure can carry the required load safely, a load test could be performed. Because of the cost and engineering expertise required, load tests are not performed routinely as part of the structural evaluation of parking structures.

The test load is specified by the governing building code. Concrete structures are tested for the load specified by the ACI building code (ACI 318),[1] which specifies that a structure has to withstand a total load of $0.85(1.4D + 1.7L)$, where D is the dead load and L is the design live load. The test load will be the difference between the total load and the dead load, which is the weight of the structure to be tested. For the usual ratio of dead to live loads, the test load usually is in the range of $1.5L$. Some building codes specify test load in a range of $2.0L$.

The required test load usually is applied by loading the structure with the required weight spread out in a pattern to achieve the load equivalent to a uniform test load. Water, sand, or concrete blocks often are used to apply the test loads, as shown in Fig. 9.23. Sometimes it is appropriate to use hydraulic jacks to load test the parking structure,[15] an example of which can be seen in Fig. 9.24. Prior to application of the test load, the structure is evaluated analytically and shoring is provided to prevent a collapse of the structure in case of overload during the test.

The test load usually is applied in several increments and is left in place for 24 hours. The portion of the structure tested should show no evidence of failure, including cracking and spalling of concrete or excessive deflections. Building codes specify the maximum acceptable deflection and the residual deflection after removal of the test load. For reinforced-concrete structures, the residual deflection should be less than 25% of the maximum deflection.

FIGURE 9.23 Load testing of a parking structure using water in a pattern loading.

FIGURE 9.24 Load testing of a ledger beam in a precast concrete parking structure using hydraulic jacks to react against the structure above as a jacking reaction block.

Load tests should be designed and performed under the supervision of structural engineers experienced in load testing parking structures. Appropriate safety provisions should be observed at all times during the load test.

EVALUATION AND REPAIR OPTIONS

Once the field work, laboratory testing, and analysis are completed, all the data must be compiled, summarized, and evaluated and appropriate repairs selected.

Correlation and Reduction of Data Obtained

All field sheets and notes should be reviewed, and the types of deterioration should be tabulated to determine the extent and amount of deterioration, with each type of deterioration summarized for each level or area of the garage. This information can be used to determine deterioration trends and the worst areas of deterioration and also can be used later to track the repair quantities and costs during the repair construction.

The summarized deterioration should be compared with the results from the materials testing and analysis to make an overall assessment of the structure. Testing such as chloride ion analysis, half-cell testing, or corrosion rate testing can provide a snapshot of where the structure is in the deterioration process, which can be confirmed by the amount of delamination detected during the condition survey. Additionally, this information regarding the ongoing corrosion and corrosion potential can assist in estimating how much the delamination quantities will increase over time.

Effect of Deterioration on the Structure

Delaminations, spalls, low concrete strength, and incorrect construction practices reduce the load-carrying capacity of a structure. In most cases, delaminations and spalls are formed by corrosion of the embedded reinforcing. As the corrosion occurs, the cross-sectional area of the reinforcement decreases, along with the bond between the reinforcing and the concrete, and this reduces the load-carrying capacity of the structure.

Cracks in the concrete contribute to the formation of delaminations by permitting water and oxygen to reach the reinforcement, allowing corrosion to begin. Cracks also can be an indication of a structural deficiency.

Deterioration resulting from ASR, DEF, high chloride ion content, insufficient air entrainment, nondurable aggregate, etc. is more of a durability or serviceability problem for the structure, and can lead to capacity loss if these items are not addressed or the structure is not protected.

Once the field work has been performed, the extent of deterioration should be reviewed to see how the structure might be affected and what repairs or protection methods need to be performed. Each structure will have different types of deterioration and issues that will have to be evaluated as part of each condition assessment. For instance, the location of the deterioration, the condition of conventional or posttensioning reinforcing, the location of the reinforcing within the member cross section, the strength of the concrete, and the analysis results can be used to determine what effect the deterioration has on the structure. If the analysis indicates that the design was proper, the field work shows that the structure basically was constructed as designed, and the deterioration is localized, then there is less concern regarding the concrete deterioration affecting the overall integrity of the structure.

However, if the design or construction of the structure was inadequate, or if the structure was designed with a reduced safety factor, then deficiencies in the structure may have a greater impact on the overall capacity of the structure. In some cases, the deterioration is significant enough that it is necessary to shore portions of the structure or reduce the loads on the affected areas until repairs can be performed.

Repair Option Development

There are several standard concrete repairs. Top-of-slab, vertical surface, and overhead surface delaminations typically are repaired by removing deteriorated portions of the member cross section, including completely removing the concrete from around the exposed existing reinforcing steel until sound concrete is encountered. This is called a *partial-depth repair.* If the slab is thin or the bottom- and top-of-slab delaminations overlap, sometimes these repairs will extend through the full depth of the slab. With respect to posttensioned or prestressed concrete, standard concrete repairs cannot be performed alone. Since the steel tendons are the main reinforcement for such structures, the load-carrying capacity of the structure is reduced if one or more tendons are broken in a given portion of the structure. Therefore, the tendons need to be repaired in addition to repairing the delaminated concrete.

Several options also should be considered regarding improving the longevity and serviceability of existing and repaired portions of the structure. For example, if the moisture content and chloride ion content in the concrete structure are high enough to promote future corrosion of the embedded reinforcing steel, then moisture ingress needs to be halted. Therefore, a sealer, waterproofing membrane system, or breathable protective coating should be applied to the surface of the structure in addition to performing the concrete repairs.

The type of waterproofing method selected depends on the parking garage's structural system and the portion of the structure that needs to be protected. For instance, if a portion of the structure that needs to be protected is on grade, then an impermeable traffic-bearing waterproofing membrane should not be installed. This is so because moisture from the soil subgrade will continue to migrate out of the slab, and the moisture will be trapped in the concrete, causing the concrete to deteriorate and the membrane to debond. For this application, a sealer should be installed because it is not impermeable. However, if a sealer is installed, it will not stop the ingress of moisture completely. All these issues need to be considered when deciding which repair and protection methods will be used.

Another method that can be used to reduce the ingress of moisture into a structure is the placement of a dense concrete overlay. If this option is selected, a structural analysis should be performed to make sure that the structure will not be overloaded with the additional dead load from the overlay.

Because new non-chloride-contaminated repair concrete is being placed adjacent to the chloride-contaminated concrete and the existing reinforcing bars connect these two concretes, a new corrosion cell can be set up between the new and existing concrete around the perimeter of the repairs. This *ring-anode effect* can be reduced by cutting off the water supply, but it cannot be stopped completely. Therefore, some future delamination formation can be expected. One method that can be used to reduce the ring-anode effect involves placing sacrificial anodes along the perimeter of the repairs and connecting them to the existing reinforcing bars.

In some cases when the chloride content is very high, all the chloride-contaminated concrete may need to be removed. This is often accomplished by either milling or hydrodemolition, along with the placement of a dense concrete overlay that also will provide future protection of the structure.

Another consideration for concrete with high chloride ion contents is chloride extraction. This involves imposing an electric current on the structure for several days or weeks to electrochemically draw chloride ions out of the concrete. Because of its expense, this option makes the most sense for structural elements that cannot be taken out of service readily.

REPORT

A written report should be prepared to describe the investigative and testing work that was performed, the findings resulting from this work, the repairs required to address the observed deterioration, the costs associated with those repairs, and the anticipated service life of the structure.

Summary of Findings

Each type of deterioration observed during the condition survey should be described, quantified, and supplemented with photographs. Detailed survey notes and test results, if included in the report, should be placed in an appendix.

Review of Repair Options, Costs, and Anticipated Service Life

The repairs necessary to address the observed deterioration need to be discussed. The anticipated quantity of these repairs needs to be increased from the condition survey to account for future growth of the repairs up to and during repair construction. These repair types and quantities should be listed in a table along with estimated repair costs based on previous repair projects of a similar size performed in the geographic area of the garage studied. All project costs, including a contingency, should be incorporated in this repair table.

Additionally, the anticipated service life of the structure can be estimated based on past projects, the amount of repairs to be performed, and the expected amount of future deterioration. Even if all the deterioration is repaired and a membrane is placed over the structure, if the chloride content is above the 0.03% by weight of concrete corrosion threshold, some corrosion may continue, particularly in areas immediately adjacent to repairs, a phenomenon known as the *ring-anode* effect. Typically, 1% to 2% of new delaminations form each year in a repaired structure, so a maintenance program and budget must be developed.

Recommendations

Depending on the amount of deterioration, the observed conditions, the overall soundness and strength of the concrete, the structure's capacity, and the budget constraints of the owner, recommendations should be given as to the repair and waterproofing options that should be performed. In some cases it may be necessary to phase the repair work, so all the repairs will need to be prioritized. If the repair work is critical to maintain the structure's integrity, then the immediacy of the repair work needs to be conveyed to the owner.

REFERENCES

1. American Concrete Institute ACI, (2002). *Building Code Requirements for Structural Concrete* (ACI 318-02) and *Commentary* (ACI 318R-02). ACI, Farmington Hills, MI.
2. ASCE Standard SEI/ASCE 11-99 (2000). *Guideline for Structural Condition Assessment of Existing Buildings.* American Society of Civil Engineers, Reston, VA.
3. American Concrete Institute (2000). *Guide for Structural Maintenance of Parking Structures* (ACI 362.2R-00). ACI, Farmington Hills, MI.
4. American Concrete Institute, (1997). *Guide for the Design of Durable Parking Structures* (ACI 362.1R-97). ACI, Farmington Hills, MI.
5. National Parking Association/Parking Consultants Council (1996). *Parking Garage Maintenance Manual.* NDA/PCC, Washington, D.C.
6. PTI Guide Specification (2001). *Design, Construction and Maintenance of Cast-in-Place Post-Tensioned Concrete Parking Structures.* Post-Tensioning Institute, Phoenix, AZ.
7. Precast/Prestressed Concrete Institute (1997). *Parking Structures: Recommended Practice for Design and Construction.* P/PCI, Chicago, IL.
8. Chrest, A. P., M. S. Smith, and S. Bhuyan (1989). *Parking Structures: Planning, Design, Construction, Maintenance, and Repair.* Van Nostrand Reinhold, New York.
9. Popovic, P. L., (1999). "Design of Parking Structures for Reduced Maintenance," in *Proceedings of IABSE Symposium.* International Association for Bridge and Structural Engineering, Zurich, Switzerland.
10. Popovic, P. L., and J. P. Donnelly (1999). "Renovation Adds Space and Value to Parking Garage," *Concrete Construction* Vol. 44, No. 2, pp. 29–33.
11. Popovic, P. L. (1994). "Maintenance and Repair of Parking Garages," *Journal of Property Management* Vol. 59, No. 6, pp. 44–47.
12. Donnelly, J. P. (Oct. 2003) "Parking Garage Maintenance and Repair," *Maintenance Solutions* Vol. 11, No. 10, pp. 14–15.
13. ASTM C876-91 (1999). *Standard Test Method for Half-Cell Potentials of Uncoated Reinforcing Steel in Concrete.* American Society for Testing and Materials, West Conshohocken, PA.
14. ASTM C42/C42M (1999). *Standard Test Method for Obtaining and Testing Drilled Cores and Sawed Beams of Concrete.* American Society for Testing and Materials, West Conshohocken, PA.
15. Popovic, P. L., P. J. Stork, and R. C. Arnold (1991). "Use of Hydraulics for Load Testing of Prestressed Concrete Inverted Tee Girder," *Prestressed/Precast Concrete Institute Journal* Vol. 36, No. 5, pp. 72–78.

CHAPTER 10

Stadiums and Arenas

LEONARD JOSEPH, P.E., S.E., THOMAS SCARANGELLO, P.E.,
and GLENN THATER, P.E.

INTRODUCTION

Many of the techniques, points of interest, and areas of focus in the investigation and assessment of more conventional buildings are applicable to sports facilities. The mechanics of building assessments in general are covered elsewhere in this book. This chapter will focus on the particular aspects of buildings that are unique to sports facilities. A reviewer who acquires a solid understanding of the performance issues for a particular element or system can better understand the intent of the original design, the elements most likely to be critical to performance, the locations most likely to exhibit distress, if any, and the significance of behaviors seen in the field. Therefore, for each building element and system, relevant design issues of strength and behavior are discussed first in some detail, followed by recommendations on the investigation, study, and field observation of the element. These elements will be discussed from the top down, i.e., from roof to foundation, both to organize the presentation and to suggest an order for an actual condition assessment. A top-down sequence aids in an understanding of the loads imposed, traces their collection and delivery to supports, and follows the path for delivery of loads to foundations and subgrade.

PREASSESSMENT ACTIVITIES

Although a sports facility includes many different components, some common activities will apply to all their reviews. Three of the common preassessment activities are described below.

Collection of Construction Documentation

First, gather all available building construction documentation.

Original documents. If at all possible, the documentation should include original construction structural contract documents, including drawings, specifications, and any documents describing the loads considered in the design. Many of these facilities are designed and built under fast-track conditions, so field and request for information (RFI) sketches can be more accurate than the original design drawings. As-built drawings that reflect field conditions are ideal. Architectural drawings of the original construction should be used for correlation with the structural drawings and confirmation of nonstructural building elements anticipated in the design. Mechanical system drawings, specifications, and "cut sheets" of manufacturers' information can provide loads and locations originally anticipated. Drainage system information also can help in evaluating the possibility of planned or inadvertent water retention on the structure, leading to the phenomenon of ponding (discussed below). Snow and wind study reports can be particularly useful because generic code forces intended for conventional buildings can be inappropriate for long-span sports facilities.

Shop drawings. Shop drawings of structural elements can be valuable, particularly for elements designed by subcontractors rather than being fully specified in the contract documents. Precast elements such as seating stadia units are often in this category. For such units, the concrete precaster's engineer develops detailed information on the geometry, mild reinforcement, prestressed reinforcement, and connection details of each piece based on performance requirements set out in the contract documents. This information appears on shop drawings and in the precaster's engineer's calculations. On steel projects, fabrication drawings may show different connection details than the contract documents if alternate methods were proposed by the contractor and accepted by the designers.

Renovation drawings. Structural, architectural, and mechanical drawings and specifications of major renovations to the facility should be reviewed because they may have altered the magnitudes and/or locations of significant loads and/or load paths from the original design.

Documenting Building Changes Over Time

The second preassessment task is to develop an understanding of building changes over time by creating a brief history of the building's construction and use. For this task, the building or facility manager should be treated as a key resource. Because of the specialized nature of these buildings, it is not unusual for the team responsible for daily operations to have considerable longevity and continuity. This "institutional memory" can be a valuable asset when the reviewer is developing the building history. The information can be as prosaic as learning about renovations and modifications through the years or as esoteric as recalling the weight and placement of special-effects rigging for the last Stones concert.

Current Building Performance

For the third common review activity, gather knowledge of the current performance of the building. Here again, the facility manager can point out locations known to require frequent attention. If a particular spot has needed repeated repairs, the building is "trying to tell you something." The reviewer should "listen" and investigate until the cause of the condition is understood.

After gathering the background information of drawings, history, and current behavior, closely examine and study the building itself. The examination is discussed here in top-down order.

ROOF MEMBRANES

As the largest single element of the facility, an arena or stadium roof is its most visible and often its most critically reviewed feature. While roofing is not a structural element, its performance can have crucial structural impacts, from reroofing loads to corrosion of framing under leaks, as well as differential thermal movements under varying levels of reflectivity and insulation efficiency. Therefore, a detailed top-down review should include the roof membrane, as well as the roof deck, drainage, structure, and suspended and supported elements.

Historical Information

A roof membrane study should start with research, gathering details of construction that include the membrane systems originally and currently in place, flashing, membrane (local) control joints, and building (structural) expansion/control joints. The research also should include information on the historical performance of the roof, including past and present leaks. The goal here is not to document and solve building envelope failures but to understand their interaction with the structure being assessed. Overlaying plan locations of leaks, joints, and structural framing often can aid in understanding roof behavior.

Roof Overview and Field Conditions

The large size and lack of landmarks on a typical arena roof can make it difficult and costly to document all conditions and to locate them in space. However, this level of detail is not necessary for a structural assessment. "Walking the roof" is still recommended for the structural reviewer to develop an understanding of the conditions over the framing. For the "big picture," infrared imaging using a helicopter also can be helpful by providing a complete picture of insulation moisture contents in one shot and guiding subsequent walks to suspicious areas. Roof walks also should concentrate on nontypical features, including through-membrane penetrations, parapets, and above-roof wiring for lighting and for lightning protection (Fig. 10.1).

FIGURE 10.1   Roofs may leak first at penetrations, such as the snow fence mounts and lightning protection cables shown here.

Flashing and Joints

Close attention to the behavior of flashing at curbs and parapets and of expansion-joint covers can lead to key information on building movements. Do they indicate roofing pulling away from walls? Do they show damage from vertical or horizontal (shearing) motions across building joints? Do they indicate roof membrane "walking" causing distress at secondary joints?

Snow Damage and Potential Exposure

In snow country, check parapets for damage from past episodes of sliding snow and ice. Miniavalanches can topple or shear off even minor projections. If snow stops or icebreaker fences are provided, check for damage to these devices and to the membrane around them. If such devices do not exist, it may be appropriate for the structural reviewer to recommend that the architect study snow conditions to see if they should be considered. If snow were to slide off the roof in large sheets, could it create a potential hazard in areas accessible to the public? Could the addition, strengthening, or modification of roof details, such as oversized gutters, intercept such material? Note that prior satisfactory performance under snowy conditions does not automatically negate such concerns; conditions may change, including changes in the thermal efficiency of insulation due to reroofing work or leaks, changes in winter operating temperatures, changes in the roofing membrane or topcoat that alter its friction value, and changes in surroundings that alter snow accumulation patterns (Fig. 10.2).

FIGURE 10.2   Snow effects should be addressed in cold climates by features such as snow bars and generous gutters designed for sliding snow impact.

Fabric Roofs

Fabric roofs are a specialized family of membrane systems. Their performance depends on the materials used for the fibers themselves; on coatings applied to the fibers; on the bias, or orientation of the fabric weave relative to the mechanical stresses; on the method of establishing geometry, such as through precut panel shapes or applied forces; on the method of maintaining shape, such as internal inflation, tensile cables, or inflation and hold-down cables; on the method of connecting panels, by sewing, lacing, or other methods; and other subtle issues. Evaluation of the performance, condition, and remaining life of a fabric roof should be performed by a specialist familiar with the particular materials and methods of installation. The subjects of structural and material assessments of tensile fabric structures are discussed in Chapters 12 and 19.

ROOF DECKS

The structural deck is the hard surface supporting the roof membrane and insulation. It spans between framing members such as beams, joists, or trusses. Structural decking options for arenas include a wide variety of materials and installations from plywood sheathing through proprietary systems such as Tectum deck (cemented wood fibers) and into formed-gauge metal decking. Some roof areas even may have decks of precast concrete in the form of plank, precast concrete channels (toes down), or teés. Others may use decking for both structural and waterproofing functions, such as standing-seam roofing systems. Because of the large structural penalty for even small increases in dead load, the long-span areas of most modern arena roofs use metal decking to save weight.

Structural Analysis Considerations

Key elements in the structural analysis portion of the assessment include the analysis of deck strength, deck stiffness, connection strength, and connection behavior.

Deck strength. Determine resistance to gravity load, including adequacy for vertical shear and for local crushing at bearing points (snow and wind loads are discussed later); resistance to uplift; and strength under horizontal shear (diaphragm action).

Deck stiffness. Evaluate flexural deflections under gravity and wind, angle changes at deck splices, and diaphragm stiffness in shear. Diaphragm stiffness can attract unanticipated in-plane forces, even when a parallel-truss system has been provided for those loads.

Connection strength. Consider forces from perpendicular loads such as uplift, forces from in-plane loads through intentional or accidental diaphragm action owing to the relative stiffness of the diaphragm compared with the in-plane bracing, forces developed as decking braces roof purlins, forces from the in-plane component of gravity loads (on steeply sloping roof areas), and force transfers of the in-plane component from the deck to the supports below. Another source of uplift is the force used to impose deck curvature when decking is "walked down" in the field to follow an arched shape (Fig. 10.3). This requires permanent uplift forces on deck panel ends and downward reactions at the first inner support from each end. Typically, curvatures of radius larger than 400 ft may be field-formed; tighter radii may require decking precurved from the factory, which would not require permanent uplift forces to hold their shape.

Connection behavior. Consider connector reliability, durability, corrosion susceptibility, and load-transfer ability for perpendicular and in-plane forces. These properties differ for puddle welds, screws, and clips. Consider their properties for in-plane diaphragm action between decking panels (side laps). Also consider the interaction of decking flexural stiffness and the connection stiffness to supports. Depending on the relative magnitude of these properties, the decking may be helpful in bracing lower flanges of purlins in uplift (Fig. 10.4). Apply current design methods, such as Load and Resistance Factor Design (LRFD) for steel, for the best understanding of capacities.

Orientation and edge conditions. A key structural aspect of roof decking is the high suction force that can occur along edges, along ridges, and at other changes in slope or elevation. Purlin spacing and/or connection spacing may be reduced to account for this effect. Nonstructural aspects of the roof deck include treatment at local and through-building expansion joints, at control joints, and at curbs. Decking orientation is important because decking can cantilever "strong way" a short distance beyond the last support but "weak way" not at all. However, depending on the curb and joint detail, considerable lateral and overturning forces may be applied at joint curbs, so it may be preferable to provide, and connect to, a structural member directly below such curbs. Deck edges are also locations where the deck, if acting as a horizontal diaphragm, requires continuous chord members and distributed connections to the chords.

Interaction of Deck and Connections

Decking features can interact in surprising ways. For example, a standing-seam deck covering a barrel vault was bent "hard way" so that the ends of the deck strips ex-

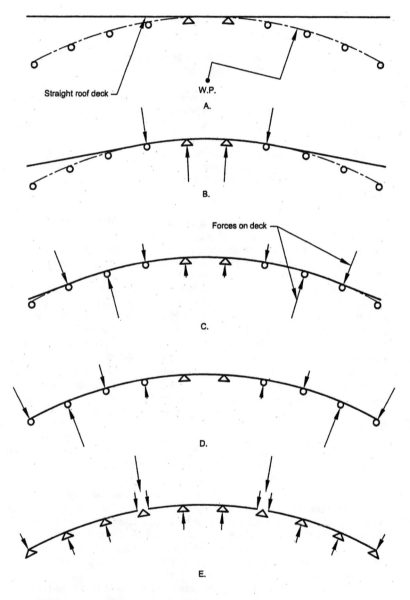

A - Initial deck anchored at crown
B - Hold down force at outer supports
C - Forces shift as deck is "walked down"
D - Eave hold-down force in final condition
E - If separate lengths of deck, hold-down at each end

FIGURE 10.3 Walking deck for curved roof.

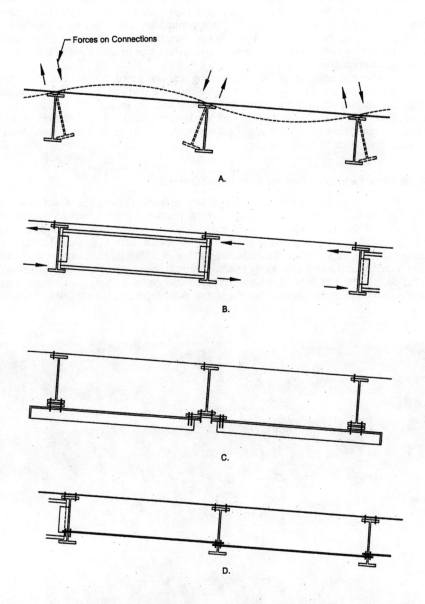

A - Deck flexure - depends on deck flexural strength, stiffness, and connections
B - Diaphragms or crossbeams
C - Main bracing system
D - Sag rod with intermittent diaphragms

FIGURE 10.4   Purlin bracing methods for uplift.

perienced sustained uplift forces. As is common for standing-seam roofing systems, the standard clips for the roof system intentionally allowed for differential in-plane movement between deck and supports to accommodate daily and seasonal thermal motion. However, because of the uplift forces, the clips "sawed through" the deck in a short time. An entirely different connection system was needed to handle this combination of conditions.

In another example, steeply sloping decking was connected to steel open-web joists acting as roof purlins. The significant in-plane component of gravity loads was required to follow an indirect load path from deck to joist to main framing through transverse loading of joist seats, but the seats had no special provision for these loads (Fig. 10.5). Supplementary bent-plate Cees and Zees were added later to deliver in-plane forces directly from deck to main framing.

Roof Deck Field Observation Recommendations

A solid understanding of the design intent and relevant performance issues can help to focus the field observation program. Typical points of interest include the general condition of the deck soffit (underside); locations of visible corrosion; locations of mechanical deck damage, such as crushed ribs, buckled surfaces, wracked areas, unsupported cuts, cuts causing discontinuities at flange connections of supporting members, unreinforced holes, and poorly executed retrofit/renovation work; deck panel side laps; deck panel end laps (if visible); types, locations, and spacing of connections; connection conditions; elements for transfer of diaphragm forces to supports; signs of

FIGURE 10.5   Sloping-roof in-plane forces would twist joist seats, so steel plates transfer shear directly to the main truss below.

connection failure, such as decking lifted off its supports; and locations of anticipated high deck shear and connection forces (e.g., near lateral-load-resisting elements and boundaries of diaphragm panels).

ROOF DRAINAGE SYSTEM

This nonstructural system should be reviewed within a structural assessment for its impact on potential structural loading. Its plumbing performance should be checked by a mechanical engineer. Impermeable surfaces, slopes that can vary from near level to relatively steep, large tributary areas, and potentially large structural deflections create challenges for sports facility roof drainage systems. This combination of conditions can result in very high short-term runoff rates during cloudbursts. Large deflections on near-flat roofs can lead to damaging ponding. And oversized roofs require commensurately large efforts to maintain good drainage.

Drainage System Design Alternatives

Runoff can be handled three ways: conventionally vented drains, siphonic drains, and detention drains. Each has different implications for structural impact.

Conventionally vented drainage lines. Rainwater conductors can be sized to carry the anticipated peak runoff rate using conventional methods and piping at normal atmospheric pressure. A vented system such as this may be used where a water trap is required between the building and the sewer because an unvented system could "blow the trap." This approach assumes that the flow will be determined by the pitch of the horizontal runs. The great length and width of sports facilities mean that horizontal runs, even with a shallow ¼ in. per foot of pitch (2% slope), must change elevation by several feet. This can affect headroom significantly. With a shallow pitch, the pipe diameter needed to handle peak flow may become impracticably large. Even where the necessary piping and headroom can be accommodated, this approach may not be acceptable where municipalities limit the flow rate into their sewers to avoid overload and backup conditions.

Siphonic drainage. Where no trap is required between roof drains and sewers, a siphonic drainage system can be used, in which smaller piping is laid with lateral runs nearly horizontal. During extreme flow conditions, the smaller piping, minimal pitch, and special inlets create a continuous column of water that can use the full change in pressure head from roof to sewer. This change in head acts on the whole line, sometimes pushing and sometimes pulling. As a result, high flow velocity is achieved. The flow rate can be limited to less than the anticipated peak rainfall rate, causing rooftop detention if required (Fig. 10.6).

On-site detention. Conventional systems also can use shallow pitches and small pipes to limit the flow rate and result in significant on-site detention. Specific provisions must be built in to permit this function. For example, the New Orleans Superdome has a very generous perimeter gutter. Where detention is anticipated to occur at rooftop level, a positive-relief method to limit detention depth must be provided, such as secondary drains or scuppers with higher inlet elevations.

Ponding

Even where rooftop detention is not anticipated, accidental water retention and subsequent ponding can be a concern. Ponding can be stiffness-related or drainage-related. On a roof with minimal pitch to drain, if structural deflection from live, snow, or hung

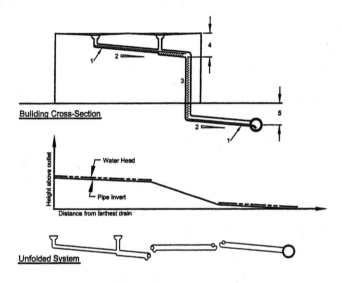

A. Conventional Roof Drainage System
1 - Horizontal runs operate partially filled at atmosphere pressure
2 - Water head change = pipe slope
3 - Vertical leaders partially filled
4 - Low point of roof runs set by distance and slope
5 -Sewer invert must be below connection elevation set by distance and slope

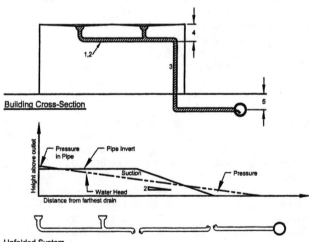

B. Siphonic Drainage System
1 - Horizontal runs full once primed
2 - Slope of pressure head drives flows; need not pitch pipe
3 - Change in pressure in vertical drives flow
4 - Pipe can run tight to roof framing
5 - Can work with shallow sewer elevation

FIGURE 10.6 Roof drainage systems.

load is large enough, the pitch can reverse, causing ponding next to the drains. This water causes added deflection. If framing is too flexible, the resulting additional water accumulation can lead to more deflection, in a potentially catastrophic cycle. Even if framing is stiff, where drainage is established by a grid of ridges and a low point in the center of each grid, one clogged drain can cause a grid square to fill until any additional water will drain across the surrounding high ridges. The extra water weight from deep ponds can overload local framing members. Several design strategies can be used to reduce these ponding risks, including the provision of adequate roof framing stiffness per American Institute of Steel Construction (AISC) *Steel Manual* criteria (Allowable Stress Design (ASD) and LRFD Specifications, Sec. K2), ample roof pitch to avoid self-perpetuating ponds between drainage points, multiple drains provided at each low point (they can supply a common leader), ridge-and-valley drainage with multiple evenly spaced drains, pitch to low points near slightly elevated edges, drains behind parapets with low relief scuppers, and pitch to external gutters so that ponding cannot occur at all (Fig. 10.7).

Maintenance

For all these systems, there is no substitute for maintaining drainage system cleanliness. Even where drains are twinned, when one drain is already clogged, the margin of protection against ponding has been eroded. Large roof expanses can attract unusual detrius; one clog was traced to shells washed into the system courtesy of neighborhood seagulls who used the hard roof and the force of gravity to shuck their clams open. Large, flat-bottomed gutter systems also can collect enough airborne soil to start their own gardens, complete with drain-clogging foliage and membrane-damaging root systems. Any site observations of roof conditions must pay special attention to this issue. In cold regions, the possibility of frozen drains also must be considered. Heat tracing and pipe insulation may be required to avoid this problem.

ROOF STRUCTURAL SYSTEMS

The long spans of arena roofs, often hundreds of feet in both directions, require large member sizes and connections for the most-loaded structural elements (Fig. 10.8). For structural efficiency and erection ease, a mix of framing systems may be used. For systems that are connected together, overall framing behavior can become quite complex. Understanding the system's behaviors, load paths, and stability issues can be a challenge. If the systems used have a wide disparity in strengths, stiffnesses, and capacities, issues of compatibility of behavior, consistency in local and overall strength values, and "weakest link" situations should be studied. Load redistribution capacity in the event of local element failure also may be of interest to the facility owner. Roof framing review should start with the loads and progress down the load paths to the supports. A review of existing conditions can proceed simultaneously.

Roof Loads

Any study of a long-span roof system should start with the loads imposed. These should include six types of loads, as discussed below.

Original design dead loads. As indicated on contract documents, these should include the self-weight of the basic structure, the roof decking, the roofing system, and superimposed dead loads from original scoreboards, signage, and equipment.

Dead loads added later. Reroofing is a prime example. Owing to the huge expanse of a roof, its limited accessibility, and the desire to reduce cost and exposure to

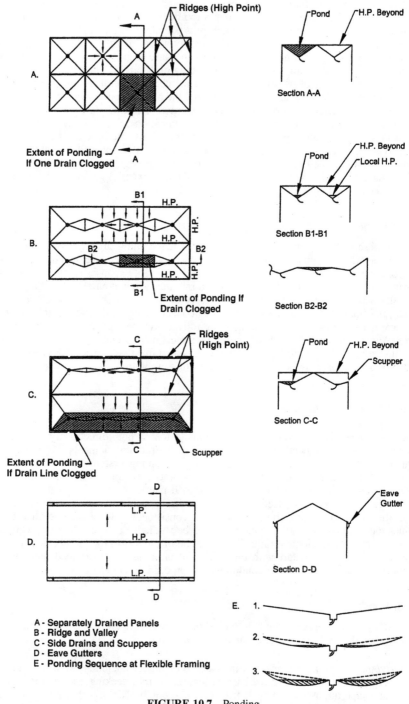

A - Separately Drained Panels
B - Ridge and Valley
C - Side Drains and Scuppers
D - Eave Gutters
E - Ponding Sequence at Flexible Framing

FIGURE 10.7   Ponding.

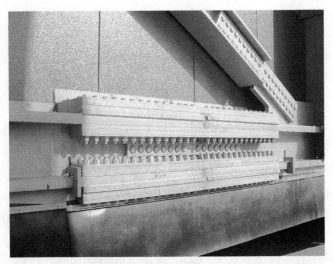

FIGURE 10.8   Long-span framing requires heavy members and connections.

damage during roofing, it is not uncommon for a second roofing membrane to be installed directly over the original roof. Even small increases on a pounds-per-square-foot basis can generate large increases of the forces in members carrying large tributary areas. Reroofing can be a particularly important structural consideration for long-span roofs in snow-free areas; the added weight can use up a large portion of live-load allowances that are already small. Sometimes future reroofing weights are considered specifically in the original design, but often they are not.

Equipment loads added later. Upgraded or newly installed heating, ventilation, and air-conditioning (HVAC) units and eye-catching new scoreboards, video screens, and "jumbotrons" bring point loads with new values and often at new locations. Roofs designed primarily for distributed loads must be checked locally for point loads.

Snow loads, where applicable. Building codes have been revised continually to reflect the latest findings on snow drift loads, so an early building design may not reflect current practice. Unbalanced snow-load conditions can be particularly critical to the performance of shallow-depth roof structures that depend on shape for strength, such as arches, barrel vaults, and domes. Changes to the roofline, such as additions or modifications to penthouses, monitors, air wells, and equipment enclosures, can change snow drift patterns for better or for worse. In heavy snow areas and for buildings with complex, cluttered, or stepped roofs, if there is a question about roof capacity, a detailed snow drift study may be advisable. Sometimes such a study was performed for the original design. Often the results of such a study are less severe than those resulting from the straightforward application of building code provisions (Figs. 10.9 and 10.10).

Wind loads. Wind pressures and suctions can be unbalanced, can change with wind direction, and can be affected by roofline changes subsequent to original construction. In addition, particularly in hurricane-prone areas, suctions can be high enough to greatly exceed available dead load, causing significant net uplift at roof edges and steps and along curved surfaces (the airplane-wing lift effect). Note that as roof spans get larger, the dynamic behavior of the roof—bouncing

FIGURE 10.9   Snow drift models help predict drift locations on large roofs; often total load is much less than from standard building code provisions.

up and down or flexing back and forth—can interact with the turbulent wind flow when generating pressures and suctions. For this situation, simple code provisions based on rigid roof framing may not be appropriate. Also note that the roof usually acts to collect top-of-wall reactions from horizontal wind loads on perimeter walls and lead them to main roof support points. If a major roof has been designed without consideration of recent code provisions or a wind tunnel test, it could benefit from project-specific testing to identify sensitive areas and, just as important, to avoid unnecessary costs by identifying areas that are not of concern.

Seismic loads. These can be quite severe compared with lateral forces from wind because arena roofs generally have minimal horizontal wind exposure but a lot of mass. Consider a 1-ft-wide "slice" through an arena. The roof may have to carry the lateral wind force from perhaps 30 ft^2 of wall area, but probably more critical are the seismic g forces acting on the mass of 200 ft^2 of roof area plus the mass of the wall area.

Load Path

The next step in reviewing a long-span roof is investigating the load path from decking to main framing. If the building is framed in steel, the secondary framing members, or purlins, may be wide-flange sections, channels, tubes, or open-web joists. Some key points to investigate include those discussed below.

Member adequacy under distributed gravity loads. Both the purlins that collect local roof deck load and the main members that act within a larger system (e.g., a truss) must be checked for capacity to resist the distributed gravity load, including snow drift effects, if applicable. Drift-pattern changes may have only

FIGURE 10.10   Aerial photograph showing actual snow drifting.

minor impacts on main members because they are affected by overall roof loads, but they may be very significant to the strength demands placed on purlins.

Member adequacy under point loads, if applicable. Trusses are particularly sensitive to point loads. The axial forces they carry can magnify moments created by transverse loads between panel points, so generally point loads should be applied only at panel points (where web members meet a chord) or where a supplementary brace was added to create a new panel point. Loads not applied in this way should be an immediate red flag. Open-web joists are designed to meet particular specified load cases very efficiently, so they are even more sensitive to unanticipated loads and load patterns.

Member adequacy under uplift loads, if applicable. Open-web joists under uplift require more bridging than Steel Joist Institute minimum requirements; closer bridging may be needed to brace the bottom chord in compression, and additional uplift bridging may be needed at bottom chord ends. Open sections (W, WT, C, Z) may have bottom flanges stabilized by continuity, end fixity through gusset plates, or torsional restraint through deck connections. Where these are not present or are insufficient, addition of diaphragms, braces, or sag rods may be required (see also Fig. 10.4).

Member adequacy under axial loads. Such loads may occur in members from acting in a load-bearing truss, as *posts* in horizontal trussed bracing systems, as *chords* along diaphragm edges, as *drag struts* gathering loads for delivery to points of lateral-load resistance, as *wind struts* propping columns or wall panels against lateral loads, or as *hoops* that resist spreading of arched or domed framing systems. With the exception of the first case, the magnitude, direction, and nature of axial forces result from complex and subtle interactions and may not be stated on the drawings, making their identification challenging. Also check for moment induced by the axial load if load is not applied at the member centerline

Connection adequacy. Connections must be adequate for the shears, moments, and axial forces being transmitted to member ends as part of the load-resisting system. Secondary effects also must be considered, such as the transfer of horizontal or in-plane loads from decking to main framing. This is particularly important where decking is in a different plane than main framing, such as atop open-web joists, where joist seats could be twisted (Fig. 10.11).

Major changes in load path. Some older arenas and field houses have been retrofitted with entirely different roof systems than provided originally. For example, some owners have replaced air-inflated, tied-down fabric roofs with hard-surface roofs and arch supports. The perimeter ring beam that was in compression is now in tension, applying very different forces on its reinforcement and internal connections. The building columns that were holding the roof down are now being asked to hold it up. The major increase in axial compression may require additional lateral bracing, column strengthening, footing enlargement, or all three.

Overall Behavior and Compatibility

The overall behavior of long-span roof framing must be understood, regardless of the particular loading values. Large spans experience large movements under gravity loads, wind loads, lateral loads, and thermal changes. These movements can cause load redistribution and redirection owing to enforced compatibility of deformations. Where such redistribution is unexpected, distress can result. Examples of conditions requiring special attention and explanations of their significance are given below (Fig. 10.12).

Long-span roof trusses parallel and close to supports. If running close to beams that are well supported by a line of columns, any framing connected to both, such as bridging or secondary trusses, must be designed and detailed to allow for large differential deformations. Otherwise, load will try to flow to the stiff beam and can fail the bridging.

Long span open-web joists parallel to stiffer framing or walls. The last bay of bridging should allow for movement by using parallel-chord bridging and avoiding cross-bridging (X's) or, if cross-bridging is required, by providing details that allow vertical slip while restraining lateral and rotational movements.

Intentional load-redistributing cross-framing. This can be an intentional part of the design to enhance system redundancy by providing alternative load paths in order to minimize local differential deformations or to spread load more evenly among supports. In such cases, the erection and loading sequences strongly influence the forces in the members. For some buildings, these sequences cannot be determined, but some members may appear highly stressed. If it is clear that they are not part of the primary load path, simply disconnecting and reconnecting the member, releasing dead load forces, may be sufficient.

Rotation and translation of long-span ends. More subtle movement patterns can occur at ends of long spans, where overall truss or girder deflection causes end rotations. These can cause truss bottoms to "kick out," tops to "pull in," or both, depending on local geometry, details, and differential strain in top and bottom chords or flanges. An ideal design minimizes this effect by supporting

FIGURE 10.11   Complex slopes can result in undesirable edge support for decking.

at an elevation where the structure does not kick or pull, such as middepth of a truss (Fig. 10.13). Where such movement is not avoided but is restricted by bearings, stiff columns, shear walls, stiff ring beams, or horizontal trusses, large forces can occur. A proper erection sequence may reduce this effect. For example, waiting to connect a diagonal strut to a bearing until after dead loads are in place can allow some structural rotation to occur without inducing a corresponding kick at the bearing.

Structural building joints. Significant vertical cross-joint movement owing to different framing behavior or different loading can affect the joint cover itself. A

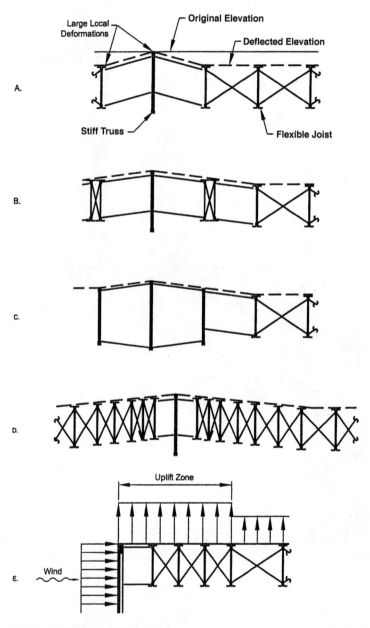

A - Parallel Bridging at Stiff Element
B - Double Up Joists to Transition Stiffness
C - Truss of Intermediate Stiffness
D - Vary Spacing of Typical Joists
E - Closer Spacing Helps at Uplift Zones

FIGURE 10.12   Differential-deflection approaches.

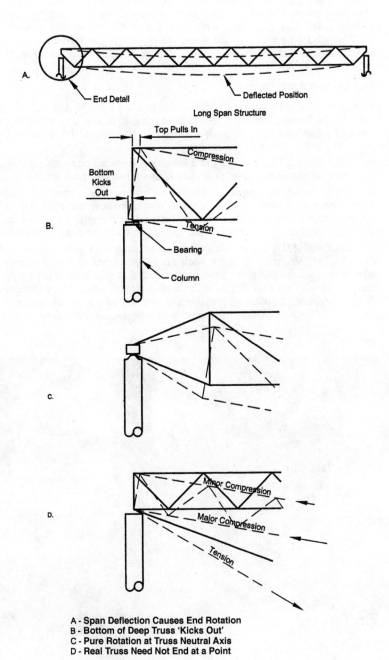

A - Span Deflection Causes End Rotation
B - Bottom of Deep Truss 'Kicks Out'
C - Pure Rotation at Truss Neutral Axis
D - Real Truss Need Not End at a Point

FIGURE 10.13   End-rotation approaches.

standard detail may not be sufficient to accommodate this motion. Similar concerns arise where horizontal cross-joint movement induces shearing forces in the joint cover. The best joint is no joint; design the whole roof to work as a unit. The second-best joint is one where the magnitudes and directions of movement are limited, perhaps by shear lugs, keepers, or special bearings (Fig. 10.14).

Main Roof System Analysis

With an adequate load path ascertained from roof to main framing and a clear understanding of overall framing behavior, the main framing itself can be studied properly. The amount of effort and level of analytic sophistication should be matched to the structural system and to any specific concerns.

Manual calculations. These may suffice for straightforward one-way framing systems, such as planar determinate trusses with simple load patterns.

Planar (2D) linear elastic computer analyses. These may be adequate for indeterminate one-way framing systems, such as planar trusses with open panels (Vierenedeel truss action). Forces induced into the framing by action of the roof-plane diaphragm or trusses would have to be added as a separate load case.

Three-dimensional (3D) linear elastic computer models. These are appropriate where load redistribution is likely, such as two-way framing systems (main and distribution trusses) and box trusses where forces in web members on opposite faces are affected by overall torsion, as well as planar loads. Three-dimensional models also permit combining gravity- and lateral-load effects. Such models are satisfactory for study of the vast majority of long-span structural systems.

FIGURE 10.14   This bearing slides left-right, whereas "keepers" restrain movement out of the page. Stainless steel top element is larger than PTFE lower element.

Three-dimensional models with geometric nonlinearity (*reflecting second-order effects*). These are appropriate where changing loads significantly affect building geometry and where changes in geometry are necessary to resist the load. For example, a roof framed like a suspension bridge with a very thin structural depth will kink under concentrated loads, so geometric nonlinearity is important. However, if the roof also includes a stiffening truss to hold its shape, second-order effects become less significant. Similarly, the use of stay cables on a roof need not require the use of a second-order analysis. If the cables are straight, taut, and not loaded transversely, they will behave as nearly linear spring supports, and linear elastic analysis may be used for structural checks (Fig. 10.15).

Three-dimensional models including geometric nonlinearity are also appropriate where small deflections can have large structural effects. For example, thin structural systems shaped to function as pure arches, barrel vaults, or domes are highly efficient in resisting load patterns that create compression forces acting along their centerlines. However, unbalanced load patterns generate lines of action that deviate from member centerlines, inducing flexural forces and flexural deflections in addition to compression. This compression, combined with member centerline deflections, can generate additional flexural forces and deflections. In unstable structures, this cycle of forces and deflections continues to grow until collapse. Checking for such instability requires modeling geometric nonlinearity.

Three-dimensional models with material nonlinearity. These are useful for study of reserve strength capacity under extreme conditions such as major seismic events, sudden failure of local elements, application of unforeseen loads, or as-built construction significantly deviating from the original design. For example, a check using design seismic forces may exhibit overload at a few elements but not necessarily an unsatisfactory overall structure. Including material nonlinearity may show that other load paths are available to resist seismic forces after those members yield. Another example is a redundancy study for the loss of a major element. Under extreme events, it is appropriate to consider postyield tensile and flexural behavior and postbuckling compressive strength for load redistribution and energy absorption. Where loads have changed or members differ from design, the ability to redistribute loads can explain realistic system behavior.

Main Roof System Acceptance Criteria

A review of member adequacy should follow the modeling of the roof structural system. In determining adequacy, six aspects should be considered, as discussed below.

Appropriate codes and standards should be used. Should acceptance be based on the building code in effect at the time of design, or can more recent criteria be applied? Concrete design began shifting from working stress to (ultimate) strength design about 40 years ago, and steel offered the alternatives of Allowable Stress Design (ASD) and Load and Resistance Factor design (LRFD) since 1986. Wood design also can be performed using either working stress or strength design criteria. Load and resistance factors were adjusted recently for both concrete and steel to better conform to the load combinations in ASCE 7, which themselves have changed over time. Strength and LRFD criteria were developed to provide a more uniform level of structural reliability. They apply lower factors to the better-known dead loads than to more variable live loads. For long-span roofs, where dead load is a major portion of the total load, systems that may not fully meet ASD criteria may still prove to be quite satisfactory using the LRFD approach. However, the acceptability of this approach should be reviewed with the owner and local building officials.

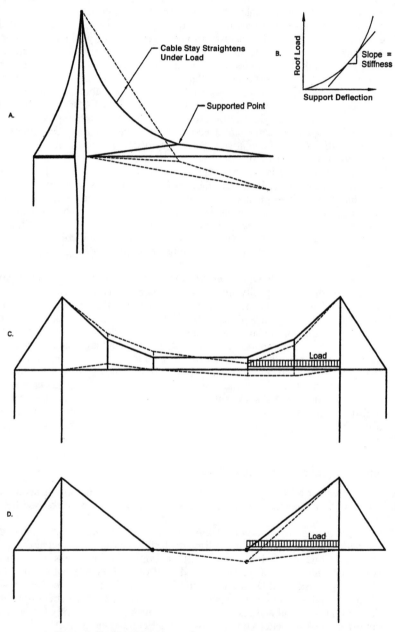

A - Heavy Cable Sag Changes with Tension - Nonlinear
B - Heavy Cable Axial Stiffness is Nonlinear
C - Catenary Suspension Cable Geometry Changes - Nonlinear
D - Simple Straight Cable Stays - Linear Analysis is Sufficient

FIGURE 10.15 Cables and nonlinear analysis.

Material properties should be appropriate. For example, concrete that has been in place for years may now have strength and elastic modulus values significantly greater than those found in 28-day cylinder tests. However, the potential benefit of applying higher values must be weighed against the cost of sampling and testing site concrete, the need to patch sample sites, and the level of confidence to use when applying values from limited samples to large members. In another example, steel mill certificates may show much higher yield and ultimate strength values than specified originally. Where particular members and mill certificates can be linked, higher values may be used. Where mill tests are not available, it may be possible to take coupons from the members in question for testing. Again, testing and patching costs should be weighed against potential benefits in the building evaluation.

Connection strengths and behaviors must be compared to forces found and assumptions used in the analyses. Connection behavior also can relate to acceptance criteria. For example, truss members usually are sized assuming pin-ended connections to ignore moment-frame action that potentially could reduce deflections and axial forces. Actual connections may have high rigidity either from member-to-member welds or from large gusset plates. The resulting secondary moments induced in members are real, but are they significant? This becomes a matter of judgement. American Association of State Highway and Transportation Officials (AASHTO) specifications for steel highway bridges have stated that some secondary bending "usually need not be considered" in slender truss elements (4 kips per square inch (ksi) in tension members, 3 ksi in compression members). Such an approach can make sense for buildings also, where a member has adequate strength and stability even with localized yielding at its ends or where improved stability from end restraint offsets the effect of induced forces.

Sensitivity to erection sequence and fit-up should be considered. If the system is highly redundant, locked-in forces and unanticipated load paths may result during erection as forces are applied to make connections fit or as joint gaps are shimmed and bolted or welded while on shores. Where the sequence is known or reasonably can be inferred, its consequences can be studied through analysis. Note that actual details from shop drawings or field observations should be used for such an analysis rather than contract document details. With ductile materials, even if forces were not planned originally, it is possible that they could redistribute before failure. Models with material nonlinearity could determine if such redistribution reasonably could be considered. Adjustment and maintenance of tension members are related issues. If simple threaded clevises are used, they should be checked for loosening and investigated for potential future tightening operations, if necessary.

Extraordinary conditions not specifically addressed in codes. What if a major member fractures or buckles? Ideally, alternative load paths could continue carrying the loads, although the members may then be acting at higher-than-allowable stresses or even yielding, with greatly increased deflections. Where such paths are not available in the system, it should be identified by the reviewer, but does not indicate an original design flaw.

"Weakest link" checks should be made. These should be done for designs that rely on a chain of interacting elements to maintain strength and stability. Apparently minor elements and connections could prove to be crucial to major systems. For example, a major truss diagonal in compression may work because a strut reduces its unbraced length. But does the strut itself rely on yet smaller members to maintain its own stability? It is not difficult to imagine a system where a chain of members ends up relying on a minor bridging line for stability. Al-

though not required by code, consider recommending discrete modification to a few members in the chain to avoid this interdependence.

Field Investigation, Findings, and Remedial Proposals

Parallel with the analytic investigation, a field investigation should be performed. Member sizes, configurations, and orientations should be spot-checked against design or fabrication drawings. Locations of obvious damage, such as members kinked by improperly rigged temporary loads, should be noted. Areas of potential damage, such as those under persistent leaks, should be accessed and documented. Sources of less obvious damage also should be investigated, such as stains that indicate roosting locations of birds and those which show that framing details have formed water-trapping pockets (Fig. 10.16).

Once the main roof framing system has been inspected and analyzed using well-defined loads and load patterns and checked for adequacy using agreed-on criteria, findings and proposals for remedial action should be presented. This may require considerable discussion with the facility owner, users, code enforcement agencies, and investigators because assessment and retrofit of existing facilities is never "black and white." For example, when individual items are found to be in noncompliance with technical criteria but the overall system still performs properly, how must that noncompliance be resolved? Spending time and money in a location that does not actually improve the system would seem wasteful and arbitrary.

FIGURE 10.16 Frozen water trapped within this box column bent its faces outward and fractured surrounding concrete.

ROOF BEARINGS AND SUPPORTS

The forces from the main roof framing system are delivered to bearing points and support framing that also warrant careful study. Bearings must accommodate a variety of forces and movements.

Bearing Motion Arrangements

Supports for main roof framing must carry the gravity loads, provide hold-down resistance against uplift loads (if applicable), provide stability against lateral loading, and allow for thermal movements and roof system rotations. This can be handled in several ways, as explained below.

Classic four-support system. Freedom for thermal expansion and restraint for lateral load is provided by having one corner fixed, the diagonally opposite corner free, the third corner slide east-west, and the fourth corner slide north-south. However, with only two bearings acting to resist lateral forces in each direction, loads are highly concentrated, and there is no redundant load path (Fig. 10.17).

Four-side bracing. A ring of axially connected members along the roof perimeter can engage brace lines or shear walls at the midpoint of each side, using their stiffness parallel to the roof edges to resist lateral loads and their flexibility at right angles to allow the roof to "breathe" laterally from a central point. In this case, gravity support points either can have sliding bearings or, if their inherent flexibility permits easy lateral movement, can be pinned to the roof framing (Fig. 10.18).

Distributed flexible restraint. An alternative approach suitable for seismic areas is support on multiple moment-connected legs with connections detailed for ductility. For thermal movements, legs are flexible enough to limit induced forces to manageable values (Fig. 10.19). For lateral loads, all legs participate to provide high redundancy and, in event of extreme conditions, high ductility.

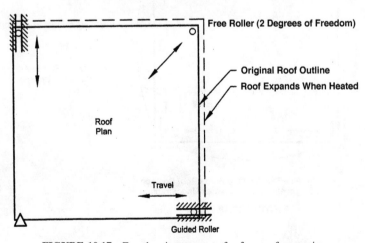

FIGURE 10.17   Four-bearing supports for free roof expansion.

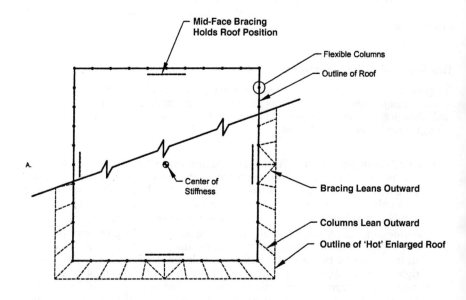

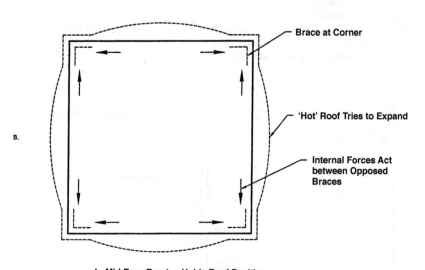

A - Mid-Face Bracing Holds Roof Position
B - Corner Bracing Creates Undesirable Restraint

FIGURE 10.18 Roof bracing considerations.

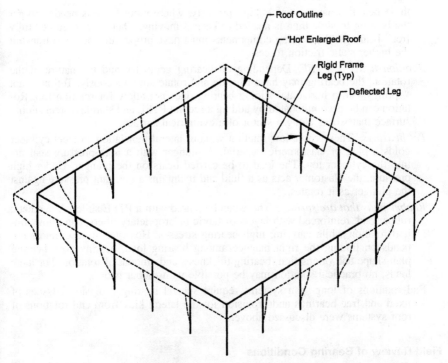

FIGURE 10.19 Roof restraint by rigid frames.

Bearing Design Considerations

Where free lateral movement is needed, current practice generally uses the technology of PTFE (polytetrafluoroethylene or Teflon) against polished stainless steel. These sliding bearings work well on clean, smooth, flat surfaces. The nature of PTFE requires certain design features outlined below.

Slide bearing protection. PTFE is soft enough that sand and grit can work its way into the pad, ruining it over time. PTFE also will be damaged by "plowing" or "shaving" if a hard edge runs across it. Therefore, usually the PTFE pad is the smaller bottom element, and the stainless steel is the top element, sized larger to keep the PTFE covered under all travel movements. In this way, no dust can settle on the PTFE, and no steel edges cross its surface, so no plowing can occur (see also Fig. 10.14).

Bearing stress limits. The softness of virgin PTFE limits its allowable bearing stress. Where loads are so large that an impractical bearing size would be needed, special formulations of PTFE and fillers can provide much higher bearing values with only a small effect on the coefficient of friction.

Guided bearing. Where a bearing must allow movement one way but provide restraint in the transverse direction, the lower bearing plate has a machined groove, and the upper plate is sized with edges to fit in the groove.

Friction values. PTFE is a low-friction material but not frictionless. First, the coefficient of friction depends on the imposed stress level. Paradoxically, high bearing stresses create a lower coefficient of friction. While higher bearing stress creates greater friction stress, friction does not increase proportionally. Second, these bearings have a "stick slip" property, where more force is needed to get the bearing to slide than is needed to keep it moving. Therefore, even at "fully free" bearings, the bearing components must have proper anchorage to transmit the higher static friction forces.

Rotation limits of PTFE. Depending on bearing geometry and the nature of the rotation, the bearing may have to slide and rotate simultaneously. Teflon is not "springy," so a horizontal sliding bearing does not allow for rotation too. Rotation can be accommodated by adding a second PTFE and stainless steel sliding surface that works along a sector of a cylindrical surface.

Pot bearings. These can allow rotation without lateral translation. A steel cylinder holds a plug of elastomeric material. A plunger with a high-pressure seal fits into the cylinder top. The load to be carried bears on the plunger. Under high pressure, the elastomer acts as a fluid and maintains a constant pressure against the plunger as it rotates.

Rubber pads that are springy. These can be paired with a PTFE slider. A synthetic rubber pad, reinforced with layers of fabric or proprietary fibers, can allow limited rotation while carrying high bearing stresses. However, its usefulness depends on finding the right balance among bearing load, bearing size, bearing plan shape and orientation, bearing thickness, and amount of rotation. For large loads, no practical solution may be possible with rubber pads.

End rotations of long-span systems require special bearing provisions. Issues of fixed and free bearings and provision for the lateral kick from end rotations of roof systems were discussed above.

Field Review of Bearing Conditions

Existing bearings should be inspected with the preceding considerations in mind, as outlined below.

Contract document comparison. Determine the bearing types used in the structure. Compare them with the contract documents. Determine if calculations are available to understand the design approach.

Bearing condition. If pot bearings are used, check for proper seals (no elastomer squeezing past the piston). If PTFE sliders are used, check for sliding surfaces free of dirt. Measure scratch marks that indicate direction and amount of travel. Measure at-rest location relative to the supporting structure while documenting environmental conditions (temperature, roof loading). Check for engagement of block-in-groove guided bearings. Check for surface conditions where guides contact. If older types of bearings such as rockers, rollers, or pins are used, check for corrosion and evidence of recent free movement.

Bearing anchorage to supports. Note any apparent shifting of bearings. Note any apparent damage to supporting structures.

Assessment of Bearing Adequacy

Comparison of analysis assumptions for bearing conditions, analysis results that show anticipated structural movements, and field inspection data can be used to determine current performance of the bearing and support system.

Lateral and rotational accommodation. Do the bearings as designed properly accommodate the anticipated lateral and rotational displacement demands of the structure? Does their current condition maintain that ability?

Vertical capacity. Do the bearings as designed properly resist the anticipated gravity and lateral loads imposed? For example, how do the existing bearings handle forces from current wind and seismic criteria? Does the current condition of bearings and supports maintain that ability?

Remedial actions. If existing bearings are not adequate by design, by current physical condition, or for current loading, what are the possible remedial actions? Are there provisions in the structure for temporarily carrying the load at different points so that the bearings can be swapped out?

Load path into main frame. Trace load paths from the bearings down into main building framing. Determine if lateral-load resistance and overall stability are provided by braced bays, shear walls, moment frames, or other methods. Compare as-built capacities with anticipated forces. Observe field conditions of the framing system, as discussed under "Framing of Concourses and Level Floors" later in this chapter.

CATWALK AND RIGGING LEVELS

Generally, catwalks and rigging points are supported by the main roof framing system, so their loads must be included in the analyses performed under "Roof Structural Systems" earlier. However, these levels deserve additional attention because they are the day-to-day working areas for the facility operator. In contrast, the roof framing is treated as a remote, rarely visited system.

Load Information and Rigging

The facility manager should be able to supply a copy of the rigging load-limit diagrams provided to riggers of the events booked into the building. This provides a good start to understanding the rigging system.

Rigging vertical capacities. Capacities of 100,000 lb and more are designed into modern arenas and similar facilities, with individual loads limited to 5000 lb and clusters of loads limited to 25,000 lb. Be alert to the doubling effect of tie-downs; a simple pulley with 1000 lb of load on the line imposes 2000 lb of force if both load and tie-down ends are anchored below it.

Rigging lateral capacities. These are important when load lines to winches are run at an angle. Modern facilities allow for simultaneous vertical and lateral forces owing to sloping tie downs by reducing vertical capacity as bridle angle increases or by allowing for the equivalent of 25% of the vertical load applied.

Alternate rigging points. If the facility being studied has significantly lower rigging limits, find out how the facility handles major shows. Do riggers bypass the rigging level and rig off the roof structure? If so, where?

Rigging methods. What method of attachment is used generally—cable chokers onto beams, pulleys mechanically clamped to beam flanges, chainfalls dropped through grates, etc. Are there specific hanging points? Are permanent fly lines available? Are they used or ignored?

Review Approach

Load path. Through hand calculations, spot checks, or incorporation into the main roof framing system analyses, review the effects of permitted rigging loads and locations on the rigging- and catwalk-level framing. Review the load path for gravity loads up to the main roof framing (if used). Review the load path for lateral loads. Is the rigging level trussed to span horizontally? Are there braces up to the roof? If braces are slender, are they in opposed pairs so that one side can always work in tension? How are building expansion joints crossed (if present)—is the framing interrupted or is rigging run across the joint? This can affect the rigging load path and the forces induced in the framing.

Field observations. Walk the rigging and catwalk levels. Note evidence of rigging points used in the past, such as scrapes or scuffs in paint or dirt and small dings at member edges. Do members show local deformations? Are handrails bent? Have rails or posts been modified to address previous overloads from rail-mounted lights? Have loads been (improperly) hung on grating? Although not a structural item, it would be appropriate to discuss informally with the facility manager any safety issues noted, such as grating surface condition and handrail/guardrail condition. Note any winches, lights and lighting panels, and other heavy equipment mounted at the catwalk and rigging levels. Note old cables and fixtures abandoned in place. Include all these loads in reviews of structural adequacy (Fig. 10.20).

Report goal. The goal of this phase of work should be to provide the owner and facility manager with a clear understanding of the capacity of the present framing and, if desired, concepts for retrofits that could enhance that capacity.

OPERABLE ROOF SYSTEMS

Although currently limited to a few facilities, operable roofs are becoming more widespread for sports facilities desiring natural grass fields. For such facilities, new issues arise in addition to those discussed earlier. If the operable roof machinery is a major concern for the client, a specialist in mechanization of large structures should be involved in this aspect of the study. In any case, the following broad outline of study items may prove helpful to provide background and perspective.

Document review. Contract documents, manufacturer's drawings and operations manual, and facility management records should be gathered and reviewed as important first steps. How was the operable roof drive system proposed to function, and how is it running now?

Conceptual review. In addition to the static, wind, and seismic loading conditions applicable to conventional framing systems, interaction of structure, drive system force application, and potential differential motion should be considered. What drive force can be applied? If some wheels or bearings are stalled and that force is applied, what is the effect on the structure? If drives become unsynchronized and a roof panel is wracked, what is the effect? Ideally, multiple levels of operating safeguards should be present to stop operation before structural damage occurs.

Field observations. A thorough field investigation can provide many opportunities to observe functionality issues for the roof operating system. These observations should be the primary responsibility of the specialist reviewing the drive system, but they also can be useful to the structural assessment team. Some key observation types are discussed below.

For rail or track conditions, plot observations on scale drawings to relate locations to operations. Are there wear patterns? Consistent rubbing on one side of a rail may indicate misalignment or mislocation (Fig. 10.21). Are track pads walking out from under the rail? If so, in which direction? Are wear marks from rail clamps visible on the rail base? If so, in which direction? These may indicate overall rail movement.

FIGURE 10.20   Watch for physical damage from mislocated rigging loads or, as shown here, errant vehicles.

State of bearing cleanliness, lubrication, and wear should be noted. Check if bearings move smoothly, stick, or make noises during operation.

Note the state of cleanliness, lubrication, and wear of motors, drive chains, drive cables (if used), and indexed motion detectors. Note if drive operates smoothly, sticks, or makes noises during operation.

Note the line of wheel contact, pattern of wheel wear, and any surface blemishes. Flat spots could indicate locked wheels, tread cracks could indicate overloaded wheels, etc.

At track-rail splices and expansion joints, check for sufficient travel to handle anticipated thermal motion and angled joints for easy crossing by wheels.

FIGURE 10.21 Wear patterns on rails can provide clues about operable roof behavior.

FRAMING OF CONCOURSES AND LEVEL FLOORS

To maintain clear sight lines from the seating bowl, roof or canopy loads generally are carried down to the structural frame of the concourses (level floors) that typically form a ring along the outer edge of a sports facility. The nature of those facilities imposes special demands on framing that otherwise might be considered conventional. The special demands are explained below.

Lateral-Load-Resisting System

Concourse framing often is used to provide lateral-load resistance and stability for the overall building. Its design must accommodate both the need for open space for circulation of crowds and the unusually tall story heights required to work with seating bowl access. Shear walls and braced bays can provide lateral resistance economically, but they may restrict free circulation and use of the concourse areas. Moment frames do not obstruct circulation, but tall stories require special treatment to provide the necessary stiffness. One approach is to provide stockier columns and stiffer beams. Another is to subdivide the story heights by adding "flying beams" above the required headroom level or by using deep trusses instead of floor beams and connecting bottom chords to the columns (Fig. 10.22).

Building Expansion Joints

Concourses forming a "donut" around a facility can have overall lengths 400 ft or more in an arena and 600 ft or more in a stadium. Over such great lengths, even small

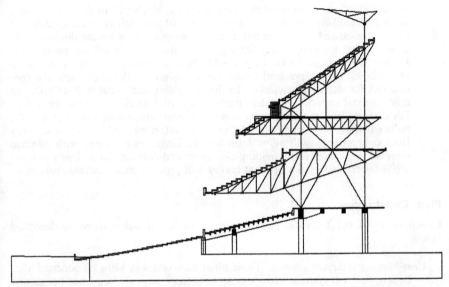

FIGURE 10.22   Column height at Petco Park is reduced by deep trusses framing into the columns.

horizontal strains can result in considerable lateral movements at each end, several inches for swings in ambient temperature (in non-climate-controlled spaces), and a similar amount for long-term shrinkage of concrete-framed concourses. These effects can be dealt with in several ways, as described below.

Multiple building expansion joints. These create short concourse segments. Joint spacing must balance the benefit of reduced lateral movement with the costs of constructing and maintaining the joints and of stabilizing individual segments. In areas of high seismic risk, because each segment created by building joints will "dance" separately, the joints must be wide to avoid pounding between segments. Mechanical systems crossing the joint must allow for this motion as well.

Open spaces or gaps that interrupt concourses. These create the effect of joints without the cost and maintenance impacts. However, gaps limit free circulation of patrons.

Pour strips. These are narrow gaps in the concourses with projecting overlapping reinforcing bars. They are left unconcreted as long as the construction schedule permits, allowing some concrete shrinkage to occur before completing the structure. Since shrinkage effects to be considered later are fewer, the permanent joints can be placed farther apart.

Jointless designs. These can avoid concourse expansion joints entirely, providing a continuous structure, if the lateral movements at extreme ends can be accommodated by the columns. In seismic areas, a continuous ring can provide better structural performance than multiple, small, odd-shaped segments while avoiding the construction and maintenance costs of joints. A continuous structure also can distribute lateral load over more columns. For this approach to work, laterally stiff elements such as shear walls and braced bays must be located towards the center of the concourse ring and never as "anchors" in the ring corners. Also, it is unlikely that the full range of thermal and shrinkage movements will be accommodated within normal design parameters, even for the double-height columns that typically occur from grade to the first framed concourse level. Pinning the ends of the columns would allow free movement, but at a considerable cost. A more practical solution is to design and detail columns and connections for ductile behavior so that local yielding can occur without affecting column axial capacity. For steel framing, compact sections provide this effect. For cast-in-place concrete, using slender column dimensions will minimize the induced strain, providing minimum vertical reinforcing, and lapping rebar away from column ends will allow formation of hinges at each end with minimal moment resistance, and providing closely spaced column ties designed to resist the shear created by the moment hinges will provide ductile column behavior.

Floor Construction

Concourse and level floors can be built using a variety of slab systems, as described below.

Cast-in-place concrete systems. These often have one-way slabs on concrete filler beams; flat-slab systems make less sense because bays are large, and beams and girders usually are needed for building frame stability anyway.

Steel-framed systems. These can use composite beams and metal deck, but relatively thick slabs are recommended to handle potentially high point loads from service vehicle wheels. Where continuous moisture exposure is possible, flexural reinforcement can be provided by rebar, and the deck can be considered as expendable. In any case, metal deck should be galvanized for reduced maintenance.

Precast on steel framing. This can be effective in some cases. The greater spanning ability of some planks can avoid the need for filler beams between main bents. A site-cast reinforced topping slab often is provided for strength to carry diaphragm forces when slabs distribute horizontal forces among frames, and for a level finished surface.

Precast on concrete framing. This is less common than cast-in-place slabs but still offers advantages in fast floor erection and prestressed long spans without the complications of site posttensioning.

Live and Superimposed Loads

Concourses typically are considered as circulation and egress areas, requiring design for a service live load of 100 lb/ft^2 or more. This sounds ample, but actual service conditions can be quite varied. Food concession vendors often deliver supplies using fork lift trucks or motorized tugs and carts. The concentrated wheel loads and impact effect of bouncing over floor irregularities may be a more severe test of the slabs than crowds are. Topping slabs over waterproofing membranes, provided when covering water-sensitive spaces, further complicate the structural picture. Topping slabs are not reinforced like structural slabs but can experience induced stresses under concentrated loads.

Walls that span between floors can create inadvertent load paths and apply unintended loads in any building type, but the presence of extensive tall masonry walls in many sports facilities makes this condition worthy of particular notice. The original design should show "soft joints" between wall tops and the floor slab above so that loads on the upper floor do not affect the walls. In addition, walls of rigid materials such as masonry should be isolated to allow building lateral drift to occur without generating damaging shear forces.

Field Observations and Maintenance Records

Careful documentation and mapping of observed leaks, cracks, joint repairs, and other damage can be very helpful in understanding the performance of the concourse framing and the causes for its behavior. Some typical field conditions worth noting are listed below.

Inoperative building expansion joints. Resulting from rusting or packing with debris, these can lead to excessive lateral movements at concourse ends because they tend to ratchet open rather than closed.

Slab cracks perpendicular to concourse length. These may also involve supporting beams. Cracks in this orientation and located at narrow points in the concourse or where less material is present owing to large openings may indicate insufficient flexibility in the structural system. Then strain changes can cause large net tension forces to occur.

Slab cracks diagonal to concourse length. These may indicate high shear stresses owing to restraint by poorly placed stiff elements such as shear walls and braced bays.

Cracks parallel to the direction of travel of service vehicles. These may indicate a vehicular load effect on flexible slab areas, such as topping slabs over membranes or insulation.

Map cracking in a relatively square grid. This may indicate general shrinkage cracking owing to insufficient curing during construction.

Wet or corroded areas. These should be checked against locations of wet services for concessions, including kitchens, bars, coolers, and beer supply pipe runs, and for locations of mechanical rooms with pumps or air-handler condensate drains.

Signs of water intrusion. These also should be reviewed with regard to the extent, treatment, and detailing of waterproofing membranes. Sometimes water that reaches the membrane pools there until it finds a weak spot in the flashing or membrane termination far from the original source. A membrane that pitches to drain, even below a level finish floor, is less likely to have this problem.

Number and spacing of topping joints, and topping joint lengths. Thin topping slabs require closely spaced joints, cut early in the curing process and extended to intersection joints.

Existing patches, previously sealed cracks, and previous joint repairs. These should be noted as indicators of building behaviors and possible recurring problems.

Cracked or crushed infill walls. These can indicate missing, inadequate, or nonfunctional isolation joints between walls and surrounding structural framing.

SEATING OR STADIA UNITS

A well-functioning seating bowl is crucial to all spectator sports facilities. The stepped surfaces that hold the seats are called *seating units* or *stadia*. In effect, they form a floor, but their design, detailing, construction, and maintenance differ from that of conventional floors, as explained below.

Seating Unit Construction Types

Seating unit construction generally falls into one of five types, each with its own design requirements and behaviors. Several construction types may occur in one facility.

Cast-in-place concrete on fill. In many large stadiums, the playing field elevation is set well below the surrounding grade level. Soil excavated from the field area is placed and compacted into a gently sloping berm. The lowest rows of seats are set on a slab on grade with a stepped surface that is cast on the berm. The cast-in-place nature of this slab means that significant shrinkage will occur as the concrete dries, so well-spaced contraction joints should be provided. In open-air stadiums, the slab is subjected to weathering and, depending on location, to freeze-thaw cycles and subgrade frost. Depending on the soil type and slope, sliding stability of the slabs may be an issue.

Cast-in-place framed slabs. These may be seen in older facilities with cast-in-place concrete concourse framing and rakers. To avoid the need for excessively thick slabs, rakers are spaced closely. Forming and shoring sloped-slab soffits and providing the open-bottomed forming for stepped top surfaces can be slow and costly. Placing and finishing the stepped concrete require skilled workers, and the resulting finish can exhibit wide variations. Durability issues for such slabs are similar to those at conventional exposed concrete. Where exposed to wide temperature swings, special care is needed to allow for strain relief through structural joints and flexible system framing.

Precast concrete stadia units. These are the most common seating elements used for modern sports facilities. Seating bowl geometry can be established with a constant step dimension for a group of seating units, permitting economical multiple reuse of precasting forms. Quality, geometry, and finish consistency can be controlled well. Units with two or, preferably, three treads speed erection. In addition to conventional reinforcing bars and wire mesh for crack control and strength, prestressing strands usually are included in long-span units for structural efficiency and cambering action. Long-span units with shallow steps can be flexible; added spanning strength or stiffness can be provided with minimal additional weight by forming *stems*, or downward-projecting fins containing prestressing strand (Fig. 10.23). The piecewise nature of precast bowls requires special care at end support and unit-to-unit connections, as discussed below.

Steel framing. Some older facilities used steel treads and risers to create the seating bowl. Depending on age, locale, raker spacing, and preference, they may use bent-steel plate for both tread and risers or may have risers of hot-rolled members with tread plates spanning between them. Because of the limited spanning ability, sealing difficulties, corrosion concerns, and noisiness of steel seating bowls, they are used rarely in construction of new major sports facilities. A notable exception is the new Invesco Field at Mile-High Stadium in Denver, where steel was used so that fans could still stomp to create the "Rocky Mountain Thunder" that intimidates visiting teams.

Aluminum framing. Aluminum can be extruded into custom shapes suitable for assembling as a "kit of parts." It also does not have the corrosion concerns of steel. However, it has one-third the stiffness and one-third the strength of steel, so it is more appropriate for relatively short spans. The most common application of aluminum seating bowl systems is as stand-alone manufactured bleachers. Bleachers are discussed later in this chapter.

Seating Bowl Loads

The design live loads for seating bowls are surprisingly varied. Therefore, judgment is required in its application. Key points to consider when evaluating or establishing appropriate live-load criteria are provided below.

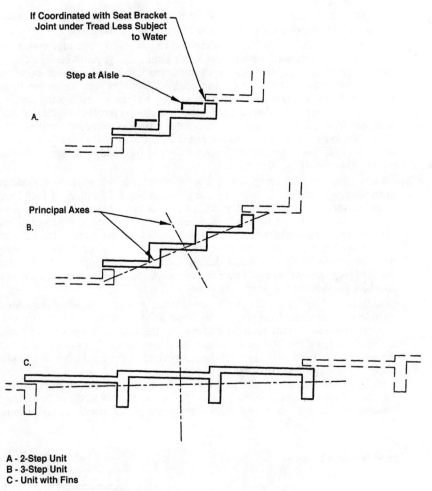

If Coordinated with Seat Bracket Joint under Tread Less Subject to Water

Step at Aisle

A.

Principal Axes

B.

C.

A - 2-Step Unit
B - 3-Step Unit
C - Unit with Fins

FIGURE 10.23   Stadia (seating bowl) units.

Realistic fixed seating filled with sports fans should average out to 30 or 40 lb/ft^2 depending on seat pitch and row spacing, with seats adding about 5 to 10 lb/ft^2.

Design loads of 40, 50, 60, 75, or 100 lb/ft^2 may be required, depending on the applicable code and, more important, the local building official's interpretation of assembly, fixed seating, and grandstand spaces. While 100 lb/ft^2 may be appropriate for the aisles (plus step weight), they cover little area and are unlikely to be fully loaded simultaneously with the seats.

Both statutory and realistic loads should be considered. For example, precast concrete units that rely on prestressed strand to create both compression and flexure to offset gravity load effects are sensitive to loads used. If strands are chosen to resist a 100 lb/ft^2 load that never occurs, the extreme amount of prestress creates excessive upward camber that will grow with time. A better approach is to prestress for the realistic load and add mild reinforcing to meet the statutory strength requirement. A review of precast shop drawings and calculations should indicate if this approach was used in the facility being assessed.

Special tub units deserve additional attention. The simplest structural system for seating has rakers running under every seating unit. This requires sufficient headroom to accommodate the raker depth under the lowest row, in addition to the seating unit tread thickness. For more compact facilities such as arenas, headroom can be tight. Then the raker is set higher, and the lowest seating row is a "*tub*," named for its U-shaped cross section that incorporates a stiffening curb along its front edge. For a proper load path, a two- or three-row unit is used, with a thicker riser between rows 1 and 2 to act as a beam. The row 1 tread cantilevers off this beam, and its ends are dapped (notched or pocketed) to bear on the raker tips (Fig. 10.24).

Secondary framing areas can generate considerable loads that must be considered in the design of the stadia units providing their support. Examples of such areas include stairs and seat steps atop cross-aisles at large breaks in seating bowl elevation and platforms for disabled access and seating areas. Often these are "frame over" construction, where a conventional stadia unit runs underneath and a short wall and precast panel floor are placed on it. The dead load certainly must be considered in stadia unit design. Establishing an appropriate live load requires judgment. Other loads may include suspended speakers, signs, monitors, and subroof systems that keep spaces below dry.

Longitudinal load values usually can be found in the governing building code. Most include a longitudinal load per foot of bleacher footboards to reflect the "start and stop" forces of fans entering and leaving the stands. While seating bowls are not bleachers, the load makes sense. Its value is unlikely to affect design of the stadia units themselves but will be transmitted to the rakers below and will affect their design for transverse loads.

If seismic requirements apply, all dead loads on the seating units contribute to mass. The contribution to mass from people (live load) realistically can be ignored because people will sway and cannot apply significant lateral forces without

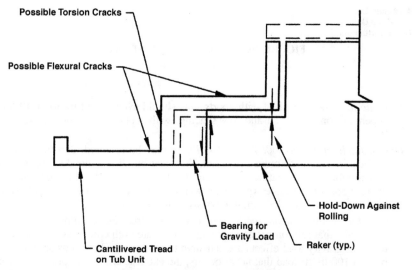

FIGURE 10.24 Tub unit for arenas.

intentionally "holding on." Their contribution to destabilizing gravity-based forces could be included. Depending on the code and structural system, it may be appropriate to apply the lateral-loading coefficient specified for diaphragm design, or modal response values from a dynamic analysis may be used if more critical. The connections may be considered equivalent to those on precast facades, with large safety factors required for calculated horizontal forces. This may control the design of the stadia unit end connections.

Transverse load values in building codes reflect bleacher loads per foot of footboards to reflect the "start and stop" forces of fans walking up or down the stands. While awkward to perform this operation in large numbers with fixed seating present, allowance for some load is still recommended. As indicated earlier, seismic requirements may govern, using diaphragm design criteria or dynamic analysis results, and connection forces may require further factoring if part of a precast system.

Stiffness and Behavior

Seating units that are too flexible or "bouncy" can be uncomfortable to seated fans. Criteria for stiffness recommendations have changed over time. A structural assessment should include information on the original design criteria and on its performance compared with current criteria.

The Precast Concrete Institute (PCI) Manual, 4th edition of 1992, recommended a high level of stiffness to provide a first-mode frequency of 5.2 to 6.5 Hz or higher to keep well clear of annoying vibration response.

The PCI Manual, 5th edition of 1999, Sec. 9.7.1, reflects subsequent testing and review of facilities that have shown satisfactory performance for frequencies between 2.8 and 3.5 Hz. Note that the lower frequency is for longer spans at angled units (arena corners), where ties to shorter spans help damp motion. At uniform-span bays, using the higher value may be prudent.

Appropriate masses for these frequency calculations should be the realistic, not statutory, loads.

Interaction of seating units and flexible supporting rakers should be considered. At cantilevered rakers that are being designed to meet a minimum frequency, we suggest designing the seating unit frequency significantly different (higher) from the raker frequency to avoid resonance. Existing facilities should be reviewed with this in mind as well.

End Details

Seating unit end conditions are an important component of overall building structural assessment because they are subject to frequent movement from thermal and loading changes and because they are so numerous. The following descriptions of intended movements can help guide that assessment.

Gravity loads. Normally these are transferred to supporting rakers by a pad at each end under each seating unit riser because the risers act as beams and are stiff in shear.

Small end rotations. The thin, stiff neoprene pads typically used under precast to avoid point loading may be sufficient to allow normal rotations.

Free longitudinal movement. To allow for thermal, long-term creep and shrinkage movements, typically one end of each unit is fixed to its supporting raker, and the other end is detailed to allow free movement along the unit length. It should resist transverse movement to keep seating units from sliding off the raker (downhill).

Freedom of wracking movement. If needed, such as where large seismic displacements can wrack seating bowl bays, one corner can be fixed, and the other three corners can be free for longitudinal movement. Sometimes there is enough play in fixed connections that the corner nearest the "fixed" point need not be detailed as free. This provides a second load path for longitudinal forces.

Constructibility and verifiability. These are key attributes of successful details. Many possible details have been used on projects, including cast-in inserts on raker tops, raker pockets to be grouted, sleeves through precast, welded nuts in precast pockets, etc. For reliability, the best systems allow the installer to see all the components in the connection. Details that are blind, such as pins thrust into concrete-filled pockets, do not offer that assurance.

Tub unit details. These differ from those of typical seating units. In addition to the gravity, longitudinal and transverse considerations discussed earlier, they must consider stability against overturning. This could be due to a dead-load center of gravity that lies beyond the support point; this is a difficult erection problem and one best avoided by careful design. Another very plausible condition is that of a fully loaded front row not counterbalanced by live load in rear rows of the unit. These conditions generally require connections in pairs to take out torsion as a force couple. Gravity loads that may be shared between these connections and normal bearing pads should be considered. For example, connections working in parallel with bearing pads could take load if the pads are not in contact or are compressible.

Side Details

Connections along stadia unit edges address vertical, transverse, and longitudinal forces. A variety of connection types can be used to handle those forces, as described below.

Vertical ties are used to avoid tearing open the horizontal sealant in joints between units and to distribute transient loads among adjacent units. If horizontal flexibility is needed (see below), slender tie rods can be used, but check for buckling in the event of compression from differential loading on adjacent units.

Horizontal-shear transfer connections are needed where diaphragm action is used to resist lateral loads and/or brace rakers. Note that for large shear loads such as can occur in severe seismic zones, many connections may be needed per unit. The field labor to make these connections may render this approach uneconomical compared with providing horizontal strength through underseating braces or transverse raker strength. Where such connections are provided, their details and installation methods must allow for varying gaps between adjacent stadia units owing to differences in camber. Forcing units closer can be time-consuming, if practical at all.

Horizontal-transverse-force transfer connections (in the absence of shear connections) are less common. Stadia units are wide compared with their span, so in-plane transverse stiffness is high, and movements in that direction are very small.

Field Observations

An informed and focused field observation program can be developed based on the preceding design and detail background. In addition to the overall observations of seating bowl structural condition, specific attention should be paid to the following conditions.

Midspan soffit cracks, indicating possible flexural overload. If found under secondary framing, these may indicate that the "frame over" loads were not considered in the design or that the secondary framing was added later.

Cracks near unit ends and corners, indicating details that do not accommodate movements. If a pattern of such cracks is found to be a widespread occurrence, a probe to see the nature of the connection is recommended, even if patching will be needed.

Cracks in units and curbs at railing posts. These could indicate mechanical damage at locations with insufficient edge distance to resist overturning forces applied to rails, expansion due to corrosion or freezing of trapped water or improper use of expansive grout to install parts. Changing mounting details may be required to bypass these problems.

Connection and tie conditions between units. Since these connections are small and usually below the units, these observations may require binoculars and/or high scaffolds.

Joint sealant damage that indicates a pattern of relative movement between units. Seals are also damaged by normal maintenance activities such as power washing, so the meaning of *damage* is subject to informed judgment.

Tub unit conditions. Note the presence of cantilevered tub units. If present, note the connections used for stability and their condition. Look for cracking at the rear edge of the bottom tread, on the second tread, and along the riser beam owing to cantilever, backup, and torsional action.

Units supporting "frame over" construction. These may exhibit water trapped under "frame over" construction that can cause deterioration, in addition to the loading cracks noted earlier.

Drainage along units may be inadequate, as indicated by standing water or stains. Drainage can be affected adversely by the inherent camber of the units and by conflict between ideal drain locations (at supports, where camber creates midspan high points) and practical locations (to the side of supports to avoid fouling the structure).

SEATING BOWL SPECIAL FRAMING

Seating units alone are not sufficient to create a functional facility. Access and egress require vomitories, cross-aisles, stairs, and platforms. Each of these components has its own loads, requires stabilization, and has the potential to act as a moisture trap or leakage path depending on the details, sealants, and maintenance actions applied.

Framing Conditions

Assessment of the seating bowl requires understanding the methods used to support it. Common framing conditions include those listed below.

Vomitories, the entry openings in the main seating bowl, are created by interrupting typical seating units. Vomitory side walls often are created by precast elements. The interrupted seating units bear on their top edges, and they load a support structure below. These walls must have connections at top and bottom sufficient to transfer gravity loads, allow longitudinal movement of seating units (their far ends should be fixed to rakers), resist movement along the wall (so that it acts like a shear wall to hold the seating units), and keep the wall itself from toppling (Fig. 10.25).

Secondary stairways and infill seating on cross-aisles, often provided at a vertical break in the seating bowl, are supported on low walls that bear on other units or the main framing below. If precast elements are used, care is required to avoid a "house of cards" with limited lateral stability. Use of cross-walls and end walls tied into the stair or tread units can create strong, stiff boxes that are stable. Compatibility of deflections between the units and supports also should be considered. An infill seating unit cast as an inverted channel may be much stiffer in flexure than the flat floor it is supposedly loading uniformly. In reality, it will bear only at the "hard spots" over each support and self-span in between.

"Frame over" platforms for disabled access and viewing were discussed earlier for their impact on the supporting typical seating units. Their connections must provide for proper load transfer and stability so that the platforms cannot "slip downhill" on the seating unit. Their layout also should avoid bridging across joints between seating units or specifically allow for movements that could occur (Fig. 10.26).

Field Observations

With a good understanding of design and construction issues, secondary framing areas can be reviewed for overall stability (are the assembled elements square and linked together), for the structural condition of elements and connections (many connections may be concealed, so indirect methods such as looking for rust stains or concrete spalls may be used), and for trapped or leaking water between layers of construction.

FIGURE 10.25   Precast panels used to frame out vomitory openings like these must be properly connected for stability and for load path.

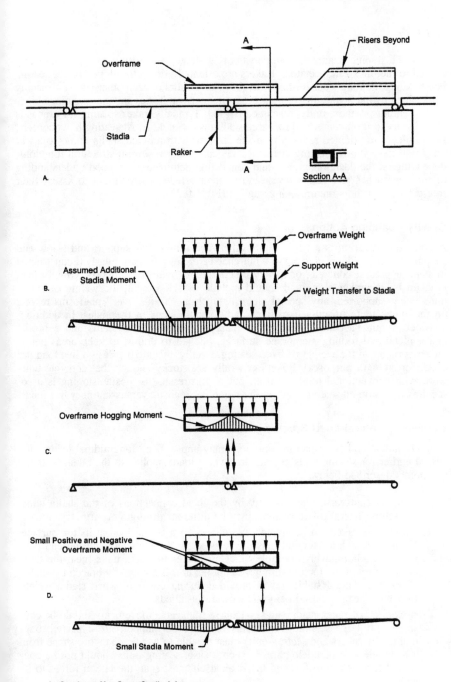

FIGURE 10.26   Stadia overframing considerations.

RAKERS

Rakers, the sloping beams that support floors with stepped seating areas, are uniquely related to facilities for spectators. Rakers are oriented "down the slope" of the seating bowl. Along the long sides of the bowl, they run parallel to each other and at a spacing determined by the layout of seating access points or vomitories and the spanning ability of the seating or stadia units (discussed below). Typically, raker spacing for arenas is about 30 to 36 ft on center and for large stadiums about 42 ft. At seating bowl corners, rakers often are oriented radially to make the turn with mitered stadia units, so lower units will have shorter seating spans. Rakers can perform several structural functions, depending on the design approach and connection details used. A good understanding of the approaches and connections, and of their effects, is important to assess their role properly in the structure being studied (Fig. 10.27).

Gravity-Spanning Action

A common condition is a beam running between a high-end support and a low-end support. It may be detailed to act as a pin-ended simple span, particularly if constructed in steel or precast concrete, or it may be continuous with the supports. Gravity load is taken by the raker alone and should include its self-weight, the weight of stadia units being supported, any special elements such as "frame over" platforms resting on the stadia units (discussed earlier), and concrete steps in the aisles. In addition, consider weight suspended below the stadia for mechanical systems and subroofs (sheet metal and roofing membrane surfaces, pitched to drains, to keep areas below the seats dry) and the weight of fixed seating. Finally, include the design live load per code (for strength purposes). Rakers generally are stocky enough that concrete units work with conventional reinforcement, but if prestressing or posttensioning is used, see the discussion on seating units earlier regarding realistic versus statutory live loads.

Transverse-Lateral-Load-Spanning Action

Several loading and deformation conditions may apply. The "longitudinal loads" discussed earlier for seating units become transverse loads applied to the raker. Note the key points described below.

Delivery of transverse loads is only by the fixed connections of the stadia units. Therefore, lateral tributary areas may be different from gravity tributary areas.

Torsion of the raker can be created by delivery of a lateral load at its top surface. If rakers act alone, torsion must be resisted by the members and the end connections. Alternatively, multiple "kicker" braces to stadia units (near the fixed end) can resist the torsion through stadia flexure. If torsionally braced by stadia, check how lateral stability was provided during erection, when stadia weight may have been imposed before braces are installed.

Compatibility of raker, stadia, and concourse transverse movements should be considered. If large seismic movements of the upper support are expected, bays defined by rakers and seating units must be able to wrack, changing shape from a rectangle to a parallelogram. In such a case, seating units should not be connected for shear transfer, and the raker itself must span the lateral forces to its ends.

Laterally self-spanning ("stand-alone") concrete rakers require rebar on both vertical faces, and steel rakers require stocky or boxed-out sections. For both materials, torsion must be addressed as discussed earlier.

Diaphragm-spanning designs can be used where wracking is not significant. Stadia units can be joined with shear-transmitting connections to form a stiff diaphragm

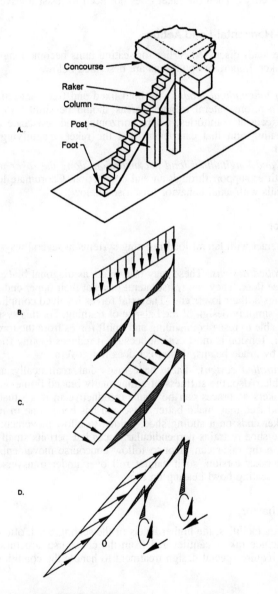

A - Isometric of Typical Condition
B - Raker Moments from Gravity Load
C - Raker Moments from Transverse Loads
D - Raker Moments from Longitudinal Load (for Foot Sliding)

FIGURE 10.27 Raker load paths.

and connected to concourse and/or foundation framing at upper and lower seating bowl edges. Then the raker does not need to resist transverse loads.

Longitudinal-Horizontal-Load Action

The transverse loads discussed earlier for seating units become longitudinal loads applied to the raker. Important conditions are described below.

Delivery of longitudinal loads is by both fixed and free connections of the stadia units. Therefore, lateral tributary areas should be similar to gravity tributary areas, except at vomitories, where horizontal load resistance may be provided by a seating unit that cantilevers past the raker, magnifying the lateral load imposed.

Drag strut and longitudinal load accumulation along the raker must be delivered to the stiffer support through an end connection. Coordinate longitudinal loads and details with axial behavior (discussed below).

Axial Behavior

Rakers can interact with lateral-load-resisting systems in several ways, as listed below.

Wind-controlled designs. These may use rakers as diagonal braces to stabilize the concourse floor. They can resist lateral loads at their upper end and deliver them to footings at their lower end. The axial forces involved complicate raker design but may simplify design of the balance of framing. For this system, the footings must be able to resist both sliding and uplift forces from the force directed along the raker. Tension is most critical because it reduces footing friction. If footings are tied by grade beams, this effect does not govern.

Seismic-controlled designs. These have loads that, realistically, are large and unpredictable. Also, the stiffness of a diagonally braced frame attracts more load. Using rakers as braces can be counterproductive in this situation. Cutting the lower end free may make better sense and has been done in modern facilities. Each raker ends on a sliding shoe, guided to allow movement along the raker axis. The shoe restrains perpendicular forces but permits small vertical-axis rotations, so the raker can pivot to follow concourse movements. The shoe also restrains raker torsion, so it cannot roll over under transverse seismic forces from the seating bowl bearing on top.

Cantilever Behavior

In modern sports facilities, the higher levels of the seating bowl, often called *club* and *upper deck*, include rakers cantilevered from the concourse structural system. Cantilevered rakers require special design treatment to handle the conditions described below.

Flexibility. A cantilever is inherently much more flexible than a simple span of the same dimension, making control of deflections and management of vibration characteristics important design criteria.

Vibration and comfort. Design criteria for audience comfort on cantilevers are not codified. However, recommended stiffness levels vary, from providing a first-mode frequency of at least 3.4 Hz for sports and soft pop venues to 6.5 Hz for hard pop venues. At the time of design, the criteria should be discussed with the facility owner. Sports facilities usually are designed for sports usage, with rock concerts recognized to be infrequent exceptions that may result in noticeable movements with an active audience. The calculation of frequency should be based on realistic dead and live loads, not full code live loads on all surfaces.

Backup and support. A cantilever requires a secure moment-resisting support. Often this takes the form of a triangular truss connecting to two concourse levels, so the resisting moment is created by interstory shear. This shear must be reflected in design of the building frame. Where freedom of interstory motion is a concern, such as in severe seismic zones, a deep, stiff, single-story cantilever backspan may be required.

Detailing. Cantilevers have stepped top surfaces to receive seating stadia units. In concrete rakers, these steps interrupt critical top reinforcing and complicate shear reinforcing. In steel rakers, flange forces must be developed each time they step. For both situations, great care is required in detailing and construction to avoid cracking or failure at corners.

Field Observations

The design issues outlined earlier suggest the following checklist for observations:

Raker design and support conditions. These should be compared with the contract documents (if available).

Distress cracks, spalls, or stains. These should be noted on all visible surfaces, generally the raker soffit and sides. Pay special attention to cantilevered raker cracking or distress originating at top-surface reentrant corners.

Relate raker behavior to the seating stadia units. Do rakers resist lateral loads, or are stadia connected to act as a diaphragm? Are rakers subjected to torsion, or are they braced to resist it?

End connection behavior. This should be related to the nature of loads being carried. Note if raker ends have slide bearings and, if so, their condition.

Dynamic behavior. If practical, determine the characteristic first-mode frequency of the rakers when the seating bowl is empty and with an audience. Relate to any complaints of excessive motion. Discuss with the facility owner the link between type of performance and audience comfort.

Raker/stadia interactions for durability. Do dissimilar metals or finishes create the potential for corrosion between raker and stadia? For example, at steel rakers, if stadia bearing points include galvanized-steel embedments, are the seat angles on rakers also galvanized or otherwise protected? Does the connection method used in the field create or avoid corrosion problems? For example, has field welding burned off galvanizing or paint with no subsequent touchup?

BLEACHERS

The dictionary definition of *bleachers* is "a set of tiered, inexpensive, unreserved open-air seats." Designers of sports facilities have their own definition, based on the differences from a main seating bowl. Typically, the main bowl has fixed seats, a closed tread-and-riser system, a solid feel underfoot, permanent construction, and open, usable areas beneath each deck. Bleachers have continuous seating boards and foot boards, possibly with gaps between them, more flexible construction, provision for easy disassembly and reassembly, and no requirement for usable space below. These differences make the design, construction, and evaluation of bleachers a separate topic from the main bowl. The descriptions below give a general idea of bleacher characteristics.

They are not intended to be comprehensive because the dimensional scale of bleachers can vary widely.

Loads. The vertical and lateral live loads listed in most codes for bleachers do indeed make sense for these structures. Without fixed, high-back seats, it is quite possible for sports fans to be squeezed tightly in the stands. This can result in high live loads, at least over small areas. It is also quite possible for fans to generate lateral loads from the start-and-stop actions of walking toward exit aisles along the foot boards or of walking down the bleacher rows toward the field. Rhythmic fan activity also could generate sway forces.

Construction materials. To satisfy the requirement for easy assembly, disassembly, and reuse, bleacher structural elements are relatively small and light. Aluminum and wood are common, with structural steel being used in selected locations where greater strength is needed. Concrete generally is limited to foundations and possibly low piers because it cannot be dismantled or reused easily.

Vertical-load-resisting framing. Aluminum and wood, while both fine structural materials, have stiffness and strength values that are a small fraction of those of steel. Therefore, spans are kept small. The "boards" of extruded aluminum with proprietary cross sections or wood planks with reinforcement are able to span between closely spaced rakers. The rakers are also relatively small, light extrusions or other sections with L-shaped brackets on top to receive the boards. To minimize peak moments and deflections, rakers often run as continuous members over the closely spaced posts. Close post spacing also keeps the load per post and the individual post sizes small.

Lateral-load-resisting framing. Triangulated bracing is the classic lateral system used for bleachers. In the transverse direction (across the boards), the rakers and posts form stable triangles. This is a very efficient and robust system as long as each raker member and connection has adequate strength to gather and deliver lateral loads to the front footing, and the footing has adequate weight and soil embedment to resist simultaneous vertical and lateral forces.

In the longitudinal direction (along the boards), each line of posts is stabilized by a series of X-braced bays. The diagonals can be rods, angles, or heavier members. By using X's, the diagonals can be designed for tension only, permitting lighter members. Connections often are made with bolts or pins for easy disassembly.

Distributed braces can restrain thermal movements. Typically, X-braced bays are distributed evenly along the length of the bleacher to limit the load per bay. Smaller forces per bay permit smaller braces, connections, posts resisting the downward component of the brace, and footings subject to potential uplift and shear. Distributing the braces means that the boards are not being asked to act as drag struts that accumulate significant lateral loads and deliver them to a central location. This permits smaller boards and connections at splices. However, the bleacher is "locked in" between each pair of braces unless specific provision is made to release longitudinal forces. For bleachers that are short in length, the forces generated by thermal strain are small and may be absorbed by play in connections and elastic strain in members. For longer bleachers, forces could grow large enough to overload braces, longitudinal members, connections, and anchor bolts.

Field Observations

If possible, begin with a comparison of design drawings with the existing field conditions and a review of the design calculations. However, many modern bleachers are provided on a "turnkey" basis, in which the manufacturer calculates, designs, fabricates, and installs the bleachers as a system. Only foundations may have been built by a contractor other than the bleacher supplier. In this situation, documentation may be sparse. Also, note that design criteria different from those for building construction may be applied to these structures, given their unconventional materials and "temporary, removable" status. For example, they may use the National Fire Protection Association's NFPA 102 Standard for Grandstands, Folding and Telescopic Seating, Tents and Membrane Structures. Observations should cover general structural conditions, signs of corrosion, locations of mechanical damage, etc. Four areas of bleacher structures deserve special attention: connections, bracing removal, thermal effects, and low-cycle fatigue.

Connections generally are designed for light loads and use mechanical fasteners. For shorter bleachers, many of the connections may be within easy reach of the ground or can be reached by climbing. Such connections may be susceptible to gradual loosening or intentional tampering.

Bracing removal and damage are potential problems. Because the bracing members are both numerous and relatively light, a facility operator may think it harmless to remove "just a couple" of them to create usable space for concessions or improve the access path. The location of all such removals must be established and analyzed to determine their acceptability. Possible effects include increased drag strut action that exceeds member or connection capacities, overload of members, connections, or foundations for the remaining braces in that line; or loss of stability of whole lines of bleacher posts. Also, low braces are subject to impact by service vehicles; a bent brace is effective.

Thermal overload effects also should be investigated. For long bleachers, endmost bays will should show the greatest effects if no internal relief joint is provided. It is possible to fail braces in outer bays without overall collapse, but the remaining system factor of safety is reduced. Other bleacher elements also could be damaged, such as longitudinal boards, their splices, and their connections to braces. Tension failure of longitudinal members can occur anywhere along the bleacher, but forces will be highest midway between joints or ends. If relief joints are provided, their performance should be checked. Look for evidence of recent free movement with temperature changes, for rust or evidence of mechanical jamming, and for signs that motions have been exceeding their design capacity. This may occur if some joints work and others do not.

Low cycle fatigue is related to the overload effects described earlier. Thermal forces may not be large enough to yield or fracture a member or connection immediately, but multiple cycles of daily or seasonal temperature changes can "work" the material and result in a failure later. Thus inspection of critical members and connections should not be a one-time exercise but rather part of a regular maintenance cycle.

SPECIAL SPORTS FACILITY FEATURES

This section is a catch-all to highlight the types of features one might encounter in a sports facility. It is neither comprehensive nor intended to reflect what one may expect in any particular type of facility.

Broadcast Camera Platforms

Sports fans at home have come to expect extensive visual coverage of their favorite teams from numerous camera positions using extreme telephoto lenses. For sports facility owners, the question is, Where to place those cameras? An ideal location is one that provides a clear view, is close to the action, is stable, and does not displace ticket-paying fans. These criteria are in conflict, so all positions are compromises. The most stable locations are those on or close to main building columns, which means near the concourse areas. These are far from the field or court, and their views are subject to interference by standing fans unless the seats in front are blocked from use. Some owners choose to use *camera baskets*, small platforms cantilevered in front of seating decks that are long cantilevers themselves. The views are great, and no seats are lost, but any motion will be magnified by the telephoto lens (Fig. 10.28). It is impractical to greatly stiffen the main seating cantilever just for TV, so the owner first should understand the limits of this position. Then the basket should be designed to minimize its own contribution to camera motion. Design measures may include stiff framing members, a torsionally rigid floor system to minimize basket twist, and a way to introduce moderate damping of vibrations that do occur. If asked to assess the structure of such platforms, the reviewer should understand, and explain to the owner, what may be improved and what level of performance is, on a practical level, "as good as it gets."

Advertisement Panels

To help defray the cost of sports facilities, ad panels usually are mounted on edges of seating decks. Structurally, their effect is limited to load. Different panel types require different handling.

Ad panels occur on the front edge of stadia units, where load-carrying capacity is limited by the strength of the front curb and/or the first interior riser and cantilever action of the first tread, depending on the location of support points (see the tub unit discussions earlier). Simple light-box panels have relatively low weight. They can be allowed for easily in the design of seating units by the precaster if the criterion is established early in construction. Even where it was not anticipated in their design, on investigation standard units may still have adequate reserve strength to carry the added load.

Heavier systems, such as matrix boards, are effectively giant flat-panel computer displays. Even though they are relatively thin, they are usually too heavy to be carried on a seating unit. They also may cause the unit to deflect more than the panel can tolerate. Such panels require separate framing that spans between raker tips and is torsionally stiff to minimize tipping deflections. Framing to rakers that are "tucked up" behind tub units can be particularly difficult.

Backstop Systems

Baseball venues are sure to require large-scale backstops to help protect fans from errant balls. Backstops generally are either self-supporting or cable-supported. Self-supporting systems are structurally simple and easy to assess. A series of posts stands behind home plate, set in sleeves embedded in concrete footings. A line of beams connects post tops, and backstop netting is stretched from beam to the ground. For a structural assessment, critical field observation points are the beam-post connections and the post conditions at grade.

Cable-supported systems are more complex. A main cable is anchored to upper deck rakers, one end above first base and the other above third base. The cable forms an open U in plan, being tugged horizontally by several other cables that follow a radial layout centered on home plate. The radial cables slope upward from the main cable. At each intersection of a radial and main cable, a hold-down cable runs down to an anchor at field level, pretensioning the system. The backstop net is laced to the main cable and each hold-down cable. Hold-down force should be large enough to permit a worker to hang on the net without the cables going slack. The system is under significant tension. Changes in geometry generate changes in tension. For example, if the main and radial cables engage different building sections, joint movement

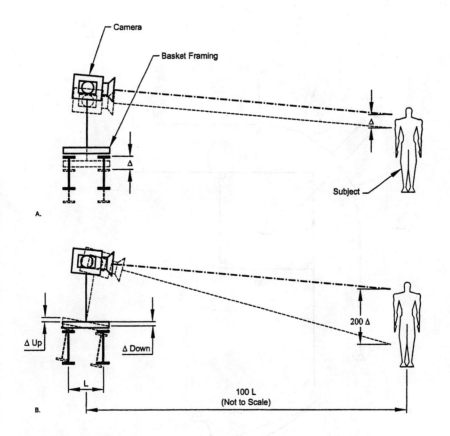

A - Translation of Basket Has Little Effect on Image
B - Twist of Basket Causes Large Movement of Telephoto Images

FIGURE 10.28 Deflections of camera baskets.

may overload cables. A structural assessment should include the condition of cables, connections, and anchor points. It also could include measurement of the geometry and size of all cables, from which tension forces can be calculated (Fig. 10.29).

Sunscreen Canopies and Windscreens

Outdoor sports facilities often have canopies above the top rows of seats, providing some sun protection to fans. The gap between canopy and top seats is often filled with fencing or open-weave synthetic fabric to provide protection against high winds in that exposed location.

Canopies have been built with everything from precast concrete to fabric and lightweight metal deck. They are carried on cantilever beams, normally projecting from main building columns. For canopies constructed with heavy materials, gravity loads may control the design. Loads may include snow (where applicable), loudspeakers, advertising banners, and other incidental suspended items. For lightweight canopies, uplift may be a critical design condition, particularly if they are framed with open-web joists or W or C shapes with unbraced bottom flanges.

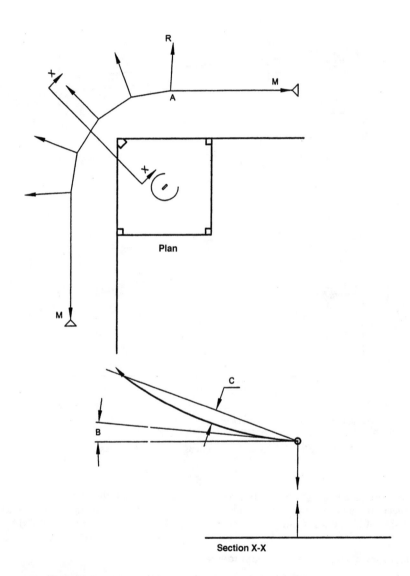

Plan

Section X-X

A - Main Cable Bends at Radial Cables - Angle < 180°
B - Radial Cable Slopes Upward to Lift Main Cables and Backstop Net, Angle > 0°
C - Sag Determines Angle B, Depending on R and Cable Weight
M - Main Cable Tension
R - Radial Cable Tension

FIGURE 10.29 Cable backstop issues.

Windscreens can impose significant loads on supporting structures, depending on the framing method. For screens freely spanning from canopy edge beam to seating unit, the catenary pull of the fabric can be several times larger than the direct wind load.

Field observations should include a review of design assumptions, a review of suspended loads, and a check of the adequacy of bracing and the deck connections and the condition of structural connections and screen connections.

Interfloor Elements

Sports facilities differ from conventional construction by the presence of large interconnecting sloped framing and slabs. Major ramps are important for the efficient circulation of tens of thousands of fans. They also are used for circulation of wheeled service vehicles.

The interaction of ramps and concourse floors should be addressed in the facility design and detailing. Parking decks with internal circulation have centrally located ramps that involve a large part of the deck area, so ramped framing can be integrated into the overall structural system. Sports facility ramps usually are located at several points along the building edge and represent only a small fraction of building area. Therefore, sports facility ramps cannot control behavior of concourses, but concourse movements can force ramp displacements. Unfavorable behavior can be controlled in several ways, as described below.

Isolation of ramps, in their own stand-alone towers, avoids the problem of imposed displacements. However, ramp framing must have sufficient strength and stiffness to stand by itself, which may carry a cost premium. Also, the expansion joint between each concourse floor and the matching ramp landing will be rather large, will require maintenance, and may affect smooth egress if damaged.

Jointing of individual flights of the ramp, while connecting landings to concourse floors, allows interfloor movement without inducing forces in the ramp framing. Joints are still required, but the required travel of each joint is smaller. The framing details at these joints add structural complexity.

Jointing of individual flights also may be needed at large precast interfloor stairs to avoid damage during severe building motions such as seismic events. Similar attention must be paid to escalators. Where large movements are anticipated, special seismic joints may be required at each end of the escalator. The possibility of finding that an escalator has "walked" to a new position after an event is recognized as the price to pay to maximize total interfloor drift capacity.

Flexible ramp framing can minimize joints and provide a smooth egress path while allowing for interstory movements. In this system, the ramp relies on the concourse framing for lateral stability. Ramp framing must be flexible enough to follow concourse movements without inducing large forces.

Field observations of ramps and stairs should note the method of isolation, if any; locations of resistance to lateral motion; and any evidence of distress where compatibility of motion would indicate large forces might be induced. If stand-alone ramps are used, the presence and condition of the self-supporting structural frame should be noted.

FACADES AND ENCLOSURES

Cladding a sports facility differs from cladding a conventional building in three ways. First, the story heights are much greater, perhaps 20 to 30 ft. Spanning these heights using the cladding system itself is often impractical or uneconomical because standard mullion sections will not be sufficient. Thus the main building frame may receive subframing to divide the story height into workable increments.

Second, the story drifts are much greater than for conventional buildings. This is due to the taller stories and the fact that interstory drift limits are not critical design criteria for these facilities. Patterns of joint movement also may be nonuniform, depending on the behavior of any subframing provided. Cladding joints require greater capacity. Also, depending on the nature of the cladding, connections back to supports may have to accommodate large differential motions as façade panels "stand tall" while subframing behind "leans" with the building, or vice versa (Fig. 10.30).

Third, there may be locations of comparatively large angular rotations at certain elevations, where roof systems meet the walls. As roof framing deflects, ends rotate and also may rise. This transition cannot be handled by standard joints. Often a change in materials and/or a change in plane is used to conceal the condition.

A structural assessment can include the condition and performance of building framing and subframing where it can be accessed and observed. Condition and performance of tieback connections from cladding to framing also can be noted.

SCOREBOARDS

In large sports facilities, scoreboards require major structural framing. They generally are framed in steel and stiffened with diagonal braces and include specialized support levels to receive video screens, "jumbotrons," matrix boards, and ad panels. The electronic elements have very specific support requirements in the geometry of pickup points, the tolerances of construction, and the deflections allowed under load.

Initial design of scoreboards should be straightforward. For known or assumed mounted weights and surface areas, the gravity, wind, and seismic loads can be developed. The framing is then sized to provided the required capacity.

FIGURE 10.30 Owing to long spans, backup framing for precast cladding can be extensive.

Structural modifications often start to creep in during final installation of electronics. Ideally, dedicated electronics support is provided in a separate plane of framing in front of the main framing, and main members do not obstruct necessary access points. When this is not the case, modifications to the main structural frame may be required. Renovation and retrofit of an existing scoreboard often add a second set of modifications because the original designer could not have anticipated the needs of devices years or decades in the future (Fig. 10.31).

Structural assessment of existing scoreboards should include a review of drawings to understand the original design intent, careful observations of the current framing system to note locations where significant modifications have been made (some of which may not have been reviewed by the structural designer), and observations of the conditions of members and connections. Complex steel-braced frames offer many potential locations to trap water, from W member webs oriented "weak way" to pockets formed by intersecting members and gussets. If the original design included drain holes, they should be checked for proper function. If the design did not, locations of trapped water or visible corrosion from intermittent trapped moisture should be noted for possible remedial drain holes.

A review of the structural design using as-built conditions may be appropriate if equipment loads, equipment locations (projecting much farther out, for example), or environmental criteria such as wind and seismic loads have changed since the original design.

FIELD LIGHT TOWERS

These stand-alone elements are major structures too. Just as for scoreboards, light towers require attention to the condition of framing members and connections, particularly at potential water traps. One design criterion unique to these features is the need to minimize angle changes at tower tops. If a stiff breeze causes the towers to twist or to rotate too much forward and backward, pointing angles of lights are affected. Pools of light that move about the field are distracting to players and fans alike. For this reason, towers are often torsionally stiff large-diameter masts, triangular trusses, or rectangular trusses. Single-plane trusses may be satisfactory if towers are linked to resist individual tower twist.

Mounting points for the lights themselves are also worthy of attention. Each light consists of a luminaire and a ballast (transformer), so the designer must be sure to get the weights for both from the electrical engineer. Each luminaire and its ballast are mounted individually to a horizontal member of a lighting rack. While open shapes such as angles had been used for this member in the past, their low torsional stiffness can allow the lights to bob in the wind. A closed tube makes a better mounting rail. An angle welded to the tube can allow for field drilling without breaching the tube wall and inviting internal corrosion.

Structural assessment of light towers can cover the same type of condition and water trap issues as discussed earlier for scoreboards. Issues of tower stiffness and pointing stability need be reviewed only if raised as concerns by the owner or facility operator.

ICE SHEETS

The construction, performance, and maintenance of ice sheets such as provided in arenas for hockey is properly a subject for a specialist mechanical engineer. However, these features do have structural implications as well. The location of the ice sheet is

FIGURE 10.31   Scoreboard framing should have separate adjustable members to receive electronic equipment, as shown here during installation.

also important. Rinks placed on grade require different treatment than those supported on a framed slab.

Ice sheets undergo significant dimensional changes as they are chilled below freezing and then allowed to thaw to room temperature. To allow this motion to occur without generating large, potentially damaging forces, the whole slab that contains the refrigeration piping "slides" on plastic sheets and stiff insulation that serves as a bond breaker over the structural concrete support slab (if present) or coarse-grained soil subgrade. The connection point where piping meets this slab must have sufficient flexibility to handle this motion too.

To provide a level surface between the ice-sheet slab and surrounding slabs when no ice is needed, the structural slab steps up to be flush. However, a cork-filled joint is still needed between the two flush surfaces to maintain an insulating gap. Otherwise, the outer slab would be chilled too.

Structural assessment items to observe include the condition of the ice slab, surrounding slab, and supporting slab and piping. Note crack patterns that indicate structural forces induced by thermal movements. Observe any leaks of refrigeration piping; older refrigeration systems used circulating ammonia, a very corrosive substance, so leaks could have serious effects on building structural integrity. Look for signs of frost or condensation outside the ice sheet, if it is operating at the time, because either could indicate missing or damaged insulation. Also note any damage or jamming of the joint around the ice sheet.

Look for signs of frost heave, indicated by uneven slab surfaces and cracking. Frost heave occurs in fine-grained soils with a nearby supply of water, such as a high water table or an undetected pipe leak. As water in soil pores freezes, more water is drawn up by capillary action, and repetition of the cycle results in ice lenses that lift the soil. This is more likely to occur at rinks on grade that are operated continuously for extended periods. The boundary line of frozen soil does not extend too far below an insulated seasonal rink as it expands during the skating season and contracts when refrigeration stops, so a gravel or coarse sand bed several feet thick can cover the likely frozen zone. However, the steady-state extent of the frozen soil line under continuously operated rinks extends far deeper. Insulation alone can reduce but cannot stop frost penetration into the subgrade. Modern installations may include heater cables or pipes below the ice sheet and insulation to warm the soil. Piped systems could even use the heat rejected by the ice rink chiller plant. Detection of heave should lead to review of the original design, original operating conditions, materials used, and current performance of chiller, insulation, and heating systems.

STADIUM AND ARENA STRUCTURAL ASSESSMENT CHECKLIST

The structural assessment of sports facilities uses the same tools and methodologies as for other building types. However, a solid understanding of the special design criteria, types of structural systems used, and likely range of behaviors are important to focus the reviewer's efforts efficiently on the likeliest areas of concern for these facilities.

Prior to field surveys

- Review available drawings on preliminary basis.
- Structural system, load schedule, typical details
- Shop drawings of any precast elements
- Available renovation and repair drawings
- Interview facility manager for background information on operational history.

Field surveys

Roof visual evaluation—walk both above and below deck.

- Waterproofing/drainage/roof penetrations/misc.
- Overall membrane condition, current/past leak locations
- Expansion joint condition/operability
- Water infiltration at roof penetrations
- Snow stop/fence conditions
- Drain number, location, condition, clogs
- Ponding evidence
- Deck type and condition
- Framing members
- Member visual condition
- Connection visual condition
- Correspondence with design drawings—spot check sizes, configuration
- Catwalk, rigging, handrail, grating, hung load conditions
- Loads present compared to allowances
- Bearing details
- Conformance with design drawings

- Evidence of motion consistent with intent
- Condition of bearing and connections to upper and lower elements

Concourse visual evaluation from above and below each level

- Structural framing configuration and condition
- Expansion joint locations and conditions
- Damage, prior repairs, leaks, cracks—nature and location

Seating/stadia units from above and below

- Framing configuration and condition
  - Cracks at midspan soffits, near unit ends, at corners
  - Connection and tie conditions between units
  - Joint sealant condition, tub unit conditions
- Adverse exposures
  - Water infiltration, ponding, entrapment
  - Mechanical damage—collisions, railing leverage, freeze-thaw, packer rust
- Load path at vomitories, secondary stairways, cross-aisle infill seating.

Rakers

- Spot check as-built sizes with design drawings
- Note cracks, spalls, and stains
- Bearing conditions—stability, unanticipated rotation
- End-connection conditions—damage, shifting pads

Bleachers

- Compare as-built with design drawings
- Damage and current condition
- Evidence of past repairs and their adequacy

Backstops

- Self-supporting or cable-supported
- Beam-post connections and post conditions at grade
- Cable and fitting conditions at upper anchors and grade

Sunscreens and windscreens

- Compare as-built condition with design drawings
- Condition of connections and braces

Ramps

- Compare with design drawings
- Member conditions, especially mechanical damage
- Connection conditions, especially at high-force locations by main stands
- Expansion joint behavior and conditions
- Condition of self-supporting structural frame for stand-alone ramps

Cladding

- Tieback and bearing support conditions between cladding and framing

Scoreboards

- Compare with design drawings and weight allowances
- Deterioration and inappropriate field modifications (holes, notches)
- Any extensive modifications made during renovations and upgrades
- Evidence of water traps along members and at joints

Field light towers

- Compare with design drawings, fixture counts and weights
- Conditions including deterioration and water traps

Ice sheets

- Slab and joint conditions adjacent to ice sheet (and below, if possible)
- Crack patterns consistent with restrained thermal movements
- Leaks of refrigeration piping and structural deterioration
- Failure of insulation, isolation or underslab heat: frost beyond ice sheet, heave

Structural analysis and assessment. Level of analysis will vary by project. Checklist items help to focus on potential problem areas but are not exhaustive.

Deck strength

- Shear, local crushing at bearing points, uplift resistance, diaphragm action
- Connection capacity along seams and to supports

Deck stiffness

- Deflections under gravity and wind, deck splices
- Interaction with supporting structure, effect of differential deflections, camber

Load review for added loads not accounted for in original design, such as

- Second roof membrane
- New scoreboards/video screens/rigging loads/etc.
- New drift or wind-load patterns from additions or owner direction

Load path—any changes due to renovations

Framing members—primary and secondary. Check for

- Gravity, including applicable snow drift and unbalanced snow
- Point loads—especially purlins, trusses loaded between panel points
- Uplift, including unbraced bottom flanges or chords
- Axial loads, thrust and tie forces, alterations that change force directions
- Connection strength
- Compatibility of deformations between members and at connections

Main roof structural system

- Match analysis approach to structural system and specific concerns.
- Follow complete load path for both vertical and lateral loads.
- Adjust results for influence of construction method on potential changes in member forces, if known.
- Consider acceptable load redistribution through self-limiting inelastic behavior.
- Check against criteria and specifications; consider those originally used and those which may now be permissible.

CHAPTER 11

Bridges

BOJIDAR S. YANEV, ENG. SC.D., P.E.

INTRODUCTION

This chapter examines bridge condition assessment as an engineering function and reviews its current practices. Engineering assessment is implied in bridge design; how-

329

ever, it has become increasingly more specific, independent, and sometimes even legally binding. Vehicular bridges on the U.S. national system have been inspected biennially since 1978. The accumulated data have been interpreted to indicate that more than 40% of the 650,000 highway bridges are structurally deficient or functionally obsolete.[1] One of three bridge crossings is over a deficient bridge. Rehabilitation and replacement costs average US $7 billion annually. The condition of approximately 101,000 railroad bridges and a comparable number of pedestrian bridges also must be evaluated. Public safety and the allocation of considerable national resources therefore are at stake.

Bridge condition assessment is administered by the Federal Highway Administration (FHWA), the American Association of State Highway Transportation Officials (AASHTO), and the American Railroad Engineering and Maintenance of Way Association (AREMA). Their publications are quoted throughout this chapter, but never in their entirety because this review is no substitute for the sources. Relevant examples are drawn from the practice in various states, the European Union, and Japan.

BRIDGE FAILURES

In contrast with many fields of human achievement where rapid change and risk taking are measures of excellence, engineering structures are stable, safe, and often monumental. Bridges in particular serve as both symbols and actual carriers of social development. History, however, records that development in terms of events, and in the case of bridges, these are the occasional accidents generally attributed (overtly, tacitly, or by default) to the structural engineer. Everyone feels competent to assess bridges after they fail. For engineers, it can be a lifetime occupation. Instructive to the profession and costly to society, failures have been the driving force in bridge condition assessment over the ages.

Reviewing the failures of the Dee Bridge at Chester, England, in 1847, the Tay bridge in Scotland on December 28, 1879, and the Quebec Bridge in 1907, Silby and Walker[2] and Petroski[3,4] found that bridge design was compelled to seek the limit of the possible by exceeding it. This viewpoint applies particularly to ambitious failures of relatively new structural types. Ignorance of the failure mode is a principal liability of new structures and materials. Throughout the nineteenth century, attempts to build long-span suspension bridges were thwarted systematically by dynamic and particularly wind loads, as was the case with the Menai Straits Bridge in 1826 and 1936 and the Wheeling Bridge in 1954.[5-7] J. Roebling brilliantly secured his record-breaking suspension spans against wind by adding stay systems[5,6]. However, the problem and the solution were stated rigorously only after the wind-induced failure of the Tacoma Narrows Suspension Bridge on November 7, 1940.[7,8] Advances in long-span aerodynamic stability remain significant today, as discussed for instance in Larsen and Esdahl.[9] Pedestrian-induced dynamic loads demonstrated their potentially fatal effects on suspension bridges as early as 1830 on Brown's Bridge at Montrose[5] and remain a concern on some recent structures, e.g., the London Millennium Pedestrian Bridge[10] and the Passerelle de Solferino in Paris.

Always noteworthy are bridge failures caused by material or construction inadequacies, such as the collapse of Navier's Pont des Invalides in Paris in 1826, the Ashtabula Creek cast iron railroad bridge in 1876 in the United States, and the Viaduct d'Evaux over the Tardes in 1881. Navier's investigation of the failure of his suspension bridge in Paris contributed significantly to the state of the art.[7]

Floods are the leading cause of bridge failures worldwide. A noteworthy recent example is the Schoharie Bridge on the New York State Thruway. Its failure during torrential flooding on April 3, 1987,[8] demonstrated the significant vulnerability of New York State bridges to scour.

Earthquakes, such as the San Fernando (1971), Loma Prieta (1989), and Northridge (1994) earthquakes, cause numerous bridge failures primarily in California. Deadliest among them, claiming 41 victims, was the failure of the elevated Route I-880 over nearly 2 miles along Cypress Boulevard in Oakland, California, during the Loma Prieta earthquake of October 17, 1989. One motorist was killed during the same event at the collapsed 50-ft (16-m) span of the East San Francisco–Oakland Bay Bridge. These 42 fatalities were the only ones sustained throughout the otherwise massively affected Bay Area. Public and professional attention focused on bridges as a result. Substantial federal and local funds were dedicated to numerous investigations, amending bridge design codes, analytic procedures, and construction and emergency response practices.[11–13]

Despite their great impact on all areas of life, however, the consequences of natural disasters were not at the origin of contemporary bridge assessment effort. Contemporary bridge condition assessment history in the United States began at about 5 P.M. EST on December 15, 1967, with the collapse of the Silver Bridge, carrying U.S. Route 36 over the Ohio River at Point Pleasant. Forty-six motorists perished, setting the twentieth-century record. Within a year, the Federal Aid Highway Act introduced the *National Bridge Inspection Standards* (*NBIS*). Mainly, the standards sought to identify the bridges on the national bridge network, eliminate potential hazards, and determine and address present and future needs.

The Silver Bridge did not collapse because of an extreme natural event or extraordinary load. Similarly inauspicious were the causes for subsequent bridge failures, e.g., at the West Side Highway in New York City in 1973 and the Mianus River Bridge in Connecticut in 1983.[8] Partial or localized failures also claimed their share of victims and publicity. A diagonal stay ruptured on the Brooklyn Bridge, in New York City, in 1981, killing a pedestrian. Concrete spalled from the underside of the Franklin D. Roosevelt Drive in Manhattan (Fig. 11.1) killing a motorist on June 1, 1989.

The causes for these failures appear dauntingly nondescript. Common to them were not particularly ambitious designs but components, both vulnerable and critical. In the absence of alternate load paths, their local malfunctions inevitably have catastrophic consequences. The Silver and Mianus bridges collapsed owing to failures of single structural elements (hence designated as *fracture-critical*[14]). Heavy rains, not unexpected for the local climate, washed a Schoharie Bridge pier away. Similarly, the user's safety was shown to depend on the condition of sometimes secondary structural components. The Brooklyn Bridge stay rupture and the West Side Highway collapse were caused by unchecked corrosion. Numerous causes contributed to the fatal deck spalling on the Franklin D. Roosevelt Drive.

In the words of Einstein,[15] "We have achieved something if we have succeeded in the formulation of a meaningful and precise question." The rational postdisaster questions are how to avoid and prevent recurrences. Engineering responds with design and inspection. By combining abstraction and application, these two activities have produced bridge condition assessment, both gaining in the process. Design has adopted performance-based criteria and includes life-cycle-cost analysis. Inspection has evolved from the identification of imminent hazards to understanding the structure's condition past, present, and future. The Federal Aid Highway Act began the integration of these tasks with the National Bridge Inventory (NBI) of 1972.

BRIDGES

The *NBIS*[16] define bridges as follows:

§650.301   A structure including supports erected over a depression or an obstruction, such as water, highway, or railway, and having a track or passageway for carrying traffic or

FIGURE 11.1   Spalling ridge deck underside.

other moving loads, and having an opening measured along the center of the roadway of more than 20 ft [6 m] between undercopings of abutments or spring lines of arches, or extreme ends of openings for multiple boxes; it may also include multiple pipes, where the clear distance between openings is less than half of the smaller contiguous opening.

The 20-ft (6-m) lower limit excludes culverts. Culverts require different expertise and may fall under the purview of other responsible owners, e.g., the Environmental Protection Agency (EPA).

Types of Bridges. The NBI[16] classifies bridges according to material (Table 11.1), main structure type (Table 11.2), feature carried (Table 11.3), and feature crossed (Table 11.4). Examples of all types of bridge structures, listed in Table 11.2, can be found in most general bridge texts.[17,18]

TABLE 11.1   Bridge Types: Material[16]

No.	Material	No.	Material
1	Concrete	6	Prestressed concrete continuous
2	Concrete continuous	7	Timber
3	Steel	8	Masonry
4	Steel continuous	9	Aluminum, wrought iron, cast iron
5	Prestressed concrete	0	Other

Although explicitly listed as no. 2 in Table 11.3, railroad bridges are part of the NBI only if they also carry vehicular traffic (no. 4 of Table 11.3). Bridges serving only train traffic are regulated by the American Railroad Engineering and Maintenance of Way Association (AREMA).

An example fitting no. 0, "Other," in Table 11.3 can be an aqueduct or utility-carrying structure. The oldest bridge in New York City, for instance, is the High Bridge over the Harlem River, dating from 1848 and modified in 1928. Modeled after the Roman Pont du Gard near Nimes, France, it is a dismantled aqueduct, to be rehabilitated as a pedestrian crossing.

Of the "Features Crossed" in Table 11.4, no. 5, "Waterway," has the greatest significance for bridge design and operation. Railroad operations under a bridge also influence management policies by limiting access for inspection, maintenance, and rehabilitation, as well as by occasionally suspending unanticipated loads from the primary structure (catenaries, signs, etc.).

Essential Parameters. Parameters are essential to a bridge if they affect management decisions. Such are key statistics and dimensions, including location, number of spans, number of approach spans, total horizontal clearance, length of maximum span, length of the structure, curb or sidewalk width, roadway width (curb to curb), deck width (out to out), minimum vertical clearance over the bridge, minimum vertical underclearance (high and low water, when relevant), channel depth (when possible), minimum lateral underclearance, and skew and curvature radii. The inventory also should include the bridge age, dates of reconstruction, and rehabilitation and maintenance records. Adequate decision support requires data on the components, elements, and members in all spans of the bridges on the network.

TABLE 11.2   Bridge Types: Main Structure[16]

No.	Main Structure	No.	Main Structure
01	Slab	13	Suspension
02	Stringer/multibeam or girder	14	Stayed girder
03	Girder and floorbeam system	15	Movable, lift
04	Tee beam	16	Movable, bascule
05	Box beam or girders, multiple	17	Movable, swing
06	Box beam or girders, single or spread	18	Tunnel
07	Frame	19	Culvert
08	Orthotropic	20	Mixed types
09	Truss, deck	21	Segmental box girder
10	Truss, thru	22	Channel beam
11	Arch, deck	00	Other
12	Arch, thru		

TABLE 11.3 Bridge Type: Feature Carried[16]

No.	Feature Carried	No.	Feature Carried
1	Highway	6	Overpass structure at an interchange or second level of a multilevel interchange
2	Railroad	7	Third level (interchange)
3	Pedestrian exclusively	8	Fourth level (interchange)
4	Highway, railroad	9	Building or plaza
5	Highway, pedestrian	0	Other

Components, Elements, Members

Most texts designate bridge *components* to include several bridge elements. A bridge *element*, in turn, may comprise several members. The span primary *members* may consist of several steel or concrete girders, as shown in Fig. 11.2. The FHWA *Coding Guide* records the number of spans and identifies the general components as in Table 11.5 but does not discretize into spans and elements for evaluation purposes. That practice was augmented in 1997. The federally sponsored bridge management package, PONTIS, introduced a new AASHTO *Guide for Commonly Recognized (CoRe) Structural Elements*.[19] The *CoRe Guide* considers certain components in greater detail, still not on a span level.

The New York State Department of Transportation (NYS DOT) bridge inventory[20,21] rates elements in every span. Condition rating is limited to the overall condition of the *element*. *Load rating* is defined by the capacity of the worst *member*, as discussed in the respective sections herein. "Erosion and scour," clearly signs of distress, are listed as bridge elements. The inventory is thus not only descriptive but also action-oriented toward safe and cost-effective bridge management.

BRIDGE MANAGEMENT

The infusion of federal funding and rigorous application of the NBIS gradually shifted the emphasis from urgent hazard mitigation to more general bridge management considerations. Section 1034 of the Intermodal Surface Transportation Efficiency Act (ISTEA) of 1991 required ". . . the development, establishment and implementation of systems for managing highway pavement of Federal-aid highways, bridges on and off Federal aid highways; highway safety; traffic congestion; public transportation facilities and equipment; and intermodal transportation facilities and systems."[22] The mandate of ISTEA was relaxed subsequently to a recommendation. By then, however, bridge management had advanced well beyond computerized inventories and, in various forms, had become the accepted practice. More than 40 states have adopted the FHWA-sponsored PONTIS. BRIDGIT, also sponsored by the FHWA, is used in Maine.

TABLE 11.4 Bridge Type: Feature Crossed[16]

No.	Feature Crossed	No.	Feature Crossed
1	Highway, with or without pedestrian	6	Highway, waterway
2	Railroad	7	Railroad, waterway
3	Pedestrian exclusively	8	Highway, waterway, railroad
4	Highway, railroad	9	Relief for waterway
5	Waterway table	0	Other

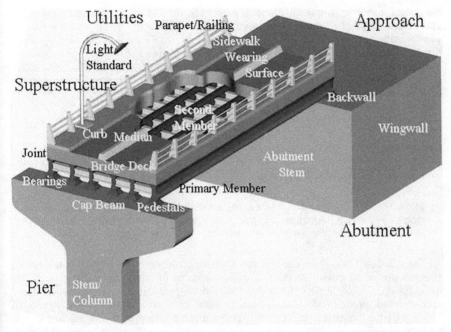

FIGURE 11.2   Typical bridge span: bridge components, elements, and members.

Pennsylvania, Louisiana, New York, and others have developed their own bridge management software.

The Organization for Economic Cooperation and Development (OECD)[23] showed an early interest in bridge management. A report (BRIME[24]) on bridge management practices in the European Union was published recently. Similar efforts are in progress in the Far East. Common to all bridge management studies[23-25] is the central role of the bridge database.

National Bridge Inspection Standards (NBIS). The *NBIS* established the national highway bridge database. Section 650 of NBIS regulates inspection procedures, fre-

TABLE 11.5   Bridge Components and Elements According to New York State DOT[21]

Components	Elements
Abutments	Joint with deck, bearings, anchor bolts, pads, bridge seat and pedestals, backwall, stem, erosion or scour, footings, piles
Wingwalls	Walls, footings, erosion or scour, piles
Approaches	Drainage, embankment, erosion, pavement, guide railing,
Stream channel	Stream alignment, erosion and scour, bank protection
Deck elements	Wearing surface, curbs, sidewalks and fascia, railings and parapets, scuppers, gratings, median, mono deck surface
Superstructure	Deck structural, primary members, secondary members, paint, joints
Pier	Bearings, anchor bolts, pads, pedestals, top of pier cap or beam, stem solid pier, cap beam, pier columns, footings, erosion or scour, piles
Utilities	Lighting standards and fixtures, sign structure, utilities and support

quency of inspections, qualifications of personnel, inspection reports, and preparation and maintenance of a state bridge inventory. The *NBIS* apply to all structures on all public roads. They are included as Appendix C in the *Recording and Coding Guide for the Structure Inventory and Appraisal of the Nation's Bridges.*[16] The National Bridge Inventory (NBI) provides the system of alphanumeric data for each structure carrying highway traffic. Both the *Coding Guide* and the *NBIS* have been the subjects of revisions and updates. The edition quoted herein is dated October 25, 1988. As a first step toward bridge inspection and management, it mandates the preparation and maintenance of an NBI by the responsible owners.

Responsibility. Identifying *maintenance* as the key responsibility of the owner is a reminder that inspection must be part of a comprehensive bridge management program. The *NBIS*[16] stipulates the following:

§650.303   Each highway department shall include a bridge inspection organization capable of performing inspections, preparing reports, and determining ratings in accordance with the provisions of the *AASHTO Manual*[26] and the Standards contained herein.

The *Coding Guide*[16] recognizes the following owners as responsible for bridge maintenance: state, county, town or township, and city or municipal highway agencies; national, state, or local park, forest, or reservation agencies; other state or local agencies; railroad, private (other than railroad), state, or local toll authorities; bureaus of Indian affairs, land management, and reclamation; military reservation/corps of engineers; U.S. Forrest Service; and unknown. Pedestrian and privately owned bridges are also subject to inspections, although their inventory is not always as well maintained. The status of overpasses crossing city streets between buildings often remains obscure, except during postaccident litigation.

Inventory. A bridge inventory must contain the information necessary for network- and project-level management decisions in a computerized form.[22] The *NBIS* defines the national highway bridge inventory as follows:

§650.311   (a) Each State shall prepare and maintain an inventory of all bridge structures subject to the Standards. Under these Standards, certain structure inventory and appraisal data must be collected and retained within the various departments of the State organization for collection by the Federal Highway Administration (FHWA) as needed [author's note: annually]. A tabulation of this data is contained in the structure inventory and appraisal sheet distributed by FHWA as part of the *Recording and Coding Guide* in January 1979. Reporting procedures have been developed by FHWA.

New and modified structures must be entered into the inventory within 90 days of their change of status if under direct state jurisdiction or 180 days if otherwise.

As a result of promptly implementing the preceding mandate, FHWA obtained a fairly detailed computerized and annually updated bridge database as early as 1982, when other developed countries were still counting their bridges.

BRIDGE CONDITION

The modeling of reality is famously described by Einstein[27]: "So far as the laws of mathematics refer to reality, they are not certain; and so far as they are certain, they do not refer to reality." The real bridge condition is neither uniquely defined nor exactly determined.

If criteria were limited to level of service, a number of qualitative and quantitative descriptors would be necessary, including

- Acceptable "as designed" / "as built" / "as is" resistance to design, current, and anticipated loads.
- Acceptable "as built" / "as is" serviceability under current and anticipated traffic volumes.

"Acceptable" quality and quantity of service and structural strength are the subject of design specifications (AASHTO for highway and AREMA for railroad bridges).

Over the years, these design specifications have undergone numerous revisions. Most bridges in use today were not designed strictly for the currently specified loads, loading combinations, and load factors. AASHTO, in turn, only provides "minimum" strength and safety requirements for spans under 500 ft (152.5 m). The owner may be in charge of a unique structure or seek superior performance. Departures from the design and imperfections of construction often are documented inadequately at the completion of a bridge project.

Emanating from a vaguely defined "as built" condition, the "as is" condition at a later date is less certain. It requires a physical investigation and analytic evaluation of the findings, two highly diverse tasks, each of them severely constrained. Parts of the substructure, such as footings, are physically inaccessible, as is concrete reinforcement, which should be entirely embedded in concrete. Hard to analyze is, for instance, the reduction of the useful life of concrete due to chemical degradation, such as alkali-silica reaction. The fatigue life of steel may be more or less exhausted, stress is not distributed according to the models, and so on. Condition evaluation therefore must, determine a number of vague structural qualities by approximate analysis of uncertain quantities.

Uncertainty and Determinism

References to uncertainty increasingly permeated design specifications during the second half of the twentieth century, culminating in the 1996 AASHTO *Load and Resistance Factor Design (LRFD) Bridge Design Specifications.*[28] They remain central to current bridge management discussions.[29,30] Three types of uncertainty are recognized in engineering assessments.[31] Events, natural or otherwise, causing bridge damage are *random.* Inaccuracies and incompleteness of measurements represent *ignorance.* Definitions, such as "bridge condition," are *vague.*

Randomness of the various causes for bridge deterioration can be modeled statistically, given ample data. Ignorance of the actual state of bridge elements can be treated probabilistically if the significant phenomena are understood. Vaguely defined bridge or element conditions can be represented by fuzzy sets.[29] Genetic algorithms and neural network models[30] take into considertaion uncertainties and vagueness, producing ranges of possibilities with perceived likelihood depending on the data and the models.

Uncertain as structural assessment might be, it is assessing structures in definite states. The resulting actions are intended to preserve or modify perceived conditions but, if effective, produce definite consequences on the actual ones. The assumptions at the origin of the assessment and the management decisions at its conclusion are deterministic, the more reliable if based on correctly estimated probabilities. This ambiguity is captured in the description of the latest Eurocode for bridge design[32] as *semiprobabilistic.*

Quantity and Quality

Design specifications gain clarity when they define the roles of uncertainty and determinism in their quantitative and qualitative assessments. According to Einstein and Infeld[33]: "One of the most important characteristics of modern physics is that the conclusions drawn from initial clues are not only qualitative but also quantitative. We wish to be able to predict events and to determine by experiment whether observation confirms these predictions and thus the initial assumptions. To draw qualitative conclusions we must use the language of mathematics."

Abstract reasoning and empirical investigations are the basic tools of assessment. To a degree, they are separated into data acquisition and analysis, as shown on Fig. 11.3. Quantities and assessments obtained on location can be analyzed independently, including by statistical methods. Inspections can compare analytic models with field findings.

There is no "fixed rate of exchange" between the quantitative and qualitative evaluations forming a bridge condition assessment. As a result, condition rating systems vary considerably. Some are used concurrently. This can be viewed as vagueness but also as a redundancy.

Redundancy

In the broadest sense, redundancy is a form of reserve. In both structural analysis and operations, however, this reserve improves reliability only if it is stored in a manner providing viable alternatives. The effect of redundancy on the reliability of a system has been modeled, for instance, by Gertsbakh.[34]

Structural Redundancy. In engineering mechanics, redundancy is static indeterminacy. Bridge assessment broadens the definition to imply an alternate load path (thus considering serviceability). The required load redistribution can be global (of the entire bridge) or local (within its elements). A structure entirely dependent on the performance of *all* its elements is nonredundant. An element whose minor defect inevitably escalates to element failure can be considered internally nonredundant. A nonredundant structure, such as the Silver Bridge[8,14] has a priori no global reserve. Its structural integrity depends on the condition of the nonredundant element, a two-eyebar chain. That element is internally nonredundant as well. A local defect in either bar will cause stress concentration, propagate into a fracture, overstress the parallel bar, and fail the member. In contrast, a suspension bridge cable consisting of many thousands of high-strength parallel wires can lose a number of them owing to corrosion and still retain its structural integrity.

These and other similar examples frequently serve to underscore the value of redundancy in bridge design and condition assessment. One of the earliest bridge inspection manuals[14] was on the inspection of fracture-critical bridge members. NCHRP Report 406[35] is recommended in the *Manual for Bridge Condition Evaluation*[36] for further quantification of superstructure load-rating calculations. The more recent NCHRP Report 458[37] addresses redundancy in highway bridge substructures. Three types of redundancy are defined[35] as follows:

Internal, implying that the failure of an element remains limited to that element (the term was used differently earlier)

Structural, implying both static indeterminacy and adequate ductility

Loadpath, requiring more than two load paths

According to the latter, a two-girder primary member system is nonredundant, but a three-member one is redundant. Thus two girders out of three would be overloaded

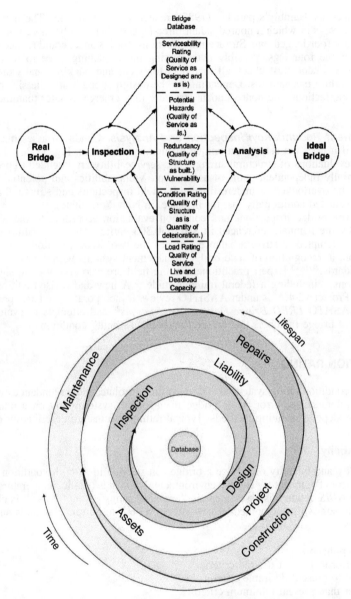

FIGURE 11.3 Bridge design, assessment, and lifecycle.

by 50%. New York State specifies that four girders supporting a span parallel to traffic are redundant, whereas three are not. The added load on three girders owing to the failure of a fourth is thus limited to roughly 133%. Such redundancy considerations reflect the governing design provisions.

The numerous definitions of redundancy and instances when they apply (load rating, hazard, and vulnerability assessments) reflect the importance and complexity of the subject. A structure is not automatically "safe" merely because it is redundant. This is illustrated by Hambly's paradox (1985), related by J. Heyman.[38] The paradox explores the ways in which a nonredundant three-legged chair can be more reliable than a redundant four-legged one. Surface imperfections, for instance, could reduce the load on one of the four legs, possibly overstressing the remaining three to failure. The correct assessment must check all members for yield and buckling and examine all possible failure mechanisms. Depending on the design strength and the size of anticipated imperfections, the nonredundant structure may emerge as safer than the redundant one.

Assessment Redundancy: Inspection and Analysis. Condition assessment is redundant if it does not rely on an unique method. As shown on Fig. 11.3a, visual and analytic evaluations of structural strength and serviceability are intended to provide at least partially independent paths of assessment. Vulnerabilities and potential hazards that may be overlooked or underestimated by field inspections and structural analysis can be evaluated additionally (as, for instance, in New York State).

Highway bridge inspections and condition evaluation according to the *NBIS* are defined by the manual, referenced in Sec. 650.303 earlier. The early editions of the manual[26] attempted to strike a balance between the two. Recent editions[36] emphasize evaluation, in recognition of state bridge management systems with their own inspection standards.[21,39,40] Expert practitioners in the field are also contributing significant publications,[41] including a federal training guide.[42] A new and revised edition under NCHRP Project 12-46[43] is under AASHTO review. It incorporates the latest provisions of the AASHTO *LRFD Bridge Design Specifications*[28] and attempts to evaluate the capacity of bridge members that are no longer in "as built" condition.

CONDITION RATING

A single condition rating system would imply an "absolute," nonredundant assessment with no tolerance for error. Most bridge management systems rely on a number of partially independent rating methods. Typical rating systems are described briefly.

Serviceability

The quality and quantity of service a bridge can provide in as-built condition can be assessed from the inventory and the environmental demands (traffic, temperature, channel). The *NBIS* define this assessment as "appraisal rating" from 0 to 9. It is obtained by comparison with a hypothetical new structure built to current standards and rated as follows:

N   Not applicable
9   Superior to present desired criteria
8   Equal to present desirable criteria
7   Better than present minimum criteria
6   Equal to present minimum criteria
5   Somewhat better than minimum adequacy to tolerate being left in place as is
4   Meets minimum tolerable limits to be left in place as is
3   Basically intolerable requiring high priority of corrective action
2   Basically intolerable requiring high priority of replacement
1   This value of rating code not used
0   Bridge closed

Tables (1 through 3B) of the *Coding Guide*[16] link the appraisal ratings to appropriate levels of average daily traffic (ADT) and bridge geometry, minimizing their subjectivity. "Serviceability" may be revised continually as traffic demands evolve. It can decline while the bridge structure may not have deteriorated significantly.

Poor condition, particularly when related to the deck elements, lowers the level of service by reducing the ADT. The resulting losses to the community are termed *user costs* and are roughly estimated as a product of the motorist-hours lost and the assumed hourly average wage. Significant phenomena, such as the relocation of businesses caused by a failing infrastructure, require more detailed investigation. The accuracy of user cost estimates can improve if users are separated into more homogeneous cate-

FIGURE 11.4   Steel column deflected by impact.

gories, such as motorists, businesses, the bridge owner, the municipality, etc. If considered properly, user costs can be decisive in life-cycle cost-benefit assessment.[29]

A central bridge management objective is to eliminate costs incurred by users and owners owing to structural malfunctions. For this task, not only structural but also environmental conditions must undergo risk analysis. M. Shinozuka defines *engineering risk* as the product of the likelihood of failure and the magnitude of the resulting penalty. Risk assessment therefore must extend beyond the purely structural and consider a broader range of potential hazards and vulnerabilities.

Potential Hazards

Depending on the condition of the bridge population, potential hazards may play a minor role or dominate management. Typical causes of potential hazards are traffic accidents (Fig. 11.4), structural deterioration (Figs. 11.5 through 11.11), and failure of nonstructural (decorative) elements (Fig. 11.12). Their identification and mitigation invariably are a highest priority. Potential hazards must be reported to the responsible owner on the day they are detected, recommending remedial action. The response ranges from emergency repair within 24 hours to engineering reevaluation.

The NYS DOT has designated potential hazards as "flags."[21] Hazard mitigation governed bridge management in the 1980s. In part, this was due to the onset of a widespread critical phase in bridge condition. However, flag numbers also grew because of "inflationary" trends in both flag definition and determination. While early flags identified general causes for potential hazards, those of recent years were issued separately for each bridge element involved. Depending on their nature, flags were divided into "structural" and "safety." According to gravity, they are currently designated as "red," "yellow," and "prompt interim action." The distinctions allow considerable overlap. A structural hazard implies a safety one (see Figs. 11.5 through 11.10), whereas the reverse may not be true (see Figs. 11.11 and 11.12). All 20 flags issued in 1982 in New York City required urgent repairs. No more than 50 of the 3000 flags in 1996 required prompt interim action (within 24 hours)—testimony to the proliferation of less than imminent hazards.

FIGURE 11.5   Failed rocker bearings and steel shoring.

FIGURE 11.6   Spalling concrete overlay.

FIGURE 11.7   Failing bridge concrete pier.

FIGURE 11.8 Corroded steel column web.

The increasingly numerous potential hazards required a reliable forecasting method. For that purpose, flag reports had to be linked to other forms of bridge condition assessment, such as condition ratings. The result[44] was a consistently accurate forecast of future needs for the repairs of potentially hazardous conditions based on current condition ratings of bridge elements. If potential hazards become the dominant bridge management priority, capital reconstruction can no longer maintain the bridge network

FIGURE 11.9   Corroded girder web and timber shoring.

in a safe operating condition. Costly emergency (and usually temporary) repairs become inevitable. As bridge condition improves, more elaborate condition assessments and forecasting techniques come to the forefront.

Vulnerability

A bridge can be considered vulnerable owing to design deficiencies or the likelihood of catastrophic events. The vulnerability of as-built structures can be estimated from the inventory without recourse to inspection results. In the database of Fig. 11.3a, vulnerabilities can be perceived as a form of "negative" serviceability. Nonredundancy, fracture-critical details, inadequate clearance, and pier footings under water are examples.

The latest AASHTO *LRFD Bridge Design Specifications*[28] recognize vulnerabilities in a number of ways, e.g., by specifying "extreme event" limit states. Possible vulnerabilities are more or less the same for all bridges, but their priority clearly depends on local conditions. Seismic vulnerability governs in California, whereas hydraulic vulnerability is dominant in New York State. Vessel collisions[18,45] are of great consequence, for instance, in Florida.

The New York State bridge management system recognizes six vulnerabilities: hydraulic, overload, steel structural details, collision, concrete structural details, and earthquake. Ratings are defined for each vulnerability, and bridges are ranked accordingly as part of the state bridge safety assurance program. The program consists of assessment, evaluation, and implementation of the selected mitigating measures. Particularly challenging is the coordination of as-built vulnerabilities with as-is structural condition ratings. Numerous algorithms have been proposed for this task, but final determination remains with the engineers in charge.

FIGURE 11.10   Piles rotted at the wet line.

Element Condition Ratings

The changes in the structure caused by service and environment require periodic re-evaluation. The process is inexact, and the available information is incomplete and imperfect. The deliberately redundant condition rating scales in use differ primarily in their manner of translating quantitative measurements into qualitative judgments. All are based on valid engineering considerations and have merits. After decades of operation, advantages and disadvantages of the different systems can be compared. All systems, however, ultimately should be evaluated in the context of local bridge management needs, resources, and constraints.

FIGURE 11.11   Failed "cushion" expansion joint and "plug" joint replacement.

NBIS. According to the *NBIS*,[16] all bridge inventory components and, ultimately, the bridges receive integer numerical condition ratings from 0 to 9 based on a mostly visual comparison of the as-is structure with a presumed as-built state. The ratings are intended to reflect general rather than local conditions as follows:

N   Not applicable
9   Excellent
8   Very good—no problems noted
7   Good—some minor problems
6   Satisfactory—structural elements show some minor deterioration
5   Fair—all primary structural elements are sound but may have minor section loss, cracking, spalling, or scour
4   Poor—advanced section loss, deterioration, spalling, or scour
3   Serious—loss of section, deterioration, spalling, or scour have seriously affected primary structural components. Local failures are possible; fatigue cracks in steel or shear cracks in concrete may be present.
2   Critical—advanced deterioration of primary structural elements. Fatigue cracks in steel or shear cracks in concrete may be present, or scour may have removed substructure support. Unless closely monitored, it may be necessary to close the bridge until corrective action is taken.
1   Imminent failure—major deterioration or serious loss present in critical structural components or obvious vertical or horizontal movement affecting structure stability. Bridge is closed to traffic, but corrective action may put back in light service.
0   Failed—out of service, beyond corrective action

The *Coding Guide* similarly rates channels and culverts.

FIGURE 11.12   Failed nonstructural cladding.

New York State DOT.[21] The conditions of all structural elements in every span, the components to which they belong, and the entire bridge are rated from 1 to 7 as follows:

9   Not accessible
8   Not applicable
7   New
6   Shade between 5 and 7
5   Minor deterioration but functioning as originally designed
4   Shade between 3 and 5
3   Serious deterioraiton or not functioning as originally designed
2   Shade between 1 and 3
1   Totally deteriorated or failed

Detailed instructions accompany the preceding ratings for all elements and for the bridge as a whole. For certain bridge elements, e.g., the primary member and the deck, the condition rating reflects the overall state for the rated span. For bearings, on the other hand, the worst one in the span is rated. Written comments, photographs, and sketches are required for all ratings that are lower than 5, changed by more than 1 point, or improved. Repair recommendations are encouraged but not strongly emphasized. The resulting inspection reports primarily are descriptive, with two very significant exceptions. One is the definition of the rating "3—Not functioning as designed." By assigning a rating equal to or lower than 3, an inspector qualitatively changes the structural assessment. The other is the requirement that, independently of all ratings, the inspecting engineer must identify and immediately report all structural hazards, essentially relieving the numerical ratings of the responsibility for decisions related to prompt remedial actions.

American Railroad Engineering and Maintenance of Way Association (AREMA). The Federal Track Safety Standards 49 CFR 213 (Appendix C) provide nonregulatory guidelines for railroad bridge inspection. The 2001 AREMA *Manual for Railway Engineering*[46] contains instructions for inspections of steel, concrete and masonry, and timber bridges independently of the detailed guidelines on rail welding and pit inspections. Inspection findings are documented in a form fundamentally different from the *NBIS*. The bridge is discretized into components: track, substructure, piers, primary elements, and retaining walls. As in the *NBIS*, there is no reference to individual spans. In contrast, however, there are no numerical condition ratings. Rather, the manual prompts the inspector to comment on anticipated defects and to report their extent. The latter determines the magnitude and urgency of the recommended repair. Examples of inspection guidelines[46] follow:

Concrete and masonry bridges
:

2. Piers and Abutments
   Material (brick, stone, concrete):_____
   Condition of backwall (plumb, clearance of structure):_____
   Condition of bridge seat:_____
   :

   b. Concrete:
   Cracks (location, size, description):_____
   Condition of reinforcing (exposed, corroded—location):_____
   Condition of waterline:_____
   :

Steel bridges
:

8. Corrosion
   Loss of section from corrosion, noting exact location and extent of such action, with measurement of remaining section if members are badly corroded, paying close attention to loss of metal in girder and beam flanges and webs and parts of lateral bracing system.
   Distortion caused by rust between rivets and built-up members.
   Damage to overhead structures from engine blast in spans.
   Pockets at bearing locations and at bottom of bearing stiffeners.

The preceding reports are defect-oriented. Implied is a confidence in prompt remedial action. Inspections are annual. The AREMA manual and the *Track Safety Standards* are also updated annually. This relatively austere practice shows the influence of nineteenth-century railroad bridge collapses, such as Ashtabula Creek in 1876 and the Forth of Tay in 1879.[4] The absence of recent incidents appears to confirm its merit.

Service d'Etudes Techniques des Routes et Autoroutes (SETRA), France. Inspections on the French national highway bridge network[47–49] combine the rating/descriptive and defect/action-oriented format. All anticipated defects, such as spalling, cracks, corrosion, etc., are described, illustrated, and assigned identification numbers in the guide. Identification numbers also are assigned to the locations on the structure where such defects can occur, span by span. Defects are rated on a variable scale (two or three levels, depending on their type) and quantified by a code designation. Comments, sketches, and photographs are attached, as in most reports.

The Laboratoire Central des Ponts et Chaussees (LCPC) manuals for reinforced concrete, prestressed, and other structures[50] subscribe to the AREMA approach. All anticipated defects are described, explained, and illustrated. The inspection must determine their presence and scope. An independent study[51] estimated the reconstruction needs for the national highway bridge network of France in 1994. Bridges were classified into the following six categories:

1   Good condition
2   Showing defects of equipment or protection components or minor structural defects without urgent need of repair
2E As above but with urgent need of repair in order to prevent more advanced structural deterioration
2S As above with urgent need of repair in order to guarantee the safety of the road user
3   Bridges with structural damage
3U As above but with urgent need of repair

DANBRO, Denmark. The bridge management system of Denmark presented by Lauridsen, for instance, in Vincentsen and Jensen[52] has been in operation for more than two decades and exports its expertise worldwide. It subscribes to the defect/action-oriented method by minimizing data collection. Routine inspections (every 2 weeks) issue work tickets for remedial action. General inspections (from 1 to 6 years apart) assign condition ratings, limited to a maximum of four levels. Special inspections are envisioned for structural or economic reasons.

From Element to Bridge Condition Rating

For highway bridges, the *NBIS*[16] compute an overall bridge "sufficiency rating" as follows:

$$0 < \text{sufficiency rating} = S_1 + S_2 + S_3 - S4 < 100\% \qquad (11.1)$$

where S_1 = 0–55%—structural adequacy and safety
           S_2 = 0–30%—serviceability and functional obsolescence
           S_3 = 0–15%—essentiality for public use
           S_4 = 0–13%—special reductions (only if $S_1 + S_2 + S_3 \geq 50\%$)

A system of weights, rules, and charts defines the computations of S_1, S_2, S_3, and S_4 using condition ratings generated from inspections, geometric and traffic data from the inventory, and the load-bearing capacity of the bridge. As defined, the system applies to all bridges on the ntional highway network. States using their own condition rating systems annually submit a numerically converted set of the ratings to FHWA. The NYS DOT defines a bridge condition rating R by the following formula:

$$R = \frac{\sum\limits_{i=1}^{n} R_i W_i}{\sum\limits_{i=1}^{n} W_i} \qquad (11.2)$$

where $i = 1$ through n, are the bridge elements considered significant to the overall bridge condition, as shown in Table 11.6
R_i = the worst condition ratings to be found on the bridge for each of the i
elements
W_i = weights, e.g., "importance factors," assigned to each R_i, as shown in
Table 11.6

Table 11.6 also shows the shortest useful life observed in New York City for each of
the n bridge elements of Eq. (11. 2). That equation uses the lowest element condition
ratings to occur in any of the bridge spans, thus rating a hypothetical "worst span."
For better estimates of rehabilitation needs, the NYS DOT[21] introduced separate condition ratings for each span. Graphically, such a condition rating history is represented
by the concave line in Fig. 11.13.

The federal "sufficiency rating" of Eq. (11. 1), the New York State "bridge condition rating" of Eq. (11. 2), and the SETRA "condition classification" are examples
of representing bridge condition by a symbol on a defined scale. The formulas used
are not unique, and some owners use more than one. New York State inspectors must
assign to the bridge a "general recommendation" integer rating between 1 and 7
independent of the "condition rating" in Eq. (11. 2). The two evaluations have been
compared[53] and appear in consistent agreement. It may be argued that despite their
limitations, different bridge condition ratings generally agree with each other because
they all converge toward the actual bridge condition.

From Bridge Condition Rating to Management

To varying degrees, the examples of rating and reporting the condition of bridges, their
components, and their elements combine three distinct stages—data collection, evaluation, and decision.

**TABLE 11.6   Components, Elements, Weights in the NYS Bridge Condition Formula,[21]
with Shortest Useful Life Observed by NYC DOT[44]**

Component	Element	Condition Rating $R_i(7 - 1)$	Weight W_i	Min. Life (Observed) L_o(years)
Abutments	Joint with deck	4		10
	Bearings, anchor bolts, pads	6		20
	Bridge seat and pedestals	6		20
	Backwall	5		35
	Stem	8		30
Wingwalls	Walls	5		50
Deck	Wearing surface	4		20
	Curbs	1		15
	Sidewalks and fascia	2		15
Superstructure	Deck structural	8		20
	Primary member	10		30
	Secondary member	5		35
	Joints	4		10
Pier	Bearings, anchor bolts, pads	6		20
	Pedestals	6		20
	Stem solid pier	8		30
	Cap beam	8		20
	Pier columns	8		30

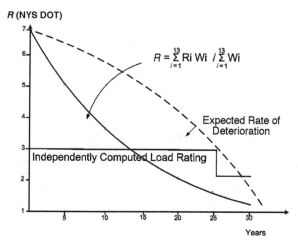

FIGURE 11.13 Bridge deterioration rates.

Rating/descriptive reports emphasize decision support. The bridge is described with its spans and elements. Qualitative evaluation consists of numeric condition ratings. Quantified data are included for independent evaluations and decisions, such as load rating and posting.

Defect/action reports take evaluation to a decision. Inspections quantify and localize precatalogued bridge malfunctions. Corrective action is recommended (in some cases also selected from a database), implying a qualitative evaluation. The available choices usually are three or four and include issuing a work ticket. Assessment is simplified, and efficiency gains. As a result, the rating/descriptive method usually reverts to defect/action for the management of potential hazards and emergency repairs.

The choice of a condition rating system depends on the range of actions available to the manager. The type, condition, and number of bridges and budget and political constraints must be considered. For bridges few in number or in excellent condition, simplicity and immediate remedial action would be preferred. Such is the choice of DANBRO, Denmark, as well as the Honshu-Shikoku and the Hanshin Expressway Authorities in Japan. For a bridge network both large (20,000 in New York State, 2200 in New York City, for instance) and in relatively poor condition, this method would not be feasible. A refined grading scale offers a greater variety of corrective actions and priority rankings. If misused, it can degenerate into "procrastination by inspection." Software packages are now available accommodating both types of database requirements.

Condition ratings support long-term bridge management decisions involving billions of dollars. Once selected, they should remain as consistent as possible. Yet an iterative development of the appropriate condition rating system seems inevitable because the optimal choice depends on a prior knowledge of the prevalent conditions. Furthermore, bridge management priorities must evolve.

SETRA[51] originally attempted to evaluate all rehabilitation needs from inspection reports. Once the available data proved insufficient, a sample group of bridges was inspected and rated on the preceding numeric scale, developed for that purpose. The alphanumeric intermediate and extreme ratings eventually were added to reconcile the findings with appropriate decision options.

The federal coding guide[16] (1988) similarly introduced the distinctions "poor," "critical," and "imminent failure" between the condition states "fair" and "failed." The AASHTO CoRe guide[19] (1997) assigns fewer condition states to a broader range of components. The CoRe guide introduces a comprehensive list of "commonly recognized structural elements" (consistent with the PONTIS bridge management package). All CoRe structural elements must be assigned to one of four condition states, where 1 is best and 4 is worst. Actions, including "Do nothing," "Rehab unit," "Replace unit," and "Perform specific repair" are associated with the condition states, expanding the number of outcomes. For instance, a reinforced-concrete element rated 4 still may be subjected to the alternatives "Do nothing," "Rehabilitate," or "Replace."

The NYS DOT rating system highlights four condition levels (1, 3, 5, and 7), with "shades" between them. New York City has used a four-level bridge rating system correlated with New York State since 1990 as follows:

New York City	New York State
Poor	1.–3.0
Fair	3.01–4.5 (changed to 5.0 in 1996)
Good	4.51–6.0
Very good	6.01–7.0

In 1996 the "good/fair" boundary was raised to 5, primarily because by then the city was able to address the rehabilitation needs of a larger "fair" bridge population. The flowchart in Fig. 11.14 (presented by Yanev, in Vincentsen and Jensen[52]) illustrates how condition ratings can guide the bridge management process.

The assessment of the Chicago mass transit rail bridges[54] combines most features discussed so far. A five-tier condition rating system is developed for the purpose, including the levels critical, poor, marginal, fair, and good. Consistent with the AREMA approach to inspection, each condition level is defined by the urgency of the needed repairs, namely, immediate, within 1 year, within 3 years, depending on further investigation, and none. The assessment determines the scope of rehabilitation required. Repairs are designed following field and laboratory tests.

The original highway bridge condition formulas were designed to determine the rehabilitation needs of a network in relatively poor condition. As conditions improve, inspections increasingly focus on the identification of maintenance tasks. An example is the expansion joint inspection form in Fig. 11.15 introduced by the NYC DOT in the mid-1990s. These inspections are designed to identify the most important early cause of bridge malfunction, e.g., the joints. The yes/no report options are particularly appropriate because joint repairs consist mostly of full replacement (see Fig. 11.11).

The AASHTO *Manual for Maintenance and Management of Roadways and Bridges*[55] contains sound and well-balanced information presented in greater detail in other AASHTO sources quoted therein. In 1999, the Government Accounting Standards Board (GASB) established new financial reporting standards (GASB 34)[56] for the management of highway bridges as part of the national infrastructure assets. An indication that bridge management has evolved over the last 25 years is the revision of the coding guide[16] due for publication in 2004.

The systems discussed so far rate bridge-related hazards, vulnerabilities, quality of service, and quality of maintenance. They rely on visual inspections and on the bridge inventory. According to Figs. 11.3, 11.13, and 11.14, bridge condition assessment also must determine the quantities of structural strength and rehabilitation work, e.g., analytic tasks falling within the scope of bridge design.

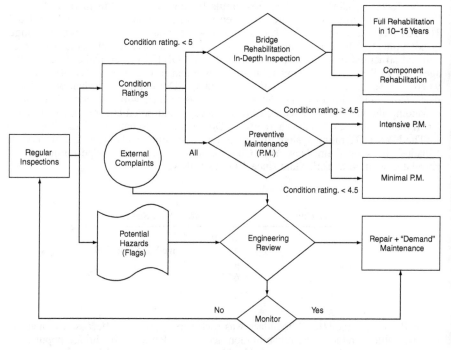

FIGURE 11.14 The role of condition ratings in bridge management.

Load Ratings

Load ratings compute the capacity of bridge elements to resist loads according to the governing design specification(s). Consistent use of terminology helps in approaching the subject. The current and proposed manuals for condition evaluation[36,43] adopt the following definitions:

Nominal resistance. Resistance of a component or connection to load effects, based on its geometry, permissible stresses, or specified strength of materials.

Safe load capacity. A live load that can safely use a bridge repeatedly over the duration of a specified inspection cycle.

Strength limit state. Safety limit state relating to strength and stability.

Serviceability limit state. Collective term for service and fatigue limit states.

For highway bridges, the coding guide[16] defines two load ratings as follows:

Item 64 operating rating. "A capacity rating resulting in the absolute maximum permissible load level to which the structure may be subjected for the vehicle type used in the rating." A bridge can be loaded to the maximum permissible level under unique circumstances, but indefinite use at that level may shorten its useful life.

Item 66 inventory rating. "A capacity rating resulting in a load level which can safely utilize an existing structure for an indefinite period of time."

NYC DOT DIVISION OF BRIDGES
BRIDGES INSPECTION / RESEARCH & DEVELOPMENT UNIT
JOINT INVENTORY AND COMPONENT CONDITION REPORT

SHEET____ OF ____

BIN: ____________ FEATURE CARRIED: ____________ FEATURE CROSSED: ____________ DATE: ________

INSPECTED BY: ____________ TITLE: ________ AMBIENT TEMPERATURE: ________ WEATHER CONDITION: ________
(CLOUDY / SUNNY / RAIN)

	CONDITION OF JOINT ABOVE DECK												CONDITION OF JOINT BELOW DECK					BEARING					DRAINAGE				
																		BEFORE JT.			AFTER JT.						
LOCATION: ABUTMENT / SPAN No.	TYPE OF JOINT	TYPE OF W.S.	COND. OF W.S./DECK (Y/N)	JT. OPENING DIM. (INCH)	VERTICAL DISPLACEMENT (INCH)	COND. OF SEAL (Y/N)	LOOSE PLATE OR MISSING PLATE (Y/N)	MISSING BOLTS OR LOOSE BOLTS (Y/N)	BROKEN WELD (Y/N)	DEBRIS (Y/N)	POOR FINGER JOINT ALIGNMENT (Y/N)	JOINT PAVED OVER (Y/N)	CONDITION OF DECK AROUND JOINT (Y/N)	CONDITION OF PRIMARY MEMBER / DIAPHRAM (Y/N)	COND. OF SEAL (Y/N)	CONDITION OF JT. ANGLE, JT. SUPPT. (Y/N)	STAINING OF SUB STRUCTURE (Y/N)	BEARING TYPE	COND. OF BEARING / A.B. (Y/N)	CLEARANCE (INCH)	BEARING TYPE	COND. OF BEARING / A.B. (Y/N)	TYPE OF DRAINAGE	CONDITION OF DRAINAGE (Y/N)	PHOTOS (YES / NO)	FLAG / NON-FLAG ISSUED (YES / NO)	RECOMMENDATION (YES / NO)
1	2	3	4	5	6	7	8	9	10	11	12	13	14	15	16	17	18	19	20	21	22	23	24	25	26	27	28

JOINTFORM.XLS Duong Tran 8/97.

N = NO COMMENT (IN GOOD CONDITION)
Y = COMMENT WRITTEN
N/A = NOT APPLICABLE
9 = NOT ACCESSIBLE DESCRIBE

FIGURE 11.15 Joint inspection form.

For railway bridges, AREMA[47] defines *normal* and *maximum* load ratings. In the most general terms, bridge rating *RT* in tons is defined as

$$RT = (RF)W \tag{11.3}$$

where W = the weight (in tons) of the nominal truck used to determine the live-load effects L
RF = rating factor

The load and resistance factor design rating calculations of refs. 28 and 43 were preceded by the AASHTO *Guide Specifications for Strength Evaluation of Existing Steel and Concrete Bridges.*[57] That guide formulates the basic structural engineering requirement that supply (of structural strength) exceed demand (of applied loads) as follows:

$$R \geq \sum_k Q_k \tag{11.4}$$

where R = resistance
Q_k = effect of load k

Design specifications stipulate the *sufficient* supply of strength. The AASHTO LRFD bridge design code[28] uses the following form (Eq. 1.3.2.1-1)[28]:

$$R_r = \phi R_n \geq \sum_i \eta_i \gamma_i Q_i \tag{11.4a}$$

where $\eta_i = \eta_D \eta_R \eta_I$ = load modifier for ductility, redundancy, and operating importance, $0.95 \leq \eta_i \leq 1.0$
γ_i = statistically based load factor applied to force effects
φ = statistically based resistance factor applied to nominal resistance

Based on Eqs. (11.4) and (11.4a), the guide[57] defines the rating factor RF as

$$RF = \frac{\varphi R_n - \sum_{i=1}^{m} \gamma_i^D D_i - \sum_{j=1}^{n} \gamma_j^L L_j (1 + I)}{\gamma_R^L L_R (1 + I)} \tag{11.5}$$

where RF = rating factor (the portion of the rating vehicle allowed on the bridge)
φ = resistance factor
m = number of elements included in the dead load
R_n = nominal resistance
n = number of live loads other than the rating vehicle
γ_i^D = dead load factor for element "i"
D_i = minimal dead load effect of element i
γ_j^L = live load factor for load j other than the rating vehicle's
L_j = nominal traffic live load effects for load j other than the rating vehicle's
γ_R^L = live load factor for rating vehicle
L_r = nominal live load effect for the rating vehicle
I = live load impact factor

Equation (11. 5) in effect states that

Rating vehicle effects = capacity − dead load effects − other live load effects

If live load effects are computed according to wheel line distribution factors, implying more than one vehicle, the other than the rating vehicle's load effects L_j can be ignored, resulting in a simplified rating factor expression:

$$RF = \frac{\varphi R_n - \gamma_D D}{\gamma_L L (1 + I)} \tag{11.5a}$$

"Load effects" can be any of the member forces entering design calculations, including bending moment, shear, and axial load. The lowest rating factor to be obtained for any of the evaluated structural members governs the structure.

The dead- and live load factors γ_D and γ_L depend on the structure and the traffic volume. These parameters are subject to design considerations and updates.

The nominal resistance R_n and the resistance factor φ pertain to existing bridges. A flowchart taking into account structural deterioration, redundancy, quality of inspection, and equality of maintenance determines φ in the following range:

$$0.55 \leq \varphi \leq 0.95$$

The condition evaluation manual[36] refers to Eq. (11. 5) in the following form:

masterpiece proves to be a manager's nightmare. Extremely limited empirical data suggest a useful life of 100 years. What, then, is the design life? When did the shay become unsafe and by what standard? Assigning 90% reliability to construction, materials, loads, and inspections each and (simplistically) assuming that they are independent would suggest salvaging at 65 years. With possibly 35% of the useful life sacrificed to safety, were the high quality and cost of construction justified? The process shown in Fig. 11.3a misses that question entirely.

LIFE-CYCLE COST ANALYSIS

According to Fig. 11.3a inspectors rate a structure's condition and designers analyze a structure's model, communicating with each other through the imperfect database. In this ostensibly fixed relationship, the engineer need not design bridges in order to inspect them and can design without ever inspecting one. The dynamic continuity of the bridge management process emerges if Fig. 11.3a is considered as a cross section and viewed in plan, as in Fig. 11.3b. So represented, design and inspection are not opposed but complementary functions. The bridge life cycle can be managed and optimized as it evolves from design to project, to asset, and finally, to liability.

Life-cycle cost analysis was recommended by the FHWA for bridge management in 1993.[64] A Bridge Life-Cycle Cost Analysis (BLCCA) software package[65] that is compatible with existing bridge management software (PONTIS, BRIDGIT) was developed to that end. The capability of the bridge database to support meaningful life-cycle cost decisions has been both questioned[66] and encouraged.[29,30,67,68] Translating past and current structural condition assessments into immediate and future cost estimates is speculative and requires continuing adjustment. While details of cost-benefit analysis are beyond the scope of structural assessment, engineers and particularly managers must be prepared to evaluate bridges in monetary terms.

The California Transportation Department (Caltrans), an early cosponsor of the federal bridge management package, PONTIS, extended the CoRe element-rating system to develop a "health index" for its 12,656 bridges.[69] The index relates bridge inspection data to an assumed asset value of the structure or the network, directly expressing structural deterioration in monetary loss as follows:

$$HI = \frac{\Sigma \, CEV}{\Sigma \, TEV} \times 100 \tag{11.7}$$

where $TEV = TEQ \times FC$
  $CEV = \Sigma(QCSi \times WFi)FC$
  $WF = [1 - (\text{condition state} - 1)(1/\text{state count} - 1)]$ (weight factor for the condition state)
  HI = health index
  CEV = current element value
  TEV = total element value
  TEQ = total element quantity
  FC = failure cost of element
  QCS = quantity in a condition state

Equation (11. 7) supplements the federal sufficiency rating of Eq. (11. 1). It aptly illustrates the role of condition ratings in pursuing bridge assessment to its management objective.

For network-level estimates, the annual replacement needs of a bridge stock worth S (U.S. dollars, USD) with an average useful life N (years) and an average age greater than $N/2$ should be expected to exceed S/N (USD/year). Bridge stock is evaluated according to recent construction costs in the particular area. During the 1990s, for instance, Caltrans reported an average bridge rehabilitation cost of approximately 1100 USD/m² of bridge deck, whereas the NYC DOT reported 4800 USD/m². In 2001, the NYC DOT recommended a full bridge replacement cost of 15,000 USD/m² for general estimates. The Williamsburg Bridge rehabilitation in New York City took roughly 15 years, included full replacement of the approaches, and cost US $900 million, by far exceeding the preceding rate. These construction estimates do not reflect the value of the bridge to the community.

On a project level, S and N are too general and can be misleading. Bridge management studies, such as refs. 52, 68, and 70 show that the shortest rather than the average useful lives of bridges and their components govern rehabilitation needs.

Bridge Condition Forecasts

"Life-cycle cost analysis does not work"[66] when life expectancy is unknown. Early management attempts to estimate the remaining bridge useful life were soon found unreliable for engineering purposes. Budget managers annually assess the funds needed for restoring the infrastructure under their responsibility to "as new" condition, a task possible only because of the vaguer definition of accuracy in finance.

Load ratings were shown to provide a more specific assessment than do condition ratings, but they offer fewer forecasting clues. As long as a bridge is in full service, the load rating should remain constant. In contrast, Fig. 11.13 shows that condition ratings may decline by more than 50% of their range before the load rating is reduced. It is precisely because they change that condition ratings are of interest, and only so long as the rate of change remains predictable. Hence remaining bridge life is deduced from condition rating forecasts. Even successful condition rating forecasts, however, merely predict bridge condition ratings at a future date, not future bridge conditions.

Condition Rating Forecasts. When condition ratings recommend repairs within a prescribed time frame, as they do in SETRA/LCPC,[51] DANBRO,[52] and the Chicago rail[54] bridge evaluation systems, deterioration forecasts are implied. The *NBIS* biennial inspection interval implies that serviceable bridges will not deteriorate significantly in less than 2 years. When the inspection interval is reduced to 1 year for bridges considered in "poor condition," that condition is essentially defined.

For a large bridge stock with more than 10 years of inspections on record, a credible general deterioration model can be developed by a number of methods. Since uncertainties are inherent to both bridge deterioration and its forecasting, a number of probabilistic models such as those in refs. 29 and 30 have been proposed. Markov chains have gained considerable popularity and are used in the forecasting routine of the federal bridge management system PONTIS, among others. Bruhwiler et al. report in Miyamoto and Frangopol[30] that the method is better suited for general forecasts concerning large populations than for modeling the behavior of specific structures. Einstein[33] already has observed that statistical laws are more helpful in explaining the behavior of crowds, not individuals. To a bridge owner, annual budgets consist precisely of individual projects.

An early comprehensive bridge management system was developed by the State of Pennsylvania Department of Transportation (PennDOT).[43] It stressed the importance of maintenance as an inventory item. PennDot models bridge useful life in terms of the *NBIS* bridge condition ratings. Equivalent and estimated ages of the inspected components are defined. A normalized convex curve results, partially governed by the following equation:

$$CNR = 9(1 - EQA/ESL)^{0.7} \qquad (11.8)$$

where EQA = equivalent age of bridge element (years)
ESL = estimated life of bridge element (years)
CNR = condition rating at equivalent age

Relating bridge age to condition rating implies knowledge of traffic, maintenance, and corresponding structural performance. Although maintenance is admittedly decisive in prolonging bridge useful life, it is absent from bridge inventories. Even when annual maintenance expenditures are documented, their benefits in retarding bridge deterioration remain unknown. Yanev and Testa[68] proposed a deterministic knowledge-based model for the effect of maintenance tasks on condition ratings. Supporting data, however, will require years of well-documented maintenance and inspections.

Data become scarcer for older bridges, but closures and failures at a known age have been reported[53] at none or negligible maintenance. In the absence of sufficient data for adequate probabilistic modeling, deterministic worst-case models based on the fastest known deterioration rates provide empirical guidance.[68,70] The results confirm known annual rehabilitation budget needs and allow setting goals for cost-effective maintenance.

A slow deterioration is expected intuitively of a new structure, accelerating toward failure as in the convex curve of Fig. 11.13 and the PennDOT model.[39] Instead, condition rating histories[53] consistently show the concave trajectory of Fig. 11.13. Two valid explanations of the concave condition rating history have been advanced:

1. Inspectors unhesitantly downgrade a new bridge but are reluctant to entirely condemn an already low-rated one.[67]
2. Undocumented repairs, such as hazard-mitigation work, do not improve conditions but delay their further decline toward the end of a bridge life.

New bridges are rarely rated "perfect," leading to a steep initial deterioration pattern. Moreover, while the top rating on the federal scale is "excellent," New York State defines it as "new," which is not necessarily the same.

Emergency repairs typically are required to mitigate potential hazards prior to a bridge closure. The useful life of a technically failed structure thus can be prolonged, contributing to the impression of a slowing rate of deterioration.

Both explanations invoke the rating's subjectivity. There is, however, an objective reason also contributing to the concavity of condition rating histories. Let the bridge deterioration rate be defined according to the New York State bridge condition formula of Eq. (11.2) as follows:

$$r = \partial R / \partial t = \frac{\sum_{i=1}^{n} (\partial R_i / \partial t) W_i}{\sum_{i=1}^{n} W_i} \qquad (11.9)$$

where r = bridge deterioration rate (points/year)
t = time (years)

If all elements i follow the steepest linear deterioration path r_i, as in Table 11.6, the resulting bridge deterioration history will have the concave curvature of Fig. 11.13. According to Eq. (11. 9), as the bridge elements with the shortest useful life fail, such as expansion joints, their contribution to r vanishes, and the rate is reduced. The bridge rating approaches 1 at the deterioration rate of the primary members and the deck, which is lower than the combined rate. In the trivial case of a failed bridge ($R = 1$), deterioration stops ($r = 0$).

As shown in Fig. 11.14, the responsible owner may initiate the rehabilitation process while the bridge is rated 5 on the premise (valid in the early 1990s) that by the beginning of construction, the rating will have reached 3 (not functioning as designed). Schedules can be adjusted over the elapsing 10 to 15 years as priorities change or conditions deviate from the forecast. Whatever the model (convex/concave/bilinear/linear), condition rating forecasts must be more conservative than load ratings, as shown on Fig. 11.13, because inspections lead in the assessment process.

Load Rating Forecasts. The links between *load* and *condition* ratings are redefined continually, reflecting the dynamic interaction between inspection and analysis. The condition rating "not functioning as designed" does not necessarily imply a reduced load rating because it is assigned without calculations, and the original structure may have been designed for higher than the current live, and dead loads and safety factors. For instance, a number of bridges have been converted from rail to automobile use, decks have been replaced with lighter ones, and so on. In fact, 5.6% of the primary bridge superstructure members in New York City were rated 3 or less in 2001 compared with 3.1% of the columns. A partial explanation may be that a visibly deteriorated girder can be rated "not functioning as designed" on the assumption that it no longer meets deflection or fatigue standards, whereas columns that are standing straight appear to be functioning as designed despite visible deterioration. This tendency to underestimate column deterioration must be offset by redundancy and risk considerations. Load redistribution is less likely in columns than in superstructure members; hence a column malfunction is more likely to precipitate a global failure. Consequently, most columns rated "not functioning as designed" are treated as potential hazards (see Fig. 11.8), unlike many similarly rated girders.

Condition ratings influence the load ratings of Eqs. (11.5a) and (11.5c) by the terms φ and φ_C. The highway condition evaluation manual[43] proposes the following values for φ_C:

Condition Rating of Member	Item No. 59[16]	φ_C
Good or satisfactory	≥6	1.00
Fair	5	0.95
Poor	≤4	0.85

The PennDOT bridge management system[39] assumes the following relationship between load capacity LC, design load capacity LC_D, EQA, and ESL of (Eq. 11.8), resulting in a convex normalized curve:

$$\frac{LC}{LC_D} = 1 - \left(\frac{EQA}{1.1ESL}\right)^5 \tag{11.10}$$

where LC = load capacity
LC_D = design load capacity

The equivalent age of the bridge element EQA is defined in terms of the condition rating, whereas its estimated life ESL is assumed. An alternative table linking LC/LC_D directly to condition rating thresholds is also available, yielding a stepwise convex relationship.

In both Eqs. (11.5) and (11.10), a bridge member load rating is not automatically equal to 0 if the respective element condition rating is 0. PennDOT makes that provision in a table, setting $LC = 0$ for condition ratings $= 0$. In general, a load rating can be 0 even if the condition rating is not, and vice versa. If implemented properly, the two ratings should form a conservative system. A bridge can be found unable to sustain the design live and dead loads by load rating calculations, or it can be termed unsafe independently by inspection. As in Hambly's paradox,[38] however, the redundancy of the ratings does not automatically guarantee a conservative assessment. Bridge closures can be forestalled by load rating calculations when condition ratings are unacceptable and by frequent inspections when the load ratings fall short. Inspections and analysis are equally responsible for preventing the latter course by rigorously adhering to their standards.

Diagnostics. Amid the described abundance of condition ratings, the term *diagnostics* increasingly signifies independent assessment. Structural and material deficiencies can be quantified and qualified directly by analytic and experimental studies. Reinforced-concrete decks, members and pavements,[71–74] orthotropic decks,[75] steel deck gratings, prestressing tendons, and suspension bridge parallel wire cables, among others, have been the subject of detailed investigations in controlled environments and under field conditions. The results can include an assessment of the member's yield and ultimate strength, deformations, resistance to fatigue and to attack by deterioration agents.

After a calibration, laboratory test results can be applied to various field conditions. The Williamsburg Bridge orthotropic deck was tested to failure under cyclic load at Lehigh University.[75] The results were compared with field stress measurements over a relatively brief period and extrapolated to obtain an expected useful life. The Chicago rail bridge assessment[54] similarly combined field inspection findings and laboratory tests of material to forecast the fatigue life of 100-year-old riveted bridge details.

Numerical condition ratings benefit from comparisons with investigations of material behavior in a controlled environment. The improved understanding of alkali-silica gel effect on the function of reinforced concrete has led to the recommendation that concrete elements exhibiting efflorescence should be rated "not functioning as designed."[21] The application of nondestructive testing and evaluation (NDT&E)[76,77] and statistical analysis to bridge diagnostics has resulted in a new type of condition assessment under the general title *structural health monitoring*.

INSPECTIONS

Inspection is bridge condition assessment by direct structural examination. Although the *NBIS* is not the sole authority on bridge inspections, the practice established by the standards is the best developed nationwide. Minimum bridge inspection requirements are broadly stipulated in the *NBIS*[16] as follows:

§650.305 (a) Each bridge is to be inspected at regular intervals not to exceed 2 years in accordance with Section 2.3 of the *AASHTO Manual*.

The following *NBIS* requirements for inspection reports are defined:

§650.309 The findings and results of bridge inspections shall be recorded on standard forms. The data required to complete the forms and the functions which must be performed to compile the data are contained in Section 3 of the *AASHTO Manual*.

In these definitions, the coding guide[16] refers to the AASHTO *Manual for Maintenance Inspection of Bridges* (1983).[78] Also relevant is the *Bridge Inspector's Training Manual* (1970,[26] 1990[42]). Fracture-critical and other elements requiring special scrutiny during inspections are addressed in a succession of manuals and related publications.[14,36,43,79]

The evolution and abundance of manuals should alert the user that they are not above interpretation. Typically, the scope of underwater inspection is described[36] as follows: "Underwater members must be inspected to the extent necessary to determine structural safety with certainty." Terms such as *to the extent necessary* place the responsibility for safe operation on the owner, who, in turn, expects the inspector to assess the bridge condition *with certainty*. Clearly, not all significant information is contained in the manual. Inspections frequently are governed by more than one set of directives and numerous updates. State highway departments, railroad companies, and other bridge owners issue their own inspection standards and manuals. Noteworthy are the manual by Park[40] of the NJ DOT, the New York State DOT *Bridge Inspection Manual,*[21] the *Bridge Inspection and Rehabilitation.*[41] A handbook for railroad bridge inspection is in preparation by AREMA Committee 10 in addition to the nonregulatory *Track Safety Standards,* 49 CFR 213, Appendix C.

Special Emphasis Details

Since a primary inspection task is to preclude failures, the details most likely to fail are specially emphasized. The following *NBIS* paragraph broadly addresses any unique and special features:

§650.303 (3) . . . Those bridges which contain unique or special features requiring additional attention during inspection to ensure the safety of such bridges and the inspection frequency and procedure for inspection of each feature.

According to Sec. 650.303(e), the inspector must identify the presence and document the condition of "unique or special features." Various designations are used by the *NBIS* and other bridge management guides for structural features requiring particular attention. In keeping with the origins of highway and railroad bridge inspection programs, foremost among them is "fracture critical."

Fracture Critical. *Fracture-critical bridge members* are defined in the *NBIS* as follows:

§650.303 (e) The individual in charge of the organizational unit that has been delegated the responsibilities for bridge inspection, reporting and inventory shall determine and designate on the individual inspection and inventory records and maintain a master list of the following:

(1) Those bridges which contain fracture critical members, the location and description of such members on the bridge and the inspection frequency and procedures for inspection of such members. (Fracture critical members are tension members of a bridge whose failure will probably cause a portion of or the entire bridge to collapse.)

The preceding definition clearly resonates with the Silver and Mianus Bridge failures. The top priority assigned to the inventory and inspection of fracture-critical bridge members is reflected in the excellent supplement to the bridge inspector's training manual entitled, *Inspection of Fracture Critical Bridge Members* (1986).[14] The supplement addresses both the general aspects of highway bridge inspection and the details prone to fatigue and fracture. Among its many important general recommendations are the rotation of bridge inspectors in order to avoid routine attitudes, the introduction of quality control review of inspection reports, and a call for quality in preference to quantity of inspections. Stress concentration owing to welds, corrosion, poor load distribution, etc. is described and classified according to the level of susceptibility.

Fatigue is a principal (but not unique) agent of metal fracture. Fatigue-prone steel details are categorized (type A through E′) in most bridge design and inspection publications.[14,79-81] Recommendations for bridge detail assessment are offered in ref. 82. For bridge inspectors, however, identifying points of stress concentration (known as *stress raisers*) in as-built details is only the beginning. An inventory of fatigue-prone details is a must for the bridge owner. Where stress-raiser details are numerous, a statistically meaningful sample must be inspected regularly.

Hard to estimate is the effect of corrosion on stress concentration and, consequently, fatigue. *Corrosion fatigue* is particularly damaging to high-strength galvanized wires of prestressing tendons, suspenders (Fig. 11.16), and suspension cables, but it affects

FIGURE 11.16   Broken wires in suspender rope.

a much broader range of steel structures, as shown in Fig. 11.17. Periodic monitoring by NDT&E techniques (discussed below) is recommended.

Less attention is paid to certain fatigue-type concrete failures. Delaminating concrete (bottom and top) deck layers (see Figs. 11.1 and 11.6) under repeated traffic loading are examples. This is attributed to the expansion of corroding reinforcement with some justification. If the concrete cover is inadequate, however, spalling also occurs when reinforcement is in relatively good condition, as in Fig. 11.7. Wearing surface and entire deck patches of higher density and strength tend to fracture in a fatigue-reminiscent manner. The concrete bridge pier of Fig. 11.7 continued to function without appreciable sag after the column-to-floor beam connection failed. The likeliest cause of damage was the cyclic nature rather than the magnitude of the live loads. The concrete of the overreinforced detail also was affected by alkali-silica reaction, making it even more susceptible to fatigue.

Concrete fatigue is not structurally critical, but it produces safety hazards and accelerates further deterioration, often necessitating structural rehabilitation.

100% Hands On. Fracture is not the only critical feature inspected. Bridge owners, e.g., the NYS DOT,[21] define critical features as "special-emphasis details." Thus "100% hands-on" inspection of all special-emphasis details must be confirmed by the signature and professional engineering license number of the inspector. As a result of inspection findings, technical advisories and engineering instructions are circulated regularly, adding new details to the special-emphasis list. The Internet has made such publications readily accessible. Consequently, it becomes essential to determine which ones of the various specifications, directives, mandates, advisories, instructions, and recommendations apply to the inspection at hand. *Hands on* is both a literal and a figurative term, stipulating the inspectors' responsibility without defining their task. Implied are *minds, eyes,* and *ears on.* The activity shown in Fig. 11.18, for instance,

FIGURE 11.17   Holes drilled to arrest fatigue crack propagation.

FIGURE 11.18   Is this a "hands-on" inspection?

would qualify as a hands-on inspection only if the inspected detail is examined with adequate knowledge of its intended function and performance and if the findings are suitably documented.

Traffic impedes hands-on inspections while allowing inspectors to observe the structure's response to live loads. At a minimum, hands-on inspections must assure the safe service of the inspected details until the next scheduled inspection. Sampling may become unavoidable when specially emphasized details are prohibitively numerous. The corners of bottom flange cover plates require hands-on inspections. Inspectors have no option but to inspect a statistically significant sample, selecting locations with most advanced deterioration and/or highest stress, and documenting their method. Concrete deck undersides similarly should be sounded for delamination. If distress is noted, the sample is increased.

Two criteria help to identify critical (fracture or not) candidates for hands-on inspection: redundancy and stability. The former (see previous discussion) gets more attention in recent manuals. The AASHTO evaluation manuals[36,43] refer to "stability" broadly, including all displacement irregularities. Structural stability, both local and global, rigorously defined as a function of compression and geometric and material properties, must be a primary consideration in every bridge inspection. Among other tasks, this implies the ability to distinguish tension from compression members, not always eminently clear, for instance, in a statically indeterminate truss.

Stability. Stability considerations occasionally are neglected during inspection owing to the false assumption that an unstable member inevitably exhibits large displacements. It is helpful to recall that the original (straight) configuration remains a possible mode of equilibrium as the axial compression approaches a bifurcation point. The columns in Fig. 11.19, for instance, are extended by approximately 25% owing to the exposure of their footings. As a result, their theoretical buckling load is reduced by a factor of $1.25^2 = 1.5625$. If the original columns were hinged at the top and fixed at the top of the footings, and if the footings could rotate around their bases, the theoretical buckling load would be reduced further by a factor of $(1/0.7)^2 = 2.04$. This assessment is extremely conservative, e.g., ignoring the rigidity of the footing blocks. Nonetheless, despite the absence of apparent displacements, stability is a realistic concern. Areas near footings are prone to accelerated corrosion, reducing the column buckling load by a factor of approximately 2.04 by eliminating the fixity of the support. Again, owing to buckling considerations, the pier column bracings shown in Fig. 11.20 are primary rather than secondary members and require 100% hands-on inspections.

Members bent by vehicular impact often are mislabeled as "buckled" (see Fig. 11.4), whereas buckled members are diagnosed as "impacted" (Fig. 11.21). The failing column shown in Fig. 11.8 appears to have buckled but is actually bent owing to an eccentric bearing on top. Steel-solid rib arches (Fig. 11.22) can be subject to buckling and are difficult to access.

Buckling is not a concern at masonry arches, but crushing and instant collapse can be, as examples repeatedly demonstrate. In France, where 75% of the bridges were built before 1900, the smallest geometric changes in masonry arches are recognized as important indications of distress and are monitored according to specific guidelines.[50]

Connections. Connections easily qualify as special-emphasis details even if they are not so designated. Design assumes that connections adequately transfer forces between elements or provide a release for them. Most bridge malfunctions prove that these assumptions are erroneous. In recently built structures, connections are most likely to suffer from construction or design defects. Prolonged bridge use adds to the possible causes of connection failures.

Connections are discrete in steel and timber structures but can be monolithic in concrete ones. Splicing and overlapping of reinforcement can be viewed as a connection's equivalent in reinforced concrete. At the Hanshin Expressway during the Hyogo-Ken Nanbu earthquake in 1995,[18] for instance, welded rebar splices failed.

For inspection purposes, connections can be divided into rigid or articulated. Unless otherwise specified, rigid connections are rated with the structural elements. Articulated ones, such as bearings, are rated as independent elements.

Connections, rigid under regular loads, may be designed to yield during extreme events. Bearings, free to adjust to thermal fluctuations, can be either free or restrained during earthquakes. After a fixed bearing failure at the San Francisco–Oakland Bay Bridge during the Loma-Prieta earthquake of 1989,[11] the expansion bearing displacement capacity proved inadequate (Fig. 11.23). As a temporary measure, the emergency rehabilitation replaced the 5-in. (127-mm) bearing pads with 3-ft (915-mm) brackets.

Since a routine visual inspection is likely to witness neither slow (thermal) nor sudden (seismic) motion, large displacements must be anticipated. Inadvertent restraints, owing to poor dimensioning, corrosion, or other causes, must be identified as a hazard. As all structures, connections are best inspected in the following sequence: Identify all loads and displacements, reduce them to their six triaxial components, and

FIGURE 11.19   Exposed column footings.

verify that they can be transmitted adequately or released. In order to recognize early symptoms of failure, inspectors must be familiar with the intended function of the connection. The 11-ft-tall (3.35-m) linkage of Fig. 11.18, for instance, "froze" owing to corrosion. This could not be observed by direct inspection, but the base of the supporting bridge pier 90 ft (27.5 m) below showed uplift. Emergency inspection and evaluation of the pier columns recommended that repairs proceed without traffic interruption. Replacement of the detail was added to the scope of the bridge rehabilitation.

FIGURE 11.20   Column bracings and girder tie-downs.

The 10-ft (3-m) tie-down links in Fig. 11.20 were designed to apply tension at the end of the nonredundant steel girders. Biennial inspections found the links buckled or loose, therefore overstressing the primary members. The condition was rated "not functioning as designed," and load rating was required. The subsequent rehabilitation replaced the links with threaded rods, allowing tension adjustment.

A number of bearings, such as pin and hanger assemblies (Fig. 11.24), are prominent on all fracture-critical lists.[14,36] Rocker and sliding metal bearings typically are eliminated during rehabilitation owing to poor seismic performance,[13] although even their routine performance (Figs. 11.5 and 11.25) is not always problem-free. Inspections continually add to the list of nonperforming details. Although all inspection and

FIGURE 11.21   Buckled truss diagonal.

evaluation manuals discuss connections at length, design texts are recommended for an understanding of the subject.

Inspecting bolted and welded connections requires knowledge of their construction specifications (e.g., shop versus field). Occasionally, welds and bolts are substituted during construction without adequate record. Widely publicized are the pedestrian walkway failure at the Kansas City Hyatt Regency Hotel[8] in 1980, caused by an

FIGURE 11.22 Hamilton and Washington Bridges, New York City.

FIGURE 11.23 Bay Bridge, San Francisco–Oakland, October 1989.

FIGURE 11.24   Pin and hanger assembly.

unauthorized hanger discontinuity, and the successful strengthening of the Citicorp building in New York City in the 1990s, following a construction change from welded to bolted connections.

Expansion joints are focal points of structural connections and the most vulnerable to early and rapid bridge deterioration. The attitude that "the only good joint is no joint" is attractive but not always practical. NCHRP Report 319[83] summarizes the national experience with various joints and their replacement. Figure 11.11 illustrates a replacement of a cushion joint with a plug joint in Manhattan. It can be argued that details are poorly designed if they require greater attention than regular inspections provide, assuming that regular inspections are optimal. The scheduling and scope of regular inspections therefore become essential.

Inspection Types

The scope and frequency of inspections vary with the type and condition of the bridges. No inspection should be undertaken without a defined set of objectives, standard doc-

FIGURE 11.25   Sliding steel bearing.

ument format, and data storage capability. The categories listed below based on the *NBIS* or other responsible sources attempt to cover most inspection needs. Manuals, guides, advisories, and instructions are available for each inspection type. The most common of them, designed primarily for vehicular concrete and steel structures, are discussed briefly below.

Regular (Biennial). For highway bridges, this is the standard bridge inspection defined by *NBIS,* Sec. 650.305, performing three essential functions: identify potential hazards, rate the condition, and update the inventory. Regular inspections are conducted according to the local[21] and federal[36,42] bridge inspection manuals and provide reports in the format prescribed therein. The numerical outputs of these reports are submitted annually to the FHWA in a form compatible with *NBIS.* New York State requires railroads to submit biennial inspection reports as well. AREMA requires annual "routine" inspections but maintains no central database.

Regularly scheduled inspections are often termed *routine,* although their execution should be anything but. Bridge inspection reports are definitive technical and legal documents signed by the engineer in charge and by the quality control reviewer with their professional engineering license numbers. Beginning in recent years, inspection reports are generated and transmitted electronically, introducing new procedures related to the legal status of the documentation.

Besides their findings, routine inspections must clearly document their own limitations. When structural changes between scheduled inspections cannot be anticipated adequately, further action must be recommended. Normally inaccessible but critical structural elements must be identified and scheduled for special inspections. These include foundations, channel, embedded reinforcement, encasement anchors, suspension cable wires, masonry arches, etc.

Interim. *NBIS* addresses the need for inspections performed between biennial inspections as follows:

650.305   (b) Certain types or groups of bridges will require inspection at less than 2-year intervals. The depth and frequency to which bridges are to be inspected will depend on such factors as age, traffic characteristics, state of maintenance, and known deficiencies. The evaluation of such factors will be the responsibility of the individual in charge of the inspection program.

Interim inspections are planned as more limited than biennials but may call for "in-depth investigations" if the need arises.

> *Monitoring.* Monitoring inspections address specific conditions and locations. Documentation is limited to describing changes in the local condition and recommending remedial action. Such inspections were found indispensable in New York City during the early 1990s when annual "flag" reports of potential bridge-related hazards reached 3000 (on roughly 800 bridges with 5000 spans).[44] The organization chart shown in Fig. 11.14 shows how monitoring of less critical conditions can alleviate the workload of emergency repairs without compromising the service.
>
> *Temporary repairs.* Temporary repairs mitigate imminent hazards without improving the condition of the structure. Their monitoring is a high priority for owners of previously neglected bridges under heavy traffic demands. Typical temporary repairs include timber and steel shorings, deck steel plates, and holes drilled at the tips of fatigue cracks (see Figs. 11.5, 11.9, and 11.17). Sandwitching a steel girder web with bolted plates to arrest crack propagation is a long-term but, ultimately, temporary repair because it alters the stiffness of the member and may cause unwelcome stress concentration elsewhere in the structure. Because of their occasionally improvised nature, such repairs place a stronger demand for judgment on the inspecting engineer.

Temporary repairs must be installed with a specified lifespan and a prescribed frequency of monitoring. Both are subject to field verification. If exceedingly frequent monitoring is required, the repair is ineffective. Timber shorings shrink, creep, rot, and split. The buckling load of a typical 14 × 14 in. (35 × 35 cm) timber post that is split in half longitudinally, is reduced by a factor of 8. Steel shorings are still temporary. Their pedestals and bearings do not conform to design specifications. Temporary columns (see Figs. 11.5 and 11.9) should not be considered effective without inspection of the wedging or other details at points of load transfer. An uninterrupted load path must replace the failed one adequately, from the supported element to an adequate footing.

Roadway steel plates bounce and shift under traffic, crushing the concrete decks. Under typical urban traffic, the fatigue life of anchoring bolts or straps is exhausted in less than a month. Holes drilled at fatigue crack tips occasionally engender multiple crack propagation paths (see Fig. 11.17). Timber shorings (Fig. 11.9) crack, shrink, and rot. Clearly, temporary repairs are best dealt with by permanent replacement.

Special. Inspections may be special owing to circumstances or the structure. Their scope varies according to their purpose. The resulting reports may not fully meet or exceed *NBIS* requirements. An inspection can be termed *interim* or *special* depending on local priorities. The expansion joint inspection, formalized according to Fig. 11.15, is considered special owing to its irregular scheduling.

In-depth. It is indicative of the nature of regular biennial inspections that they can serve to prioritize rehabilitation projects but cannot be used to determine the scope of rehabilitation work. This is the task of in-depth inspections. Such inspections include field verification of as-built drawings. Destructive testing, such as coring of concrete decks, avoided during most inspections, is standard during in-depth inspections. Videotaping of inspection highlights has become common practice. The NYS DOT has issued a guide for in-depth inspections. An example of an in-depth inspection is the Chicago transit rail bridge assessment.[54]

Essential Completion. *Essential completion* inspections serve two purposes. First, they ascertain whether the work on the bridge is essentially complete under the current contract(s). If appropriate, a "punch list" of items to be completed is created and verified by a subsequent inspection. The scope of rehabilitation work is particularly hard to assess because bridges are not always restored to their original condition, and significant items may have been added by contract change orders. Second, new as-built drawings are field-verified and added to the bridge file. The bridge inventory is updated to reflect the new condition (according to the owner's manual) and becomes a reference source for future inspections.

Essential completion inspections should recognize that construction occasionally deviates from design and that design is not always flawless. Condition ratings assigned to bridges after rehabilitation, for a variety of reasons, are rarely the highest. Large bridges in western Europe are inspected 10 years after construction completion by commissions representing the contractor, the owner, and government auditors in order to determine the responsibility for any nonperformance. What constitutes a completed project can become a matter of litigious dispute. A manual of professional practice on the "quality of the constructed project" was prepared by the American Society of Civil Engineers (ASCE) in 1988 as a guideline for owners, designers, and constructors, but it remained only a recommendation. The lack of uniform standards in this field underscores the importance of construction supervision.

Construction Supervision/Quality Control. Construction supervision focuses on the process but requires a thorough familiarity with the product. It is defined in standard specifications of construction and materials[84-86] and in contract documents. The subject is also treated comprehensively in general texts on bridges[18] and industry reports.[87] A number of essential construction tasks, such as welding, high-strength bolting, prestressing, reinforcing, excavating, etc., are subject to quality control regulated by ICC, ACI, AWS, etc. Safety and environmental protection may be monitored by government agencies (OSHA, DEP). Conformity with government regulations typically is supervised by independent qualified professionals under contract agreement.

Competent construction supervision can extend considerably the useful life of a bridge by eliminating the numerous flaws typical of most construction projects, such as inadequate sampling of materials, unacceptable pile driving, poor water-cement ratio in concrete, poor compaction of concrete, poor temperature control of concrete during curing, substandard welding (shop and field), substandard bolt connections, misalignment of bridge elements, inadequate placement of reinforcement, inadequate steel surface preparation prior to painting, inadequate waterproofing, and nonconformity with OSHA and DEP regulations.

Construction quality may be verified by field tests. Figure 11.26 shows a load test by hydraulic jacks anchored below a bridge. The purpose is to determine the capacity of the deck primary structure. Alert supervision must avert the particularly pernicious departures from the stipulated construction sequence. The footings in Fig. 11.18, for instance, should have been exposed for only brief periods and under special conditions.

FIGURE 11.26   Construction load test.

Design also can be improved during construction. *Value engineering* reviews accepted designs, seeking (life-cycle) cost-effective modifications.

Maintenance. When maintenance is a continuous process, as on some unique bridges, inspections can be incorporated in it. Recurring (routine) maintenance tasks, such as spot painting, caulking, and drain cleaning, have been combined with the condition rating and recordkeeping required by inspections and conducted concurrently. Certain components, such as all movable parts of movable bridges, travelers of suspension bridges, electric equipment, etc., may require frequent inspections (including by electrical and mechanical engineers), whereas others, such as, for instance, the wrapped wires of suspension cables, are beyond the biennial scope. Recent designs of unique bridges include bridge-specific maintenance manuals. If future inspections are able to compare field findings with maintenance records, the actual effect of maintenance on bridge condition finally will emerge.

Long and Unique Spans. From a management point of view, a bridge may be unique either because regular inspections do not fully capture its condition or because it provides unique service and a condition rating of "not functioning as designed" is unacceptable. *Long* is a vague term. Inspections need not rigidly adhere to the 500-ft (152.5-m) AASHTO design specification limit. The number of spans, rather than the bridges, best quantifies the scope of inspection. Structures comprised of multiple relatively short spans are typical for urban areas and sometimes are not even perceived as bridges. Multispan bridges often have unique details, but their magnitude remains the main challenge, adequately handled by *NBIS* and state inspection practices.

Long spans combine critical structural details with the demand for rapid assessment of voluminous data. In the inventory, they are at a disadvantage if they are treated as average spans. Discretizing long trusses into fictitious spans between panel points improves data management. Approaches typically are multispan structures with important features of their own.

The scope of biennial inspections had to be exceeded at New York's Williamsburg Bridge in 1988. Critical details, including portions of the suspension cables and the cantilever roadways, were so deteriorated that inspectors closed the bridge to all but pedestrians until a detailed evaluation could allow a reopening. A subsequent in-depth investigation resulted in a 16-year rehabilitation. A bridge-specific maintenance manual is part of the as-built documentation.

Eye-Bar Chain, Suspension, Cable-Supported, Prestressed Bridges. The Silver Bridge collapse has discredited the internally nonredundant two eye-bar chain suspension systems, but multiple eye-bar chains are present on many railroad and highway trusses and on a few suspension bridges. When the parallel eye-bars between two panel points are numerous, a single one of them is not as critical, but the analysis must consider the probability that if one has failed, the others may be close to failure for similar reasons, in addition to experiencing overstress. Eye-bar chains and their pinned connections are very well suited to nondestructive testing and evaluation techniques, as has been the case at the Queensboro Bridge (Fig. 11.27) and the Brooklyn Bridge (Fig. 11.28) in New York City.

Twenty-nine suspension bridges with aerially spun parallel wire cables had been built in North America prior to the year 2000; 3 have prefabricated parallel wire strands, and 21 have helical strands. Most spans are greater than 700 ft (213 m). The number of cable-stayed bridges is growing rapidly. The Cincinnati-Covington Bridge and the Brooklyn Bridge are examples of John Roebling's trademark hybrids of suspension and stay bridges.

The details peculiar to cable-supporting systems are recognized as critical[14,36,43] Included are the main cables and the suspenders of cable-stayed bridges and the stays of cable-stayed bridges, saddles, and anchorages (Fig. 11.29). The high-strength steel wires used in cable-supported bridges are grouped in parallel or helical strands. They usually are galvanized, although there are exceptions, such as the Williamsburg Bridge, New York City (parallel wires), and the original cables on Pont de Tancarville, France (helical strands, now replaced with galvanized ones). Helical strands or wire ropes are typical for suspenders (see Fig. 11.16) and stays. The latest record holders, Akashi-

FIGURE 11.27   Eye-bar inspection, Queensboro Bridge, New York City.

FIGURE 11.28 Franklin Square Truss, Brooklyn Bridge, New York City.

FIGURE 11.29 X-ray diffraction test at suspension bridge anchorage.

Kaikyo for suspension and Tatara for cable-stayed spans, in Japan, use parallel wire strands for suspenders and stays, respectively, requiring new inspection and assessment.

Suspension and stay system details are mostly inaccessible during regular inspections. Conventional condition ratings capture neither the rate of their deterioration nor the consequences to overall bridge safety. Adequate assessment of suspension bridges must take into account the mechanisms of material degradation, the structural failure mechanisms, and the techniques of detecting malfunctions. A general understanding of cable-supported bridges is a prerequisite.

The corrosion mechanism of high-strength, highly stressed, galvanized (or not) steel is the subject of many detailed investigations.[88,89] Pitting and stress corrosion can cause square breaks of suspension cable wires. When such phenomena are suspected, the cable must be unwrapped and wedged (Fig. 11.30) to determine the extent of deterioration. Acoustic monitoring of wire breaks is a noninvasive alternative becoming standard practice for monitoring suspension and stay cables, as well as prestressing tendons.

If confirmed by in-depth inspection, the deterioration of a suspension system triggers a detailed evaluation of the bridge load-bearing capacity and life expectancy, including laboratory tests of wire samples, analysis of the as-built and as-is structure, and cost-benefit assessment. The Williamsburg Bridge investigation concluded that the main cables could be rehabilitated, whereas the suspenders had to be replaced during the 1990s. All suspenders and stays of the Brooklyn Bridge were replaced after one stay ruptured owing to corrosion in 1981, killing a pedestrian. The main cables of the Pont de Tankarville were replaced with galvanized ones[90] after one helical strand ruptured owing to corrosion. The similar cables of the Pont d'Aquitaine in Bordeaux were replaced without awaiting analogous developments. A number of texts[18,88,89] describe cable-supported bridge assessment. NCHRP Project 10-57 is expected to produce a

FIGURE 11.30 In-depth suspension cable inspection.

manual for the inventory, inspection, and assessment of parallel-wire suspension bridge cables in 2004.

Although cable-stayed bridges appear closest to suspension bridges in general aspects, their function has a lot in common with prestressed-concrete structures. Extrados bridges represent a transition between the two. The primary target of inspection, mostly by nondestructive techniques, is the stays. The same applies to segmental posttensioned bridges with external or embedded tendons. Manuals specifically address prestressed bridge condition assessment.[91,92] Most recent is the NCHRP Report 496, *Prestress Losses in Pretensioned High-Strength Concrete Bridge Girders* (2003). Loss of prestressing tension owing to shrinkage, creep, relaxation, construction imperfections, corrosion, or combinations of factors is critical to the integrity of pretensioned concrete structures. Inspections should be aware of the bridge construction sequence (pre- or posttensioned), as well as the prestressing tendons' protection systems (grouted or not). Access is often entirely obstructed. An example are the many three-span bridges with inclined-pier columns prestressed by tie-downs at the abutments. Inspections are meaningless if they ignore the existence of the tie-downs or their condition.

The need to investigate and evaluate the condition of embedded or encased high-strength tendons and rods[93,94] is universally recognized. During the early 1990s, the United Kingdom halted all new construction of posttensioned segmental bridges until a reliable method of tension verification during their use could be found. This halt has now been relaxed, although the reliability of inspections remains a concern. The collapse of a 5-year-old prestressed concrete pedestrian bridge in Concord, North Carolina, on May 21, 2000, apparently due to tendon corrosion, is a reminder of the sensitivity of prestressed concrete bridges.

High-precision monitoring of overall structural geometry becomes particularly important in prestressed concrete structures, as well as close observation of any local indication of distress, such as cracking and pigmentation of concrete. The crushing of the concrete at the joints between the segments in Fig. 11.31 prompted in-depth inspections and eventually a rehabilitation aimed at restoring the estimated loss of pre-

FIGURE 11.31   Posttensioned segmental bridge.

stressing forces. Wire breaks are monitored acoustically, as in cable-supported bridges. Acoustic monitoring of corrosion has been attempted but not confirmed at this writing.

Movable Bridges. Movable bridges uniquely require functionality assessment. Their mechanical, hydraulic, electrical, interlocking, control, and other special components must be inspected by appropriately licensed engineers. Inspections are guided by the FHWA manual,[95] currently in its second edition. Also essential are the AASHTO movable bridge design specifications.[96] The cables supporting counterweights on lift bridges must be certified by the manufacturer and inspected according to the American National Standards Institute (ANSI).[95] To the bascule, vertical lift, and swing bridges discussed in ref. 95, the rarer retractile ones could be added for completeness.

Movable bridges must operate at the request of the U.S. Coast Guard. As a result, owners not only must maintain and inspect but also operate them. A number of details are unique and prone to specific modes of failure that must be recognized and anticipated (see Fig. 11.32). Movable bridge decks typically are open steel gratings suffer-

FIGURE 11.32 Ruptured auxiliary weight cable on lift bridge.

ing from corrosion fatigue and fractures. Structural repairs, particularly temporary ones, should not alter the ballance of the movable parts.

Underwater (Diving) Inspections. The *NBIS* specify underwater inspection frequency as follows:

§650.303　(2) Those bridges with underwater members which cannot be visually evaluated during periods of low flow or examined by feel for condition, integrity and safe load capacity due to excessive water depth or turbidity . . . shall be described, the inspection frequency stated, not to exceed five years, and the inspection procedure specified.

The condition evaluation manual[36] classifies underwater inspections in three levels according to their scope. The *NBIS* mandate above refers to a level III inspection, requiring hands-on verification of the entire underwater substructure by qualified divers. State departments of transportation issue specific directives for diving inspections.[97]

Access to underwater bridge elements, a professional feat in itself, must be followed by collection and correct interpretation of meaningful data. Scour is particularly hard to detect even during direct access to a susceptible pier footing. Infrared camera photography allows for visual inspection, but scour still can go unnoticed. The early stages of marine borer attacks on timber piles are only detectable by sample testing. Underwater inspections must be conducted by specialized consultants under contract to the bridge owner. The same does not apply to bridge details located at the highly vulnerable *wet line*. Figure 11.10 shows timber piles rotted between 50% and 100%, entirely submerged at high tide but in full view 6 hours later.

Extreme Event/Emergency. Two types of assessments are developing to meet the demand: pre- and postevent. Preassessment seeks to avoid emergencies; postassessment must resolve them. Owing to the randomness of extreme events, adequate response requires a core team of permanent employees to be supplemented by contracted forces as needed. Personnel charged with emergency inspections must be specially trained in safety and communications by the responsible agency, as well as by law enforcement and federal (FEMA) and local emergency management authorities (e.g., the Office of Emergency Management). Bridge inspectors must be clearly advised of their responsibilities and options for remedial action, e.g., to maintain or close traffic, request emergency repairs, etc. Emergency procedures must be documented, coordinated with related agencies, and tested. Means of contact with all responsible personnel and access to any location must be provided.

Triage. Management, including that of bridges, must resort to triage when allocating its scarce resources. Three categories are usually identified in the process. Two are excluded: the "good" and the "unsafe" (closed and awaiting replacement). The available funding is spent on keeping the middle third in service. Condition assessment is suspect when it is constrained by limited resources. Triage is the crudest form of management, and life-cycle considerations seek to eliminate the need for it.

Post–extreme event assessments have no option but to perform triage on demand. Emergency response depends on limited workforce, time, and material. For expediency, bridge owners occasionally combine the inspection and repair function by assigning engineers competent to assess the damage, select the appropriate repair, and supervise its completion. Independent inspections should determine the useful life of emergency repairs. The bridge inventory and the condition ratings may have to be updated accordingly.

All factors relevant to an emergency inspection, including parties present, possible causes of accidents, atmospheric conditions, types and licenses of vehicles involved, etc., must be documented in addition to the structural assessment. The most recent inspection report should be available at the site as early as possible. Repair recommendations should include measures to mitigate the cause. The most common causes for emergency inspections are discussed briefly below.

Traffic Accidents. Traffic accidents on the bridge deck level typically damage parapets and guide rails and light and sign structures. Under the bridge, vehicular impact can destroy columns (Figs. 11.4 and 11.33) and primary structure (Fig. 11.34). The nature and location of the damage often are repetitive. A steel signpost or overhead floor beam may sustain numerous impacts with minor distortions but eventually could crack. Complacency, always a threat to bridge inspections, is particularly likely to affect investigations of traffic-induced damage as accidents recur at the same locations and create the impression of routine.

Train impact (see Fig. 11.4) is much greater than vehicular impact and requires more effective prevention. Even more destructive are vessel collisions (see Fig. 11.33), discussed in numerous reports.[45] Protection systems, such as crash walls, fenders, dolphins, etc., are inspected as part of the bridge. Traffic must be considered unsafe unless such systems function reliably. After the collision shown in Fig. 11.33, traffic was suspended until completion of emergency repairs.

Reliable forecasts of accident-prone locations and frequency of occurrence can be obtained from local traffic departments. Inspections of such locations should recommend appropriate mitigation, possibly including changes to the bridge geometry. After

FIGURE 11.33   Collision damage to concrete bridge pier.

FIGURE 11.34 Impact damage to superstructure.

numerous collisions, bridge inspections recommended the removal of the structure in Fig. 11.34. The NYS DOT collision vulnerability manual is an attempt to address this hazard systematically.

Earthquakes. Earthquakes have caused repeated vast destruction of bridges. Nationwide, California has suffered the most from recent earthquakes and, accordingly, has a well-developed postevent response capability. The Applied Technology Council offers a *Field Manual for Postearthquake Safety Evaluation of Buildings* (ATC 20-1, 1989) and conducts courses on the subject. CalTrans has issued manuals for postearthquake structural inspections and has emergency procedures tested during the Loma-Prieta[11] (1989) and Northridge (1994) earthquakes. The Japanese experience, particularly after the Hyogo-Ken Nanbu earthquake, in Kobe, in January 1995, is another important source of information on postearthquake response. Numerous bearings and restrainers failed during that event. The San Francisco–Oakland East Bay Bridge span collapsed owing to failure of both the fixed and expansion bearings (see Fig. 11.23). Rocker and sliding steel plate bearings must be inventoried for replacement during bridge retrofits in seismic zones. The bearings in Figs. 11.25 and 11.35 did not fail in an earthquake but would have performed similarly. Such considerations inform the preevent vulnerability assessments.

In the words of professor R. Clough of the University of California, Berkeley, earthquake-induced loads, unlike any others, depend on the properties of the structure. The analytic model of the structure is subjected to hypothetical earthquakes with estimated return periods and likelihood of occurrence. The uniquely speculative nature of the assessment is best summarized in the sobering reminder of professor G. Housner of Caltech that "earthquakes repeatedly prove us stupid."

When a relatively low seismic hazard is coupled with a relatively poor average bridge condition, as in New York State during the 1980s and 1990s, the seismic vulnerability program replaces all substandard details during planned bridge rehabil-

FIGURE 11.35 Tilted rocker bearing and cracked pedestal due to fire.

itations. Nationwide, the FHWA is funding a considerable effort to develop a comprehensive pre- and postearthquake bridge safety program through its three national centers. A number of publications on seismic bridge design and retrofit are available on federal[12,13,98] and state levels. The current FHWA Project DTFH61-98-C-00094 will update the seismic bridge retrofit manual and issue new vulnerability guidelines for the existing highway infrastructure.

Floods. Floods are recognized historically as the primary cause for bridge failures. The Schoharie Bridge[8] in New York State is an example. The NYS DOT hydraulic vulnerability is one systematic management approach to mitigating this hazard. During the 1994 floods in Delaware County, New York, New York City inspection teams were deployed by helicopter to inspect the affected areas. Even though emergency inspections can be helpful in allowing limited loads on structures, underwater inspections are required after torrential floods. Scour can be monitored sonically. It has been noted that the undermining of footings can occur very quickly during high water and offers no early warnings during underwater inspections.

Hurricanes. Most bridges are designed for 100-mi/h (160 km/h) wind loads, but the corresponding force is applied statically. The Tacoma Bridge collapse more than any other demonstrated the destructive effect of the dynamic nature of wind loads. The wind response of the Whitestone Bridge has been adjusted by numerous retrofits, the latest one still in progress at this writing.

The wind response of suspension structures is assessed both analytically and by testing models in wind tunnels. Ambient vibrations are monitored continually on a number of bridges in Japan, such as the Akashi-Kaikyo Bridge and the Hakucho Bridge, as reported in several publications by Professor Y. Fujino of the Univerity of Tokyo and by the Honshu-Shikoku Bridge Authority. The behavior of long suspenders and diagonal stays remains a matter of concern[10] and can be monitored continually thanks to increasingly available accelerometers.

Crowds. Crowds have caused structural failures as far back as 1830 on Brown's Bridge at Montrose.[5] More recently, the Millennium Bridge in London[10] and Passerelle de Solferino in Paris were retrofitted to correct their response to crowd-induced motion. The sag of the Golden Gate Bridge in San Francisco under a crowd celebrating its fiftieth anniversary (1987) reminded engineers that pedestrians can exceed design live loads. For the AASHTO design lane load of 640 lb/ft (9.4 kN/m) over a 10-ft (3-m) traffic lane and with more than two 12-ft-wide (3.7-m) lanes, allowing multiple-use reduction, the equivalent unit load is $(640/12)(3/4) = 40$ lb/ft^2 (1.9 kN/m^2). A crowd uniformly packed closer than 4 ft^2 (0.8 m^2) per person would be heavier. As in all dynamic phenomena, this static model is superficial. It has been demonstrated[10] repeatedly that crowds tend to excite relatively flexible bridges laterally by synchronizing their gait with the dominant bridge frequency. In such cases, the comfort of the pedestrians, rather than concern for the structure, may govern. It can be recalled that on May 30, 1883, a week after the opening of the Brooklyn Bridge in New York, an estimated crowd of 20,000 panicked, possibly due to bridge sway, and rushed to the exits, trampling 12 people to death. Pedestrians using the Manhattan-bound traffic lanes during the blackout on August 14, 2003 reported lateral motion at 7 P.M. At 8 P.M., under roughly similar live load, I could not confirm the reports. A number of experts and I, interviewed by the *Village Voice* (August 27, 2003), including professor M. Shinozuka (currently of the University of California, Irvine), considered the movement a discomfort but not a structural hazard. Nonetheless, the article was entitled "Point of Collapse," contributing little to public confidence. The NYC DOT promptly initiated an emergency investigation of the dynamic response of the bridge to lateral excitations. Public safety demands not only accurate but also convincing assessment.

Structural Malfunction (Nonperformance). If there were no marked correlation between structural malfunctions and reports of potential hazards (e.g., the NYS DOT "flags"), the inspection system would have to be corrected. Regardless of the systematic precautions, however, unanticipated structural malfunctions cannot be ruled out. The term *nonperformance* is applied occasionally to the extreme-event failures discussed earlier, but it also applies to malfunctions under routine circumstances. A typical example is the spalling of the underside of a concrete bridge deck in New York City on June 1, 1989 (see Fig. 11.1), leading to the death of a motorist. Within days, in-house staff identified 400 city bridges as potentially hazardous based on reviews of inspection reports. Temporary make-safe measures, including shielding and netting of bridge undersides, were implemented while capital deck replacement contracts were processed. Bridge deck undersides have been treated as special-emphasis details ever since, promoting the use of stay-in-place deck forms.

The sliding bearing malfunction shown in Fig. 11.25 resulted from a misalignment of the expansion joints of a track rail and a supporting girder at a suspension bridge tower. Identifying the malfunction and averting a potential accident revealed the need for better communication between the train service and bridge management.

Movable bridges can suffer costly malfunctions owing to electrical and mechanical failures and the occasional operator's error. Figure 11.32 shows the socket of a ruptured auxiliary counterweight cable on a lift bridge. The incident occurred during manual operation of the lift span and required around-the-clock repairs. Condition evaluation in such cases is invariably an emergency owing to the demand for bridge service.

Nonstructural bridge appurtenances, such as masonry cladding (see Fig. 11.12), can be extremely hazardous to the public, the more so because they do not contribute to the structure's function and can be overlooked easily. Complete removal of cladding is a typical emergency measure. Purely decorative elements have maintenance needs, requiring cost-benefit analysis during design. Nonperformance culminates with a bridge closure, but that does not eliminate all need for condition assessment.

Closed Bridges. Closed bridges require inspections. As a general rule, dead loads (although static) amount to 90% of the design loads on vehicular bridges (beyond a certain length) and 80% on railroad bridges. Combining that load with the poor condition usually causing a closure can have catastrophic consequences. Figure 11.8 shows a deflected 45-ft (14-m) steel pier column with a 100% corroded web. The structure had been closed for 12 years before the condition was declared an emergency. Temporary supports were erected. Closed bridges frequently become dumping grounds and are more than commonly exposed to fire hazards.

Fire. Fire can be due to a traffic accident, illegal dumping, arson, or other source. Bridges cannot be fireproof to the extent that buildings are. When a fire occurs under the bridge deck, it can be catastrophic, as was the case on the New Jersey Turnpike in 1989 (see Fig. 11.35). Consequences include spalling and cracking of concrete, buckling of steel, and bearing and pedestal failures. Flammable materials, including all debris, are bridge-related hazards, and their unprotected storage in the bridge vicinity must be prohibited.

Sabotage. Bridge sabotage is an act of war, but peacetime concern is justified by the events of September 11, 2001, in New York City. In that and other instances, the experience of the Bridge Inspection Unit of the NYC DOT shows that qualified bridge inspectors have all the key assets of training and equipment required for a reliable structural assessment. Coordination with the responsible agency, e.g., the federal and local offices of emergency management, is of primary importance. Federal and local guidelines are under development.

Inspection Reliability and Optimization

The numerous inspection types and objectives described earlier seek to match the diverse causes and forms of bridge distress. Inspections are generously funded after structural failures. Once the level of risk is perceived as acceptable, management shifts focus to minimizing inspection costs. Biennial inspections were a national mandate in 1980, but in 2003, a more confident bridge management contemplates relaxing that rule to a recommendation. An alternative view holds that inspections, as well as maintenance, should be regular, as are the seasonal climate changes and traffic. The two positions can be reconciled by regular inspections with a scope that is varied according to the structural condition.

While bridge management cannot be optimized in isolation from other social activities, inspection frequencies and scope can be, depending on structures, traffic, climate, budget, and other constraints. Originally, inspections were designed and optimized to suit the demands of structural design and maintenance, but the process can be reversed. A structural assessment unable to meet the demands of bridge management is essentially recommending revisions of maintenance and design practices. Consequently, a "shoot the messenger" reception is among the inspector's occupational hazards. Given this responsibility, are inspections reliable?

"Experience is never at fault," Leonardo da Vinci[99] wrote, "it is our interpretation that is in error." The interpretation of experiences obtained from a structure errs owing to vagueness and ignorance in the forms of subjectivity and lack of quantification.[24] Vagueness is always present in less than fully informed opinions. "Everything," B. Russell cautioned, "is vague to a degree you do not realize until you have tried to make it precise,"[100] Ignorance, on the other hand, is more transparent but indecisive. The rotation of inspectors at inspection sites is intended to compensate for the inevitable subjectivity of condition ratings. Unfortunately, the opposite tendency is also present: The old reports increasingly influence the new. (Withholding previous reports is considered cost-ineffective.) Inspectors are reminded regularly to quantify their findings using the expanding NDT&E capabilities.

Consistency with measurements promotes objectivity but does not reduce the demand for subjective judgment. The wrong experience is also possible. Meaningless measurement could be as likely as misinterpretation. Loss of steel cross section to corrosion is typically exaggerated unless it is measured exactly. Even if it is quantified accurately, however, it does not reveal the risk of cracking owing to stress concentration (see Figs. 11.9 and 11.17). The correlation among hollow-sounding concrete, the corrosion rate in reinforcement, and deck spalling is hard to quantify. It is safe to assume that quantification will remain incomplete to a degree, allowing sufficient need for subjective (qualitative) judgment. As a special case of deducing truth from unreliable experiences, condition assessment inherits vast intellectual resources, none of them in user manual format. Plato[101] attributed to Socrates a dialogue on obtaining reliable knowledge from transient opinions, circa 420 B.C. Four centuries later, Seneca declared knowledge "not justified unless shared with others." Fifteen centuries after him, Leonardo da Vinci[99] wrote, "All knowledge is based on opinion." In the seventeenth century, Descartes began screening opinions by accepting only confirmed truths, reducing all problems to their smallest components, proceeding from the simplest to the more complex, and maintaining the input readily retrievable. The slightly younger Pascal found the Cartesian method "useless and uncertain," but it still informs decision support and knowledge-based systems. In mid-twentieth century, Einstein[15] allowed, "The whole of science is nothing more than a refinement of everyday thinking."

Following in this tradition, late-twentieth-century condition assessment arrives at the thinking that bridges are "deteriorated." "Damage" is refined into "minor," "serious," or "advanced," the latter two adjectives being used interchangeably. Recommended are the decisive "repair as necessary" and "restore to as new." Seeking further clarification, the FHWA investigated the quality of visual inspections in the early twenty-first century.

The resulting report[102] consists of literature review, survey of existing practice, and results of field tests with 49 state bridge inspectors. Among the important conclusions are the following: Condition ratings produced during routine inspections vary significantly and do not show a systematic approach. In total, 95% of the condition ratings spread over five contiguous condition levels, with 68% varying within one level. In-depth inspections were found unlikely to identify significant specific defects correctly, such as welding cracks in steel primary members. (Use of the term *in-depth* suggests that one owner's routine inspection is another's in-depth inspection.) The inability to recognize structurally significant features, such as support condition, bridge skew, fracture-critical members, and fatigue-sensitive details, was common.

Among the factors contributing to these deficiencies were fear of traffic, (lack of) formal bridge inspection training, reported structural maintenance level, accessibility, visibility, time constraints, and wind. Significantly, licensed professional engineers were less likely to misdiagnose a bridge condition. The findings suggest a shortage of both expert systems and experts. Recommendations include better quantification of the findings, possibly by expanded use of nondestructive testing and evaluation methods.

Nondestructive Testing and Evaluation (NDT&E)

So far, regular inspections tend to recommend NDT&E applications beyond their own scope. Rapid technologic advances are changing this practice. The theoretical basis for NDT applications to material quality control was addressed, for instance, in refs. 103 and 104. The following general NDT categories are identified[105]: visual, radiologic, ultrasonic, magnetic, electrical, penetrant flaw detection, acoustic emission, and others. Nondestructive techniques were eventually extended to structural evaluation. NDE is divided[105,106] into primary, secondary, and tertiary depending on whether the sought parameters are related to the primary function of the structure, to a fault that might lead to a failure thereof, or directly to a failure symptom.

The American Society for Non-Destructive Testing (ASNDT) has produced a large number of important publications. The ASTM[107] and CRC Press[108] specify a number of nondestructive methods. Specifically for bridges, the FHWA has conducted a number of strategic highway research projects (SHRP) to develop and assess nondestructive testing and evaluation (NDT&E) techniques. One of many examples is FHWA Demonstration Project No. 84, *Corrosion Detection Equipment.*

Based on the type of wave propagation involved, NDE methods can be grouped into mechanical, sonic, electromagnetic, optical, x-ray, and thermal. Specific applications include strain gauging, corrosion surveying by half-cell potential, magnetic particle evaluation of welds, ultrasonic evaluation of steel, acoustic emission and other methods for evaluation of fatigue crack propagation, infrared thermography for concrete decks, nuclear measuring of the water-cement ratio in fresh concrete, impact-echo thickness and flaw evaluation, fiberoptic sensors for concrete, radar survey for roadway surfaces, the dynamic characteristic method, long-range remote monitoring using electronic clinometers, ultrasonic testing, etc. Pile and foundation testing are gaining interest. Nationally, the FHWA Nondestructive Evaluation Validation Center (NDEVC) is the source of information on available techniques and their performance. Worldwide, a number of international events are held to discuss NDE developments annually, and various Web sites carry related information (SPIE, SMARTEC). The abundance of information adds to the need for a sound understanding of each method's range of applications. A brief description of the most common nondestructive techniques follows with examples of their applications.

Enhanced Methods of Observation. Enhanced methods of observation, such as dye-penetrant testing and thickness meters, have long been common practice; however, bridge inspectors find the equipment excessively sensitive or cumbersome. Steel thickness meters, for instance, are largely underused despite the importance of the information they provide primarily because of their need for repeated calibration and the unrealistic demand for surface preparation. Dye-penetrant testing is extremely helpful in determining the sizes of fatigue cracks once their location is suspected.

The capabilities of electronic boroscopes are constantly improving, and thus previously inaccessible locations can be viewed and photographed. Training is required before the relatively expensive equipment can be used effectively in the field. High-resolution digital photography is combined with advanced software into Quick Time Virtual Reality (QTVR) and panoramic image-creation utilities for recording of field observations. This is a noteworthy database enhancement as long as it does not become a substitute for hands-on inspections by qualified engineers.

Surveying. Laser technology provides highly accurate surveying with excellent field performance. Laser beam distance gauges have replaced the telescopic poles and have facilitated the essential (but previously deficient) bridge inspection task of measuring clearances. Using such equipment to monitor changes in bridge geometry can be extremely useful. Essential is the appropriate choice of monitored locations and frequency of sightings, as well as adequate recordkeeping (including ambient temperature, etc.). Typical applications include long-term monitoring of movable bridges for indications of approach settlement and temperature effects.

Global Positioning Systems (GPS). Currently, GPS systems can monitor structural displacements with accuracy within 5 mm in the horizontal plane and attempting to match that in the vertical plane. Results can be transmitted by telephone to any computer station.

Load-Deflection Tests. Load testing by calibrated vehicles is gaining recognition as a method of bridge load rating.[43] On a number of occasions bridges have been subjected to controlled loadings, e.g., with hydraulic jacks (see Fig. 11.26) or calibrated trucks, and the response has been measured with extensometers and strain gauges, as in Fig. 11.36. Load cells are used during the replacements of suspenders on suspension bridges. Strain gauges can be used to evaluate the live load response of the eye-bars, whereas x-ray diffraction can give some estimates of the total load, as was the case at the Franklin Square truss of the Brooklyn Bridge approach in New York City (see Fig. 11.28). Total load on a steel member also can be evaluated by electromagnetic methods.[109]

Strain and Displacement Gauges. In a functioning structure, strain gauges determine mainly the response to transient loads, such as those owing to traffic and temperature variation. Since the amplitude and frequency of live load cycles are essential

FIGURE 11.36 Strain gauges on exposed rebars.

in fatigue life estimates, gauges have been employed in a number of in-depth and special inspections of steel structures supporting automobile and train traffic.[54,75] In exceptional cases, strain gauging may yield data on the structural response to the total load. Such opportunities arose on both the Manhattan and Williamsburg Bridges in New York City during the reanchoring of suspension cable strands. The new anchorage rods were strain gauged before the load transfer, and the progress of the operation was monitored.

The exposed reinforcing bars shown in Fig. 11.36 were strain gauged to demonstrate that they still respond to live load. The results allowed the structure to remain open to traffic despite the low condition rating based on visual inspections.

Most common are the familiar electrical resistance gauges. New technologies, however, allow the use of acoustic or other gauge types. They can be more robust, allowing attachment to the structure by magnets (in the case of steel) or by clamps and bolts, eliminating the sensitive gluing process. Inevitably, such gauges have a larger base and average measurements over a finite length. As a result, they can be regarded as "displacement" rather than "strain" gauges and should be used when that limitation is acceptable. Figure 11.37 shows a cracked anchorage monolith monitored by gauges with a 6-in. (150-mm) base.

Fiberoptic Sensors. Fiberoptic sensors are relatively new to the field but are gaining rapidly for strain, temperature, and pressure measurement applications. Owing to their exceptionally adaptable length and durability, they frequently are embedded in prestressing tendons, although other applications are also reported. Data are transmitted easily by telephone for online monitoring.

X-ray Diffraction. X-ray diffraction is used for the determination of residual stress in steel. When the stress distribution across the section of a structural member is relatively uniform or well known, this method can estimate the stresses in a loaded structure. Good results were obtained in New York City at the Williamsburg Bridge anchorage eye-bars (see Fig. 11.29) and on the Brooklyn Bridge Franklin Square truss (see Fig. 11.28). Stresses in posttensioning tendons were measured at the La Guardia Airport runways, owned by the Port Authorities of New York and New Jersey.[110]

The method can be used to compare the stress levels in structural members intended to carry equal loads, e.g., two legs of a bridge pier tower. A reliable calibration of a stress-free sample, comparable with the in situ tested material, is highly desirable but not always possible. A good knowledge of the elastic and plastic stress-strain relationships in steel[111] is indispensable.

Acoustic Emission. Acoustic emission (AE) has been used for the detection of fatigue or corrosion-fatigue cracks in steel bridge members for decades. The State of Virginia Transportation Council reported results of acoustic monitoring of steel bridge members in 1997.[112] Once crack locations are known, AE can monitor their propagation. Difficulties arise from the inability of vehicular traffic to generate a meaningful response from the structure and from the need to filter out noise. Monitoring of wire breaks is used widely for prestressed and cable-supported structures. AE detection of corrosion in inaccessible high-strength wires is attempted currently.

Ultrasonic Testing. This method has been considered for the inspection of nickel-alloy steel pins [diameter 0.4 m, length 2.0 m (15.75 in. 78.75 in) approx.] with exposed face sections. The main reservation is that crack initiation cannot be distinguished readily from surface roughness. The California Department of Transportation reported self-compensating techniques for improved results. The method was used at the Williamsburg Bridge in the mid-1990s for evaluating the condition of nickel alloy steel pins as part of the structural rehabilitation. A calibration allowing for visual

FIGURE 11.37   Crack monitoring in suspension bridge anchorage.

verification of the sonograms was demonstrated on test pins. The results were found to be satisfactory. A statistically significant sample of the pins and eye-bars on the Queensboro Bridge (see Fig. 11.27) in New York City will be inspected by this method with accelerated scanning.

Corrosion Detector. An early corrosion detector developed at the ATLSS Center, Lehigh University, consists of a sample probe implanted in normally inaccessible areas of a structure, e.g., under the wrapping of a suspension bridge cable, where the level of corrosion is of interest. An estimate can be made of the ambient corrosion by measuring the resistance of the probe to direct current (dc). Corrosion sensors of this or other types would be highly desirable implants under the protective wrappings or the grouting of cable-supported high-strength steel tension members. The FHWA is conducting research on this subject.

Half-Cell Potential. Potential readings have been used to assess the corrosion of deck reinforcing bars on several bridges prior to their demolition. This demonstrated that the method, when applied adequately, yields accurate results. Currently, similar techniques are used to evaluate the effect of a deck waterproofing membrane. One reservation in applying this method is that corrosion in rebars embedded in concrete decks usually is localized in certain areas, whereas other areas remain free of it. This highly nonuniform condition is difficult to evaluate. The erratic nature of the results can (sometimes unduly) cast doubt on the procedure.

Dynamic Signature Tests. Dynamic signature investigations measure structural natural frequencies and compare them with theoretical values or earlier measurements. Deviations from the expected value can be interpreted as indications of modeling error or structural damage, respectively. The method has had some success in components with strongly pronounced lowest vibration modes, such as the vertical suspenders of the George Washington Bridge, where tests were conducted by the Port Authority of New York and New Jersey and Columbia University in the 1980s.

The dominant natural frequencies of bridges are of great interest during analyses of seismic performance. Response to ambient excitations is recorded by a number of transducers located on the structures and synchronized by satellite. Natural modes and frequencies thus are provided for the dynamic analysis of seismic and wind load response. Typically, results are compared with values obtained analytically. It is noted that computerized structural modeling tends to underestimate the stiffness of large-truss structures. One possible explanation is that a number of rotation and translation release devices in the investigated structures act rigidly under small traffic excitations.

Ultraviolet Rays for the Detection of Alkali-Silica Gel in Concrete. Alkali-silica reaction in concrete can be detected quickly by applying 5% uranil acetate solution to a sample. Alkali-silica gel exhibits a yellow pigmentation under ultraviolet light. The method is not quantitative but helps to determine if alkali-silica reaction is the reason for spalling, e.g., in a concrete deck or pier.

Magnetic Particle Flux. Magnetic particle flux tests can detect cracks in steel under the cover of paint. The method has been under review at the FHWA Fairbank Validation Center. Field tests suggest that it can be very useful, e.g., in investigating painted-steel members deformed by impact. The Honshu-Shikoku Bridge Authority successfully inspected suspender ropes at the Inoshima Bridge[113] with this method.

Structural Health Monitoring. As NDE technology advances, the lists of available techniques become obsolete within months. The online and intermittent surveillance of environmental and structural conditions has evolved into structural health monitoring,[76] changing the relationship between structural design and assessment. The *International Journal on Structural Health Monitoring,* published quarterly since July 2002, and events such as the First International Conference on Structural Health Monitoring and Intelligent Infrastructure in Tokyo, in November 2003, are representative of the intensified professional interest in the subject. Einstein's concern that "perfection of means and confusion of goals seem to characterize our age"[59] occasionally may apply.

Cost-benefit estimates of health monitoring by NDT&E technology depend on the stage of the respective bridge life cycle, as shown on Fig. 11.3*b*. So far, most successful

NDT&E applications have either maintained a bridge in operation, thus benefiting the traveling public and the local economy to an inestimable degree, or avoided potentially fatal accidents by identifying unsafe structures (another incalculable cost). This mode of use can be associated with the liability stage. At that point, NDT&E applications often are expected to determine levels of potential hazard. While this purpose may indeed be served, it must be approached with caution. NDT&E applications occasionally have indicated no significant structural distress, but existing codes and safety considerations have overruled them in favor of emergency remedial work (as in the support arch shown in Fig. 11.28). Surveying alone, for instance, does not fully guarantee the safety of a deteriorating brick masonry arch. The absence of acoustic emission signals is not conclusive proof that wires are not breaking in a suspension cable.

At the asset stage, NDT&E applications can extend bridge life by recommending optimal maintenance levels. Monitoring structural parameters for serviceability rather than "health" also provides life-cycle cost benefits. At many bridges in Japan (see report by Sumitro[76]), at the Tsing-Ma Bridge in Hong Kong, and at the future Woodrow Wilson Bridge in Washington, D.C.,[76] health monitoring systems are designed along with the structure. Interest is growing in monitoring of maintenance effectiveness. Certain maintenance tasks depend on advanced sensor systems. Anti-icing systems, such as, for instance, on the Brooklyn Bridge in New York City, can be equipped with weather-activated control. Also in evidence are attempts to incorporate NDT&E applications at the project stage, while the bridge is in construction and design. A design alternative can become feasible only because of advanced health monitoring capabilities but should be evaluated on the merit of its life-cycle costs and benfits.

NDT&E applications always must be preceded by analysis. Anticipated results and the appropriate response actions must be defined. Data invariably will be wasted unless they are part of a systematic structural evaluation plan. The independent monitoring of different structural response parameters, such as strain and acoustic emission, strain and displacement, and acceleration and velocity, provides a welcome redundancy. Approval of any proposed technique by ASTM, ASCE, and AASHTO is essential.

The inspections described herein and the NDT&E technologies they can employ are oriented primarily toward steel and concrete bridges. As carbon, glass, and other materials gain application, management will have to adapt to their needs.

INSPECTION MANAGEMENT

Bridges can be inspected by the owner or, as in most transportation departments, by consultants. Authorities in charge of toll bridges (which are also unique structures) find the *NBIS* biennial inspections too restrictive and design bridge-specific maintenance inspections, as discussed earlier. Management may wish to avoid "conflict of interest" by separating the staff conducting assessment from that performing maintenance tasks.

A comparison of bridge-related expenditures and conditions suggests that maximizing toll revenues requires maintenance intensity well beyond the needs of safe traffic. As a side effect, improved maintenance minimizes the need for inspection. Inspection efficiency evolves under the contradictory demands for quantity and quality. The manager must determine and maintain the highest average productivity that does not compromise results. Productivity is usually measured in inspector/span/day and depends on a number of variables describing structures, climate, access, experience, etc. Inspection teams typically consist of a leader (licensed P.E.), one or more assistants (civil engineers), and the required equipment operators. The FHWA report[102] discussed earlier recommends a number of procedural changes aimed at improving quality. Inspection reports must pass quality control and show consistency with load ratings. The staff regularly attends inspection and safety refresher courses. Rotation of personnel on bridges offers the equivalent of a peer review.

Personnel

The basic qualifications of inspectors are defined by the *NBIS* as follows:

§650.307 (a) The individual in charge of the organizational unit that has been delegated the responsibilities for bridge inspection, reporting, and inventory shall possess the following minimum qualifications:
(1) Be a registered professional engineer; or
(2) Be qualified for registration as a professional engineer under the laws of the State; or
(3) Have a minimum of 10 years experience in bridge inspection assignments in a responsible capacity and have completed a comprehensive training course based on the "Bridge Inspector's Training Manual," which has been developed by a joint Federal-State task force, and subsequent additions to the manual.
(b) An individual in charge of a bridge inspection team shall possess the following minimum qualifications:
(1) Have the qualifications specified in paragraph (a) of this section; or
(2) Have a minimum of 5 years experience in bridge inspection assignments in a responsible capacity and have completed a comprehensive training course based on the "Bridge Inspector's Manual."
(3) Current certification as a Level III or IV Bridge Safety Inspector under the National Society of Professional Engineer's program for National Certification in Engineering Technologies (NICET) is an alternative acceptable means for establishing that a bridge inspection team leader is qualified.

The *Bridge Inspection Training Manual*[26] quoted earlier has been updated repeatedly. The most recent edition is dated July 1991.[42] FHWA periodically offers training courses, as do the states. A civil engineering license is not uniformly mandatory, reflecting the two approaches to assessment discussed earlier. Rating/descriptive evaluations require experts (e.g. licensed engineers) who can qualify and quantify both routine and unprecedented findings. Defect/action reports can be prepared by experienced technicians given clear and comprehensive instructions (as by an expert system). Both methods suffer from the shortages of their main assets, the experts in the former and the expert systems in the latter.

Expert systems[30] integrate the bridge database with existing experience, service requirements, and repair options. Information can be treated as fuzzy sets and modeled by neural networks. Portable computers facilitate on-site data access and processing. Heuristic guides can prompt the inspector to seek and identify catalogued defects. These new developments will enhance inspection quality if implemented without relaxing personnel qualification requirements. Emergencies, such as the lost bearing in Fig. 11.25 or the damaged pier in Fig. 11.33, must be assessed promptly by qualified engineers capable of assuming responsibility for decisions affecting traffic and workforce deployment. Periodical inspections lose significance if they do not reflect engineering expertise. A database can be homogeneous yet impenetrable if it is obtained mechanically.

Safety and Equipment

All bridge inspection manuals instruct inspectors to comply with safety rules and regulations "as required." This is unavoidably vague because safety rules often are revised by a number of responsible agencies with occasionally overlapping jurisdictions.

The Occupational Safety and Health Administration (OSHA), U.S. Department of Labor, is the federal authority on construction work standards.[114,115] The standards are updated periodically. Bridge inspection differs from bridge construction in the level of exertion and the time spent at locations requiring safety precautions. The extent of this difference is ill-defined but significant because it is argued that inspection is not the type of labor OSHA always envisions. An example is the requirement that anyone working higher than 6 ft (1.835 m) from the ground should be tethered. The rigid enforcement of this requirement effectively would abolish the widespread and mostly uneventful inspections of bridge abutments by ladders. The existing ambivalence makes it imperative to define clear safety rules for any specific inspection operation, as well as a line of responsibility for their implementation and adherence. The rules should be reviewed at the start of a job regardless of past experience. All inspections must be supervised by a responsible, qualified, and duly licensed professional.

Traffic Management. Most inspection accidents are traffic-related. Traffic conditions, rules, and needs vary for different communities and municipalities. Temporary embargos often are imposed on the traffic closures required for inspections. Transportation agencies normally have construction (and other operations) coordination centers that authorize lane closures and issue permits. Lane closures in heavy traffic (see Fig. 11.38) may require recourse to specialized professionals. Inspection personnel must regularly attend the traffic management courses offered by local law enforcement or transportation agencies.

Night Work. Night inspection may be required owing to lack of daytime access, e.g., over railroad tracks or in emergencies. All personnel must have adequate vision for the tasks. Lights powered by generators must be available. Infrared vision aids are available. Light-reflecting safety gear is mandatory. The responsible inspector must determine if the findings are satisfactory or, alternatively, if daytime inspection remains necessary.

Railroad Track Work. Inspections of highway or pedestrian bridges over railroad tracks are distinct from those of railroad bridges. Work performed from or above railroad tracks must comply with OSHA *and* the safety standards of the particular railroad company. All personnel working on the tracks must complete a training course offered annually by the respective company and obtain a certificate. No bridge inspection is to be conducted from train tracks without the supervision of railroad personnel. Before proceeding with the inspection, the engineer in charge must obtain a clear statement of responsibility for the safety of his or her staff from a responsible railroad representative.

Access Equipment. All access equipment, such as bucket trucks, snoopers, scissor lifts, ladders, etc., must be appropriately certified for the work intended. A bucket truck may be appropriate for inspection but not for repair. Aluminum ladders are inadequate in the proximity of power lines. Thirty-foot (10-m) boom trucks can be used by qualified bridge inspectors only after completing a course of instruction. Other equipment, such as 80-ft (25-m) boom trucks (Fig. 11.38) and snoopers (Fig. 11.39), require a licensed operator. Bucket trucks operated from barges (Fig. 11.40) are swayed easily by waves. Vessels should be kept at a safe distance. Inspections conducted from waterborne vessels in a navigable channel must comply with U.S. Coast Guard regulations.

Contemporary design is required to provide access for bridge inspection to all hands-on or critical locations. Travelers or fixed inspection platforms are built along with new bridges or added to older ones. Such nonstructural appurtenances must be inspected along with the rest of bridge elements and rated most appropriately as "utilities." Operation must be certified by qualified mechanical and electrical engineers.

Owing to their relatively light cross sections, maintenance and inspection platforms are the first to lose structural integrity under corrosion attack. Fatal inspection accidents

FIGURE 11.38 Lane closure for inspection with 80-ft (25-m) bucket truck.

have been caused by unsafe platforms. Pans built under bridge decks for collecting debris may not be designed for any live loads and often are unsafe. In the absence of access platforms and adequate equipment, scaffolding must be erected by a qualified contractor.

Safety Gear. All helmets, light-reflecting jackets, harnesses, lanyards, belts, goggles, flotation vests, respirators, and other personal safety gear must be certified, inspected

FIGURE 11.39   Inspection with snooper.

periodically, and replaced according to specifications. Incomplete or inadequate safety equipment is cause for work cancellation. Mountain-climbing techniques are effective on a number of bridges where scaffolding would be the only alternative. Successful use of mountain-climbing techniques for bridge inspections was reported, for instance, at the New River Gorge Arch and in California.

Frostbite is a common inspection hazard in winter. Gloves, clothing, and shoes must provide adequate protection in low temperatures and high winds.

FIGURE 11.40   Inspection with bucket truck from a barge.

Portable computers are standard for many bridge inspections. In addition to the inspector's computer skills, managers must ascertain that operating this equipment in field conditions does not pose new hazards. With the availability of portable telephones, procedures must be established for prompt communications in the event of accidents. Most inspection reports are generated and transmitted electronically. Data filing and management must be adapted to the new and less cumbersome technology.

Environmental Hazards. Inspectors may come in contact with toxic materials, particularly waste, disposed under bridges. When such a hazard is suspected, the area must be vacated immediately and declared unsafe. The sanitation department must be notified. Respirators must be used in the presence of pigeon droppings, frequently abundant under bridges, and in cellular abutment structures. Heavy shoes provide protection against sharp objects (sneakers are never appropriate).

Lead paint is a toxic waste. Lead contamination is not likely to result from chipping paint during inspection, but the generated debris should not be allowed to pollute the environment. A lead disposal protocol usually exists for the specific area and must be followed.

SUMMARY

Bridge condition assessment is necessary because design and maintenance are imperfect. It evaluates, just as imperfectly in its way, structural performance under every known type of uncertainty. While many performance aspects can remain a matter of interpretation, safety must be ensured. Yet no safety margin can be uniquely defined, determined, or universally accepted. Inspectors could be easily tempted to assume that everything *possible* will go wrong. This (optimistic) version of Murphy's law is attributed to U.S. aircraft inspection during World War II. Applied to bridges, it appears conservative but offers little decision support. A "zero tolerance" to deterioration abdicates all responsibility and invites the suspicion that the assessment causes more problems than it solves. Beyond the misuse of funding, the immediate repair of all perceived defects could cause greater losses by interrupting essential traffic.

Under the contradictory constraints of safety and efficiency, a new field of engineering has evolved. Professor E. Brühwiler of Ecole Polytechnique de Lausanne proposes the term "examineering." New and significant is the implied cyclic coexistence of design and inspection. Design specifies how a bridge should function, whereas inspection evaluates its performance, design, and maintenance. Determinism and probability similarly complement each other in these tasks. Prescriptive specifications cannot determine every pertinent loading condition and structural response without the support of performance-based probabilistic ones. The uncertain and inconclusive inspection data offer opportunities for interpretation to both expert systems and inspection experts.

Ultimately, "experience alone can decide on the truth."[59] Formalized algorithms do not determine budget policies based on virtual reality databases without direct personal knowledge, and neither can they arrive at engineering decisions. Bridge design, construction, and maintenance will change, and so will condition assessment. Experts in both abstraction and reality will be needed to reconcile assumption with fact, completing the cycle from engineering design to its product.

ACKNOWLEDGMENTS

This chapter expresses the views of the author and not those of any organization. Establishing and guiding the Bridge Inspection and Bridge Management Unit at New York City Department of Transportation and teaching at Columbia University Civil Engineering Department have formed these views. Mr. J. Shah, P.E., director of bridge inspection, Mr. K. McAnulty, P.E., director of bridge management, and their highly capable staff have provided expert support at the NYC DOT and earlier at IKW, PC.

REFERENCES

1. Chase, S. B., C. Nutakor, and E. P. Small (1999). "An In-depth Analysis of the National Bridge Inventory Database Utilizing Data Mining, GIS and Advanced Statistical Methods,"

in *International Bridge Management Conference, Boulder.* TRB, FHWA, Washington, p. C-6 (IBMC-047).

2. Silby, P., and A. S. Walker (1997). "Structural Accidents and Their Causes," *Proceedings of the Institution of Civil Engineers* **62**(1):191–208.

3. Petroski, H. (1993). "Predicting Disaster," *American Scientist* **81**:110–113.

4. Petroski, H. (1994). "Success Syndrome: The Collapse of the Dee Bridge," *Civil Engineering* **64**(4):52–55.

5. Hopkins, H. J. (1970). *A Span of Bridges.* Praeger, New York.

6. Billington, D. P. (1983). *The Tower and the Bridge.* Basic Books, New York.

7. Kranakis, E. (1997). *Constructing a Bridge.* MIT Press, Cambridge, MA.

8. Levy, M., and M. Salvadori (1992). *Why Buildings Fall Down.* Norton, New York.

9. Larsen, A., and S. Esdahl (eds.) (1998). *Bridge Aerodynamics.* Balkema, Rotterdam.

10. Dallard, P., et al. (2001). "London Millennium Bridge: Pedestrian-Induced Lateral Vibration," *ASCE Journal of Bridge Engineering* **6**:398–411.

11. Housner, G. W. (1990). "Competing Against Time: The Governor's Board of Inquiry on the 1989 Loma-Prieta Earthquake." Sacramento, CA.

12. *Seismic Retrofitting Manual for Highway Structures.* MCEER, FHWA, Washington, 2002.

13. *Guide Specifications for Seismic Isolation Design,* 2d ed., AASHTO, Washington, 1999, interim, 2000.

14. Harland, J. W., et al. (1996). *Inspection of Fracture Critical Bridge Members.* FHWA-IP-86-26, U.S. Department of Transportation, Washington.

15. Einstein, A. (1941). "The Common Language of Science," *Advancement of Science* **2**(5).

16. *Recording and Coding of the Structure Inventory and Appraisal of the Nation's Bridges.* FHWA, ED-89-044, Washington, 1988.

17. Taly, N. (1998). *Design of Modern Highway Bridges.* McGraw-Hill, New York.

18. Chen, W. F., and L. Duan (eds.) (1999). *Bridge Engineering Handbook.* CRC Press, Boca Raton, FL.

19. *Guide for Commonly Recognized (CoRe) Structural Elements.* AASHTO, Washington, 1998, interim 2002.

20. *Bridge Inventory and Inspection System.* New York State Department of Transportation, Albany, 1991.

21. *Bridge Inspection Manual.* Department of Transportation, State of New York, Albany, 1997.

22. *Federal Register* **58**(229), 1993.

23. Organization for Economic Cooperation and Development (1992). *Bridge Management.* Paris, OECD.

24. Woodward, R. J., et al. (1997). *BRIME: Bridge Management in Europe.* European Commission Project RO-97-SC.2220, Brussels.

25. Hudson, S. W., et al. (1987). *Bridge Management Systems.* NCHRP Report 300, TRB, NRC, Washington.

26. *Bridge Inspector's Training Manual 70.* U.S. DOT, FHWA, Washington, 1979

27. Einstein, A. (1921). "Geometry and Experience," lecture before the Prussian Academy of Science, Jan. 27, 1921.

28. *LRFD Bridge Design Specifications,* 2d ed. AASHTO, Washington, 1998.

29. Frangopol, D. M., and H. Furuta (eds.) (2000). *First International Workshop on Life-Cycle Cost Analysis and Design of Civil Infrastructure Systems.* ASCE, Reston, VA.

30. Miyamoto, A., and D. Frangopol (eds.) (2001). *Second International Workshop on Life-Cycle Cost Analysis and Design of Civil Infrastructure Systems.* Yamaguchi, Japan.

31. International Organization for Standardization (ISO) (1995). *Guide to the Expression of Uncertainty in Measurement,* 2d ed. ISO, Geneva.

32. Calgaro, J.-A. (1996). *Introduction aux Eurocodes.* Presses de l'Ecole Nationale des Ponts et Chaussees, Paris.

33. Einstein, A., and L. Infeld (1942). *The Evolution of Physics.* Simon and Schuster, New York.

34. Gertsbakh, A. (2000). *Reliability Theory with Application to Preventive Maintenance.* Springer, Berlin.

35. Ghosn, M., and F. Moses (1998). *Redundancy in Highway Bridge Superstructures.* NCHRP Report 406, TRB, NRC, Washington.

36. *Manual for Condition Evaluation of Bridges,* 2d ed. AASHTO, Washington, 2000.

37. Liu, W. D., et al. (2001). *Redundancy in Highway Bridge Substructures.* NCHRP Report 458, TRB, NRC, Washington.

38. Heyman, J. (1998). *Structural Analysis: A Historical Approach.* Cambridge University Press, Cambridge, England.

39. The Pennsylvania Bridge Management System. FHWA-PA-86-036-84-28A, Harrisburg, PA, February 1987.

40. Park, S. H. (1980). *Bridge Inspection and Structural Analysis.* NJ DOT, Trenton, N.J.

41. Silano, L. G., (ed.) (1993). *Bridge Inspection and Rehabilitation.* Wiley, New York.

42. Hartle, R. A., et al. (1991). *Bridge Inspector's Training Manual90.* FHWA-PD-91-015, Washington.

43. *Manual for Condition Evaluation and Load and Resistance Factor Rating of Highway Bridges.* NCHRP Report 12-46, Washington, March 2000.

44. Yanev, B. (1994). "Emergency Repair Needs Assessment for the New York City Bridges." in *Maintenance of Bridges and Civil Structures, Colloque International.* Presses de l'Ecole National des Ponts et Chaussees, Paris, pp. 501–516.

45. Gluver, H., and D. Olsen (eds.) (1998). *Ship Collision Analysis.* Balkema, Rotterdam.

46. AREMA, (2001). *Manual for Railway Engineering.* AREMA, Washington, DC.

47. *Service d'Etudes Techniques des Routes et Autoroutes (SETRA)* (1994). Ponts a Poutres sous Chaussee en Beton Arme, Proces Verbal de Visite, Paris.

48. Direction des Routes, (1984). *Instruction technique pour la surveillance et l'entretien des ouvrages d'art, Ponts en beton precontraint.* Ministere des Transports, Paris, 1984.

49. Direction des Routes, (1981). *Defauts Apparents des Ouvrages d'Art Metalliques.* Ministere des Transport, Paris.

50. *Auscultation, Surveillance Renforcee, Haute Surveillance, Mesures de Securite immediate ou de Sauvegarde.* Instruction Technique du 19 Octobre 1979, Laboratoire Central des Ponts et Chaussees, Paris.

51. Robichon, Y., C. Binet, and B. Godart, (1995). "Evaluation of Bridge Condition for Improved Maintenance Policy," in *Extending the Lifespan of Bridges.* IABSE Symposium, San Francisco, IABSE-AIPC-IVBH, Zurich.

52. Vincentsen, L. J., and J. S. Jensen, (eds.) (1998). *Operation and Maintenance of Large Infrastructure Projects.* Balkema, Rotterdam.

53. Yanev, B., and X. Chen, (1993). Life-Cycle Performance of New York City Bridges," in *Transportation Research Record No. 1389, Materials and Construction.* Transportation Research Board, Washington, pp. 17–24.

54. Walther, R. A., and M. J. Koob, (2002). *Condition Assessment of Chicago's 100-Year-Old Elevated Mass Transit System.* Stahlbau, Berlin, pp. 117–124.

55. AASHTO, (1999). *The Maintenance and Management of Roadway Bridges.* AASHTO, Washington.

56. Primer, (2000). *GASB 34.* Office of Asset Management, U.S. Department of Transportation, FHWA, Washington.

57. AASHTO, (1989). *Guide Specifications for Strength Evaluation of Existing Steel and Concrete Bridges.* AASHTO, Washington.

58. AASHTO, (2000). *Standard Specifications for Highway Bridges,* 17th ed., AASHTO, Washington.

59. Einstein, A. (1950). "On the Generalized Theory of Gravitation," *Scientific American* **182**(4).

60. Barker, R. M., and J. A. Puckett, (1997). *Design of Highway Bridges.* Wiley, New York.

61. Shinozuka, M. (1969). "Methods of Safety and Reliability Analysis," in *International Conference on Safety and Reliability.* Pergamon Press, New York, pp. 11–45.

62. Boller, P. (1895). *Construction of Iron Highway Bridges, for the Use of Town Committees.* Wiley, New York.

63. Holmes, O. W. (1895). *The Complete Poetical Works.* Houghton Mifflin, Boston.

64. *Life Cycle Cost Analysis, Searching for Solutions—A Policy Discussion Series.* FHWA, U.S. DOT, Number 12, Washington, 1994.

65. Hawk, H. *Bridge Life-Cycle Cost Analysis* (BLCCA) (2003). NCHRP Report 483, TRB 2003, Washington, DC.

66. Veshosky, D. (1992). "Life-Cycle Cost Analysis Doesn't Work for Bridges," *Civil Engineering,* **62**(7):6.

67. Veshosky, D., et al. (1994). "Comparative Analysis of Bridge Superstructure Deterioration," *ASCE Journal of Structural Engineering* **120**(7):2123–2136.

68. Yanev, B., and R. B. Testa, (2000). "Annualized Life-Cycle Costs of Maintenance Options for New York City Bridges," in *Bridge Management 4.* Thomas Telford, London, pp. 400–407.

69. Shepard, R. W., and M. B. Johnson, (2001). "Health Index: A Diagnostic Tool to Maximize Bridge Longevity, Investment," *TR News* **215**:6–11.

70. Yanev, B. (1997). "Life-Cycle Performance of Bridge Components in New York City," in *Proceedings, Recent Advances in Bridge Engineering* EMPA-Columbia University, Zurich, New York, pp. 385–392.

71. *Strength Evaluation of Existing Reinforced Concrete Bridges.* (1987). NCHRP Report 292, TRB, NRC, Washington.

72. Vesikari, E. (1988). *Service Life of Concrete Structures with Regard to Corrosion of Reinforcement.* Technical Research Centre of Finland, Research Report 533, Espoo.

73. Clifton, J. R. (1991). *Predicting the Remaining Service Life of Concrete.* NISTRIP 4712, U.S. Department of Commerce, Washington.

74. Wadia-Fascetti, S., et al., (2000). *Subsurface Sensing for Highway Infrastructure Condition Diagnostics.* TRB, Washington.

75. *ATLSS Evaluates Williamsburg Bridge Orthotropic Deck Prototype.* Lehigh University Center for Advanced Technology for Large Structural Systems, Bethlehem, PA, August 1995.

76. Aktan, E., et al. (2001). *Health Monitoring of Long Span Bridges.* National Science Foundation Workshop, University of California, Irvine.

77. Smith, F. C. (2001). "Increasing Knowledge of Structural Performance," *Structural Engineering International* **3**:191–195.

78. AASHTO (1983). *Manual for Maintenance Inspection of Bridges.* Including revisions from Interim Specifications for Bridges 1984, 1985, 1986, 1987–1988, 1989, 1990, Washington.

79. *Fatigue Evaluation Procedures for Steel Bridges.* NCHRP Report 299, TRB, NRC, Washington, September 1987.

80. Yen, T., et al. (1990). *Manual for Inspecting Bridges for Fatigue Damage Conditions.* Lehigh University Report No. 511-1, PennDOT Project 85-02, Bethlehem, PA.

81. Fisher, J. W., et al. (1997). *A Fatigue Primer for Structural Engineers.* Lehigh University, ATLSS Report No. 97-11, Bethlehem, PA.

82. Miki, C. (1992). "Lessons to Be Learned from Fatigue Damage Accidents of Steel Bridges," in Drdacky, M. (ed.)," *Lessons from Structural Failures 2.* ed. Aristocrat, Prague.

83. Purvis, R. (2003). *Bridge Deck Joint Performance.* NCHRP Synthesis 319, Transportation Research Board, Washington.

84. *Bridge Welding Code.* ANSI/AASHTO/AWS, Washington, 1996, interim revision, 1999.

85. *LRFD Bridge Construction Specifications,* 1st ed., AASHTO, Washington, 1998, interim, 2002.

86. *Standard Specifications for Transportation Materials and Methods of Sampling and Testing,* 21st ed. AASHTO, Washington, 2001.

87. *ACI Manual of Concrete Inspection.* (1992). Report by ACI Committee 311, Publication SP-2(92) Detroit.

88. *Construction: Pavement, Bridge, Quality Control/Quality Assurance, and Management.* Transportation Research Record 1654, Transportation Research Board, Washington, 1999.

89. Stahl, L., and C. P. Gagnon, (1995). *Cable Corrosion in Bridges and Other Structures.* ASCE Press, New York.

90. Virlogeux, M. (1999). *Replacement of the Suspension System on the Tancarville Bridge.* Paper No. 99-0604, TRR No. 1654, TRB, NRC, National Academy Press, Washington, pp. 113–120.

91. *Guide Specifications for Design and Construction of Segmental Concrete Bridges,* 2d ed. AASHTO, Washington, 1999.

92. Budelmann, H., et al. (2000). "Monitoring of Reinforced and Prestressed Concrete Structures," in *Present and Future of Health Monitoring.* Aedificatio, Freiburg, Bauhaus University, pp. 135–146.

93. *Post-Tensioned Concrete Bridges,* (1999). Anglo-French Liaison Report, Thomas Telford, London, 1999.

94. Burdekin, F. M., et al. (1991). *Non-Destructive Methods for Field Inspection of Embedded or Encased High Strength Steel Rods and Cables.* CAPCIS, Ltd., NCHRP 10-30 (3), University of Manchester, U.K.

95. *Movable Bridge Inspection, Evaluation and Maintenance Manual.* AASHTO, Washington, 1998.

96. *Standard Specifications for Movable Highway Bridges.* AASHTO, Washington, 1988.

97. *Contractual Requirements for Diving Inspections for Bridges.* New York State Department of Transportation, GD/DCR01102, Albany, 1993.

98. *Seismic Retrofitting Manual for Highway Bridges.* Publication No. FHWA-RD-94-052, Washington, May 1995.

99. McCurdy, E. (1935). *Leonardo da Vinci's Notebooks.* Empire State Book Co., New York.

100. Russell, B. (1985). *The Philosophy of Logical Atomism.* Open Court, La Salle, IL.

101. Rouse, W. H. D. (trans.) *The Great Dialogues of Plato.* New American Library, New York.

102. Moore, M., et al. (2001). *Reliability of Visual Inspections for Highway Bridges.* FHWA-RD-01-020, Washington.

103. Hull, B., and J. Vernon, (1988). *Non-Destructive Testing.* Macmillan Education, London.

104. Halmshaw, R. (1987). *Non-destructive Testing.* Edward Arnold, London.

105. Collacott, R. A. (1985). *Structural Integrity Monitoring.* Chapman and Hall, London.

106. Agbabian, M. S., and S. F. Masri, (eds.) (1988). *Nondestructive Evaluation for Performance of Civil Structures.* USC Press, Los Angeles.

107. Bush, J., and G. Y. Baladi, (1989). *Nondestructive Testing of Pavements and Backcalculation of Moduli.* ASTM 89-38726, New York.

108. Malhorta, V. M., and N. J. Carino, (eds.) (1991). *Handbook on Nondestructive Testing of Concrete.* CRC Press, Boca Raton, FL.

109. Lozev, M., G. Clemena, J. Duke, Jr., M. Sison, Jr., and M. Horne (1994). *Acoustic Emission Monitoring of Steel Bridge Members.* Report No. FHWA/VTRC 97-R13, Washington.

110. Carfagno, M. G., F. S. Noorai, M. Brauss, and J. Pineault, (1995). "X-Ray Diffraction Measurement of Stresses in Post-Tensioning Tendons," in *Extending the Lifespan of Structures, IABSE Symposium, San Francisco.* ETH Honggerberg, CH-8093 Zurich, pp. 201–206.

111. Noyan, C., and J. B. Cohen, (1987). *Residual Stress.* Springer-Verlag, Berlin.

112. Schwesinger, P., and F. H. Wittmann, (eds.) (2000). *Present and Future of Health Monitoring.* AEDIFICATIO Publishers, Freiburg.

113. Honshu-Shikoku Bridge Authority, (2002). "Nondestructive Testing of Hanger Rope," *Newsletter on Long-Span Bridges* **11**:1.

114. *Fall Protection in Construction.* OSHA 3146, U.S. Department of Labor, Washington, 1995.

115. *Final Rule Revising OSHA Safety Standards for Scaffolds Used in the Construction Industry.* U.S. Department of Labor, 61 FR 46026, Washington, August 30, 1996.

CHAPTER 12

Tensile Fabric Structures

TIAN-FANG JING, P.E., and WESLEY R. TERRY, P.E.

INTRODUCTION

During the past 50 years, architects and engineers have had many calls for innovation in new structural forms with lightweight structures to accommodate large unobstructed spaces. Advances in structural analysis and construction technology allowed architects and engineers to make more use of fabric materials in their design applications and to create more lightweight load-carrying structural systems with exciting architectural expressions. Various tensile fabric structures have been developed that benefited from new construction materials and new building techniques.

The forms of early tensile fabric structures were influenced directly by traditional tents. They were used mainly as temporary structures where portability, lightness, forms of flexibility, and ease of construction were critical features. These forms are still used widely in many new tensile fabric structures, such as field shelters and bandstands for circus and exhibition events. Improvements in the strength and durability of industrial fabric have expanded the use of tensile fabric structures dramatically, especially for long-span roofs.

Modern tensile fabric structures have been used in a broad range of building types, such as sports stadiums, exhibition pavilions, transportation terminals, airplane hangars, performing arts facilities, and shopping malls. Besides these long-span roof applications, they also are being used on mid- and short-span roofs, exterior wall closures, and canopies, most of which are permanent structures. In the past three decades, significant progress has been made in the process of developing tensile fabric structures. Many notable projects have been designed and constructed with tensioned fabric structures in a variety of configurations and applications all over the world.

Although most of the existing fabric structures today are in sound condition, some have experienced weathering problems, material degradations, and structural distress. These structures have been repaired, modified, or altered in order to continue their services. The existing tensile fabric structures are also faced with a growing need for more flexible use and additional load-carrying capacity. The structural evaluation and rehabilitation of existing tensile fabric structures have become new matters to be resolved by architects and engineers.

The purpose of structural condition assessment of existing buildings is to evaluate the general condition of building structures and identify their areas of deficiency. Engineers often complain about the lack of original construction documentation, which can lead to uncertainties and speculations about the characterization of existing structural conditions. To remedy this problem, extensive field investigation is to be conducted in conjunction with the knowledge of existing structural systems. Since tensile fabric structures are relatively new designs, most architects and engineers are not yet familiar with their design and behavior. It is therefore necessary to present a simple review of the basic types of tensile fabric structures in this chapter before discussing their structural condition assessment.

BASIC TYPES OF TENSILE FABRIC STRUCTURES

In general terms, a tensile fabric structure consists of a tensioned fabric surface membrane and a structural support system such as masts, arches, columns, beams, and cables. The stresses developed in the tensioned membrane interact with those in the structural support system, and the entire assembly acts together to resist the imposed loads.

A fabric membrane is a thin, flexible skin material that has little or no bending and shear stiffness. As a surface structure, fabric membrane must rely on its shape and the prestress alone to achieve stability and to carry loads. The ideal shape of the surface structures to resist the imposed loads is the doubly curved form (Fig. 12.1). According to structural mechanics, two primary shapes are available with double curvatures: the hyperbolic paraboloid–shaped form and the conical-shaped form (Fig. 12.2).

All tensile fabric structures are developed from these two primary forms and their combinations. They offer a variety of possibilities for unique configurations to meet

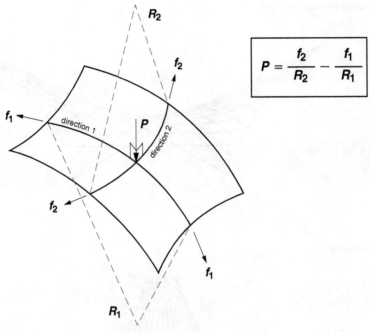

$$P = \frac{f_2}{R_2} - \frac{f_1}{R_1}$$

FIGURE 12.1   Doubly curved form.

application requirements. From the structural system point of view, tensile fabric structures usually are classified according to their support structural systems. The following basic types of tensile fabric structures have been used in the design practice: mast supported, arch supported, cable-dome or cable-truss supported, air supported, and air-inflated fabric structures. There are many other types that are derived from these basic types.

Mast-Supported Fabric Structures

Many fabric structures are supported by large masts at the center. As an example, the computer model and the plan layout of the fabric atrium roof at the Guaranty Bank in Hiawatha, Iowa (Fig. 12.3) show the principal structural features of this type of fabric structure. The upper support points are provided by a steel tension ring, which is supported at the top of the mast. The lower support points are anchored in the steel framing around the center opening at the low roof. Between the upper tension ring and the low roof framing are a series radial cables that follow the core-shaped radial lines. They originate from the top tension ring and are anchored at the low roof framing. This fabric roof shows the ideal conically shaped tent form. This system uses limited compression elements, and it can be erected easily by jacking the mast. The imposed downward and uplift forces are resisted by the mast and the low roof framing, respectively. The lateral loads are transmitted to the low roof structure through fabric panels and radial cables. This concept can be applied to multiple tents by introducing multiple masts. The fabric roof at the Denver Airport terminal building (Fig. 12.4) is an interesting example of the multiple mast-supported fabric roofs. A pair of central masts acts as the main compression columns spaced 150 ft apart, and they occur every

Hyperbolic Paraboloid Form

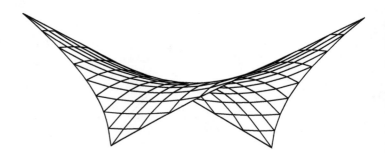

Conical Form

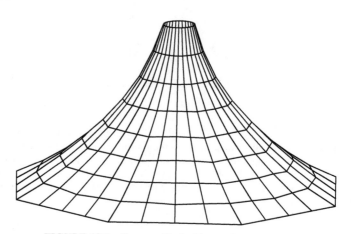

FIGURE 12.2   Shapes with double curvatures.

60 ft along the building length. There are two different mast heights arranged in an alternating sequence. The ridge cables draped over the top of a pair of masts and anchored to the low roof act as the main cables of a suspension bridge and are designed to carry the downward loads and transmit them to the masts. Between any two-ridge cables are arch-formed valley cables that run parallel with each other. These valley cables are designed to resist the upward loads and are tied down to the anchor points at the low roof. The edges of the fabric roof are formed by edge caternary cables.

The fabric structure supported by masts at the edges is another popular structural system. A hyperbolic paraboloid-shaped surface can be formed by a minimum of four points consisting of opposite pairs of support and anchor at different elevations. The supports are the high points, and the anchors are the low points. The curvatures of fabric surface are controlled by the placement of high points and low points. This principle is shown in a tent structure supported at four points in Fig. 12.5. The edges of this tent are formed by the caternary cables. The same concept can be applied to

3D Computer Model

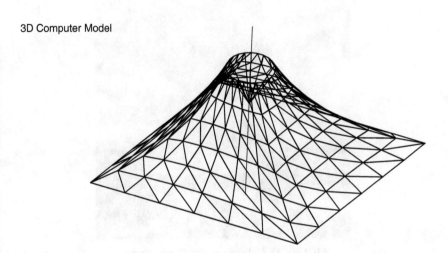

Plan Layout

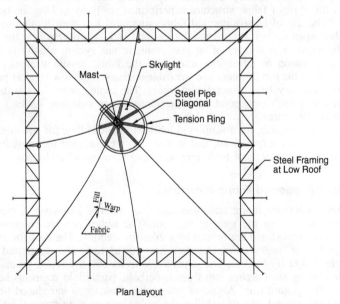

Plan Layout

FIGURE 12.3 Fabric atrium roof.

FIGURE 12.4 Fabric roof at Denver airport.

more complex geometries by introducing more high points and more low points. The support points usually are formed by sloping masts and tie-down cables. The erection of this type of fabric structure is performed easily by jacking the tie-down cables.

The use of mast-supported fabric structures has been limited mostly to mid- and short-span roofs owing to the flatness and lack of stiffness near the midspan between the masts. The number of support points in this system usually is determined by the configuration of fabric curvatures and the fabric tensile stresses. The high support points at the top of masts and the cable anchors at the low support points usually have to resist very large concentrated forces. The cable connections at the mast and at the tie-down anchoring points are the major design detailing issues for mast-supported fabric structures.

For architects, mast-supported fabric structures allow for structural forms with dramatic artistic expressions, and in addition, they provide the most flexible options for plan configurations of both temporary and permanent buildings.

Arch-Supported Fabric Structures

Arch-supported fabric structures use one or more long-span arches as compression members across the span of the roof. The arches provide continuous lines of high points (ridges) and are braced by cables for stability. The roof for the Tropical Forest Pavilion at Boston's Franklin Park Zoo was an early arch-supported fabric roof structure design (Fig. 12.6). The structure, with its 300-ft circular ground plan, has a roof divided by steel arches into three hyperbolic paraboloid segments between the arches and the ground ring. A grid of orthogonal cables was introduced between the arches and the ground ring to stabilize the arches and provide support for the fabric. Arch-supported fabric structures can be developed from two basic arch arrangements: the parallel-arch system and the cross-arch system. The parallel-arch structure consists of a series of parallel arches located some distance apart. These arches need to be braced by cables or struts for overall structural stability. The cross-arch system provides a self-stable structure by virtue of the arches crossing each other.

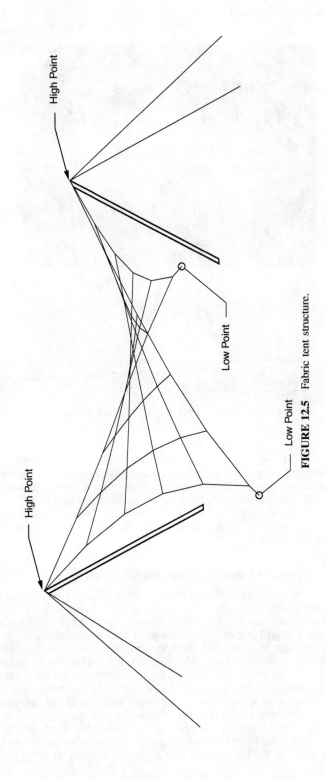

FIGURE 12.5 Fabric tent structure.

FIGURE 12.6   Fabric roof at Franklin Park Zoo.

Arch-supported fabric structures generally are used for long- and midspan roofs. They offer clear, uninterrupted space without interior supports, such as the masts. Arch-supported forms can provide a much shallower space for a similar plan distance, in comparison with the mast-supported forms. With the arches being the main support structural elements and having a direct impact on fabric curvatures, key design considerations such as determining the optimal shape of the arches and their arrangement must be taken into account.

Cable-Dome- or Cable-Truss-Supported Fabric Structures

The first cable net roof structure was built in 1951 for the J. S. Dorton Arena in Raleigh, North Carolina (Fig. 12.7). The roof structure is a 300-ft diameter suspended cable net using a single layer of cables and is supported by a pair of opposing slanted concrete arches. The grid of roof cables consists of 47 main cables and 47 cross cables running perpendicular to each other. Both are suspended from the opposing slanted arches. Additional vertical tie-down cables are introduced between the slanted arches and the foundation structure. The whole cable net forms a parabolic shape of roof surface like a saddle. This roof is not covered by fabric, but its saddle-shaped form in response to the nature of surface tension had direct influence on the new cable-dome- and cable-truss-supported tensile fabric structures.

The first long-span cable-dome-supported fabric roof based on the "tensegrity" concept was constructed in 1986 for the Gymnastics Arena for the Asian Games in Seoul, Korea (Fig. 12.8). The circular roof is 400 ft in diameter using radial cable

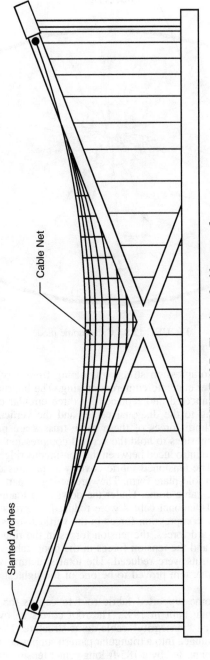

Cable Net

Slanted Arches

FIGURE 12.7 Elevation of cable net roof.

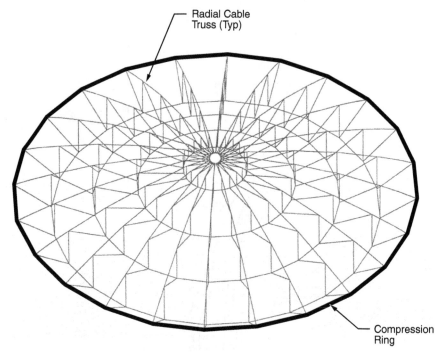

Radial Cable
Truss (Typ)

Compression
Ring

FIGURE 12.8   Cable-dome model.

trusses (with vertical compression struts) originating from a center tension ring and supported by a perimeter concrete compression ring. The bottom chords of the radial cable trusses are disconnected and replaced by three circular cable hoops acting as tension rings connecting to the diagonal cables and the vertical compression struts. The required bottom chord forces of these cable trusses are provided by the radial forces from these tension rings to hold the vertical compression struts in position. The radial valley cables were introduced between the continuous ridge cables (the top chord of the cable trusses). The tensioned fabric cover was prestressed by the ridge cable and valley cables into a fold-plate form. They are acting as part of structural elements in order to stabilize the cable dome. Under the downward loading, the tension forces in the ridge cables and diagonal cables were reduced, whereas the tension forces in the hoop cables and the compression forces in the vertical struts were increased. Vice versa, to resist the upward forces, the tension forces in the ridge cables and diagonal cables were increased, and the tension forces in the hoop cables and the compression forces in the vertical struts were reduced. The total structural weight of this roof is less than 3 lb/ft². This system proved to be one of the lightest long-span roof structures.

The largest cable-dome-supported fabric roof to date is the Georgia Dome (Fig. 12.9) in Atlanta, Georgia, built in 1992. The roof covers an oval plan of 780 ft by 650 ft. It is a triangulated "tensegrity" cable-dome structure. The ridge cables of each radial cable truss are arranged into a triangular pattern supported by compression struts. The radial trusses are connected by a 185-ft-long center tension truss and are supported on a perimeter concrete compression ring. Three cable hoops are introduced to replace the disconnected bottom chord of the radial trusses and to support the vertical compression struts. To hold the vertical struts in position, diagonal cables connect the top

FIGURE 12.9   Georgia Dome, inside view.

joints of the ridge cables and the vertical struts to the bottom joints of the hoop cables
and vertical struts. The fabric panels are in diamond shape with hyperbolic paraboloid
surfaces. They act as roof cover only to transfer the vertical and lateral loads to the
cable-roof system.

Besides the cable-dome structures, several cable-truss-supported fabric structures
have been constructed on large soccer stadium roofs. Most of them are designed as
the roof-canopy structures over the spectator seat area. The fabric roof at the Gerry
Weber Stadium in Germany uses cable trusses in a spoked-wheel configuration sup-

ported by the perimeter steel-trussed compression ring. This arrangement offers the option of a retractable roof portion over the center field.

Design of cable-dome- or cable-truss-supported fabric structures requires sophisticated full model analysis to determine overall structural behavior. Structural detailing design is an important element of the design process of cable-dome- or cable-truss-supported fabric roof structures. The tension forces residing in the cable-dome and cable-truss structures have to be resisted by a perimeter compression ring. Most compression rings are made of concrete. The ring structures and the cable anchors at the compression rings have become important design issues for cable-dome- and cable-truss-supported fabric structures.

Air-Supported Fabric Structures

The concept of air-supported fabric structures is that the fabric roof is supported by the outward pressure of an inflation system. The first major application of a low-profile long-span air-supported fabric roof was built for the U.S. Pavilion at the 1970 World's Fair, in Osaka, Japan. A grid of cable net was used to limit the fabric spans. The single-fabric enclosure, acting as a curtain, is supported and prestressed by the enclosed air pressure that exceeds the atmosphere. The fabric enclosure has to be anchored along its perimeter to a concrete compression ring to prevent being lifted. This air-supported fabric roof provided the best initial cost-saving solution to meet the reduced budget of roof construction. The development of new technology is always driven and related to human needs and the economy. Following the success at the 1970 World's Fair, a dozen low-profile air-supported large fabric roofs, together with hundreds of medium-sized and small air-supported fabric structures, were built to cover stadiums, arenas, tennis courts, swimming pools, and storage facilities. Figure 12.10 shows the Carrier Dome at Syracuse University in New York.

The ability to address budgetary concerns, along with its quick construction, distinguished this system from the conventional structural systems at that time.

Many of these structures have experienced problems with the melting and removal of snow and ice. The additional cost expenditures and the inconvenient maintenance of highly pressurized buildings have led their owners to look for improvements or new solutions for this type of fabric structure.

FIGURE 12.10 Carrier Dome at Syracuse University.

Air-Inflated Fabric Structures

Another type of fabric structure is called an *air-inflated structure*. It differs from an air-supported structure in that only members are pressurized, not the enclosed space. This system uses tubular fabric as members that are inflated by high pressure and set to behave like traditional rigid-line elements such as columns, beams, and arches or surface elements made of double-fabric envelops that are pressurized to form walls, shells, and slabs. Theoretically, this system has no limit to its applicability. However, from a practical standpoint, such structures are used mainly for either temporary fabric shade structures or unconventionally shaped fabric sculptures. The pressure required for this system is much higher than for the air-supported structures, and thus, high-strength fabric materials are needed to contain the high pressures. This system requires periodic adjustments to restore any change in pressure caused by leakage and changes in temperature. The most common applications of this system are for small-span structures.

CODES AND STANDARDS

To evaluate the existing structures, criteria must be established in accordance with local building codes. Fabric membrane structures are classified under the category of "Special Construction" in most of the local building codes in the United States. The criteria for fabric structures specified by these building codes are very limited. Several technical publications are available by American Society of Civil Engineers (ASCE) that discuss the industry standards related to fabric structures.

Typically, fabric structures are designed to carry all loading requirements as specified by the building codes. Aside from the weight of the structure itself, wind, snow, live loads, and other environmental loadings are specified by the building codes. The following is a list of building codes and standards used to determine the design-load criteria:

International Building Code[1]
Uniform Building Code[2]
BOCA National Building Code[3]
Standard Building Code[4]
ASCE Standard, *Minimum Design Loads for Buildings and Other Structures*[5]

For some specially shaped large fabric structures, wind tunnel tests may be required to determine more accurate design wind loads.

In general, tensile fabric structures can be built using various construction systems for their support structures. Each type of construction has to follow the code provisions and specifications published by the technical trade associations for the materials used. They are as follows:

ASCE Publication, *Air Supported Structures*[6]
ASCE Publication, *Tensioned Fabric Structures: A Practical Guide*[7]
ASCE Publication, *Structural Applications of Steel Cables for Buildings*[8]
AISC, *Manual of Steel Construction (ASD)*[9]
AISC, *Manual of Steel Construction (LRFD)*[10]
ACI, *Building Code Requirements for Structural Concrete* (ACI 318)[11]

Several specifications relating to fabric membrane materials are published by the American Society for Testing and Materials (ASTM). These specifications summarize the material testing methods and required results and set the material standards for fabric membranes. The following is a partial list of these standards:

ASTM D4851, *Coated and Laminated Fabrics for Architectural Use*[12]

ASTM E84, *Tested Method for Surface Burning Characteristics of Building Materials*[13]

ASTM E108, *Test Method for Fire Tests of Roof Covering*[14]

ASTM E136, *Test Method for Behavior of Materials in Vertical Tube Furnace at 750°C*[15]

These codes and standards provide the requirements for design of tensile fabric structures. They can and indeed should be used as the standards for the structural condition evaluation of tensile fabric structures. The ASCE also has published a document entitled, *Guideline for Structural Condition Assessment of Existing Buildings*. The detailed procedures and methods specified in this standard can apply to all types of existing buildings, including tensile fabric structures.

COMPONENTS OF FABRIC STRUCTURES

Structural Fabric Membranes

As the primary structural element used for fabric structures, structural fabric materials must have enough strength to span between the supporting structural elements and carry the imposed loads. In order to meet the architectural requirements, they must behave as enclosure elements and be airtight, waterproof, fire resistant, and durable. Besides these basic requirements, transmitting daylight, reflecting heat, controlling sound, and self-cleaning are also desired characteristics for some of the long-span fabric roof structures.

Structural fabrics consist of a structural base material and a surface coating. The two most commonly used structural base materials are fiberglass and polyester cloths. The available surface coatings are polyvinyl chloride (PVC), Teflon, and silicone. The surface coatings can improve the fabric strength and durability and act as a self-cleaning feature for the fabric surface.

PVC-coated polyester fabrics generally are used for temporary structures, and Teflon-coated fiberglass fabrics are used widely in permanent building applications. Silicone-coated polyester fabrics have been used for some building applications and have been found to develop problems at the seam joints. Most of the structural fabrics used are PVC-coated polyester and Teflon-coated fiberglass. The structural fabrics are produced by a limited number of suppliers in United States. The detailed specifications for each fabric can be obtained from its supplier. Table 12.1 shows the comparison between some of the fabric materials. Design safety factors of 3 to 8 typically are used for fabric membranes.

The strength of structural fabrics is described by three main uniaxial strength properties: (1) uniaxial strip tensile strength in the warp and fill directions, (2) trapezoidal tear strength in the warp and fill directions, and (3) the coating adhesion strength. The actual ultimate strength of most fabric materials depends on their biaxial behavior. Fabric materials made of a woven substrate demonstrate orthotropic material properties because the interaction between the warp and fill threads can result in a loss of ultimate strength in biaxial stress loading.

All structural fabric materials exhibit degradation with time owing to ultraviolet (UV) radiation exposure, water exposure, and the weathering impact of wind, rain,

TABLE 12.1 Fabric Material Comparison

Manufacturer	Chemfab		Ferrari		Shelter Rite	
Product designation	Sheerfill I	Sheerfill II	Précontraint 1202S	Précontraint 1202T	8028	8028
Substrate material	Fiberglass (beta yarn)	Fiberglass (beta yarn)	Polyester	Polyester	Polyester	Polyester
Principal coating	PTFE (Teflon)	PTFE (Teflon)	PVC	PVC	PVC	PVC
Top coat	None	None	"Blended" PVDF	100% PVDF	Urethane	Tedlar laminate
Bottom coat	None	None	Acrylic lacquer	Acrylic lacquer	Urethane	Urethane
Roll width (ft)	13	13	6	6	5	5
Physical Data						
Weight (oz/y^2)	45.5	38.5	39.0	39.0	28.0	28.0
Strip tensile capacity, warp	975	785	710	710	490	490
Fill (lb/in.)	900	560	615	615	490	490
Tear strength warp	95	70	90	90	83	83
Fill (lb/in.)	120	65	93	93	125	125
Estimated Life						
Aesthetic life, years	25–40	25–40	4–6	8–12	1–3	7–10
Serviceable life, years	25–40	25–40	7–12	10–15	7–12	10–15
Cleanability	Excellent	Excellent	Fair	Good	Fair	Very good
Mildew resistance	Excellent	Excellent	Fair	Good	Good	Good
Fire Performance						
Combustibility	Noncombustible	Noncombustible	Combustible	Combustible	Combustible	Combustible
Smoke generation	5	5	N/A	N/A	415	415
Flame Spread	5	10	N/A	N/A	25	25
Application						
High visibility (architectural)	Excellent	Excellent	Fair	Good	Poor	Very good
Medium visibility (commercial)	Excellent	Excellent	Good	Very good	Fair	Excellent
Low Visibility (Industrial)	Excellent	Excellent	Excellent	Excellent	Very good	Excellent

snow, and other environmental effects. One of the most common questions for fabric structures is related to their lifespan. Some testing results of the strip tensile retention from existing fabric structures are listed in Table 12.2. In general, polyester-based fabrics are more susceptible to UV damage than fiberglass-based fabrics. PVC-coated polyester fabric is a low-cost material with limited durability owing to the soiling problem on its surface. The lifespan of this fabric is approximately 5 to 10 years. Although Teflon-coated fiberglass fabric is more expensive, it is an incombustible, strong, and durable material that provides a self-cleaning surface. This fabric is recognized as a permanent building material. Some roofs constructed with this fabric have maintained sufficient strength for almost 30 years.

Fabric Connections

There are two types of fabric connections: (1) fabric-to-fabric and (2) fabric-to-support structural elements. The connections in a tensile fabric structure ensure that the forces are transferred within the fabric enclosure and from the fabric to the support structure. These connections need to have adequate flexibility to accommodate the expected large displacements and rotations. In some cases, to prevent the development of stress concentrations in the fabric, connections have to be detailed carefully. Besides the structural requirements, the connection joint may have to be watertight for its intended architectural applications.

Fabric-to-fabric connections can be made by several methods: heat sealing, gluing, and sewing. Fabric membranes are produced as sheets of different widths and are supplied in rolls. The fabric patterns then are cut from the roll goods into the assembly pieces, and finally, these pieces are assembled together in the field to produce the desired fabric configuration.

The fabric-to-fabric connection joints usually are called *seams,* and these seams must provide adequate strength to transmit the fabric stresses. The seam strength depends primarily on the coating adhesion for glued or welded connections and on the seam width. The standard seam widths are 1 to 2 in. for PVC-coated polyester fabrics

TABLE 12.2   Strip Tensile Retention from Existing Structures

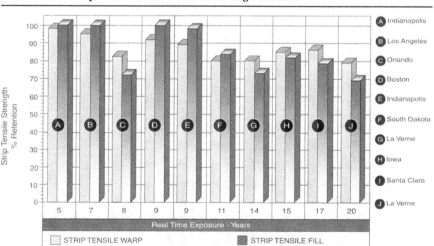

and 2 to 3 in. for Teflon-coated fiberglass fabrics. A typical lap seam details is shown in Fig. 12.11.

Fabrics-to-support-structure connections vary with the type of construction. One typical configuration is the termination of fabric membrane at a rigid edge, such as a concrete or steel curb. The common connection for this condition is called *clamping detail* (Fig. 12.12). A fabric roped edge is compressed between an aluminum clamping bar and a base plate anchored to the rigid edge. The clamping force is provided by tightening the fasteners. This type of connection should be continuous around the fabric edges in order to transfer the fabric stresses uniformly to the support structure.

In some cases, the fabric relies on a structural member for support but is not attached to it. Usually a seam is located to coincide with the member to improve appearance if the member is a steel pipe. When fabric is supported on a cable, a fabric bias wear strip should be introduced to reinforce the fabric. The use of a plastic coated cable is recommended in this case.

The termination of fabric membrane at a cable location has two typical conditions: the one-side connection that usually occurs at a cable caternary edge and the two-side connection that is used to sectionalize the fabric membrane. For both connections, the fabric should be secured continuously to avoid the free edges. Several typical fabric-to-cable and fabric-to-pipe connections are shown in Fig. 12.13.

Cables and Other Support Structural Elements

High-strength steel cables are important structural elements in tensile fabric structures. Fabric membranes often are supported and reinforced by flexible steel cables to form their shapes. The common types of cables are steel strand and wire rope (Fig. 12.14). The minimum ultimate tensile strength of cables is in the range of 200 to 220 ksi depending on the coating class. Three classes of coating weights are available to meet the range of corrosion-resistance requirements.

A strand consists of wires laid helically around a center wire. Wire rope is made of strands laid helically around a core. Since a strand has more metallic area than a rope of the same diameter, strands are stronger and stiffer than rope. It is important to know that the effective elastic modulus for cables is slightly less than that of solid

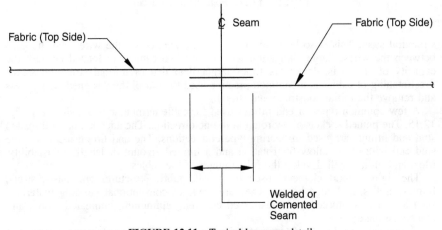

FIGURE 12.11   Typical lap seam detail.

Section

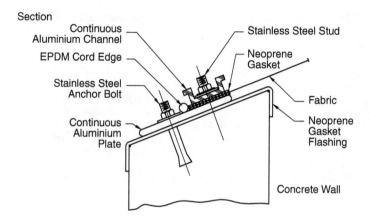

Continuous Aluminium Channel

Stainless Steel Stud

EPDM Cord Edge

Neoprene Gasket

Stainless Steel Anchor Bolt

Fabric

Continuous Aluminium Plate

Neoprene Gasket Flashing

Concrete Wall

Plan View

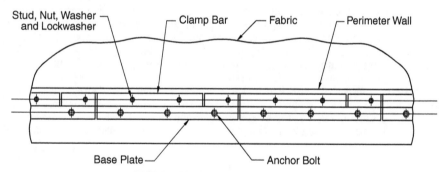

Stud, Nut, Washer and Lockwasher

Clamp Bar

Fabric

Perimeter Wall

Base Plate

Anchor Bolt

FIGURE 12.12 Edge clamping detail.

structural steel. This is so because of the coating applied on the wires and the gaps between the wires. Large elongation will occur when cables are loaded because the capacity of cables is almost six times larger than that of normal structural steels. Prestretching of cables is a common practice used to control the designed cable stress and remove the initial constructional stretch.

A few common types of end fittings used for cable terminations are shown in Fig. 12.15. The pinned sockets allow rotation in one direction. The anchor stud or threaded stud end fittings are used in bearing-type connections. The end fitting using a cable stud end with clevis allow for rotation and a limited amount of length adjustability. Most end fittings will develop the full strength of the cables.

The rigid support elements used for tensile fabric structures are masts, struts, frames, arches, and edge beams. They are made of conventional building materials, such as steel, reinforced and prestressed concrete, aluminum, laminated wood, and composite materials.

The connection details of tensile fabric structures are critical elements. Because of large displacements under the imposed loads, the connection details have to be flexible enough to accommodate the expected deflections and rotations. Most of these connec-

A. Fabric to Cable Connections

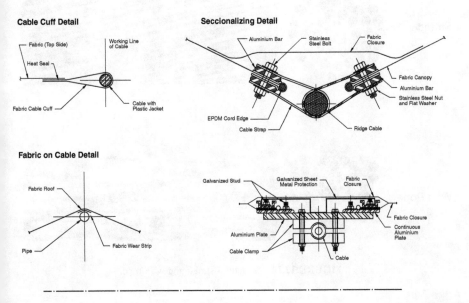

B. Fabric to Steel Pipe Details

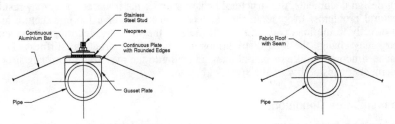

FIGURE 12.13   Typical fabric connection details.

tions are exposed, and their appearance is an important architectural design feature. The geometry of these details is much more complicated than that for conventional buildings owing to the unique shapes of tensile fabric structures. In cases where many members are to be connected at one working point, it is important to minimize or eliminate the eccentricities.

COMMON TYPES, CAUSES, AND CONSEQUENCES OF DEGRADATIONS AND DEFECTS

Understanding the common types, cause, and consequences of degradations and defects is a basic necessity for assessing tensile fabric structures. They are discussed in four parts.

Strand

Wire Rope

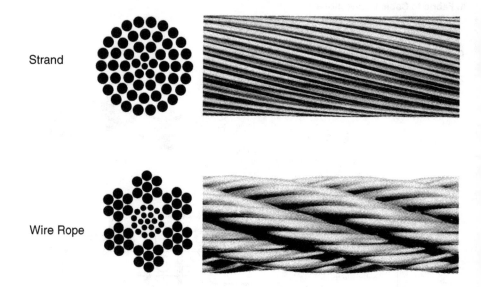

FIGURE 12.14 Structural strand and wire rope.

Overall Structural Conditions

The shape and geometry of fabric structures vary and are influenced by design re-quirements and the structural system. If the as-built conditions are different from the original design documents, the structural behavior under the imposed loads may result in structural problems. In general, the shape and geometry of a fabric structure are related directly to the load path and the stiffness of the structure. Incorrect shape and geometry may change the load distributions and the overall structural stiffness. This may cause structural failure or partial damage owing to the overstressing of key struc-tural elements or instability of the structural system.

Fabric Membrane Conditions

With fabric as a main structural element, there are three types of fabric failures that often occur in tensile fabric structures: tensile failure, tear failure, and seam failure. An overloading of fabric membrane usually causes *tensile failure*. This occurs when the accumulation of water or snow results in ponding on the fabric membrane, and the fabric stress exceeds the ultimate strength. *Tear failure* can occur when fabric is cut by external hard objects or is blown during windstorms or hurricane events. A *seam failure* occurs when a joint between two fabric panels opens up and causes a loss of fabric prestress and shape, eventually resulting in fabric tear failure.

Degradations and defects of fabric membrane materials will have a direct effect on fabric failures over time. The loss of structural strength is one such degradation, which is usually caused by UV degradation, loss of coating owing to abrasion, and water absorption and migration into the base substrate. Loss of prestress owing to creep in the fabric materials is a major issue for PVC-coated polyester fabrics because it causes wrinkles, bagginess, and potential water ponding. Loss of physical appearance is an-other form of degradation. Dirt accumulation on the coated fabrics is either an indi-cation of extreme air pollution or a failure in the coated fabrics. Teflon-coated fabrics

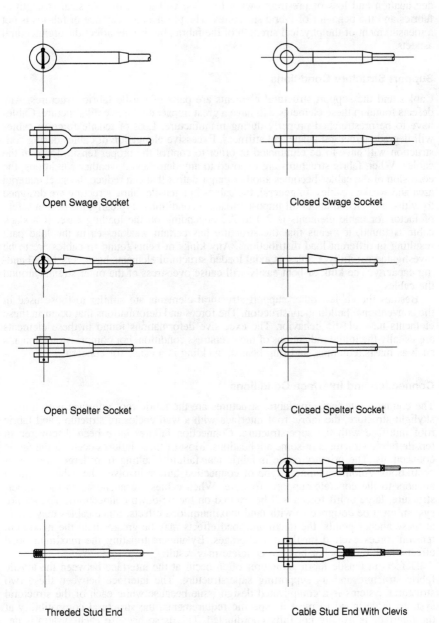

FIGURE 12.15   Typical cable fittings.

tend to retain their appearance over time, but the PVC-coated fabrics show a wide range of dirt accumulation depending on the coating composition and top coating. UV degradation and loss of prestress owing to creep will affect the physical strength of fabrics and the behavior of fabric structures. The physical appearance of fabrics is not a measurement of the physical strength of the fabric, but it may affect the architectural aspects.

Support Structure Conditions

Cables and the support structural elements are parts of tensile fabric structures. Any defects found in these elements will have a great impact on the overall structure. Cables have to be prestretched properly during manufacture. Loss of tensile force in cables will degrade structural shape and stiffness. Excessive elongation of cables during construction will have to be eliminated in order to control the proper tensile force in the cables. When fabric structures are exposed to any detrimental weather conditions, the corrosion of the cables becomes another major defect that can reduce the steel material area and weaken cables. In general, the cables in a tensile fabric structure are designed to withstand tension under all imposed loading conditions. Typically, the design safety of factor for cable elements is 2.0 to 2.2 depending on the loading case. If a slack cable is found, it means that the structure has certain weaknesses in the load path resulting in different load distributions. Any kinks or bents found in cables are to be investigated carefully. Cables as axial loaded structural elements have very little bending capacity. The kink or bent easily will cause overstress of the outer strands around the cables.

Besides the cables, other support structural elements are similar to those used in the conventional building construction. The forces and deformations that occur in these elements have elastic behavior. The excessive deformations found in these elements are usually the initial indications of the overstress condition. For compression elements such as masts, columns, and ring beams, buckling is a cause for concern.

Connection and Interface Conditions

The common interfaces for fabric structures are the fabric roof interface with a glass skylight structure, the fabric roof interface with a wall enclosure structure, and fabric roof interface with the superstructure. Connection failures have been discovered in tensile fabric structures in some applications. Most of these failures occur at the fabric connections. The primary cause is fabric tear failure resulting from large stress concentrations. Another common type of connection failure involves the cable tie-down anchors to the concrete support structure. When cables are anchored to the support structure, large uplift forces will be exerted on the tie-down connections. The anchor system must be designed to withstand maximum load effects. Since cables may rotate at these anchor points, the maximum load effects may be greater than the maximum tension forces exerted on these anchorages. By underestimating the maximum load effects, failure of anchor bolts or concrete may result.

Defects in tensile fabric structures often occur at the interface between the tensile fabric structure and its supporting superstructure. The interface between these two structural systems is a complicated design issue because while each of the structural systems is designed to meet its specific requirements, the structural compatibility at the interface is usually not fully coordinated. This is so because each system is designed by a different engineer. The structural incompatibilities at interfaces are caused mainly by force distributions and structural deformations. Tensile fabric structures make limited use of structural elements to achieve the lightness. The structural elements in this system are dealing with a relatively large magnitude of concentrated forces, and conventional building systems are designed to distribute the forces in a more uniform pattern. The deformation of tensile fabric structures is usually much larger than the limits set for conventional building systems. In most cases, the incompatibilities at the interface affect the architectural details more than the structural details. Water and air leakage are the common defects. If these defects are not resolved properly, subsequent structural defects may occur.

STRUCTURAL EVALUATION AND REPORTING

The structural condition assessment of existing tensile fabric structures should follow the same guidelines used for conventional structures as specified by SEI/ASCE 11, *Guideline for Structural Condition Assessment of Existing Buildings*.[16] It should be conducted according to the following procedures:

- Review available construction documents.
- Conduct site inspection to identify the structural deficiencies.
- Establish loading and performance criteria.
- Perform structural analysis.
- Perform material testing, if necessary.
- Integrate the results from the inspection and the analysis.
- Determine the findings regarding the structural conditions.
- Make recommendations.

The results of structural evaluation are to be summarized in a report. In the report, the purpose of this structural evaluation and the procedures should be introduced first. The integration of the inspection findings and an analysis results should be discussed in detail. The recommendations concerning the actions for strengthening, retrofitting, and future adoptions are presented.

INSPECTION OF TENSILE FABRIC STRUCTURES

The inspection of existing structures is an important part of structural condition assessment. The inspection provides engineers with the opportunity to confirm the correctness of the existing design information and to assess the visible condition of exiting tensile fabric structures. During the inspection, evidence of structural modification, deterioration of materials, discrepancies in documentations, and signs of weakness in structural elements or connections should be noted and recorded. For large and important fabric structures, inspections are scheduled and conducted once every 2 years. The scope of the inspection for tensile fabric structures is usually divided into several phases, as listed below.

Fabric Membrane Conditions. The fabric membrane is to be inspected carefully and thoroughly for the following items:

- Check fabric membrane for holes and tears.
- Check fabric membrane for scratches and abrasions. All scratches and abrasions are to be examined carefully for evidence of exposed yarns.
- Inspect all seams and splices for tears and separations.
- Check all fabric reinforcements for tears, delaminations, and other damages.

- Check the overall configuration of the fabric membrane for any signs of unusual looseness, bagginess, and other loss of tension.
- Check the entire fabric membrane for evidence of unusual dirt accumulation, soiling, etc.
- Note and record all areas of standing or ponded water on the fabric membrane.

Fabric Clamping Conditions. The fabric membrane attachments are to be inspected carefully and thoroughly as follows:

- Check all fasteners, gaskets, and aluminum clamping bars for damage, corrosion, and other deterioration.
- Check that all aluminum clamping bars are fastened securely.

Cables and End Fittings. The cables and end fittings are to be inspected carefully and thoroughly for the following items:

- Inspect all cables for corrosion.
- Check all accessible cables for looseness or apparent loss of tension.
- Check all cable end fittings for corrosion.
- Check all cable end fittings and pins to ensure that pins are in their proper locations and cotters/retainers are in place.

Structural Support System. The structural support system (masts, arches, frames, columns, beams, perimeter ring and plates, etc.) is to be inspected carefully and thoroughly inspected for the following items:

- Inspect all related structural steel for paint damage, corrosion, evidence of deformation, and other signs of deterioration.
- Inspect visible welds for obvious evidence of cracking.
- Inspect cable attachment lugs for cracking or corrosion.
- Inspect all attachment points and bolted connections for evidence of loose bolts, movement.
- Inspect all related structural concrete for evidence of cracking, deterioration.

The inspection should be carried out by a team with at least one structural engineer experienced in structural evaluation. The inspection should be conducted according to a detailed inspection checklist (see "Tensile Fabric Structure Inspection Checklist" at end of chapter) prepared by the engineer of record, who is responsible for the assessment evaluation. All inspection findings and results are to be summarized into a report. This information will be used as part of the structural condition assessment of existing tensile fabric structures.

STRUCTURAL ANALYSIS

Depending on visible condition, age, and history of the facility, structural analyses of parts or the whole may be warranted. A full structural analysis is performed to evaluate the entire structural system. Partial structural analyses and specific structural element analyses usually are conducted only on the critical or defective members in the system. The purpose of these analyses is to estimate the structural capacity to meet the specified structural criteria and to identify the structural deficiencies and their impact on the minimum life of the structure and public safety standards.

Prior to the structural analyses, the loading and performance criteria must be established in accordance with the governing codes, regulations, and owner-prescribed requirements. The original design criteria and the code requirements in effect when the structure was built also should be reviewed. The differences between these two criteria are to be determined, and those which govern should be noted.

Full Structural Analysis

Full structural models often are used to determine the capacity and behavior of the overall existing structure for building retrofitting and future adoption. To perform a full structure analysis, the structural model should be developed according to the as-built condition, as shown on the final as-built construction documents. If any uncertain conditions exist, then a detailed field survey and required probe testing must be conducted to determine the actual as-built condition.

Once the model is completed, the engineer should establish the loading criteria that meet the specific performance requirements of building retrofitting or future adoption. The first analysis to be conducted is the model under the prestress and as-built self-weight load conditions. The results of this analysis will allow the engineer to compare the model with the actual conditions and to confirm the accuracy of the full structural model.

The following analysis will be the model under the specific loading conditions, and the engineer should check the results for load distributions, structural deformations, support reactions, and member capacities. Once the engineer finds that these results are acceptable, the detailed evaluation of these results in comparison with the required performance criteria will be conducted finally. The findings of the evaluation will allow the engineer to make a professional judgment and recommendations about the structural capacity and behavior of the overall existing structure.

Many long-span fabric structures have been built to cover multiuse facilities. Some of these existing structures have been evaluated for their structural assessment under the request of the building owners. Georgia Dome is one of these buildings. After the Georgia Dome was built, the owner realized that there would be additional and different rigging loads to be hung from the fabric roof structure for the different events. A detailed evaluation study was performed by the architect and engineering design team to determine the overall structural capacities of the existing fabric roof to adopt these rigging loads. A full analysis was performed by the engineer (Fig. 12.16). The results indicated that this fabric structure was capable of hanging an additional 200 kips of rigging loads without reinforcing the existing structure. The engineer prepared the detailed maps to summarize the acceptable model of hanging load criteria based on the analysis results. These maps have been used by the owner as guidelines to present to the different event organizers.

Partial Analysis and Specific Structural Member Analysis

When evaluating the critical members and connections in a tensile fabric structure, using partial analysis and specific structural member analysis will provide the most accurate and detailed results with respect to resistant capacities associated with specified loading and performance criteria. The partial model is the part of the structure that includes the critical specific structural members to be investigated. In order to reflect the actual behavior of the partial model in relationship to the overall structure, proper boundary conditions of the partial model must be introduced. The detailed analysis procedure should follow that listed for a full structural analysis. The advantage of using the partial analysis is that not only will the amount of work be reduced, but the results relating to the critical members and their connections remain viable.

Since the fabric membrane is the main element of tensile fabric structures, it is nearly always analyzed in a condition assessment. The notable stipulation for this type

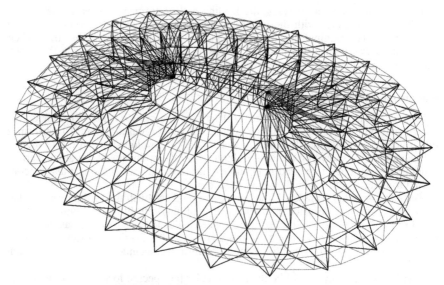

FIGURE 12.16 Full model of Georgia Dome.

of analysis is that the actual shape of the fabric panel and its boundary condition must be modeled properly. In general, a large fabric enclosure is constructed by jointing the individual fabric panels together. If water ponding or loss of tension has been seen at individual panel locations, the fabric panel model analysis should be used to evaluate the stresses. The results of the fabric model analysis will provide detailed information regarding fabric warp and fill stresses, stress concentrations, and the deformation contours.

Tensile fabric structures are unique structures consisting of many different structural systems and varied configurations. The type of structural model analysis used to assess the structural conditions is an important procedure. This decision is to be made by engineers depending on the specific application. As a general rule, the partial structural models and the specific structural member models are primary choices in the preliminary analysis. This will provide engineers with useful results related to the structural deficiencies associated with the critical structural members for their strengthening.

TENSILE FABRIC STRUCTURE INSPECTION CHECKLIST

1. Fabric membrane condition
   The fabric membrane is to be inspected carefully and completely for the following items:
   _____ Check membrane for holes and tears.
   _____ Check membrane for scratches and abrasions. All scratches and abrasions are to be carefully inspected for evidence of exposed yarns. If any exposed yarns are present and repair must be made.
   _____ Inspect all seams and splices for tears or separations.
   _____ Check all fabric reinforcements for tears, delaminations or other damage.
   _____ Check overall configuration of the membrane for any signs of unusual looseness, bagginess or other loss of tension.

_____ Check entire membrane for evidence of unusual dirt accumulation, soiling, etc.

_____ Note and record all areas of standing or ponded water on the membrane.

Comments: _____

_____

2. Perimeter clamping condition

The attachment of the fabric membrane and the perimeter is to be inspected carefully and completely for the following items:

_____ Check all fabric reinforcements for tears, delaminations, or other damage.

_____ Check all fasteners, gaskets, and aluminum clamp bars for damage, corrosion, or deterioration.

_____ Check that all clamp bars are fastened securely.

Comments: _____

_____

3. Cables and end fittings

Cables and end fittings are to be carefully and completely inspected for the following items:

_____ Inspect all cables for corrosion.

_____ Check all accessible cables for looseness or apparent loss of tension.

_____ Check cable end fittings for corrosion.

_____ Check cable end fittings and pins to ensure that pins are in their proper locations and cotters/retainers are in place.

Comments: _____

_____

4. Structural steel support system

The structural steel support system (flying masts, perimeter pipe and plates, etc.) is to be carefully and completely inspected for the following items:

_____ Inspect visible welds for obvious evidence of cracking.

_____ Inspect all related structural steel for paint damage, corrosion, evidence of deformation, etc.

_____ Inspect cable attachment lugs for cracking or corrosion.

_____ Ensure that welds that come into contact with fabric membrane are ground smooth and touched up with the correct paint.

_____ Inspect all attachment points and bolted connections for evidence of loose bolts, movement, etc.

Comments: _____

_____

5. Miscellaneous

_____ Check that all lightning rods and ground cables are installed properly in accordance with the manufacture's instructions.

_____ Inspect and inventory the contents of the fabric repair kit.

Comments: _____

_____

REFERENCES

1. International Code Council, Inc. (ICC) (2003). *International Building Code.* ICC, Country Club Hill, IL.
2. International Conference of Building Officials (ICBO) (1997). *Uniform Building Code.* ICBO, Whittier, CA.
3. Building Officials and Code Administrators International, Inc. (BOCA) (1999). *National Building Code.* BOCA, Chicago.
4. Southern Building Congress International, Inc. (SBCCI) (1997). *Standard Building Code.* SBCCI, Birmingham, AL.
5. SEI/ASCE 7 (2002). *Minimum Design Loads for Buildings and Other Structures.* American Society of Civil Engineers/Structural Engineering Institute, Reston, VA.
6. ASCE 17-96 (1996). *Air Supported Structures.* American Society of Civil Engineers, Reston, VA.
7. ASCE (1996). *Tensioned Fabric Structures.* American Society of Civil Engineers, Reston, VA.
8. ASCE 17-96 (1996). *Structural Applications of Steel Cables for Building.* American Society of Civil Engineers, Reston, VA.
9. AISC (1989). *Manual of Steel Construction (ASD).* American Institute of Steel Construction, Inc., Chicago.
10. AISC (2001). *Manual of Steel Construction (LRFD).* American Institute of Steel Construction, Inc., Chicago.
11. ACI (2002). *Building Code Requirements for Structural Concrete* (ACI 318), American Concrete Institute, Farmington Hills, MI.
12. ASTM D4851 (1997). *Test Methods for Coated and Laminated Fabrics for Architectural Use.* American Society for Testing and Materials, West Conshohocken, PA.
13. ASTM E84 (1994). *Test Methods for Surface Burning Characteristics of Building Materials.* American Society for Testing and Materials, West Conshohocken, PA.
14. ASTM E108 (1993). *Test Methods for Fire Tests of Roof Covering.* American Society for Testing and Materials, West Conshohocken, PA.
15. ASTM E136 (1994). *Test Methods for Behavior of Materials in Vertical Tube Furnace at 750°C.* American Society for Testing and Materials, West Conshohocken, PA.
16. SEI/ASCE 11 (1990). *Guidelines for Structural Condition Assessment of Existing Buildings.* American Society of Civil Engineers, Reston, VA.

CHAPTER 13

Broadcast and Transmission Towers

JOHN L. WINDLE, P.E.

INTRODUCTION

The development of wireless communications created a need for tower structures. As new forms of communication were developed, the demand for towers increased dramatically to the point where there is a proliferation of tower structures for radio and television broadcasting, emergency communications, data transmission, and cellular telephone systems. Assessment of the structural condition of these towers is important not only to prevent interruptions of the communication systems but also to protect the safety of the general public.

The installation of new antennas and transmission lines on a tower that differ from those considered in the original design; a lack of proper maintenance resulting in deterioration; defects in the original design, fabrication, or installation; settlement or lateral movement of the foundations and guy anchorages; changes in building codes; improvements in design and analysis methods; and a general lack of structural redundancy are some of the reasons why the structural integrity of a tower may be threatened. Therefore, it is important that a periodic structural condition assessment program be developed for every tower structure.

TOWER TYPES AND CONSTRUCTION

Towers are classified as either *self-supporting* or *guyed*. Those which are supported laterally by a system of cables extending to ground anchors are designated as *guyed* (Fig. 13.1), and those which are free-standing are designated as *self-supporting* (Fig.

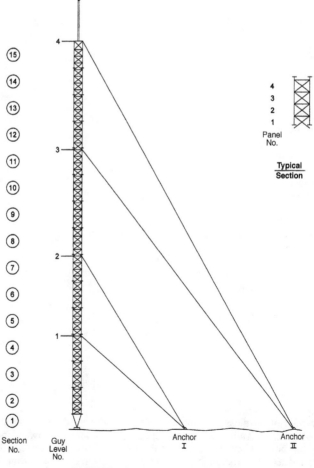

FIGURE 13.1 Typical guyed tower.

13.2). The tower structure may be either a single structural element, such as a solid or tubular shaft, or a lattice column. Nearly all guyed towers are lattice columns. Self-supporting towers used for wireless communications are often single-element steel monopoles, whereas those used for radio and television broadcasting and microwave transmission usually are open-faced structural steel space trusses.

The tower structure consists of modular sections usually between 20 and 30 ft (6 and 9 m) in length. In place, most have an equilateral triangular cross section, although some have a square cross section. The principal components are the vertical legs and the horizontal and diagonal braces. The bracing may be any number of styles, with the more common being an X, Z, or K configuration. There also may be horizontal diaphragm members between the faces of the tower (see Fig. 13.2). Leg members may be made from structural-steel angles, formed plates, or structural pipe, tubes, or solid round bars with welded connections and splice plates. Bracing members may be made

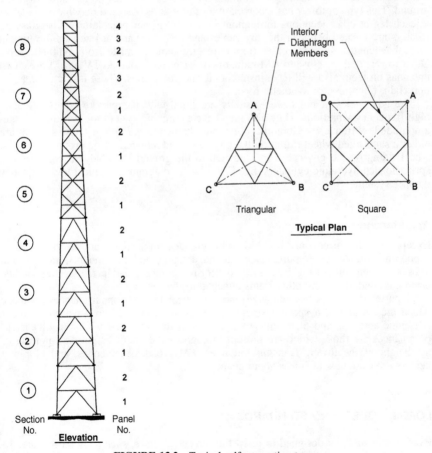

FIGURE 13.2   Typical self-supporting tower.

from solid round bars with welded end plates, structural pipes or tubes with either flattened ends or welded end plates, structural angles, or channels. Sections with a face width (distance between legs) of less than 5 ft (1.5 m) frequently are completely shop-welded.

Monopoles consist of sections of hollow tubes of varying lengths. Those having a circular cross section are usually straight cylinders with welded flanges on the ends of their sections for bolting them together. Those having a polygonal cross section usually are tapered, with the sections telescoping into one another.

Guy Systems

Three cables, placed at 120-degree angular spacing, are used at each guy level for triangular guyed towers. For square towers, four cables at 90-degree angular spacing are used. At those levels where significant torsion forces occur, frames are attached to each face of the tower, and two cables are used in each direction. See Fig. 13.3 for typical guy arrangements. The guy cables usually are made from extra-high-strength or structural-grade galvanized strand.

Some towers, such as those used for AM radio broadcasting and ground wave transmission, actually function as radiators themselves and must be insulated from the ground. This type of tower has a porcelain insulator at its base and may have insulators at each leg at other locations throughout its height. Either porcelain or fiberglass insulators are located between the guy cables and the tower and at intervals throughout the cable lengths. In some cases the entire guy cable is made from synthetic rope. Other towers that are close to AM radiators or have side-mounted TV or FM broadcast antennas on them also will have insulators in the guy cables. These insulators may be porcelain, fiberglass, or synthetic rope.

The guy cables of tall towers usually are fitted with dampers to control Aeolian high-frequency vibrations. They also may have a damping system for preventing large-amplitude low-frequency vibrations commonly known as *galloping*. These systems are either a sand-filled wheel riding on the guy cable and tethered to the ground or a solid wheel riding on the guy cable and tethered to the ground by a halyard connected to spring-loaded hydraulic cylinders. See Fig. 13.4 for diagrams of typical low-frequency damper systems.

Appurtenances

Except for those used only for AM radio broadcasting, all towers are designed to support appurtenances. Appurtenances include antennas and their mounts, transmission lines and their supporting structures, platforms, climbing ladders, elevators, lights, electrical conduit, or any other items attached to the tower.

Appurtenances are classified as *discrete* or *linear*. Discrete appurtenances are those which are located at a specific single height on the tower. Microwave and cellular telephone antennas and platforms are examples of discrete appurtenances. Linear appurtenances are those which are present throughout the entire height or a portion of the height of the tower. Transmission lines, FM broadcast antennas, and climbing ladders are examples of linear appurtenances.

LOADS, CODES, AND STANDARDS

Every tower must be designed to resist the forces of nature, such as wind and ice. To provide standard methods for calculating these forces and determining the capacity of

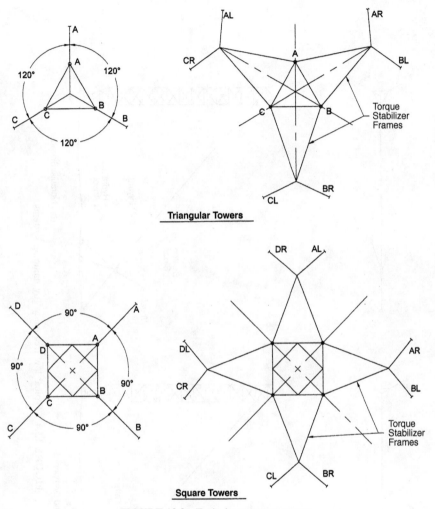

FIGURE 13.3 Typical guy arrangements.

a tower to resist them, standards and codes have been developed for their design, fabrication, installation, and maintenance.

Loads

The loads considered in the design of a tower structure in addition to its own weight are the weight of any appurtenances supported by the tower, the weight of ice accumulations on the tower and the appurtenances, and horizontal forces caused by wind acting on the tower and the appurtenances, including the thickness of ice accumulations.

Since almost no redundancy is included in the design of these structures, the failure of any one component may result in a catastrophic collapse. Therefore, it is especially important that their structural condition be assessed periodically.

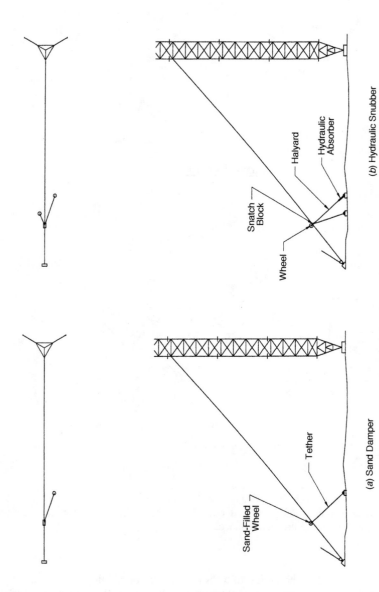

(a) Sand Damper

Sand-Filled Wheel

Tether

Snatch Block

Halyard

Wheel

Hydraulic Absorber

(b) Hydraulic Snubber

FIGURE 13.4 Typical low-frequency guy damping systems.

Industry Standard ANSI/TIA/EIA 222

The most comprehensive standard for tower design, fabrication, erection, and maintenance is ANSI/TIA/EIA 222-F-1996, *Structural Standards for Steel Antenna Towers and Antenna Supporting Structures,* published by the Telecommunications Industry Association (TIA). Most of the towers in the United States have been designed and constructed in accordance with this standard.

This standard was published in 1949 with the designation RETMA TR116. It was expanded and revised in 1959 and designated as EIA RS-222. A major revision, designated ANSI/EIA 222-D, was made in 1987 when the methods of specifying and calculating wind loads were changed significantly, and greater emphasis was placed on the importance of considering ice loads. Since then, there have been two revisions of the standard. The effects of the various revisions on the magnitude and profile of wind loads are discussed and illustrated in an article entitled, "Recent Changes in Structural Standards for Communications Towers," in the *Proceedings of the 1991 SBE and Broadcast Engineering Conference, Journal of the Society of Broadcast Engineers,* 1991.

ANSI/TIA/EIA 222-F-1996 requires the tower to be adequate to resist the following combinations of loads:

Dead plus wind without ice
Dead and weight of ice, plus 75% of wind with ice

Wind loads are determined from a basic fastest-mile wind speed, v mi/h, (m/s), at a height of 33 ft (10 m) above ground level. The standard provides minimum recommended values of basic wind speed based on a 50-year recurrence (0.02 probability) for every county in the United States. This basic wind speed is converted to a horizontal wind pressure, q lb/ft², (Pa) by the relationship

$$q = 0.00256 \times v^2 \quad \text{for } v \text{ in mi/h or}$$

$$q = 0.613 \times v^2 \quad \text{for } v \text{ in m/s}$$

This pressure q, modified by various factors, is applied to the aerodynamic areas of each section of the tower and linear appurtenances to determine the horizontal force F lb (N) applied to the tower in the direction of the wind. For latticed towers,

$$F = qK_zG_h(C_fA_e + \Sigma\ C_aA_a) \leq 2.0qK_zG_hA_g$$

where K_z = exposure coefficient = $(Z/33)^{2/7}$ for Z in ft or $K_z = (z/10)^{2/7}$ for Z in m

Z = height above ground level, ft (m)

G_h = gust coefficient = $0.65 + 0.60\ (H/33)^{1/7} \geq 1.0$ but ≤ 1.25 for H in ft or $G_h = 0.65 + 0.60\ (H/10)^{1/7}$ for H in m

H = total height of the tower structure, ft (m)

C_f = force coefficient = $3.4e^2 - 4.7e + 3.4$ for triangular towers and $4.0e - 5.9e + 4.0$ for square towers

e = solidity ratio = $(Af + Ar)/Ag$

A_f = projected area of flat structural components in one face of the tower, ft² (m²)

A_r = projected area of round structural components in one face of the tower, ft² (m²)

A_g = gross area of one face of the tower as if it were solid, ft² (m²)

A_e = $DfAf + DrArRr$, ft² (m²)

D_f = wind direction factor for flat structural components
D_r = wind direction factor for round structural components
R_r = reduction factor for round structural components = $0.51e^2 + 0.57 \leq 1.0$
C_a = linear or discrete appurtenance force coefficient
A_a = projected area of a linear appurtenance, ft² (m²)

For discrete appurtenances,

$$F = qK_zG_h \left[\Sigma(C_aA_c)\right]$$

where A_c = projected area of a discrete appurtenance, ft² (m²)

Antenna manufacturers often will provide the aerodynamic area, C_aA_c, as well as the weight of the antenna.

For a parabolic microwave antenna (dish), it is necessary to calculate two horizontal vectors of the wind force acting through its apex and the torque moment about its apex. These values are determined from the following equations:

The force vector along the axis of the antenna:

$$F_a = C_aAK_zG_hv^2$$

The force vector normal to the axis of the antenna:

$$F_sC_sAK_zG_hv^2$$

The torque moment about the apex of the antenna:

$$M = C_mADK_zG_hv^2$$

Values for C_a, C_s, and C_m can be obtained from Annex B of the standard or from the manufacturer. D is the diameter of the antenna, and A is its frontal area. K_z, G_h, and v are as defined previously.

The standard states that linear appurtenances attached to a face of the tower and not extending beyond the width of the face may be considered as structural components for the purpose of calculating wind loads. When linear appurtenances are on more than one face of the tower, this provision usually results in smaller wind forces.

Ice loads are based on a coating of uniform thickness having a density of 56 lb/ft³ (8.8 kN/m³) on all tower members, guys, and appurtenances. The standard recommends a minimum thickness of 1/2 in. (12.7 mm) for towers in areas subject to ice accumulation. It does not provide a specific recommended thickness for each county. When calculating wind loads with ice, the increase in exposed area resulting from the ice accumulation must be considered.

This standard references the following specifications and codes for determining allowable stresses and minimum safety factors:

American Institute of Steel Construction (AISC), *Specification for Structural Buildings—Allowable Stress Design and Plastic Design*
ANSI/ASCE 10-90, *Design of Latticed Steel Transmission Towers*
American Concrete Institute ACI-89, *Building Code Requirements for Reinforced Concrete*

The standard has special provisions that modify or supersede certain portions of the referenced codes and specifications. These include the determination of k factors for calculating the slenderness ratios for single-angle compression members, load factors for calculating design loads for reinforced-concrete foundations and guy anchorages, allowable stress-increase factors for structural-steel components, and the tightening methods and washer and locking-device requirements for high-strength bolts.

Statutory Codes

Most municipal and state governments have statutory building codes. Many of these are patterned after one of several model building codes. The most common of these are

Building Officals and Code Administrators International (BOCA), *Basic Building Code*
International Conference of Building Officials (ICBO), *Uniform Building Code*
Southern Building Code Congress (SBC), *Standard Building Code*

The industry standard ANSI/TIA/EIA 222-F-1996 is compatible with these codes and is included by reference in some of them. Since it is necessary to comply with the statutory code, it is important to consult with the appropriate building official to determine what the statutory requirements are.

DEGRADATION AND DEFECTS

All components of the structure are subject to degradation and defects. Degradation may be a gradual process over time, such as corrosion, or it may be damage caused by a severe storm or carelessness by workers installing new equipment on the tower. Some defects may be the result of design errors, defective materials, and poor workmanship and may have existed throughout the entire life of the tower. This section outlines the common types, causes, and consequences of degradation and defects.

Common Types

The common types of degradation and defects are the following.

In the tower structure:

- Missing, damaged, deformed, or corroded members
- Missing or blocked drain holes in hollow tubular members
- Missing, loose, or corroded fasteners
- Cracked, incomplete, or otherwise defective welds
- Damaged insulators
- Incorrectly sized components
- Peeling or faded paint

In the guy system:

- Damaged or corroded strand
- Damaged, loose, or corroded end connectors
- Improperly installed end connectors
- Incorrect strand or hardware size

- Damaged insulators
- Missing or damaged dampers
- Incorrect pretension

In the tower/guy system as a unit:

- Out-of-plumbness tolerance
- Out-of-twist tolerance
- Out-of-straightness tolerance

In the tower foundation:

- Damaged or deteriorated concrete
- Missing or deteriorated grout
- Damaged, loose, or corroded anchor bolts
- Excessive or differential settlement
- Excavations or other soil disturbances in close proximity
- Broken or missing grounding connections

In the guy anchorage:

- Damaged or corroded anchor shafts
- Excessive lateral creep or settlement
- Excavations or other soil disturbances in close proximity
- Broken or missing grounding connections

At the appurtenances:

- Improperly located appurtenances
- Appurtenances not considered in the original design
- Improper or damaged anchorage of appurtenances

Causes

Degradation and defects are the results of human, natural, and normal causes.

Human Causes. Among the human causes of defects are design errors, use of defective or incorrect types and grades of materials, fabrication (especially welding) errors, assembly and installation errors, and damage occurring during erection or the installation or removal of equipment on the tower. Among the human causes of degradation are vandalism and lack of proper maintenance. Proper maintenance requires inspection of the tower at regular intervals and after any severe wind or ice storms and taking corrective measures such as cleaning blocked drain holes, repairing or replacing damaged or corroded components, tightening loose connections, adjusting guy tensions to their correct values, and adjusting the alignment and plumbness to permissible tolerances.

Natural Causes. Wind speeds and ice accumulations exceeding the design values, earthquakes, corrosive atmospheres, and electrolytic action are natural causes of defects and degradation.

Normal Causes. Repeated cyclic-load-induced motion leading to fatigue.

Consequences

Many towers with defects have been standing for years. However, this is probably so because they have not yet experienced their design wind and/or ice loads. When they do experience these loads, there may be no noticeable consequence but perhaps only a degradation or loss of signal. More likely, however, there will be a catastrophic collapse, resulting in extensive property loss and damage, as well as possible personal injuries or deaths. Additionally, there will be significant loss of revenue to the companies having antennas on the tower.

WHAT TO LOOK FOR

In this section, guidance is given on what to look for in the various parts of a tower when conducting a structural condition assessment.

Tower Structure

Aside from inspecting the parts and components, it is important to check the tower structure as a unit throughout its height for straightness, plumbness, and twist.

Leg Members. When inspecting towers with round leg members, look carefully at the welds for the bracing connection plates. Cracks are most likely to occur in the heat-affected zones at the ends of these plates on solid round legs. Check also for any signs of undercutting of welds. Look for any bent or torn plates.

For legs made from hollow tubes, look for proper drain holes. Drain holes must be provided to prevent the accumulation and possible freezing of water inside the legs. These holes may be either at the bottom of each leg or only at the bottom of the lowest hollow leg. Some manufacturers of self-supporting towers do not provide drain holes in the legs themselves but require a drainage slot in the grout at the tower base. Note any blockages of drain holes or slots. Check these legs carefully for any cracks.

For towers with legs made from structural angles or formed plates, look for any local buckling or tears in individual legs. Check the straightness of all legs. The permissible deviation is 1:500.

Bracing Members. Look for any bowed, kinked, or bent members. For those made from structural angles or formed plates, look for signs of local buckling or tears. Look for abrasions or cuts. These sometimes occur when rigging on the tower has been positioned such that a moving line rubs against a member, particularly the outstanding leg of an angle.

Look for loose tension rods. For towers with X-braced diagonal tension rods, look carefully at their points of intersection. The two rods should be fastened securely together, usually with stainless steel strapping or U-bolts and plates. These connections often become loose or broken, thereby permitting the two rods to rub against each other as the tower sways. For towers serving as radiators and for other towers at locations in the radiation field of side-mounted FM or TV broadcast antennas, electrical arcing will occur and cause serious damage to the rods if this connection is not tight.

For tall towers equipped with elevators, it is important to check the bracing members, including any horizontal diaphragm members, for damage from the elevator lift cables striking them. This is especially important in the areas of side-mounted TV or FM broadcast antennas where electrical arcing can occur.

Connections. Check all bolted connections for loose or missing bolts and missing washers. Verify the grades of bolts and nuts and the size and quantity for each connection.

Some leg-splice connections in the upper spans of a guyed tower, all leg-splice connections in any cantilevered portion of a guyed tower, and all leg splices in self-supporting towers are subject to both tension and compression forces at various times. At these locations the bolts in lapped splices for legs made from angles or formed plates and those in flanged splices for round member legs (unless not subject to prying action) must be provided with hardened washers and tightened to the full specified pretension. Verification of this tension should be in accordance with the provisions of the AISC, *Specification for Joints Using ASTM A325 and A490 Bolts.*

All other leg splices and most bolted connections for bracing members are designed as bearing rather than friction-type connections. Therefore, it is only necessary that they be tightened to the snug-tight condition.

All bolts other than A325 high-strength bolts tightened to the full pretension must have nut-locking devices. Galvanized A490 bolts should not be used in any connection.

Visually check all welds for cracks, undercutting, or other defects. If cracks are suspected, remove the zinc coating and perform dye-penetrant testing.

Guy System

Strand. Check all guy cables for individual broken wires in the strand. Check for "birdcaging" of the strand. This is a condition caused by the strand being subjected to a compression load that causes the individual strands to separate and balloon outward to form what appears to be a nearly spherical cage.

For guys or segments of guys made from synthetic rope, look for any tears in the outer jacket that protects the rope from deterioration caused by ultraviolet (UV) rays.

Connections. For connections made with bolted clips, check for tightness. U-bolt clips should be installed with the saddle on the main side of the strand. For connections made with grip-type dead-end connectors, check that the ends are completely snapped in place and that an end cap is in place for those grips pointing up. All turnbuckles should be wired safety or fitted with lock nuts to prevent turning.

For connections made with poured sockets, check for any deterioration of the zinc or epoxy resin material. Where closed-bridge sockets are used at the anchors, there should be a nut and a lock nut on each leg of the hairpin. These nuts should be positioned so that the bowl bears evenly against them and the load is distributed equally between the two legs of the hairpin. Check that all pins are secured properly with cotter pins.

Check the holes in the guy attachment plates at the tower and the anchor for any signs of elongation or undue wear.

Pretension. All guyed towers are designed on the basis of each guy having a specific value of pretension. These values are usually 10% of the breaking strength of the strand, with a suggested minimum value of 8% and a maximum value of 15%. The design values of pretension are given in the erection drawings for the tower.

Insulators

Check all porcelain insulators at the tower base and in any other locations on the tower and in the tower guys for chips and scratches in the glaze coating and for cracks in the porcelain itself. Some porcelain insulators are hollow and filled with oil. Look for any signs of oil leaks at these locations.

Check all fiberglass insulators in the guys for scratches or other deterioration of the outer coating that protects the fiberglass from deterioration caused by UV rays.

Tower Foundation

Check for loose nuts on the anchor bolts. Look for missing or deteriorated grout beneath the base plate. Check the visible portion of the concrete for spalling, cracking, or other deterioration. Look for signs of settlement and any abrupt changes in contour of the soil around the foundation. Be sure that the grounding conductors are not damaged or broken and that their connections are intact and tightened properly.

Guy Anchorages

Check the visible portions of the guy anchor shafts for bends, cracks, cuts, abrasions, or other damage. Check the visible portion of the concrete for spalling, cracking, or other deterioration. Look for signs of creep and settlement and any abrupt changes in contour of the soil in the vicinity of the anchor. Note any vegetation immediately around the anchor and beneath the guy cables that might pose a fire hazard. Be sure that the grounding conductors are not damaged or broken and that their connections are intact and tightened properly.

Corrosion

Check all metallic components of the tower structure, guy system, base foundation, and guy anchorages for corrosion, especially severe rusting with pitted surfaces. Nearly all tower components are zinc-coated for corrosion protection. Small scratches or nicks in this coating usually are not detrimental because the underlying steel is protected by galvanic action.

For hollow members, look at the areas around drain holes and the end connections for any stains that would indicate corrosion on the inside of the member. Try to locate the source of any obvious stains. Many times an unprotected appurtenance connection or hardware item is the source of the stain on the structural member.

Components made from steels with high silicon content often will become discolored even though they have been zinc-coated. This discoloration is not detrimental.

Look for places where dissimilar metals may be in contact. These usually will be at grounding connections for appurtenances and at the tower base and guy anchorages. Copper conductors or connectors fastened to zinc-coated steel components will show galvanic corrosion.

Guy anchor shafts, even if zinc-coated, in direct contact with the soil should be checked for corrosion owing to electrolytic as well as galvanic action. This is especially important in areas where an outside source of electric current is present in a soil having a resistivity of less than 2000 Ω·cm.

Appurtenances

All towers are designed to support a certain arrangement of appurtenances. The actual quantity, types, and locations of these appurtenances may vary from those used as the basis for the tower design and thus may impose a different magnitude and distribution of loads on the tower. Therefore, it is important to identify all appurtenances and their locations in both elevation and plan on the tower.

HOW TO LOOK AND HOW TO RECORD

This section provides information for preparing a comprehensive program for planning, executing, and recording a tower inspection.

Preparation

Proper planning and preparation will make the condition assessment more efficient, more economical, and even more reliable. It should include the acquisition of existing documents, definition of the scope of work, identification and acquisition of the applicable codes and standards, development of identification systems, preparation of preliminary diagrams, and attention to safety. These are discussed below.

Documentation. Collect and review all documentation available for the tower. This should include the original design calculations, installation and shop drawings, geotechnical reports, as-built drawings, structural analyses, modification drawings made subsequent to the original installation, and reports from previous inspections.

Scope. The scope of inspection may vary depending on the documentation that is available. If the documentation includes the original design criteria, the sizes and types of materials for each component, and the construction details and design capacities of the tower base and guy anchorages, it will be necessary only to determine what defects or degradations exist. If this information is not available, it also will be necessary to obtain it as part of the inspection process. The scope of the inspection should be discussed and agreed on with the owner before proceeding.

Applicable Codes and Standards. Since it is probable that there have been revisions to the standards and codes used originally, it is necessary to determine what design and construction codes are applicable to the tower at the time of the structural condition assessment. Annex F of ANSI/TIA/EIA Standard 222-F-1996 states that the latest revision of the standard should be used if there is a change in antennas, transmission lines, or other appurtenances; a change in deflection limitations for operational performance; or a need to increase the magnitude of the wind or ice loading from the original design criteria.

It is advisable to inquire from the building officials of the municipality in which the tower is located as to what the statutory requirements are and what revisions of the required codes and standards are applicable before beginning the assessment.

Identification System. Develop an identification system for the tower. A typical system would include the following:

For the tower structure:

- Number the sections beginning with one for the bottom section.
- Number the bracing panels in each section beginning with one for the bottom.
- Identify the legs of the tower in plan view with uppercase letters.
- Identify the faces of the tower with the letter designations for the legs in them, e.g., face *AB*, *BC*, *CA* for a triangular tower. Be sure to relate the faces of the tower to magnetic north. See Figs. 13.1 and 13.2 for illustrations of tower structure identification systems.

For the guy system:

- Number each guy level beginning with one for the lowest.
- Identify each guy at a level with the letter designation of the leg to which it is attached. Where more than one guy extends in the same direction, such as levels with torque stabilizers or face-guyed towers (two guys extending from each leg of the face and perpendicular to it), add an *L* or *R* suffix to indicate if it lies to

the left or right of the guy path looking from the tower toward the anchor. For example, guy 3AL would be the guy on the left side of the guy path in the direction of the A leg at the third level up. See Fig. 13.3 for illustrations of guy identification systems.

For the guy anchorages:

- Identify each guy anchorage with the letter designation of the leg that is in the same direction as the anchorage.
- For towers that are face-guyed, identify each anchorage with the letters of the face from which the guys extend.
- Use roman numeral suffixes, beginning with I for the anchorage nearest the tower base, when there are more than one anchorage in a direction. For example, guy anchorage BII would be the second guy anchorage from the tower base in the direction of leg B.

Diagrams. Prepare a set of line diagrams for use in recording the observations from the inspection. These should include an elevation of each face and a cross section of each tower section, an elevation of the guy cables in each direction, and an elevation of the tower as a unit with the section numbers and guy levels identified. If the available documentation for the tower is incomplete, it will be necessary to develop all or some of these diagrams during the field inspection. See Figs. 13.1, 13.2, and 13.3 for examples of these diagrams.

Safety Considerations. Identify what type of ladder safety climbing device, if any, is on the tower and what type of belt will be required to interface with it. Always be tied to the tower structure with a belt and lanyard.

Identify what electrical power and/or radiation hazards exist. For towers used for AM radio broadcasting and low-frequency ground-wave transmission, consult with the station engineer to determine when and how you can board the tower safely. For towers supporting FM and TV broadcast antennas, find out if these antennas can be off the air during the inspection. If not, protective clothing must be worn while working in the areas of these antennas.

Do not work alone. Have a "walkie talkie" system available for communication between personnel on the tower and on the ground.

Field Inspection

At a minimum, the inspection should include the activities described below:

Visually inspect each tower section. Record the location and description of any observed defect or degradation on the appropriate diagram. If there is reason to suspect the quality of the material in any component, tests should be made using methods described in Chap. 17. Photograph each observed defect or degradation, recording the photograph number on the appropriate diagram.

Visually inspect the guy system. Record the location and description of any defect or degradation observed on the appropriate diagram. Photograph each observed defect or degradation, recording the photograph number on the appropriate diagram. Photograph typical connections at both the tower and the anchorage and any observed defect or degradation, recording the photograph number on the appropriate diagram

Visually inspect the tower base and guy anchorages. Record the location and description of any defect or degradation observed on the appropriate diagram. If there is reason to suspect the quality of the concrete, further tests should be made using methods described in Chap. 16. For anchorages with the anchor shaft in direct contact with the soil, it will be necessary to excavate around the shaft to the point where it enters the concrete to detect any galvanic or electro-lytic corrosion. This excavation must be done carefully and kept to a minimum. Photograph the tower base and each guy anchorage, recording the photograph number on the appropriate diagram.

Locate and identify all appurtenances. Record the type, model number, size, lo-cation in both elevation and plan, and azimuth orientation of each appurtenance on the tower elevation and plan diagrams. It is especially important to note where linear appurtenances such as transmission lines, conduits, ladders, etc. are lo-cated in the plan and if they are attached to a face. Photograph each appurte-nance, and record the photograph number next to the appurtenance on the tower diagrams.

Determine the pretension in the guy cables. It is only necessary to measure the pretension in the guys in one direction for towers with guys in three directions and in two mutually perpendicular directions for towers with guys in four di-rections. The direction to be measured may be noted on the original installation drawings. If not, select the direction that has the least difference in elevation between the tower base and the guy anchorages.

It is important to record the ambient temperature at the time the tensions are mea-sured. The values of pretension used in the design are for a specific reference tem-perature. This temperature should be shown on the original erection drawings along with a chart of pretension values for other temperatures. If not, a value of 60°F should be used as a reference, and the measured values should be adjusted for the difference between this value and the recorded ambient temperature.

Measurements should not be taken if the ground wind velocity is greater than 10 mi/h. If this is not possible, record the wind speed and direction so that proper allow-ances for wind-induced increases or decreases in tension can be made.

There are several methods of measuring the pretension; three of them are described below.

For smaller towers with guys up to 1 in. in diameter, the pretension may be mea-sured directly using a *tensiometer*. This instrument must be calibrated for each specific type and size strand on which it is used. Figure 13.5 shows a typical tensiometer.

For larger towers, the most common method used is the *tangent intercept method*. A sighting device is placed on the guy at the anchor end, and a line of sight tangent to the cable at the anchor is established. The vertical distance from the intercept of this sight line along the tower leg measured upward to the guy attachment point on that tower leg is identified as the intercept distance I in feet. This distance may be measured directly, but usually it is estimated based on the number of bracing panels within it. The pretension T_a in pounds (N), is related to this distance by the relationship

$$T_a = \frac{WC \sqrt{H^2 + (V - I)^2}}{HI}$$

where W = the weight of the entire length of the guy, lb (N)
           C = the horizontal distance form the tower attachment point to the center of gravity of the guy, ft (m).
           H = the horizontal projection of the guy, ft (m)
           V = the vertical projection of the guy, ft (m)

This method is illustrated in Fig. 13.6.

FIGURE 13.5   Tensiometer for measuring pretensions in guy cables. (*Courtesy of Penn Tech International, Inc., West Chester, PA.*)

The pretension also may be determined indirectly by the *pulse method*. A sharp jerk is applied to the guy near the anchor end, thereby initiating a wave traveling up and down the guy. On the first return of the wave to the anchor, a stopwatch is started. The elapsed time P in seconds for a number of returns N is measured. The tensions in pounds (N) at midguy T_m and at the anchor T_a then may be calculated from the relationships

$$T_m = \frac{WLN^2}{8.05P^2}$$

$$T_a = \sqrt{\left(T_m - \frac{WV}{2L}\right)^2 + \left(\frac{WH}{2L}\right)^2}$$

where W = the weight of the guy, lb (N)
H = the horizontal projection of the guy, ft (m)
V = the vertical projection of the guy, ft (m)
L = the chord length of the guy (ft) (m) = $(H^2 + V^2]^{1/2}$

Another direct method for measuring the pretension in guys with closed bridge sockets at the anchor end is to use a *calibrated hydraulic ram system*. To use this method, it is necessary to have fixtures designed to symmetrically mount two rams between the bowl of the socket and nuts placed on each leg of the socket hairpin. The rams are connected through a manifold to a hydraulic pump fitted with a pressure

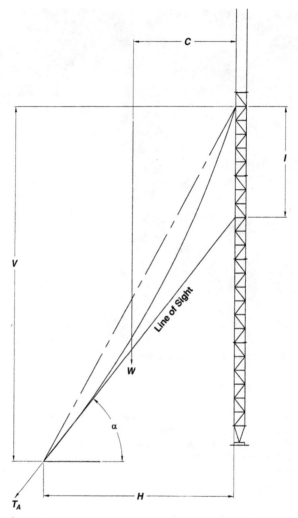

FIGURE 13.6 Tangent intercept method for measuring pretensions in guy cables. (*From ANSI/TIA/EIA Standard 222-F-1996, published by Telecommunications Industry Association.*)

gauge. The pressure in the system is increased until the instant the socket bowl breaks free of the nuts restraining it. The pressure reading at this time then can be compared with the calibration chart to obtain the tension at the anchor T_a in pounds (N). This method is illustrated in Fig. 13.7.

> *Measure the alignment and twist of the tower.* The alignment and twist of a tower may be determined by using a transit to measure the offsets of the tower legs from true vertical from the base up throughout its height. For a self-supporting tower, it is necessary to use one transit position for each leg. While it is possible

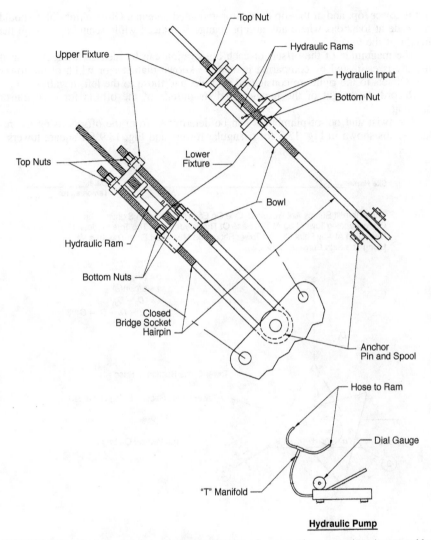

FIGURE 13.7   Typical arrangement of hydraulic rams for measuring pretensions in guy cables.

to use only two transit locations for a guyed tower, it is better to use a different position for each leg of the tower.

The transit position for each leg should be nearly on a line extending from the center of the tower through the center of the leg at a distance that will permit observation of the entire tower. The vertical crosshair of the transit should be aligned on an edge of the leg at the tower base. Observations are then made at intervals throughout the height of the tower. For guyed towers, observations should be made at each guy level, at the tower top, and at the top of any top-mounted antenna. For self-supporting towers, observations should be made at intervals equal to the height of two sections, at the tower top, and at the top of any top-mounted antenna. Observations also should be made at locations where any abrupt change is noticed while scanning through the height of the tower.

The magnitude of the offset for each observation can be measured directly, but it usually is estimated by comparing it with the known diameter or width of the tower leg. Offsets to the right are considered positive, and those to the left, negative.

Record the height of the location and magnitude of the offsets for each transit sighting.

The twist and out-of-plumbness can be determined from the offsets using the relationships shown in Fig. 13.8 for triangular towers and Fig. 13.9 for square towers.

Site Name:_____ Date:_____
                                    Wind:_____ Temperature:_____

Three Transit Setups Are Required, One On Each Leg Azimuth Sighting The Corresponding Tower Leg At The Base Of The Tower To Set The True Vertical. The Deflection At Any Point On The Tower Should Be Measured From This True Vertical, Using The Sign Convention Below.

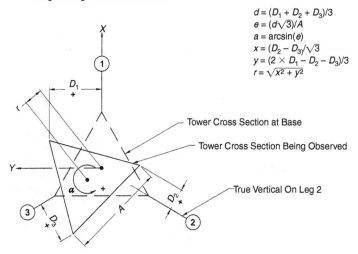

$d = (D_1 + D_2 + D_3)/3$
$e = (d\sqrt{3})/A$
$a = \arcsin(e)$
$x = (D_2 - D_3)/\sqrt{3}$
$y = (2 \times D_1 - D_2 - D_3)/3$
$r = \sqrt{x^2 + y^2}$

Tower Cross Section at Base

Tower Cross Section Being Observed

True Vertical On Leg 2

Observed Mast Data					Calculated Twist			Calculated Out-of-Plumb		
Mast Elev. Ft.	A In.	D_1 In.	D_2 In.	D_3 In.	d In.	e	a Deg.	x In.	y In.	r In.

FIGURE 13.8 Twist and out-of-plumb determination for triangular towers. (*From ANSI/TIA/ EIA Standard 222-F-1996, published by Telecommunications Industry Association.*) Reproduced under written permission from Telecommunications Industry Association

Site Name:_____   Date:_____
                                  Wind:_____ Temperature:_____

Four Transit Setups Are Required, One On Each Leg Azimuth Sighting The Corresponding
Tower Leg At The Base Of The Tower To Set The True Vertical. The Deflection At Any
Point On The Tower Should Be Measured From This True Vertical, Using The Sign
Convention Below.

$$d = (D_1 + D_2 + D_3 + D_4)/4$$
$$e = (d\sqrt{2})/A$$
$$a = \arcsin(e)$$
$$x = (D_2 - D_4)/2$$
$$y = (D_1 - D_3)/2$$
$$r = \sqrt{x^2 + y^2}$$

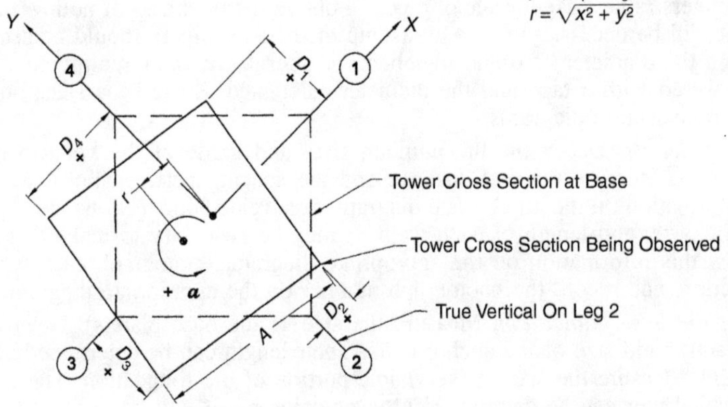

Tower Cross Section at Base

Tower Cross Section Being Observed

True Vertical On Leg 2

	Observed Mast Data				Calculated Twist			Calculated Out-of-Plumb		
Mast Elev. Ft.	A In.	D_1 In.	D_2 In.	D_3 In.	D_4 In.	d In.	e	a Deg.	x In.	y In.	r In.

FIGURE 13.9   Twist and out-of-plumb determination for square towers. (*From ANSI/TIA/EIA Standard 222-F-1996, published by Telecommunications Industry Association.*) Reproduced under written permission from Telecommunications Industry Association.

The permissible tolerance for out of plumbness is that the horizontal distance between the vertical centerlines of a leg at two locations should be no more than 0.25% of the vertical distance between the locations (or 1/400). The permissible tolerance for twist is 0.5 degrees in any 10 ft of height, with the total twist in the tower not exceeding 5 degrees.

Data for New Documentation

When the available documentation for the tower is incomplete, some or all of the information discussed below must be obtained during the inspection.

Configuration and dimensions. Measure the width at the top and bottom and the height of each section of the tower structure. Make a diagram of each section showing these dimensions and the arrangement of the members. Make a diagram of the tower as a unit showing the section numbers and the elevation of each guy level. Count the number of cables at each level, and make a diagram of their arrangement in plan and elevation. Determine the position of each guy anchorage referenced to the tower centerline at the top of the tower base foundation in both plan and elevation. Record these data on the appropriate guy diagrams. See Figs. 13.1, 13.2, and 13.3 for illustrations of these diagrams.

Member sizes. When measuring the thickness of angles, it is preferable to use micrometers rather than a scale or tape. To obtain the thickness of hollow members, it will be necessary to use ultrasonic methods. Calipers should be used to measure the diameter of round members, or alternately, the circumference can be measured with a tape and the diameter calculated. Record the sizes on the appropriate section diagrams.

Connection details. Determine the number, size, and grade of the bolts in each connection. Measure the end distance and the spacing between bolts. Record this information on the appropriate diagram. For welded connections, determine the type, size, and length of each weld. It may be necessary to make a sketch to show this information on the appropriate diagram. Photograph each typical connection, and record the photograph number on the appropriate diagram.

Details of the base foundation. Measure the size of the base plate(s). Determine the quantity and size of the anchor bolts. Their length can be determined ultrasonically. Measure the size of the visible portion of the foundation. The depth of a drilled pier can be determined ultrasonically, but it will be difficult to tell if it is belled at the bottom. The size of a spread footing and the distance to its top can be determined by probing in the area around the base. Its thickness can be determined ultrasonically or by excavating along one of its sides. Make a sketch of the foundation, and record with the base foundation diagram.

Details of the guy anchorages. Measure the sizes of the anchor shaft and the plates to which the guys are connected. Measure any visible concrete around the shaft. The depth of a drilled pier can be determined ultrasonically, but it will be difficult to tell if it is belled at the bottom. The size and depth to the top of a buried anchor block can be determined by probing in the area behind, i.e., away from the tower base, the anchor shaft's head. Its thickness can be determined ultrasonically or by excavating along one of its sides. Make a sketch of the guy anchorage, and record with the guy anchorage diagram.

Material strengths. The yield strengths of the steel components of the tower and the compressive strength of the concrete in the tower base foundation may have to be determined. See Chaps. 16 and 17 for methods of obtaining these data. Record them on the appropriate diagrams.

Soil conditions. If no geotechnical information or design capacities of the base foundation and guy anchorages are available, a subsurface investigation must be made to determine the parameters required to evaluate the stability of the tower foundation and guy anchorages. Test borings should be taken adjacent to the tower base and each guy anchorage. The actual locations should be discussed with the owner prior to making them in order to prevent any damage to underground conduits or grounding systems.

Inspection Checklist

A checklist of inspection items and the recommended frequency for performing them are shown in Table 13.1.

TABLE 13.1 Inspection Checklist

Item	1 Year	3 Years	After Major Storm
Tower structure			
Members		X	X
Connections		X	X
Condition of drain holes	X		
Insulators		X	X
Corrosion		X	
Alignment and twist		X	X
Guy system			
Strand		X	X
Connections		X	X
Insulators		X	X
Corrosion		X	
Pretensions		X	X
Foundation and guy anchorages			
Exposed structural steel		X	X
Corrosion		X	
Concrete condition		X	X
Evidence of creep or settlement		X	X
Grounding connections and conductors		X	X
Appurtenances			
Type, size and location		X	X
Connection to tower		X	X

Report

A report of the inspection should be prepared. This report should include the following:

- Authorization for and the scope of the inspection
- Date and time of the inspection
- Names of the personnel performing the inspection
- Brief description of the tower
- Diagrams of the tower and guy system
- Photographs of typical connections, the base foundation, and the guy anchorages
- A table listing all appurtenances
- Photographs of all appurtenances with cross-references to the table
- A table listing all defects and degradations found
- Photographs of all defects and degradations with cross-references to the table
- A list of any further tests that should be performed
- A list of any recommended corrective actions, along with the degree of urgency for making them

STRUCTURAL ANALYSIS

Detailed discussion of structural analysis is not intended here, but the following considerations are highlighted.

An evaluation of each component with an observed defect or degradation should be made to determine its safe load capacity. If there is a structural analysis of the tower with the appurtenances as it presently exists, the maximum calculated load in the component by that analysis can be compared with this capacity to determine if any corrective action is necessary. If no such structural analysis exists, it will be necessary to perform one.

If a structural analysis of the tower as it exists with the appurtenances is required, it should be performed by a qualified professional engineer in accordance with the applicable standards or codes. This analysis must include the following:

- Calculation of the forces acting on the tower. For wind forces, the wind must be considered as acting in any direction with respect to the tower's orientation in plan. For a triangular tower, a minimum of three directions are required. These are normal to one face in both directions and parallel to one face. For a square tower, a minimum of two directions are required. These are normal to one face and along a diagonal axis of the tower. For towers with asymmetric guy arrangements, large differences in anchor distances or elevations from the tower base, and those supporting microwave antennas, it may be necessary to consider additional directions to ensure that the maximum stresses and deflections are determined.
- Application of those forces to the tower.
- Calculation, in accordance with the principles of structural mechanics, of the displacement, sway, and twist of the tower.
- Calculation of the maximum internal stress for each member.
- Calculation of the allowable internal stress for each member.
- A comparison of the maximum calculated internal stress with the calculated allowable value.
- A comparison of the maximum calculated displacement, sway, and twist with any specified operational limitations. This is particularly important at the location of microwave antennas.

There are several different software packages suitable for the structural analysis of towers. For a self-supporting tower, a generic structural analysis program for a space frame or truss is adequate. For a guyed tower, it is necessary to use a program that can handle the nonlinear aspects of the guy system. It is important to model the tower and apply all the loads in accordance with the requirements of the program being used. When using a generic structural analysis program, this can be a very time-consuming task, as well as a likely source for error. Therefore, tower manufacturers and engineering firms that specialize in evaluating towers have customized software that greatly simplifies the modeling of the structure and determination of the capacity and adequacy of the components. While most of these are proprietary, there are some that can be purchased.

A report of the structural analysis should be prepared. This report should include the following:

- Authorization for and scope of the analysis
- History of the structure, including all known modifications
- Conditions for which the analysis was performed, including all the appurtenances considered and the sources of all data used
- Applied wind and ice loads used and the criteria for calculating and applying them
- Standards or codes used for determining allowable stresses and load, resistance, and safety factors

- Table showing the maximum calculated loads and the safe capacity of each component with any overstresses noted
- Table showing the maximum calculated deflections and their specified limits where applicable
- Conclusions with respect to the adequacy of the tower
- List of recommended corrective actions required to make the tower adequate
- Certification by the responsible professional engineer

CHAPTER 14

Foundations and Retaining Walls

ROBERT W. DAY, P.E., G.E.

INTRODUCTION

A *foundation* is defined as that part of a structure that supports the self-weight and loads acting on the structure and transmits this load to underlying soil or rock. Foundations are commonly divided into two categories: shallow and deep foundations. Table 14.1 presents a list of common types of foundations together with descriptions of their characteristics. The most frequently encountered conditions that cause damage to foundations and structures are ground movement such as settlement, expansive soil, and slope movement. Other conditions include deterioration and earthquake damage.

Visible damages to buildings owing to ground movements traditionally have been divided into three general categories: architectural, functional, and structural.[1-5]

- *Architectural damage.* This type of damage affects the appearance of the building and is usually related to minor cracks in the walls, floors, and finishes. Cracks in plaster walls greater than 0.02 in. (0.5 mm) wide and cracks in masonry walls greater than 0.04 in. (1 mm) wide are considered to be typical threshold values that would be noticed by building occupants.

461

TABLE 14.1 Common Types of Foundations

Category	Common Types	Comments
Shallow foundations	Spread footings (also called pad footings)	Spread footings are often square in plan view, are of uniform reinforced-concrete thickness, and are used to support a single-column load located directly in the center of the footing.
	Strip footings (also called wall footings)	Strip or wall footings are often used for load-bearing walls. They are usually long reinforced-concrete members of uniform width and shallow depth.
	Combined footings	Reinforced-concrete combined footings are often rectangular or trapezoidal in plan view and carry more than one column load.
	Conventional slab on grade	A continuous reinforced-concrete foundation consisting of bearing-wall footings and a slab on grade. Concrete reinforcement often consists of steel rebar in the footings and wire mesh in the concrete slab.
	Posttensioned slab on grade	A continuous posttensioned concrete foundation. Tensioning steel tendons or cables embedded within the concrete create the posttensioning effect. Common posttensioned foundations are the ribbed foundation, California slab, and PTI foundation.
	Raised wood floor	Perimeter footings that support wood beams and a floor system. Interior support is provided by pad or strip footings. There is a crawl space below the wood floor.
	Mat foundation	A large and thick reinforced-concrete foundation, often of uniform thickness, that is continuous and supports the entire structure. A mat foundation is considered to be a shallow foundation if it is constructed at or near ground surface.
Deep foundations	Driven piles	Driven piles are slender members, made of wood, steel, or precast concrete, that are driven into place by using pile-driving equipment.
	Other types of piles	There are many other types of piles, such as bored piles, cast-in-place piles, and composite piles.
	Piers	Similar to cast-in-place piles, piers are often of large diameter and contain reinforced concrete. Pier and grade beam support is often used for foundation support on expansive soil.
	Caissons	Large piers are sometimes referred to as caissons. A caisson also can be a watertight underground structure within which construction work is carried on.
	Mat or raft foundation	If a mat or raft foundation is constructed below ground surface, or if the mat or raft foundation is supported by piles or piers, then it should be considered to be a deep foundation system.
	Floating foundation	A special foundation type where the weight of the structure is balanced by the removal of soil and construction of an underground basement.
	Basement-type foundation	A common foundation for houses and other buildings in frost-prone areas. The foundation consists of perimeter footings and basement walls that support a wood floor system. The basement floor is usually a concrete slab.

Note: Shallow and deep foundations in this table are based on the depth of the soil or rock support of the foundation.

- *Functional (or serviceability) damage.* This type of damage affects use of the building. Examples include jammed doors and windows, extensively cracked and falling plaster, and tilting of walls and floors. Ground movements may cause cracking that leads to premature deterioration of materials or leaking roofs and facades.
- *Structural damage.* This type of damage affects the stability of the building. Examples include cracking or distortions to support members such as beams, columns, or load-bearing walls. This category also would include complete collapse of the structure.

In addition to visible damage to structures, there could be other conditions such as hidden (or latent) damage and monetary (ancillary) losses. A *latent condition* refers to a hidden weakness in the structure. There may be no visible evidence of damage. A latent condition could be discovered by reviewing the design calculations, by the testing of substandard construction materials, or by closely inspecting or testing a struc-

FIGURE 14.1   Niguel Summit landslide. The photograph was taken at the head of the landslide with the ground surface having dropped down and away from the house. The smaller arrow points to the corner of the house, which is suspended in midair, and the larger arrow points to the main landslide scarp.

tural member that is suspected of being defective. Latent damage is referred to as a *hidden problem*, with damage waiting to happen.[6]

An *ancillary condition* does not involve actual damage to the structure itself but rather refers to damage in the form of monetary losses. For example, the owner or contractor may file a claim because of cost overruns. Another example is the failure to complete the project on time, which could result in a lawsuit because of lost revenues.

In summary, there are five general categories of damage. Three categories (architectural, functional, and structural) refer to visual damage of the structure; the fourth category is latent damage, which is hidden damage; and the fifth category refers to monetary losses (ancillary damages).

INVESTIGATION AND ASSESSMENT

After accepting the assignment of assessing the condition of a foundation, the first step is the field investigation, which could include such items as document review, interviews, and site visits to document visible distress or deterioration, if any, of the foundation. A main objective of the field investigation is to determine the cause of foundation distress. In some cases the cause may be obvious, such as foundation damage owing to massive landslide movement, such as shown in Figs. 14.1 through 14.3. Two additional examples where the cause of damage is readily apparent are rock falls (Figs. 14.4 and 14.5) and earthquake damage (Figs. 14.6 through 14.9). In other cases it may be very difficult to determine the cause or even the existence of foundation damage, and an extensive investigation may be required. The assessment may include

FIGURE 14.2   Niguel Summit landslide. The photograph also was taken at the head of the landslide, with the ground surface having dropped down and away. The vertical distance between the two arrows is the distance that the landslide dropped downward.

FIGURE 14.3 Niguel Summit landslide. This photograph was taken at the toe of the landslide. The arrow indicates the original level of this area, but the landslide toe has uplifted both a portion of the road and the condominium.

an initial site review, document search, subsurface exploration, laboratory testing, and monitoring.

Initial Site Review

The purpose of the initial site visit is to evaluate the scope and nature of the failure. The project engineer, who may be accompanied by assistants, should perform the initial site visit. For assessment projects where the structure has had sudden damage or collapse, it is important to perform the initial site visit immediately after the assignment has been accepted. This is so because assessment details may be lost or disturbed as time goes by. At the initial site visit, it may be appropriate to perform interviews with persons knowledgeable about the damage. It is also important to take photographs of the observed damage and collect samples if it is likely that the samples will be lost or destroyed before the investigation can be completed.

Numerous photographs should be taken of the damaged structure. It may be appropriate to take a professional photographer who has the proper equipment for long-range and close-up photographs at the site. Sequentially numbering and marking the

FIGURE 14.4   Rock fall. This photograph shows a large rock that landed on the roof of a house.

FIGURE 14.5   Rock fall. For the rock fall shown in Fig. 14.4, this photograph shows the interior damage to the house. Note that the door frames were originally rectangular but became distorted on impact from the rock fall.

FIGURE 14.6 Earthquake damage. This building collapsed during the California Northridge earthquake on January 17, 1994.

FIGURE 14.7 Earthquake damage. Same building as shown in Fig. 14.6. This photograph shows a view of the inside of the collapsed first-floor parking garage. The arrows point to the columns that had inadequate shear resistance and thus collapsed during the ground shaking.

FIGURE 14.8   Earthquake damage. This building suffered a liquefaction-induced bearing-capacity failure during the Izmit earthquake in Turkey on August 17, 1999. (*Photograph from the Izmit Collection, EERC, University of California, Berkeley.*)

location of the photographs on a site-plan sketch may help to refresh recollection of where the pictures were taken. In the excitement of the first visit, there is a natural tendency to forget about photographs. For example, I was involved in one project where an airport runway was damaged by a flood. The damage consisted of sand deposits on both sides of runway cracks that were attributed to a loss of runway base material. Although eight engineers visited the site and observed the damage, no photographs or samples were taken of the sand deposits along the runway cracks. The next day the maintenance crew swept the runway, and all evidence of the sand deposits was lost. Without actual samples or photographs of the sand deposits, it was more difficult to convince the insurance company that the runway had been damaged during the flood.

Document Review

Table 14.2 presents a list of typical documents that may need to be reviewed in order to assess the damage, if any, to a foundation. The following is a brief summary of these types of documents:

- *Project reports and plans.* The reports and plans that were generated during the design and construction of the project may need to be reviewed. The reports and plans can provide specific information on the history, design, and construction of the project. These documents also may provide information on maintenance or alterations at the site.

FIGURE 14.9   Earthquake damage. The photograph shows the Kawagishi-cho apartment buildings located in Niigata, Japan. The buildings suffered liquefaction-induced bearing-capacity failures during the Niigata earthquake on June 16, 1964. (*Photograph from the Godden Collection, EERC, University of California, Berkeley.*)

- *Building codes.* A copy of the applicable building code in effect at the time of construction should be reviewed.
- *Technical documents.* During the course of the damage assessment, it may be necessary to check reference materials, such as geologic maps or aerial photographs. Other useful technical documents can include journal articles that may describe a failure similar to the one under investigation. For example, the ASCE *Journal of Performance of Constructed Facilities* deals specifically with construction-related failures or deterioration.

Manometer Survey

A manometer survey is also referred to as a *floor-level survey.* It is a nondestructive means of finding damage or flaws in foundations. A manometer survey is used commonly to determine the relative elevation of a concrete slab on grade or other foundation element. The survey consists of taking elevations at relatively close intervals throughout the surface of the floor slab. These elevation points are then contoured, much like a topographic map, to provide a graphic rendition of the deformation condition of the foundation. Soil movement can cause displacement of the foundation, and a manometer survey can detect this deformation. For example, if one side or corner of a concrete slab is significantly lower than the rest of the slab, this could indicate settlement or slope movement in that area. Likewise, if the center of the slab is bulged upward, it would be detected by the manometer survey, and this could indicate expansive soil. Other soil phenomena or slab conditions also can be detected by manometer survey. For example, close deformation contours indicate a high angular distortion,

TABLE 14.2 Typical Documents That May Need to Be Reviewed for the Damage Assessment

Project Phase	Type of Documents
Design	Design reports such as geotechnical reports, planning reports, feasibility studies, etc.
	Design calculations and analyses
	Computer programs used for the design of the project
	Design specifications
	Applicable building codes
	Shop drawings and design plans
Construction	Construction reports such as inspection reports, field memos, laboratory test reports, mill certificates, etc.
	Contract documents (contract agreements, provisions, etc.)
	Construction specifications
	Project payment data or certificates
	Field change orders
	Information bulletins used during construction
	Project correspondence between different parties
	As-built drawings, such as as-built grading plans, foundation plans, etc.
	Photographs or videos
	Building department permits and certificate of occupancy
Postconstruction	Postconstruction reports such as maintenance reports, modification documents, reports on specific problems, repair reports, etc.
	Photographs or videos
Technical data	Available records such as weather reports, seismic activity, etc.
	Reference materials such as geologic and topographic maps, aerial photographs, etc.
	Technical publications such as journal articles that describe similar failures

which in many cases corresponds to the location of foundation cracks. Throughout the chapter, examples of manometer surveys that depict foundation displacement owing to different soil mechanisms will be presented.

Subsurface Exploration

The subsurface exploration can be accomplished by excavating test pits and borings. Table 14.3[7-10] (ref. 7 is based on the work by ASTM) summarizes the boring, core drilling, sampling, and other exploratory techniques that can be used by the engineer. The test pits or borings are used to determine the thickness of soil and rock strata, estimate the depth to groundwater, obtain soil or rock specimens, and perform field tests such as sand cone tests or standard penetration tests (SPT). The Unified Soil Classification System (USCS) can be used to classify the soil exposed in the borings or test pits.[11] The subsurface exploration and field sampling should be performed in accordance with standard procedures, such as those specified by the American Society for Testing and Materials (ASTM)[12-14] or other recognized sources (e.g., refs. 15 to 18).

Another common type of destructive testing is the coring of foundations. By coring the foundation, the thickness of concrete, reinforcement condition, and any deterioration can be observed. Also, soil samples can be obtained directly beneath the foundation. Figure 14.10 shows a portion of a foundation that has been chipped away to reveal that the welded wire mesh actually was placed below the concrete rather than positioned within the foundation.

TABLE 14.3 Boring, Core Drilling, Sampling, and Other Exploratory Techniques

Method (1)	Procedure (2)	Type of Sample (3)	Applications (4)	Limitations (5)
Auger boring, ASTM D1452	Dry hole drilled with hand or power auger; samples preferably recovered from auger flutes	Auger cuttings, disturbed, ground up, partially dried from drill heat in hard materials	In soil and soft rock; to identify geologic units and water content above water table	Soil and rock stratification destroyed; sample mixed with water below the water table
Test boring, ASTM D1586	Hole drilled with auger or rotary drill; at intervals samples taken 36 mm ID and 50 mm OD driven 0.45 m in three 150-mm increments by 64-kg hammer falling 0.76 m; hydrostatic balance of fluid maintained below water level	Intact but partially disturbed (number of hammer blows for second plus third increment of driving is standard penetration resistance or N)	To identify soil or soft rock; to determine water content; in classification tests and crude shear test of sample (N value a crude index to density of cohesionless soil and undrained shear strength of cohesive soil)	Gaps between samples, 30 to 120 cm; sample too distorted for accurate shear and consolidation tests; sample limited by gravel; N-value subject to variations depending on free fall of hammer
Test boring of large samples	50 to 75 mm ID and 63 to 89 mm OD samplers driven by hammers up to 160 kg	Intact but partially disturbed (no. of hammer blows for 2nd plus 3rd increment of driving is penetration resistance)	In gravelly soils	Sample limited by larger gravel
Test boring through hollow stem auger	Hole advanced by hollow stem auger; soil sampled below auger as in test boring above	Intact but partially disturbed (no. of hammer blows for 2nd plus 3rd increment of driving is N value)	In gravelly soils (not well adapted to harder soils or soft rock)	Sample limited by larger gravel; maintaining hydrostatic balance in hole below water table is difficult
Rotary coring of soil or soft rock	Outer tube with teeth rotated; soil protected and held stationary inner tube; cuttings flushed upward by drill fluid (examples, Denison, Pitcher, and Acker samplers)	Relatively undisturbed sample, 50 to 200 mm wide and 0.3 to 1.5 m long in liner tube	In firm to stiff cohesive soils and soft but coherent rock	Sample may twist in soft clays; sampling loose sand below water table is difficult; success in gravel seldom occurs
Rotary coring of swelling clay, soft rock	Similar to rotary coring of rock; swelling core retained by third inner plastic liner	Soil cylinder 28.5 to 53.2 mm wide and 600 to 1500 mm long encased in plastic tube	In soils and soft rocks that swell or disintegrate rapidly in air (protected by plastic tube)	Sample smaller; equipment more complex

472

TABLE 14.3 *(Continued)*

Method (1)	Procedure (2)	Type of Sample (3)	Applications (4)	Limitations (5)
Rotary coring of rock, ASTM D2113	Outer tube with diamond bit on lower end rotated to cut annular hole in rock; core protected by stationary inner tube; cuttings flushed upward by drill fluid	Rock cylinder 22 to 100 mm wide and as long as 6 m depending on rock soundness	To obtain continuous core in sound rock (% of core recovered depends on fractures, rock variability, equipment, and driller skill)	Core lost in fracture or variable rock; blockage prevents drilling in badly fractured rock; dip of bedding and joint evident but not strike
Rotary coring of rock, oriented core	Similar to rotary coring of rock above; continuous grooves scribed on rock core with compass direction	Rock cylinder, typically 54 mm wide and 1.5 m long with compass orientation	To determine strike of joints and bedding	Method may not be effective in fractured rock
Rotary coring of rock, wire line	Outer tube with diamond bit on lower end rotated to cut annular hole in rock; core protected by stationary inner tube; cuttings flushed upward by drill fluid; core and stationary inner tube retrieved from outer core barrel by lifting device or "overshot" suspended on thin cable (wire line) through special large dia. drill rods and outer core barrel	Rock cylinder 36.5 to 85 mm wide and 1.5 to 4.6 m long	To recover core better in fractured rock, which has less tendency for caving during core removal; to obtain much faster cycle of core recovery and resumption of drilling in deep holes	Same as ASTM D2113 but to lesser degree
Rotary coring of rock, integral sampling method	22-mm hole drilled for length of proposed core; steel rod grouted into hole; core drilled around grouted rod with 100 to 150 mm rock coring drill (same as for ASTM D2113)	Continuous core reinforced by grouted steel rod	To obtain continuous core in badly fractured, soft, or weathered rock in which recovery is low by ASTM D2113	Grout may not adhere in some badly weathered rock; fractures sometimes cause drift of diamond bit and cutting rod
Thin-walled tube, ASTM D1587	75- to 1250-mm thin-walled tube forced into soil with static force (or driven in soft rock); retention of sample helped by drilling mud	Relatively undisturbed sample, length 10 to 20 diameters	In soft to firm clays, short (5 diameters) samples of stiff cohesive soil, soft rock and, with aid of drilling mud, in firm to dense sands	Cutting edge wrinkled by gravel; samples lost in loose sand or very soft clay below water table; more disturbance occurs if driven with hammer

tube, fixed piston	which has internal piston controlled by rod and keeps loose cuttings from tube, remains stationary while outer thin wall tube forced ahead into soil; sample in tube is held in tube by aid of piston	length 10 to 20 diameters	soft clays (drilling mud aids in holding samples in loose sand below water table)	Samples sometimes damaged by coarse sand and fine gravel
Swedish foil	Samples surrounded by thin strips of stainless steel, stored above cutter, to prevent contact of soil with tube as it is forced into soil	Continuous samples 50 mm wide and as long as 12 m	In soft, sensitive clays	Misleading in gravel or loose saturated fine cohesionless soils
Dynamic sounding	Enlarged disposable point on end of rod driven by weight falling fixed distance in increments of 100 to 300 mm	None	To identify significant differences in soil strength or density	Stopped by gravel or hard seams
Static penetration	Enlarged cone, 36 mm diameter and 60-degree angle forced into soil; force measured at regular intervals	None	To identify significant differences in soil strength or density; to identify soil by resistance of friction sleeve	
Borehole camera	Inside of core hole viewed by circular photograph or scan	Visual representation	To examine stratification, fractures, and cavities in hole walls	Best above water table or when hole can be stabilized by clear water
Pits and trenches	Pit or trench excavated to expose soils and rocks	Chunks cut from walls of trench; size not limited	To determine structure of complex formations; to obtain samples of thin critical seams such as failure surface	Moving excavation equipment to site, stabilizing excavation walls, and controlling groundwater may be difficult
Rotary or cable tool well drill	Toothed cutter rotated or chisel bit pounded and churned	Ground	To penetrate boulders, coarse gravel; to identify hardness from drilling rates	Identifying soils or rocks difficult
Percussion drilling (jack hammer or air track)	Impact drill used; cuttings removed by compressed air	Rock dust	To locate rock, soft seams, or cavities in sound rock	Drill becomes plugged by wet soil

Source: Adapted with permission from Sowers and Royster.[7]

FIGURE 14.10 A jackhammer has been used to chip away the concrete foundation and reveal that the welded wire mesh actually was placed below the concrete rather than within the concrete.

There can be numerous other types of subsurface explorations performed by the engineer. Examples include field load tests of the foundation or underlying soil, cone penetration testing of the soil, and in-place strength testing, such as determining the shear strength of in-place soil or rock.

Laboratory Testing

Soil, groundwater, or foundation material samples recovered from the site visits and subsurface explorations can be taken to the laboratory for testing. Laboratory tests on soils are used commonly to determine the classification, moisture content, density, index properties, shear strength, compressibility, and hydraulic conductivity of the soil. Tests on foundation samples include standard compression tests and petrographic analysis.

Usually at the time of the laboratory testing, the engineer will have developed one or more hypotheses concerning the cause of failure. The objective of the laboratory testing should be to further investigate these failure hypotheses. It is important that the engineer develop a logical laboratory testing program with this objective in mind. The laboratory tests should be performed in accordance with standard procedures, such as those recommended by the ASTM or those procedures listed in standard textbooks or specification manuals (e.g., refs. 19 to 23).

For some foundation assessments, it may be important to determine the future potential for soil movement and damage. In these cases, the laboratory testing should model future expected conditions so that the amount of movement or stability of the ground can be analyzed.

Monitoring

Many types of monitoring devices can be used to assess and/or quantify the damage to a foundation. Some of the more common monitoring devices are as follows:

- *Inclinometers.* The horizontal movement preceding or during, the movement of slopes can be investigated by successive surveys of the shape and position of flexible vertical casings installed in the ground. The surveys are performed by lowering an instrument in flexible vertical casings that is capable of measuring the deviation of the instrument from the vertical. An initial survey (base reading) is performed, and then successive surveys are compared with the base reading to obtain the horizontal movement of the slope since the time of the base reading.
- *Piezometers.* Piezometers are installed routinely to monitor pore water pressures in the ground. Several different types are available commercially, including borehole, embankment, or push-in piezometers. In their simplest form, piezometers can consist of a standpipe that can be used to monitor groundwater levels and obtain groundwater samples.
- *Settlement monuments or cells.* Settlement monuments or settlement cells can be used to monitor settlement or heave of the foundation. More advanced equipment includes settlement systems installed in borings that not only can measure total settlement but also can indicate the incremental settlement at different depths.
- *Crack pins.* A simple method to measure the widening of a crack in a concrete foundation is to install crack pins on both sides of the crack. By periodically measuring the distance between the pins, the amount of opening or closing of the crack can be determined. Other crack monitoring devices are available commercially, such as the Avongard crack monitoring device.
- *Other monitoring devices.* Many other types of monitoring devices can be used to assess foundation damage. Some commercially available devices include pressure and load cells, boreholes and tape extensometers, soil strain meters, beam sensors and tiltmeters, and strain gauges.[24]

TYPES OF FOUNDATION PROBLEMS

Many conditions can cause damage to foundations. Some of the more common causes of foundation damage are settlement, expansive soil, slope movement, and concrete deterioration.

Settlement

Settlement can be defined as the permanent downward displacement of a foundation. Common causes of settlement are consolidation of soft and/or organic soil, settlement from uncontrolled or deep fill, and the development of limestone cavities or sinkholes.[25,26] Foundations also can experience settlement owing to natural disasters, such as earthquakes or the undermining of the foundation resulting from floods. There also have been reports of widespread ground subsidence caused by the extraction of oil or groundwater, as well as the collapse of underground mines and tunnels.

There are two general types of settlement: that due to the weight of the structure and those due to secondary influences.

Settlement due directly to the weight of the structure. The first type of settlement is caused directly by the weight of the structure. For example, the weight of a building may cause compression of an underlying sand deposit or consolidation of an underlying clay layer.

Settlement owing to secondary influences. The second general type of settlement of a building is caused by secondary influences, which may develop at a time long after the completion of the structure. This type of settlement is not caused directly by the weight of the structure. For example, the foundation may settle as water infiltrates the ground and causes unstable soils to collapse (i.e., collapsible soil). The foundation also may settle owing to yielding of adjacent excavations or the collapse of limestone cavities and underground mines and tunnels. Other causes of settlement that would be included in this category are soil erosion or collapse caused by water main breaks, natural disasters such as settlement caused by earthquakes, and undermining of the foundation from floods.

Subsidence usually is defined as a sinking down of a large area of the ground surface. Subsidence could be caused by the extraction of oil or groundwater, which leads to a compression of the underlying porous soil or rock structure. Since subsidence is due to a secondary influence (extraction of oil or groundwater), its effect on the structure would be included in the second general type of settlement described above.

Most structures can tolerate a certain amount of settlement. The allowable settlement depends on many different factors, such as the following[27]:

- *The type of construction.* For example, wood-frame buildings with wood siding would be much more tolerant than unreinforced brick buildings.
- *The use of the structure.* Even small cracks in a house might be considered unacceptable, whereas much larger cracks in an industrial building might not even be noticed.
- *The presence of sensitive finishes.* Tile or other sensitive finishes are much less tolerant of movements.
- *The rigidity of the structure.* If a footing beneath part of a very rigid structure settles more than other foundation elements, the structure will transfer some of the load away from the footing. However, footings beneath flexible structures generally settle more before any significant load transfer occurs. Therefore, a rigid structure will have less differential settlement than a flexible one but is more prone to show cracks.

It has been stated that the acceptable settlement for most structures, especially buildings, will be governed by aesthetic and serviceability requirements, not structural requirements.[27] Unsightly cracks, jamming doors and windows, and other similar problems will develop long before the integrity of the structure is in danger.

In terms of the damage assessment of foundations subjected to settlement, there often will be a distinct pattern of foundation displacement and corresponding building cracking. For example, Fig. 14.11 presents a manometer survey for a house that has a concrete slab on grade. Most of the house is relatively level, but at the southwest corner the foundation has settled about 3.5 in. Associated with the foundation settlement are foundation cracks, wallboard cracks, and exterior stucco cracks. For example, Fig. 14.12 shows a foundation crack located at the southwest corner of a house. Consistent with the settlement of the southwest corner of the house, there was differential movement on the two sides of the cracks, with the part of the foundation closer to the corner (i.e., left of the crack in the picture) being down relative to that part of the foundation further away from the corner. The settlement at the corner of the house also was visible from the exterior, where the roof actually was observed to deform downward, as shown in Fig. 14.13. The cause of settlement of the southwest corner

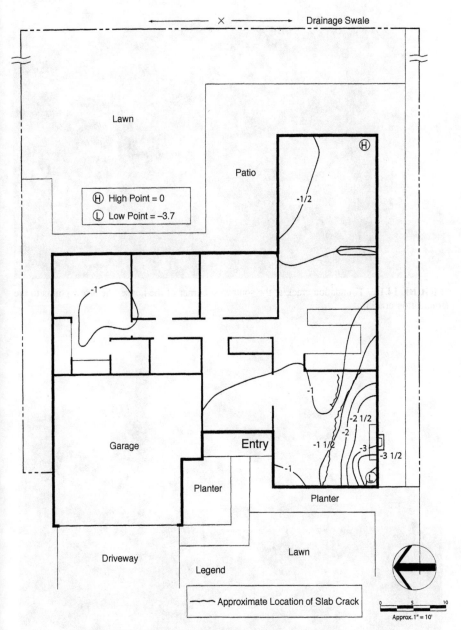

FIGURE 14.11   House foundation experiencing settlement at one corner.

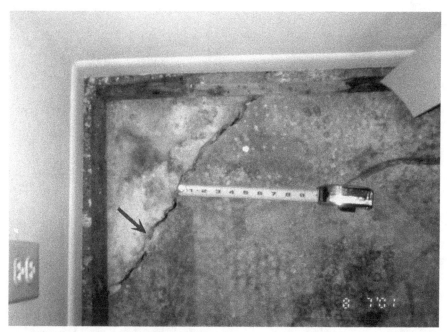

FIGURE 14.12 Foundation crack at the southwest corner of the house (the arrow points to the foundation crack).

FIGURE 14.13 Exterior view of the southwest corner of the house (the arrow points to the corner of the house that is settling).

of the house was believed to be the construction of large planters adjacent to the west wall, which caused compression of the underlying soft alluvial soils.

In damage assessment of structures subjected to settlement, it is often observed that the cracks in walls are wider at the top than the bottom, due to the downward rotation of the settling portion of the wall.

Expansive Soil

Expansive soils are a worldwide problem, causing extensive damage to civil engineering structures. Although most states have expansive soil, certain areas of the United States, such as Colorado, Texas, Wyoming, and California, are more susceptible to damage from expansive soils than others.[28] These areas have large surface deposits of clay and have climates characterized by alternating periods of rainfall and drought.

The engineer often can identify desiccated clay by visual inspection because of the numerous ground surface cracks, such as shown in Fig. 14.14. Examples of damage mechanisms owing to expansive soil are as follows:

Moisture migration beneath slab-on-grade foundations. Figure 14.15 shows the common expansive soil damage progression caused by moisture migration underneath a slab-on-grade foundation. Damage often develops first when the foundation is subjected to edge lift. For example, Fig. 14.16 shows the manometer survey of a foundation that has experienced edge lift at the front entry of the house. Typical damage often consists of interior wallboard cracks, ceiling cracks, and exterior cracks that often develop at window corners, such as shown in Fig. 14.17. Eventually, the moisture will migrate underneath the foundation, resulting in center-lift (Fig. 14.15).

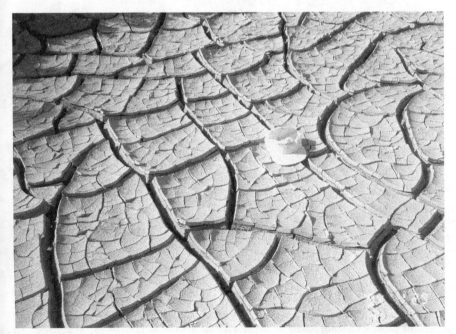

FIGURE 14.14 Deposit of desiccated clay located in Death Valley, California. Note that the hat in the center of the photograph provides a scale for the size of the desiccation cracks.

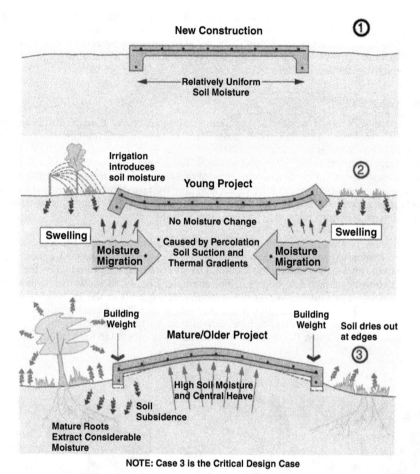

FIGURE 14.15 Expansive soil damage progression. (1) New construction of the foundation on expansive soil, which often has relatively uniform soil moisture. (2) Water from irrigation and rainfall migrates underneath the perimeter of the foundation, resulting in edge lift. (3) For older projects, the water eventually will migrate beneath the center of the foundation, resulting in center lift.

Uplifting of foundation elements. Figure 14.18 is a view from the crawl space of a raised wood floor foundation. At this site, expansive soil has uplifted the lightly loaded concrete pad footing, causing distortion of the wood post supporting the floor beam. Note the wet condition in the crawl space.

Differential movement of flatwork: Flatwork can be defined as appurtenant structure that surrounds a building or house such as concrete walkways, patios, driveways, and pool decks. It is the lightly loaded structures, such as pavements or lightly loaded foundations, that are commonly damaged by expansive soil. Because flatwork usually only supports its own weight, it can be especially susceptible to expansive soil-related damage, such as shown in Fig. 14.19. The arrows in Fig. 14.19 indicate the amount of uplift of the lightly loaded exterior sidewalk

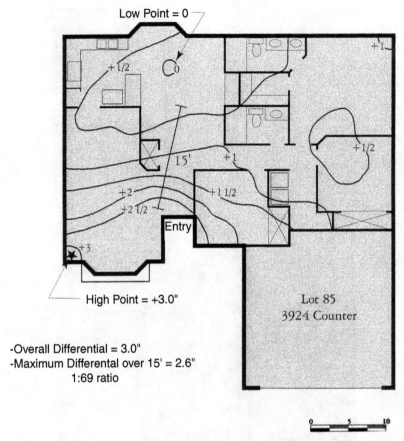

Low Point = 0

+1/2   C 0

+1

+1/2

15'   +1

+2

+1 1/2

+2 1/2

Entry

+3

High Point = +3.0"

Lot 85
3924 Counter

-Overall Differential = 3.0"
-Maximum Differential over 15' = 2.6"
    1:69 ratio

0   5   10

FIGURE 14.16   House foundation experiencing expansive soil uplift at the entryway.

relative to the heavily loaded exterior tilt-up wall of the building. The sidewalk heaved to such an extent that the door could not be opened, and the door had to be rehung so that it opened into rather than out of the building.

Additional examples are shown in Figs. 14.20 and 14.21. In Fig. 14.20, there had been expansive soil uplift that cracked the concrete driveway. The expansive soil uplift tends to produce a distinct crack pattern, which has been termed a "spider" or X type of crack pattern. In Fig 14.21, expansive soils uplifted the concrete patio, causing differential movement and cracking.

Slope Movement

Slope movement is often divided into two different categories: falls and slides. Rock falls are distinguished by their relatively free-falling nature, where the rocks will detach themselves from a cliff, steep slope, cave, arch, or tunnel.[29] An example of damage caused by a rock fall is shown in Figs. 14.4 and 14.5.

Slides are different from falls in that there is shear displacement along a distinct failure (or slip) surface. Common types of slides are rock block slides, soil slumps,

FIGURE 14.17   Typical damage caused by expansive soil heave. Cracks often develop at the corners of windows or doors.

and landslides. Figures 14.1 through 14.3 show damage caused by the Niguel Summit Landslide in California. Slides often develop when there is an increase in the driving forces, such as owing to a rise in the groundwater table or the placement of fill at the top of the slope. Slides also can develop when the resisting forces are reduced, such as when the toe of a slope is removed during grading operations or by erosion.

Another common cause of slope movement is the seismic shaking from earthquakes.[30] For example, Fig. 14.22 shows the Government Hill School, located in Anchorage, Alaska, which was severely damaged by earthquake-induced landslide movement. The school building straddled the head of the landslide. When the landslide moved during the earthquake, it caused both lateral and vertical displacement of the building.

These and other types of earthquake-induced slope movements are listed in Tables 14.4 and 14.5.[31,32] The minimum slope angle listed in column 4 of Tables 14.4 and 14.5 refers to the minimum slope inclination that usually will initiate a specific type of earthquake-induced slope movement. Note that for an earthquake-induced rock fall, the slope inclination typically must be 40 degrees or greater, whereas for liquefaction-

FIGURE 14.18   Expansive soil damage of a raised wood floor foundation.

FIGURE 14.19   Expansive soil uplift of a sidewalk.

FIGURE 14.20 Expansive soils that have caused driveway cracking.

FIGURE 14.21 Expansive soils that have caused cracking and differential movement of a concrete patio.

FIGURE 14.22   Close-up view of damage to the Government Hill School located at the head of a landslide caused by the Prince William Sound earthquake in Alaska on March 27, 1964. (*Photograph from the Steinbrugge Collection, EERC, University of California, Berkeley.*)

induced lateral spreading, the earthquake-induced movement can occur on an essentially flat surface (i.e., minimum angle of inclination = 0.3 degrees).

Deterioration and Shrinkage

Concrete foundations also can be susceptible to damage from deterioration since construction and from excessive shrinkage of the concrete during construction. For example, if the concrete mix had a high water content, then large shrinkage cracks can develop. Figure 14.23 shows an example of a crack in a concrete floor that developed owing to shrinkage. Cracked and porous concrete can enhance moisture migration through the foundation, which, if it is a floor, can cause damage to floor coverings (Fig. 14.24) and growth of mold on the surface (Fig. 14.25).

A common cause of deterioration of concrete foundations is sulfate attack. In this case, there is a chemical reaction between the sulfate in the soil or groundwater and the cement paste.[33] Figures 14.26 through 14.28 show damage to concrete flatwork caused by the sulfate attack.

RETAINING WALLS

A *retaining wall* is defined as a structure whose primary purpose is to provide lateral support for soil or rock. In some cases a retaining wall also may support vertical loads.

TABLE 14.4 Types of Earthquake-Induced Slope Movement in Rock

Main Type of Slope Movement	Subdivisions	Material Type	Minimum Slope Inclination	Comments
Falls	Rock falls	Rocks weakly cemented, intensely fractured, or weathered; contain conspicuous planes of weakness dipping out of slope or contain boulders in a weak matrix.	40° (1.7:1)	Particularly common near ridge crests and on spurs, ledges, artificially cut slops, and slopes undercut by active erosion.
Slides	Rock slides	Rocks weakly cemented, intensely fractured, or weathered; contain conspicuous planes of weakness dipping out of slope or contain boulders in a weak matrix.	35° (1.4:1)	Particularly common in hillside flutes and channels, on artificially cut slopes, and on slopes undercut by active erosion. Occasionally reactivate preexisting rock slide deposits.
	Rock avalanches	Rocks intensely fractured and exhibiting one of the following properties: significant weathering, planes of weakness dipping out of slope, weak cementation, or evidence of previous landsliding.	25° (2.1:1)	Usually restricted to slopes of greater than 500 feet (150 m) relief that have been undercut by erosion. May be accompanied by a blast of air that can knock down trees and structures beyond the limits of the deposited debris.
	Rock slumps	Intensely fractured rocks, preexisting rock slump deposits, shale, and other rocks containing layers of weakly cemented or intensely weathered material.	15° (3.7:1)	Often circular or curved slip surface as compared to a planar slip surface for block slides.
	Rock block slides	Rocks having conspicuous bedding planes or similar planes of weakness dipping out of slopes.	15° (3.7:1)	Similar to rock slides.

486

Main Type of Slope Movement	Subdivisions	Material Type	Minimum Slope Inclination	Comments
Falls	Soil falls	Granular soils that are slightly cemented or contain clay binder.	40° (1.7:1)	Particularly common on stream banks, terrace faces, coastal bluffs, and artificially cut slopes.
Slides	Soil avalanches	Loose, unsaturated sands.	25° (2.1:1)	Occasionally reactivation of preexisting soil avalanche deposits.
	Disrupted soil slides	Loose, unsaturated sands.	15° (3.7:1)	Often described as "running soil" or "running ground."
Slides	Soil slumps	Loose, partly to completely saturated sand or silt; uncompacted or poorly compacted manmade fill composed of sand, silt, or clay, preexisting soil slump deposits.	10° (5.7:1)	Particularly common on embankments built on soft, saturated foundation materials, in hillside cut-and-fill areas, and on river and coastal flood plains.
Slides	Soil block slides	Loose, partly or completely saturated sand or silt; uncompacted or slightly compacted manmade fill composed of sand or silt, bluffs containing horizontal or subhorizontal layers of loose, saturated sand or silt.	5° (11:1)	Particularly common in areas of preexisting landslides along river and coastal flood plains, and on embankments built of soft, saturated foundation materials.
Flow slides and lateral spreading	Slow earth flows	Stiff, partly to completely saturated clay and preexisting earth-flow deposits.	10° (5.7:1)	An example would be sensitive clay.
	Flow slides	Saturated, uncompacted or slightly compacted manmade fill composed of sand or sandy silt (including hydraulic fill earth dams and tailings dams); loose, saturated granular soils.	2.3° (25:1)	Includes debris flows that typically originate in hollows at heads of streams and adjacent hillsides; typically travel at tens of miles per hour or more and may cause damage miles from the source area.
	Subaqueous flows	Loose, saturated granular soils.	0.5° (110:1)	Particularly common on delta margins.
	Lateral spreading	Loose, partly or completely saturated silt or sand, uncompacted or slightly compacted manmade fill composed of sand.	0.3° (190:1)	Particularly common on river and coastal flood plains, embankments built on soft, saturated foundation materials, delta margins, sand spits, alluvial fans, lake shores, and beaches.

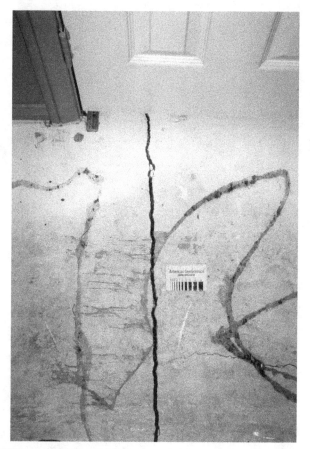

FIGURE 14.23   Foundation crack caused by the shrinkage of concrete.

Examples include basement walls and certain types of bridge abutments. They may be constructed of reinforced concrete, masonry, stone, or timber.

Some of the more common types of retaining walls are gravity walls, counterfort walls, cantilevered walls, and crib walls. Gravity retaining walls are built routinely of plain concrete or stone, and the wall depends primarily on its massive weight to resist failure from overturning and sliding. Counterfort walls consist of a footing, a wall stem, and intermittent vertical ribs (called *counterforts*) that tie the footing and wall stem together. Crib walls consist of interlocking concrete members that form cells, which are then filled with compacted soil.

Although mechanically stabilized earth retaining walls (MSE walls) have become popular in the past decade, cantilever retaining walls are still probably the most common type of retaining structure. There are many different types of cantilevered walls, with the common features being a footing that supports the vertical wall stem. Typical cantilevered walls are T-shaped, L-shaped, or reverse-L-shaped.[34]

To prevent the buildup of hydrostatic water pressure on the back of a retaining wall, clean granular material (no silt or clay) is the standard recommendation for backfill material. Import granular backfill generally has a more predictable behavior

FIGURE 14.24 Damage to wood flooring caused by excessive moisture migration through the concrete foundation. The arrow points to a moist stain, and the asterisk indicates the upward warping of the wood floor.

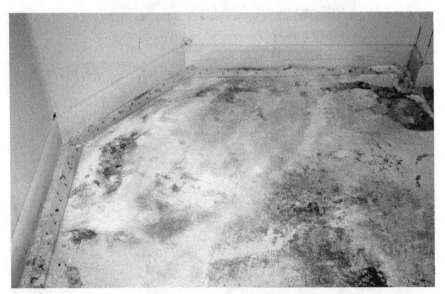

FIGURE 14.25 Salt deposits and the growth of mold caused by moisture migration through a concrete floor slab. Note that the carpets have been removed.

FIGURE 14.26   Sulfate attack of a concrete sidewalk.

FIGURE 14.27   Sulfate attack of a concrete patio.

FIGURE 14.28 Expansion of concrete owing to sulfate attack, which has caused the upward movement of the concrete.

FIGURE 14.29 Collapse of a retaining wall that has a clay backfill.

FIGURE 14.30 The liquefaction-induced retaining wall movement was caused by the Kobe Earthquake on January 17, 1995. Both lateral spreading fissures and sporadic sand boils were observed behind the retaining wall. The site is near the pier of Nishihomiya Bridge, which previously supported the collapsed expressway section. (*Photograph from the Kobe Geotechnical Collection, EERC, University of California, Berkeley.*)

in terms of earth pressure exerted on the wall. A backdrain system often is constructed at the heel of the wall to intercept and dispose of any water seepage in the granular backfill. Probably the most common reason for a retaining wall failure is the use of an inferior backfill material. For example, Fig. 14.29 shows the collapse of a retaining wall owing to the increased pressure from a clay backfill. For many retaining and basement walls, especially if the clay backfill is compacted below optimal moisture content, seepage of water into the clay backfill causes horizontal swelling pressures well in excess of the active earth pressure values, and the pressure may even exceed at-rest values. Besides the swelling pressure induced by the clayey soil, there also can be groundwater or perched water pressure on the retaining or basement wall because of poor drainage of clayey soils. In those cases where the retaining wall has failed, such as shown in Fig. 14.29, it is common practice to excavate test pits behind the wall to evaluate the earth pressures acting on the wall as compared with those values assumed during the design stage.

The ground shaking associated with earthquakes also often damages retaining walls. For example, Fig. 14.30 shows liquefaction-induced retaining wall movement that was caused by the Kobe earthquake on January 17, 1995. Both lateral spreading fissures and sporadic sand boils were observed behind this retaining wall.

REFERENCES

1. Skempton, A. W., and D. H. MacDonald (1956). "The Allowable Settlement of Buildings," in *Proceedings of the Institution of Civil Engineers,* Part III. Institution of Civil Engineers, London, England, no. 5, pp. 727–768.

2. Bromhead, E. N. (1984). "Slopes and Embankments," in Attewelb P. B., and Tylor, R. K. (eds.), *Ground Movements and Their Effects on Structures.* Surrey University Press, London, p. 63.

3. Boscardin, M. D., and E. J. Cording (1989). "Building Response to Excavation-Induced Settlement," *Journal of Geotechnical Engineering,* ASCE, **115**(1):1–21.

4. Feld, J., and K. L. Carper (1997). *Construction Failure,* 2d ed., Wiley, New York.

5. Burland, J. B., B. B. Broms, and V. F. B. de Mello (1997). "Behavior of Foundations and Structures: State of the Art Report," in *Proceedings of the 9th International Conference on Soil Mechanics and Foundation Engineering.* Japanese Geotechnical Society, Tokyo, pp. 495–546.

6. Greenspan, H. F., J. A. O'Kon, K. J. Beasley, and J. S. Ward (1989). *Guidelines for Failure Investigation.* ASCE, New York.

7. Sowers, G. F., and D. L. Royster (1978). "Field Investigation," in Schuster, R. L., and Krizek, R. J. (eds.), *Landslides, Analysis and Control,* Special Report 176. Transportation Research Board, National Academy of Sciences, Washington, pp. 81–111.

8. Lambe, T. W. (1951). *Soil Testing for Engineers.* Wiley, New York, p. 165.

9. Sanglerat, G. (1972). *The Penetrometer and Soil Exploration.* Elsevier Scientific Publishing, New York.

10. Sowers, G. B., and G. F. Sowers (1970). *Introductory Soil Mechanics and Foundations,* 3d ed., Macmillan, New York.

11. Casagrande, A. (1948). "Classification and Identification of Soils," *Transactions,* ASCE, **113:** 901–930.

12. American Society for Testing and Materials (ASTM) (1970). *Special Procedures for Testing Soil and Rock for Engineering Purposes.* ASTM Special Technical Publication 479, Philadelphia.

13. American Society for Testing and Materials (ASTM) (1971). *Sampling of Soil and Rock.* ASTM Special Technical Publication 483, Philadelphia.

14. American Society for Testing and Materials (ASTM) (2000). *Annual Book of ASTM Standards,* Vol. 04.08: *Soil and Rock,* Vol. 1. Standard No. D 420-98, *Standard Guide to Site*

Characterization for Engineering Design and Construction Purposes. West Conshohocken, PA.

15. Hvorslev, M. J. (1949). *Subsurface Exploration and Sampling of Soils for Civil Engineering Purposes.* Waterways Experiment Station, Vicksburg, MI.

16. American Society of Civil Engineers (ASCE) (1972). *Subsurface Investigation for Design and Construction of Foundations of Buildings.* Task Committee for Foundation Design Manual, Part I, *Journal of the Soil Mechanics and Foundations Division,* ASCE, 98(SM5):481–490; Part II, (SM6):557–578; Parts III and IV, (SM7):749–764.

17. American Society of Civil Engineers (ASCE) (1976). *Subsurface Investigation for Design and Construction of Foundations of Buildings.* Manual No. 56, American Society of Civil Engineers, New York.

18. American Society of Civil Engineers (ASCE) (1978). "Site Characterization and Exploration," in Dowding, C. H. (ed.), *Proceedings of the Specialty Workshop at Northwestern University.* ASCE, New York.

19. Lambe, T. W. (1951). *Soil Testing for Engineers.* Wiley, New York.

20. Bishop, A. W., and D. J. Henkel (1962). *The Measurement of Soil Properties in the Triaxial Test,* 2d ed. Edward Arnold, London.

21. Department of the Army (1970). *Engineering and Design, Laboratory Soils Testing (Engineer Manual EM 1110-2-1906).* Prepared at the U.S. Army Engineer Waterways Experiment Station, Published by the Department of the Army, Washington.

22. *Standard Specifications for Public Works Construction,* 11th ed. BNi Building News, Anaheim, CA, 1991; commonly known as the "Greenbook."

23. Day, R. W. (2001). *Soil Testing Manual: Procedures, Classification Data, and Sampling Practices.* McGraw-Hill, New York.

24. Slope Indicator Company (1998). *Geotechnical and Structural Instrumentation.* Prepared by Slope Indicator Company, Bothell, WA.

25. Greenfield, S. J., and C. K. Shen (1992). *Foundations in Problem Soils.* Prentice-Hall, Englewood Cliffs, NJ.

26. Day, R. W. (1999). *Forensic Geotechnical and Foundation Engineering.* McGraw-Hill, New York.

27. Coduto, D. P. (1994). *Foundation Design, Principles and Practices.* Prentice-Hall, Englewood Cliffs, NJ.

28. Chen, F. H. (1998). *Foundations on Expansive Soils,* 2d ed., Elsevier Scientific Publishing, New York.

29. Stokes, W. L., and D. J. Varnes (1955). "Glossary of Selected Geologic Terms with Special Reference to Their Use in Engineering." *Colorado Scientific Society Proceedings* **16**:1–165.

30. Day, R. W. (2002). *Geotechnical Earthquake Engineering Handbook.* Sponsored by the International Conference of Building Officials (ICBO). McGraw-Hill, New York.

31. Keefer, D. K. (1984). "Landslides Caused by Earthquakes," *Geological Society of America Bulletin* **95**(2):406–421.

32. Division of Mines and Geology (1997). *Guidelines for Evaluating and Mitigating Seismic Hazards in California.* Special Publication 114, Department of Conservation, Division of Mines and Geology, Stockton, CA.

33. Day, R. W. (1999). *Geotechnical and Foundation Engineering: Design and Construction.* McGraw-Hill, New York.

34. Cernica, J. N. (1995). *Geotechnical Engineering: Soil Mechanics.* Wiley, New York.

CHAPTER 15

Vulnerability to Malevolent Explosions

ROBERT SMILOWITZ, Ph.D., P.E., and PAUL F. MLAKAR, Ph.D., P.E.

INTRODUCTION

Among the myriad of possible malevolent actions against buildings and other structures, explosions remain the most insidious, requiring the least sophisticated materials and expertise to assemble and deploy. Without arousing any suspicion, the principal components of an explosive device may be obtained at a variety of retail outlets. For this reason, the federal government has developed guidelines and criteria for the assessment of their facilities based on the threat of explosion. These criteria recognize the risks and hazards associated with large-scale terrorist explosive events and prescribe different levels of protection based on the nature of the structure and its criticality, occupancy, and symbolic importance.

Many local jurisdictions and commercial property owners do not have similar criteria for their properties. This chapter provides guidelines for their engineers—based on the hazards associated with the site conditions, building layout, and control of access, as well as the structural framing and facade components—for assessment of the vulnerability of their structures. This is a specialized discipline in which the expertise and experience of the engineer strongly influence the validity and utility of the assessment. If the owner and engineer retain a qualified consultant for this, this chapter will aid in establishing an effective working relationship.

Although each successive major domestic terrorist event exceeded the intensity of the predecessor, the vast majority of structures should be evaluated relative to the credible blast threat established by intelligence and law enforcement sources. Only a

495

few iconic structures may need to resist the more extreme threat at the tail end of the distribution. The evaluation of the magnitude of explosive charge weight that a structure may sustain depends on the extent of damage that the facility may tolerate. In general, the greater the extent of deformation, the greater will be the damage to the structure and assets.

While it is very difficult to correlate precisely the extent of damage with calculated deformations, experience has provided some useful guidelines to help quantify these levels. Three qualitative levels of protection—high, moderate, and low—are related to the casualties and extent of repair or renovation that would be required subsequent to a detonation. In a concrete slab, as an example, a *high level of protection* would shield the occupants from all but a spall of concrete off of the slab surface, and the slab would be cracked but require minimal repair subsequent to the event. A *moderate level of protection* would result in larger deformations with the potential for greater injury to occupants, and the slab would require extensive repair following the event. A *low level of protection* would result in extensive deformation with severe injury to occupants, requiring total replacement of the slab. These qualitative assessments are very imprecise; however, they correspond to the descriptions identified in the government documents discussed in the next section.

While life safety is the fundamental objective, casualties may occur, and assets may be damaged. Continued operations are the most difficult level of protection to achieve. Continuity of services requires the minimization of debris and, in some cases, the control of accelerations that otherwise might disable equipment.

GOVERNMENT CRITERIA

Following the 1995 Oklahoma City bombing and other malevolent acts, the Interagency Security Committee (ISC) was formed to incorporate the protective design needs of the community of federal agencies and bureaus. The current *ISC Security Criteria for New Federal Office Buildings and Major Modernization Projects* was issued in 2001. These criteria attempt to facilitate a risk-based approach. The document endeavors to take a flexible and realistic approach to the reliability, safety, and security of federal office buildings. It also tries to consider urban design principles, cost-effectiveness, and geographic locations. These criteria have evolved over a decade and can be expected to develop further as the profession gains further experience in their practical application. Thus use and interpretation of the security criteria depend on the expertise and experience of the engineer in charge.

The *ISC Security Design Criteria*[1] identify several levels of threat and response for government buildings. The criteria use designations ranging from low to higher to indicate the severity of the risk to a facility and to designate the appropriate degree to which the building should offer protection against specific tactics. The classification of buildings into their respective levels depends on the symbolic importance, the criticality, the consequences, and the perceived threat.

The assessment of physical security involves not only the structural resistance to blast and progressive collapse but also the exclusion capability of the perimeter, the debris hazard from nonstructural items, the isolation of occupied spaces from internal explosive threats, and the robustness of systems for evacuation and continuation of critical functions. While the architectural and structural details determine the resistance to dynamic loading, the operational security procedures and technical security devices significantly affect the locations to which threats may enter the facility. Since nothing can be done to limit the access of vehicles of all sizes around the perimeter of the building, the exterior facade components establish the hazards associated with a nominal level of blast loading.

It is important to note that security alone does not determine level of protection. A combination of technical and operational security does influence the parcels entering the facility. It is then structural engineering that defines the consequences from these malevolent parcels.

EXPLOSIVE THREAT

The density of common high-energy explosives, such as ammonium nitrate and fuel oil (ANFO), tri-nitro-tolulene (TNT), and composition 4 (C4), is on the order of 100 lb/ft³. Significant weights can be transported in vehicles. It is possible to package substantial amounts within a large briefcase that unobtrusively fits on a luggage carrier. Smaller weights of explosives may be packaged within backpacks and shopping bags, all of which arouse little suspicion by experienced security guards. Explosive devices weighing anywhere from several ounces to several pounds may be delivered through the mail or private courier services.

A parcel bomb may be delivered wherever packages and visitors are allowed access prior to inspection. While there is no limit to the magnitude of an explosive device that may be detonated on the public roads beyond the controlled perimeter, only a relatively small vehicular explosion associated with a surreptitiously placed bomb may occur within the controlled perimeter. The extent and thoroughness of the screening and vetting processes therefore will determine the size of an explosive threat and the location it may be placed within a facility. While visual inspections will be able to identify large or suspicious parcels within vehicles, a visual inspection will not detect explosives hidden within the wheel wells or door panels of vehicles. Although the use of bomb sniffing canines relies extensively on the experience of the handlers and working conditions, it is perhaps the most effective means of vetting truckloads of deliveries. However, there are limitations to the effectiveness of all operational security measures; the sensitivity of canines to the smell of explosives diminishes with time, and their noses lose sensitivity and effectiveness over the course of the day.

Thus one must consider all the preceding in assessing the credible size and location of an explosive threat. Some systematic procedures have been presented for this.[2,3] Since the assessment deals with the vulnerability to the inherently uncertain field of irrational human behavior, these procedures should not be regarded as prescriptive algorithms. They are instead useful guides for experienced experts.

BLAST PRESSURES

The dynamic blast loads from an explosion consist of a short-duration shock wave that is illustrated in Fig. 15.1 as a plot of overpressure p as a function of time t. For structural assessments, this load is conveniently characterized by a peak pressure, P_{so} and an impulse i_s. The peak pressures indicate the intensity of the applied loading. The impulse is the temporal integral of the overpressure and considers the duration of the loading pulse. Both parameters are important to the response of structural systems and mechanical elements.

The blast overpressure for a particular explosion is a function of the type of explosive material, the amount of explosive, and the distance to the detonation. The effectiveness of various explosive types is expressed practically as an equivalent amount of TNT. Table 15.1 lists these effectiveness factors for some of the materials that have been used in malevolent acts. Thus, for example, the 5000 lb of ANFO employed in the 1995 Oklahoma City bombing was equivalent to the detonation of $0.8 \times 5000 = 4000$ lb of TNT.

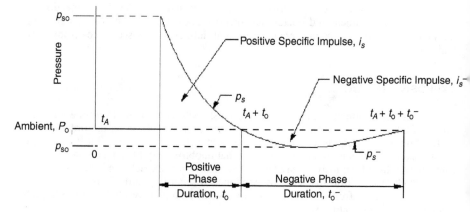

FIGURE 15.1 Blast overpressure.[4]

The blast load increases with the weight of explosive W and decreases with the range R from the detonation. In practice P_{so} and i_s are approximated by a function of scaled range $R/W^{1/3}$, as shown in Fig. 15.2.

To continue the example of the Oklahoma City bombing, the center of the building face adjacent to the bomb in Fig. 15.3 was at a range of 71 ft or a scaled range of $71/4000^{1/3} = 4.5$ ft/lb$^{1/3}$. Here the blast peaked at 54 lb/in.^{2} and had an impulse of 267 lb/in.^{2} · msec. If this loading were approximated by a suddenly applied triangular pulse (see Fig. 15.1), the duration would be 9.8 msec.

When a blast wave encounters an obstruction, the wave is reflected. The increased intensity of the peak reflected pressure P_{ra} as compared with the peak-incident pressure P_{so} is a function of the angles of incidence, as given in Fig. 15.4. Note that these reflection factors do not vary monotonically with the angle and are a function of the intensity of the peak pressures. Continuing the example of the Oklahoma City bombing, the pressure incident at the center of the building face reflected at an angle of 34 degrees. For this angle and incident pressure, the reflection factor is 3.2, and the reflected pressure thus is $3.2 \times 54 = 170$ lb/in.^{2}.

While Fig. 15.4 represents the reflected pressure that is directly incident, it does not include the effect of reflections from neighboring structures. Reflections from neighboring structures typically do not increase the initial peak pressure, but they

TABLE 15.1 Effectiveness of Explosive Materials

Explosive Name	Effectiveness
ANFO	0.80
Composition B	1.09
Composition C4	1.13
Octol 70/30	1.12
Pentolite 50/50	1.09
PETN	1.17
RDX	1.15
TNT	1.00

Source: From ref. 4.

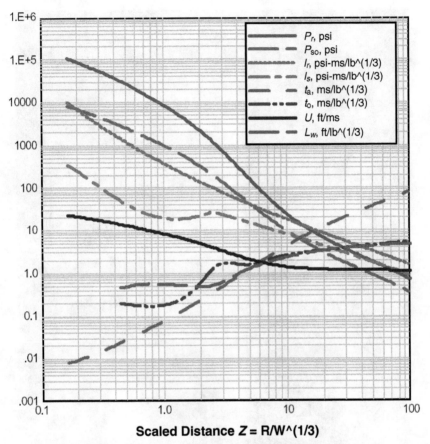

FIGURE 15.2 Blast parameters for TNT surface burst.[4]

usually amplify the accumulated impulse. In a dense urban setting, the reflected shock pulse may amplify the impulse by factors ranging from 1.5 to 2.0, and a nominal amplification factor of 1.75 has been suggested.[4]

On the other hand, the blast loading on specific components of a building may be reduced by virtue of their location. For example, the blast waves resulting from an explosion at the outer perimeter or internal parking areas may not have a direct line of sight to these locations and must sweep over or around structural components in order to load these surfaces. This disrupted path will diffuse the blast waves, thereby reducing the intensity of the pressure applied to the roofs and skylights.

As shown in Fig. 15.5, the blast pressure from an internal explosion is even more complex. In addition to the shock waves discussed earlier, there is a less intense but longer-duration gas pressure from the confinement of the gaseous products of the detonation in the enclosure. The amplitude of this load is a function of the quantity of explosive and the volume of the enclosure. The duration depends on these factors, as well as the vent area. Practical approximations of these relationships have been tabulated for design and assessment.[4]

Thus the first step in a building evaluation is to tabulate the relevant charge weights, along with the standoff distances and the building configuration. These parameters will be used to develop air blast models that produced the incident and reflected pressure-

FIGURE 15.3 Murrah Federal Building in Oklahoma City.

time histories at various locations around the building. A parametric study therefore must be performed in which the vehicle threat is assumed at various locations around the site, and the resulting pressures and impulses must be catalogued on all faces of the building. If vehicular access to the site is restricted, an explosive satchel at close range instead could be considered for blasts within the compound. Contour plots de-

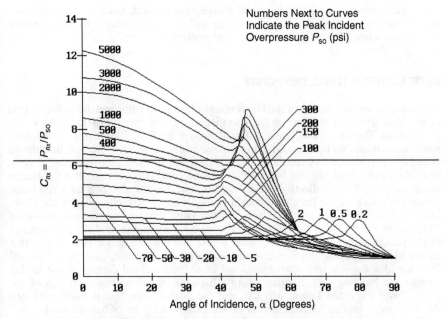

FIGURE 15.4 Reflected pressure coefficient versus angle of incidence.

picting the peak pressures and corresponding impulses must be developed to identify the zones of hazard along the building faces. These peak pressures and corresponding impulses define the blast loading on the individual structural components at their respective locations around the building. Once the entire blast environment is determined, each building component is bounded by a maximum pressure and corre-

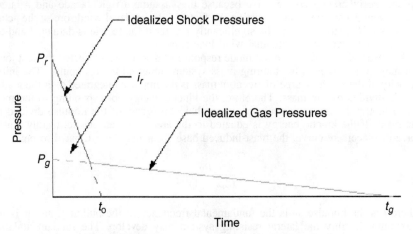

FIGURE 15.5 Overpressure from internal explosion.[4]

sponding impulse based on each critical charge location as dictated by site conditions and engineering judgment. This information will be used subsequently to determine the various hazard levels associated with the existing structure.

GLOBAL STRUCTURAL RESPONSE

The lateral resisting systems of the building must be able to withstand the global effects of the blast forces in response to the specified threat and transfer these forces to the foundations. These forces will be transferred through the floor diaphragms to the lateral resisting elements such as braced frames, shear walls, or moment frames. In order to determine the intensity of the blast loads transferred to the lateral resisting system, equivalent static blast-induced base shears may be generated assuming that the facade is infinitely rigid and collects all blast impulse and that the slab acts as a rigid diaphragm to distribute the impulse throughout the structure. The equivalent static blast-induced base shear depends primarily on the building size, the natural period of vibration of the building, the impulse collected over the facade, and the performance of the facade. Although it is conservative to assume that the facade is rigid and that all the blast impulse is transferred to the lateral resisting system, glazing failure will limit the load the structure must resist, and localized structural damage at the loaded face of the building will dissipate energy. In actuality, a significant portion of the glazing over the entire facade will fail, and the total impulse that is transferred into the structure is limited by the capacity of the facade and exterior bay materials.

The base shears owing to wind loading, for which the building was designed, may be used to estimate the lateral resistance of the building. By this approach, the wind load can be calculated for the building using the procedures specified in the building code in effect at the time of construction. In order to convert these forces into ultimate base shears, the forces must be summed over the height of the building and amplified by the factors of safety used in conventional lateral design for wind forces. This corresponds to amplifying the wind base shear by the typical load and resistance factors used in Load and Resistance Factor Design (LRFD) design. This factored wind base shear may be compared with the blast-induced base shear for a given intensity blast loading, and by a process of iteration, the minimum charge weights associated with failure of the lateral system may be determined. These thresholds for structural failure are inherently conservative because they assume a rigid facade and a lateral system that is capable of just resisting the factored design-level wind forces; the actual capacity of the structure may be significantly greater if the facade is damaged and the lateral system exceeds its nominal wind load capacity.

For example, the fundamental mode response of a structure subjected to blast loading may be represented by a spring mass system subjected to an impulse. The initial velocity of the single-degree-of-freedom mass is the impulse integrated over the loaded facade divided by the mass. Therefore, the kinetic energy corresponding to the mass set in motion by the initial velocity is one-half the square of the impulse divided by the mass. If the kinetic energy is equated to the area under an elastic, rigidly plastic force-displacement curve, the blast-induced base shear may be estimated to be:

$$\frac{i\omega}{(2\mu - 1)^{1/2}}$$

where i is the impulse, ω is the fundamental frequency of the building, and μ is the maximum ductility the lateral resisting system may develop. The fundamental frequency and the maximum ductility the lateral resisting system may develop may be obtained from descriptions of building systems published in the seismic code provisions.

For example, 200 kip-seconds of impulse will be applied to a 200-ft-wide, 50-ft-tall concrete frame building owing to a 1000-lb detonation of TNT at a standoff distance of 100 ft. If the fundamental period of this building is approximately 0.5 sec, and it may sustain a maximum ductility of 5 in response to the impulsive loading, the equivalent lateral load of 56.6 lb/ft^2 may be compared with the ultimate lateral-load-resisting capacity of the building. This ultimate lateral-load-resisting capacity may be estimated by multiplying the design wind load by the load factors and dividing by the understrength factor for the materials.

In practice the global effects of blast forces do not usually exceed the capacity of the structural system intended to resist the lateral forces owing to wind or earthquake. However, it is always prudent to perform this check as described herein.

RESPONSE OF STRUCTURAL ELEMENTS

Beams and columns that may be subjected to blast pressures from either exterior or interior explosive threats must be evaluated to determine their limiting capacities. The different structural elements therefore will be analyzed to determine the intensity of tributary loads that will produce an acceptable level of deformation. These analyses are necessarily dynamic and inelastic, and the loads to which they are subjected are a combination of the reaction forces from the adjacent wall or slab and the load that is applied directly to the beam or column. Although the technology for evaluating existing structures is relatively straightforward, information regarding the material properties and construction details may be difficult to obtain. Even if the original design drawings are still available, the corrosive effects of time and structural modifications may require nondestructive and possibly destructive testing to verify conditions. Lacking accurate information, the investigation is best served by bracketing the likely response of the structure by assuming different degrees of member fixity and different extents of internal reinforcement. With less reliable information, it is prudent to set relatively conservative deformation and ductility limits for defining the onset of structural damage and failure. It is therefore the engineer's experience and judgment that establishes the limiting capacities of the components.

While single-degree-of-freedom analyses accurately represent the fundamental flexural mode response to short-duration impulsive loads, these methods are less accurate when considering the high-frequency effects of direct shear. The effects of direct shear are most pronounced for near-contact explosive threats, and for these cases, transient dynamic finite-element analyses provide the most accurate evaluation of structural performance.

The exterior columns and spandrel beams are exposed directly to the effects of an exterior vehicle threat, whereas the underside of the exterior bay floor slabs will be loaded only if the exterior facade components are overwhelmed by the blast loading. The net uplift pressures that may be applied to the underside of the slabs will have to overcome the self-weight of the floor system before the structure is subjected to upward deformation. While the inertial resistance increases the apparent capacity of the floor system, the lack of top reinforcement at midspan will provide little additional capacity. The rebounding floor system once again will be required to carry its self-weight, and the capacity of the damaged structure will be seriously compromised.

Supporting beam structures must be able to resist the slab uplift reaction forces, and the magnitude of these forces is limited by the plastic capacity of the slabs. Unless vulnerable to punching shear, this is not as likely to cause collapse as it is to require extensive repair or replacement. If multidegree-of-freedom analyses are performed instead of sequential single-degree-of-freedom analyses, using the dynamic reactions of one analysis to serve as the loading function for the next, then the accurate phasing of the loads for the different frequency systems may mitigate some of the dynamic effects. Three-dimensional analyses of the structural systems using inelastic dynamic finite-element methods provide the most accurate interaction among the structural members; however, unless accurate information regarding the structure is known to the analyst, the additional accuracy may not be justified.

Although the magnitude of the hand carried satchel threat within the lobby or mailroom and the surreptitiously hidden vehicle threat within the loading docks or underground parking structures is relatively small compared with the exterior vehicle threat, their proximity to critical structure increases the potential for hazard. A contact or near-contact explosion will produce an extremely short duration but extremely high-intensity pressure pulse that will shatter most construction materials and cause a breach of the structure. Flanged steel sections are likely to suffer extensive local deformations that can diminish the section properties and leave the members incapable of developing their plastic capacity. Since the semiempirical blast loading methods[4] do not calculate the high intensity but short-duration near-contact loading functions and only approximately represent the multiple reflections within a confined space, a computational fluid dynamics (CFD) analysis provides the most accurate loading function. Furthermore, the by-products of the internal detonations produce relatively low-intensity but long-duration quasi-static gas pressures that must be vented to the atmosphere. The magnitude and duration of these gas pressures are also approximated by the semiempirical methods. Unless the structure provides adequate venting, these quasi-static pressures will fail structure that is previously weakened by the shock wave. While energy methods may be used to consider the yield line capacity of the slabs and the potential for punching shear failures, inelastic dynamic finite element analyses may provide the most accurate response of the system. The self-weight of ceiling slabs will provide an apparent increased capacity to uplift pressures, but, the slab will experience intense upward accelerations that may induce unacceptable vibrations and may propel occupants and lighter equipment into the air. These spaces therefore may be lethal environments despite the apparent survival of the structure.

For example, 10 lb of TNT may be detonated on a slab on grade (zero height of burst) within a fully vented loading dock, and the blast pressures will apply an uplift blast loading to the underside of the ceiling slab. If the ceiling height is 10 ft, the uplift shock loading averaged over a 20 by 20 ft area may be found to be 130 lb/in.^{2} with a corresponding impulse of 73 lb/in.$^2 \cdot$ msec. The 8-in.-thick doubly reinforced two-way reinforced concrete ceiling slab must be supported by a doubly reinforced concrete beam to withstand the uplift pressures and withstand the subsequent rebound. If the reinforced slab contains $\frac{1}{2}$ reinforcement each way on each face, the slab will be lifted 0.6 in. at approximately 11 msec and will rebound to a deformation of approximately 0.7 in. at approximately 33 msec. This corresponds to a ductility of approximately 1.15. If the reinforcement were reduced in half to $\frac{1}{4}\%$ each way on each face, the slab will be lifted 0.75 in. at approximately 15 msec and will rebound to a deformation of approximately 0.5 in. at approximately 47 msec. This corresponds to a ductility of approximately 2.37. The rebound for the greater percentage of reinforcement, for which the slab essentially remained elastic, was more pronounced than for the lesser percentage of reinforcement, for which the slab dissipated a greater amount of energy through inelastic deformation. The shear stress in the slab is in both cases less than the nominal capacity of the concrete; however, the greater percentage of reinforcement corresponds to a greater equivalent static capacity, which corresponds to greater slab reaction forces.

PROGRESSIVE COLLAPSE

Evaluation of a building's vulnerability to collapse and building component failure is the primary focus of the structural evaluation. *Progressive collapse* is defined as the spread of an initial local failure from element to element, resulting in the collapse of a disproportionately large part of a structure. For example, a progressive collapse occurs when the loss of a single column results in the destruction of a large portion of a building. A properly designed structure will be able to absorb large displacements, redistribute the loads, and although damaged, remain standing. This evaluation therefore must assess the potential for a progressive collapse by evaluating the structural redundancy, ductility, and detailing.

The ISC criteria for new facilities require the structure to be designed for "the loss of a column for one floor above grade at the building perimeter without progressive collapse." This design and analysis requirement for progressive collapse is intended to be threat-independent, which will "ensure adequate redundant load paths in the structure should damage occur for whatever reason." It is intended that "the structure will not collapse or be damaged to an extent disproportionate to the original cause of the damage" should a larger than design explosive event cause a localized partial collapse. Analysts are permitted to apply static and/or dynamic methods of analysis and take advantage of ultimate load capacities of the structural members to meet this requirement.

The American Society of Civil Engineers (ASCE) *Minimum Design Loads for Buildings and Other Structures,* SEI/ASCE 7-02, describes protection through "an arrangement of the structural elements that provides stability to the entire structural system by transferring loads from any locally damaged region to adjacent regions capable of resisting these loads without collapse." From this approach, ASCE 7-02 discusses three design alternatives that may be part of a multi-hazard design approach. The alternatives are the indirect design approach, the alternate-path direct design approach, and the specific local resistance direct design approach. The alternate-path approach presumes that a critical element is removed from the structure owing to an abnormal loading and that the structure is required to redistribute the gravity loads to the remaining undamaged structural elements. The method of specific local resistance requires all critical gravity-load-bearing members to be designed and detailed to be resistant to a postulated abnormal loading. Each design approach is based on assumptions and conditions that offer technical advantages and disadvantages. However, the intention is to determine the capacity of a structure either to resist an abnormal loading, thereby preserving the load-carrying capacity of the critical elements, or to redistribute gravity loads if a critical load-bearing element is removed.

The response of either the elements or the structure to abnormal loading conditions is most likely to be dynamic and nonlinear, both geometrically and in the material behavior. Therefore, the analytic methods that are required to determine the response of the structure must represent the sudden application of the abnormal loading, the dynamic behavior of the materials under very high strain rates, the inelastic post-damage behavior of the materials, and the geometric nonlinearity resulting from large deformations. Further, the ability of the structural elements to withstand the abnormal loading or the structural system to redistribute the loads depends to a great extent on the behavior of the structural details that define the connections. As a result, the computational tools and modeling constructs that are used to analyze the damage response of structures is often critical to the success of the design approach. Perhaps most critical to the success of the design is the experience of the engineers modeling the structure and materials. By assessing the consequences of different damage mechanisms, the engineer can determine whether the structure has a high potential for progressive collapse.

The alternate-path direct design approach has the appearance of being threat-independent. This approach explicitly considers the resistance to progressive collapse when a primary load-bearing member is removed. While this does not specify a threat or a cause for the damaged state, it is limited in the applicability to abnormal loading conditions that would fail only one load-bearing member. An advantage of this approach is that it promotes structural systems with ductility, continuity, and energy-absorbing properties that are desirable in preventing progressive collapse. This method is also very consistent with the seismic design approach used in many building codes throughout the world. The seismic codes promote regular structures that are well tied together. They also require ductile details so that large plastic rotations can take place. This would be essential in designing structures that could resist progressive collapse when a primary load-bearing member was lost.

A first approximation for evaluating progressive collapse can be obtained using linear elastic analysis techniques. Linear elastic analyses are incapable of accounting for the redistribution of forces, P-delta instability, nonlinear material properties including rate effects, and the development of membrane modes of resistance. Therefore, engineers using linear elastic analysis approaches use judgment, which they believe generally results in a more conservative design, and then independently check for P-delta instability after the initial design is complete. The elastic analysis does not account for the greater capacity resulting from plastic hinge/yield-line formation, membrane action, or enhanced strength owing to rate-dependent material properties.

The forensic study of the Oklahoma City bombing illustrated this linear alternate-load-path analysis.[6] As shown in Fig. 15.3, the structural frame included a transfer girder at the third floor that enabled the elimination of alternate columns along the exterior line of the building in the structure below. A two-dimensional linear elastic analysis of this reinforced-concrete frame demonstrated that the destruction of a remaining column will result in a collapse of the structure above. The bombing directly removed three of these columns and triggered a general collapse, as shown in Fig. 15.6.

The inclusion of geometric nonlinearity resulting from large deformations accounts for the redistribution of loads from a flexural response to a membrane response as a

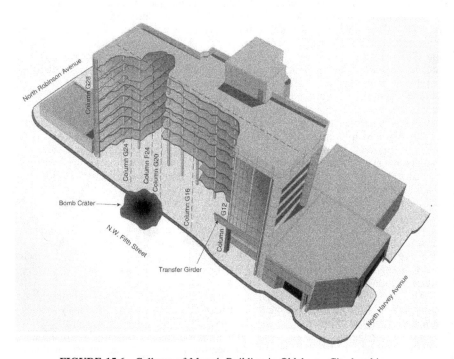

FIGURE 15.6 Collapse of Murrah Building in Oklahoma City bombing.

column is removed and the structure attempts to reequilibrate to the larger spans. The members that originally spanned a single bay must now span two bays, and the center span will be at the location of the damaged column, where the connection details may have limited capacity to develop positive moments. The inclusion of geometric non-linearity will enable the analyst to account for the tension-membrane stiffening of the slabs and spandrel girders as they sag and develop catenary resistance. These membrane forces must be compared with the tensile capacity of the members and their connections to make sure that they are capable of developing the axial forces.

Since the alternate-path approach is essentially an academic exercise that ignores all other damage to the structure that may accompany the removal of the critical column support, a girder spanning a single bay is instantaneously transformed into a girder spanning two bays. The transition from the original structural configuration to the damaged state is assumed to be instantaneous, and therefore, the structure is exposed to a dynamic effect. The dynamic amplification factor represents the effect of this dynamic phenomenon, which, for a single-degree-of-freedom system, is a function of the duration of the load and the extent of inelastic deformation. The instantaneous removal of a column and corresponding transfer of the gravity load to the double-bay span is best represented by a suddenly applied step pulse of infinite duration. The most conservative condition corresponds to an elastic single-degree-of-freedom system subjected to a step pulse of infinite duration, for which the dynamic amplification factor is twice the static load.[5] The development of plastic hinges in response to removal of the column will dissipate energy, and the corresponding dynamic amplification factor will be reduced.

The specific local-resistance direct design approach explicitly considers resistance to progressive collapse by requiring the structure to withstand the intensity of a given abnormal loading and avert collapse altogether. This approach is often the only rational approach when considering the capacity of an existing building. The cost of bringing the building into compliance with the alternate-path direct design method may be so great as to make the approach impractical. The specific local-resistance direct design approach might allow the owner to increase the resistance of key structural elements, such as exterior columns, to withstand any reasonable threat that would be levied against the structure. The specific local-resistance direct design approach averts progressive collapse by hardening the vulnerable load-bearing members to a given threat using selected upgrades. These upgrades may be selective, and only the most critical elements need to be retrofitted.

The specific local-resistance direct design method requires a high-fidelity explicit, nonlinear, large-deformation transient analysis finite-element analysis to demonstrate the resistance of the key elements in response to the effects of blast loading. However, since this method is only applied to an individual part of the structural system at one time, the finite-element models are relatively small and efficiently analyzed. Individual columns may be analyzed to determine their capacity to resist explosive loading, and where the capacity is found to be insufficient, these same columns may be analyzed to determine the increased capacity provided by a retrofit.

The results of this study may demonstrate the ability of the retrofitted columns to sustain the blast environment that would severely damage two columns, thereby violating the basic premise of the alternate-path direct design method. Furthermore, the retrofitted column may be capable of sustaining a charge weight at the defensible perimeter that would damage the surrounding spandrel girders and floor slabs. The specific local-resistance direct design method actually may improve the resistance of the existing structure to a larger threat than the alternate-path method would have done. In some cases, the specific local-resistance direct design approach provides more protection than an upgrade with the alternate-path method. Furthermore, the column retrofits at the first-floor level may be less intrusive than the alternate-path upgrades, and this can have a significant impact on the cost and viability in upgrading existing structures.

RESPONSE OF FACADE

In explosive events in which no building collapse occurs, a very high number of casualties and fatalities may result from the impact of flying glass fragments and nonarchitectural features. Conventional glass used in most windows breaks at very low blast pressures, resulting in hazardous dagger-like shards. Minimizing these hazards has a major effect on limiting mass casualties following an explosive event. Glazing response is categorized into different performance levels based on the behavior of the glass while either within the frame or any postbreakage debris impact into the interior space. The hazard level owing to debris associated with window damage also may be attributed to the failure of the frames in retaining the glass or failure of the anchorage in holding the mullions and frames to the structure. These various modes of window hazard may be evaluated separately to determine the lowest blast environment producing the governing mode of failure. The blast environments associated with these failure modes may then be compared with the pressures and impulses associated with the postulated threat. The hazard level determined by the calculations only considers debris associated with glazing damage, but the hazard to occupants can be amplified by the placement of personal items on the windowsill. In the case of a blast event, these items can become lethal projectiles. Window and door glazing evaluations therefore must consider glazing, frames, connections, and the structural components to which they are attached as an integrated system. The exterior facade must be capable of withstanding both the directly applied blast pressures and the dynamic reactions from the windows.

The United Kingdom pioneered the categorization of hazard levels and relates these to the distance the fragments of glass debris will be propelled in a standard cubicle test. This is shown in Fig. 15.7 and Table 15.2. The U.S. government generally adopted the U.K. hazard rating by converting the distances to English units and creating several intermediate categories. According to the ISC, a safe level of protection corresponding to no hazard occurs if the glass does not break (level 1). A very high level of protection corresponding to very low hazard occurs if the glass cracks but remains in the frame (level 2). The ISC assigns a high level of protection corresponding to a low hazard if the glass exits the frame but the debris flies no further than 3 ft (level 3A) or 10 ft (level 3B) from the face of the glass. A medium level of protection corresponding to a medium hazard occurs if the debris impacts a witness panel that is 10 ft behind the plane of the glass at a height no greater than 2 ft above the floor (level 4). Finally, a low level of protection corresponding to a high hazard occurs if the debris impacts the witness panel at a height greater than 2 ft above the floor.

The evaluation of a facade's hazard thresholds requires detailed description of the materials, thicknesses, and dimensions. Ideally, the investigators have access to the design drawings and specifications; however, this is rarely the case. Most often the investigators need to make educated guesses based on the standard practice for a particular location or construction type. Glazing thickness and nominal strength may be estimated using a laser gauge; however, the accuracy of these devices may be limited to simple glazing makeups. Alternatively, a range of capacities may be developed by evaluating the performance of the window system using a range of practical glazing makeups. This range provides the owner with the best estimate of the hazard thresholds.

Parametric studies may be performed in order to develop isodamage plots in the form of pressure-impulse $(P\text{-}I)$ curves or range-to-effect curves. These logarithmic

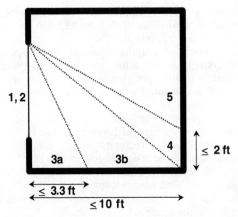

FIGURE 15.7 Schematic indication of fragment impact locations.

TABLE 15.2 Glazing Protection Levels Based on Fragment Impact Locations

Performance Condition	Protection Level	Hazard Level	Description of Window Glazing Response
1	Safe	None	Glazing does not break. No visible damage to glazing or frame.
2	Very high	None	Glazing cracks but is retained by the frame. Dusting or very small fragments near sill or on floor acceptable.
3a	High	Very low	Glazing cracks. Fragments enter space and land on floor no further than 3.3 ft from the window.
3b	High	Low	Glazing cracks. Fragments enter space and land on floor no further than 10 ft from the window.
4	Medium	Medium	Glazing cracks. Fragments enter space and land on floor and impact a vertical witness panel at a distance of no more than 10 ft from the window at a height no greater than 2 ft above the floor.
5	Low	High	Glazing cracks and window system fails catastrophically. Fragments enter space impacting a vertical witness panel at a distance of no more than 10 ft from the window at a height greater than 2 ft above the floor.

plots indicate the combination of peak pressure and impulse or the combination of weight of explosive and standoff distance that produce a particular hazard level for a given size of glass and makeup. Although there is no unique combination of peak pressures and impulses associated with the threshold, the shapes of the curves provide pressure and impulse asymptotes that define limiting capacities of the glazing. Using the *P-I* isodamage chart, the limiting capacities of glazing at the specified elevations can be catalogued, and the associated weights of explosives may be used to identify the vulnerability (Fig. 15.8).

In this manner, the response of the glass is presented in the form of three lines defining the transition from nodamage to break safe, break safe to low hazard, and low hazard to high hazard. Each curve represents the combination of peak pressure and impulse at the threshold of the specified hazard level. All combinations of peak pressure and impulse lying below and to the left of the curves indicate the performance to be less severe than the specified hazard level. Conversely, all combinations of peak pressure and impulse just lying above and to the right of the curves indicate the performance of the glass to the specified hazard level.

In order to determine the extent of a glazing hazard, the investigation must indicate the explosive threat associated with the onset of a particular level of hazard and the increased threat associated with widespread occurrence of the particular level of hazard. As might be expected, a relatively small satchel threat in close proximity to the structure is likely to produce a high level of hazard in immediate proximity to the

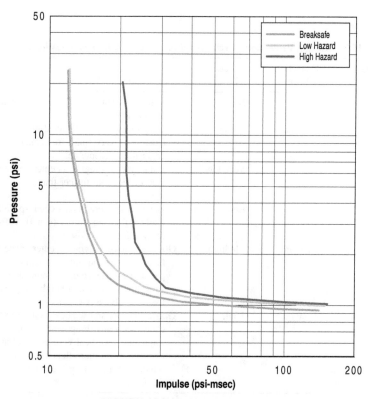

FIGURE 15.8   *P-I* isodamage chart.

detonation, but the extent of hazard is limited to a relatively small area of the facade. Alternatively, if the evaluation is to consider the glazing performance in response to a specified threat, the investigation should tabulate the total area of glazing that responds to the different categories of hazard. This will provide the best indication of the extent of hazard to occupants in response to a specified threat. In either case, several evaluations will be required, positioning the threat at the locus of possible threat locations, in order to determine the worst-case conditions. As a guide, the ISC recommends that no more than 10% of the total glazed area of the building produce high-hazard debris in response to the specified threats.

Although the glazing-response calculations identify the hazard associated with glass failure, the facade is only as strong as the framing system and anchorages that support it. If the framing system is weaker than the glazing and fails first, it becomes the weak link and does not allow the glass to achieve the maximum capacities indicated earlier. The properties of the mullions and anchorages often are unavailable when performing an investigation of an existing building. In order to estimate these properties, the engineer therefore must use a combination of measurements and engineering judgment based on standard practice for the location and type of construction. Observations must be made as to whether the facade is glazed from the exterior or the interior. This may indicate whether the glass bears against the strength portion of the mullion or relies on pressure plates to transfer the load. In some interior glazed systems, the glass bears against the snap-on cover during exterior loading, and the snap-on cover will not be capable of resisting large glazing-edge reactions.

To determine the adequacy of the mullions to support the glass, single-degree-of-freedom models must be developed and loaded with the dynamic edge reactions of the glazing at maximum capacity. This will determine the maximum force the glazing and any laminate interlayer may impart onto the mullions. Often the facade systems contain both vertical and horizontal free-spanning mullions that interact with each other as the load is transferred from the glazing to the anchorages. Where primary mullions are loaded by small discontinuous mullions, the resulting reaction-point loads were incorporated into the analysis. Dynamic analyses of these models indicate whether the responses of the mullions are within acceptable ductility and/or rotation limits. A multidegree-of-freedom model of the facade will determine the accurate interaction of the individual mullions and the phasing of the interconnecting forces. Since all response calculations must be dynamic and inelastic, the accurate representation of the phasing of these forces may have a profound effect on the calculated performance. Where mullions are attached within punched-out openings, the spacing of the anchorages will determine the span of the mullions and the force each anchorage is required to resist.

The anchorage of the facade to the structure must be able to resist the ultimate capacity of the mullions and framing members. Often the details of the anchorage can neither be observed nor inferred. At best, the capacity of the anchorage may be estimated on the basis of the design-level wind forces amplified by the factors of safety required by the building codes. While this may understate the capacity of the facade, it may be the only practical way of determining the capacity of the anchorage. Some glazing systems that are not attached directly to the structure must rely on the backup wall system to support the windows and withstand the reactions from the window framing. In this case, the backup wall system must either span continuously between the floor slabs or cantilever up and down from each floor slab to transfer the window reactions back to the structure. These wall systems may be masonry, precast, or metal stud systems. Once again, the anchorage of the mullions to these backup walls may limit the forces that may be transferred; however, in addition to the mullion reaction forces, these walls also must resist the dynamic blast loads that may be applied directly to them. This combination of loading is necessarily dynamic and the response of the backup wall elements is necessarily inelastic. The accurate representation of the phasing of the load and the material response is critical to determination of the facade capacity. Often the wall systems are capable of providing significant inertial or structural resistance, but their connections represent the weak link in the load transfer.

Facade systems may contain combinations of glazing, metal panels, and stone panels. Metal panels provide little inertial resistance but are capable of developing large inelastic deformations; however, the screws that attach these panels to the mullions or metal studs often are incapable of transferring the large membrane forces. Stone panels provide significant inertial resistance but are relatively brittle and have little strength beyond their modulus of rupture. Stud wall systems that restrain these facade panels may deform within acceptable levels and develop a membrane-stiffening capacity, and strain-energy methods may be used to calculate their response. However, the anchorages of the studs to the floor and ceiling slabs are likely to limit the forces they can develop.

Since a significant surface of the exterior walls enclosing the mechanical spaces must be louvered to provide the required ventilation, these facade materials allow infill pressures to load the interior equipment and interior partition walls. These interior partitions must be analyzed to determine whether they protect occupants adequately from the effects of infill blast loading. Furthermore, the response of electrical and mechanical equipment to blast environments has been studied in support lethality and survivability algorithms that were developed for the U.S. Department of Defense. However, short of a detailed analysis of all components comprising the critical electrical and mechanical systems, it may not be possible to determine whether the infill pressures or the in-structure shock motions will result in a disruption of services. Therefore, if any of the equipment within this space is deemed to be critical for life safety and evacuation, the equipment must be protected with a hardened plenum or located on a higher elevation.

For example, a curtainwall spans a 12-ft slab-to-slab floor height and consists of primary vertical mullions spaced 5-ft apart that are stacked just above the floor slab and four horizontal mullions that span between vertical members. The facade is exposed to a 400-lb TNT threat at a standoff distance of 200 ft, that produces a nearly uniform peak pressure of 4 lb/in.^{2} and a corresponding impulse of 45 lb/in.^{2} · msec over the surface of the building. Each lite of insulated (double-pane) glass, 36 in. tall by 60 in. wide, is supported on all four edges by either a vertical or horizontal mullion. If the glass is to crack but fall to the floor directly behind the facade, performance condition 3a, it must consist of two plies of $3/16$-in.-thick plies of annealed glass with a 0.030 PVB interlayer for the inner lite, a $1/2$-in. airspace and a $1/4$-in.-thick lite of monolithic heat strengthened glass for the outer lite. This lite of glass will resist nearly 10% more impulse before it exits the frame, and the peak edge shear reaction loads along the long span is 69 lb/in.^{2}, and the corresponding magnitude along the short span is 49 lb/in.^{2} The variation of the glazing reaction forces as a function of time may be idealized by a linear increase to a peak magnitude at 10 msec, at which time the glass cracks, followed by a sudden drop to zero and a nearly sinusoidal variation to 25 lb/in.^{2} along both long and short edges, with a period of approximately 90 msec.

In an existing building, the inner lite is likely to be $1/4$-in.-thick monolithic glass; therefore, a 0.070 antishatter film mechanically attached on two (long) sides will be required to provide an equivalent level of protection. The edge reactions will oscillate with a peak force of 61 lb/in.^{2} at a period of approximately 50 msec.

The calculated reaction forces may be applied to the horizontal and vertical mullions as line loads, and the dynamic inelastic analysis may be used to determine the section properties that are required to limit the peak mullion end rotations to less than 2 degrees.

NONSTRUCTURAL CONSIDERATIONS

In most cases no uninspected vehicles may be assumed to approach closer than the curbline surrounding the site. Although a curb is a legitimate protective boundary for stationary (parked) vehicles, antiram barriers are required around the site perimeter to prevent a moving vehicle from approaching the building. To ensure the maximum standoff distance, a defined security perimeter typically is established with bollards, hardened street architecture, or retaining walls. These barriers must be capable of withstanding the impact from a moving vehicle to be effective in establishing a protected perimeter. Operable bollards, either in the form of a retractable piston, cable-reinforced beams, or plate wedges, are required at entrances. Two rows of barriers at entrances are required to prevent tailgating, but this requirement is considered to be excessive for most facilities, particularly at the employee parking and loading dock entrances.

There is no comprehensive methodology for designing barrier systems for a wide variety of foundation conditions, barrier systems, and materials. Traffic studies are required to evaluate the approach vectors and determine, from the available turning radii, the maximum attainable impact velocities. Impact physics considers the kinetic energy and momentum of the vehicle, but realistic analyses also must consider the amount of energy that may be dissipated through the crushing of the vehicle. Nonlinear dynamic analyses are required to evaluate the performance of these devices and their ability to maintain the required standoff distances. Furthermore, dense urban streets and sidewalks are cluttered with underground utilities, vaults, subways, etc., and none of the barriers that have been tested by the U.S. government or commercial manufacturers have considered foundation designs that accommodate these difficult subgrade conditions.

REFERENCES

1. Interagency Security Committee (2001). *ISC Security Design Criteria.* ISC, Washington, D.C.
2. *Protecting Buildings from Bomb Damage, Transfer of Blast Effects Mitigation Technologies from Military to Civilian Applications.* National Research Council, Washington, D.C., 1995
3. ASCE (1999). *Structural Design for Physical Security: State of the Practice.* ASCE, Reston, VA.
4. *Structures to Resist the Effects of Accidental Explosions TM5-1300.* Departments of the Army, the Navy, and the Air Force, Washington, D.C., December 1991.
5. Biggs, J. M. (1964). *Introduction to Structural Dynamics.* McGraw-Hill, New York.
6. "The Oklahoma City Bombing: Improving Building Performance through Multi-Hazard Mitigation." Federal Emergency Management Agency 277, Washington, D.C., 1996.

PART IV

EVALUATION AND TESTING OF STRUCTURAL MATERIALS AND ASSEMBLIES

Concrete

SCOTT F. WOLTER, P.G.

INTRODUCTION

When attempting to evaluate the structural condition of a concrete structure to develop an appropriate plan for future use, a thorough understanding of the concrete is mandatory. The first phase of any evaluation is to identify a qualified team of investigators. The person in charge should be a structural engineer who has passed his or her professional engineers examination and is in good standing within his or her profession. The lead engineer first should perform an extensive review of all existing information about the structure, which, among other materials, includes the following three items:

Design documents. A thorough review of the original design documents will aid in attaining an understanding of the size and nature of the structure.

Construction records. Any and all construction documents will help further in understanding the type of construction inherent to build the structure.

Service history. Service records will help to understand how the structure has performed over time and further aid in a general understanding of the overall condition.

CONDITION SURVEY

Once an understanding of the structure has been reached after a thorough review of all existing information, a field condition survey then should be performed. An initial walk-through is appropriate with careful documentation of the following activities:

Visual inspection. An overall appreciation of the physical aspects of the structure can be obtained by performing a walk-through. Various observations should be noted, and photographs should be taken of good visual examples of specific conditions.

Crack mapping. Careful and detailed mapping of cracks is important in learning about the health of a particular structure. Various aspects of cracks should be documented, such as length, width, and pattern, as well as any evidence of water movement. Secondary deposits and staining propagating from cracks are obvious indicators of water movement. Additionally, it is important to document if any movement has occurred relative to observed cracking, such as shrinkage, displacement, or volumetric expansion.

Deflection, flatness, and alignment. Other physical aspects of a structure that yield important information on quality and condition are deflection, flatness, and alignment. These parameters can be measured and are considered part of the overall condition survey of various concrete structures.

Once the condition survey has been completed, a more detailed analysis of the concrete may be necessary. A quality assessment of the concrete and/or the need to solve a particular problem(s) can be pursued through an appropriate testing plan. Several invasive and noninvasive test procedures can be implemented to determine and document the physical properties of the concrete.

Petrographic analysis is extremely useful in diagnosing common concrete problems, such as low strength, cracking, tearing, surface delamination, blisters, scaling, dusting, popouts, "D" cracking, alkali-aggregate reaction, and ettringite distress.

COMMON CONCRETE PROBLEMS

Low Strength

There are several causes of low strength in concrete. A few examples that can be diagnosed through microscopic analysis include those discussed below.

Delayed Set Time. A delay in set time will slow the strength development of concrete. Thin-section analysis can document the size of reaction rims around cement grains to determine the relative degree of hydration. The degree of hydration can be compared with the age of the concrete to gain an understanding of where along the hydration–strength-gain curve the material is. Several factors, alone or in combination, can delay set time, including an overdose of chemical and/or pozzolanic admixture, an incompatibility of chemical and/or pozzolanic admixture, exposure to cool temperatures during curing, and ineffective curing (Fig. 16.1).

High Water-to-Cementitious Ratio. Water-to-cementitious (w/cm) ratio is a good predictor of strength. A ratio in excess of roughly 0.55 will adversely affect anticipated compressive strength. As water is added to a concrete mixture, there is a reduction in cement particle concentration, resulting in an increase in paste porosity.

High Entrained-Air Content. The most important aspect of entrained air that can adversely affect cement strength is when the smallest air voids are spaced too closely together. When this happens, microcracking will develop at lower stress levels. The spacing factor is a measure of this parameter and will begin to cause a problem at about 0.003 in. or less. Even early-age concrete with low strength attributed to low spacing factor will not see appreciable strength gain with time and additional hydration. Entrained air content is also an important parameter when assessing concrete durability. Typically, the volume of entrained air should fall in the range of 3.0% to 6.0% with a spacing factor of 0.008 in. or less. Synthetic air-entraining admixtures typically generate the smallest average-sized air bubbles and are most prone to low spacing factors.

Poor Paste-to-Aggregate Bond. There are at least three situations where bond between the paste and aggregate can be adversely affected. Rounded gravel aggregates will not develop as good of a bond with the paste as a crushed aggregate. The highly

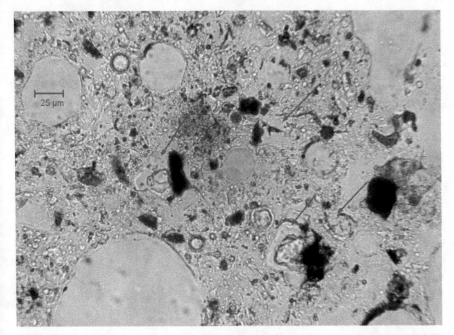

FIGURE 16.1 Relatively well to fully hydrated alite cement grains (*arrows*) indicating relatively little strength gain potential. Also notice the spherical fly ash pozzolan particles (plane polarized light).

angular surface of a crushed particle will provide a much better mechanical bond with the paste. The geology of the aggregate also will affect the bond. A dense, low-porosity aggregate, such as quartz or an igneous type, will resist penetration of the paste. Limestone and other carbonates have a higher porosity, which will allow slight ingress of the paste through the surface of the particle, greatly improving bond.

Dirty aggregate with a high percentage of fines can have a significantly negative impact on paste-to-aggregate bond. This condition can be diagnosed through thin-section review, where numerous fine particles are observed within the paste.

Air-entrained concretes can exhibit numerous entrained sized air bubbles concentrated along the paste-aggregate interface. In compression, microcracking tends to follow these weakened planes around the aggregate and reduce strength. This condition, sometimes called a *string of pearls,* is usually caused by dirty aggregate. Rock fines and mud often are problematic and also can be easily diagnosed through thin-section review (Fig. 16.2).

Cracking

Perhaps the most important condition to examine, document, and understand in concrete is cracking. Proper assessment of cracking is critical in structures, and various techniques are used. In addition to crack mapping, detailed information about the type and nature of cracking is important in helping to determine the cause. Petrographic analysis can document several important aspects related to various types of cracking, such as those discussed below.

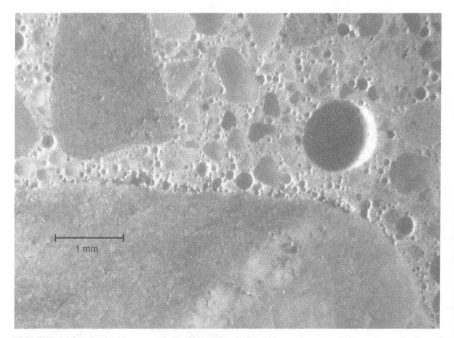

FIGURE 16.2   A significant amount of entrained-air voids concentrated along the perimeter of a coarse aggregate particle often is caused by dirty aggregate; microcracking will develop along these "strings of pearls" and reduce compressive strength.

Drying Shrinkage Cracking. These cracks are relatively large in scale (typically easily visible to the naked eye) and will propagate around the coarse aggregate in a subvertical orientation. The width of the crack usually will lessen with depth and often will penetrate the full depth of the member. The basic cause of drying shrinkage cracking is a reduction in volume of the concrete owing to water loss as the concrete dries. As the water content in the concrete increases, the shrinkage potential also increases, and the greater is the volume change. As the shrinkage potential increases, so will the degree of cracking. Other factors that affect drying shrinkage cracking include water-to-cement ratio, control-joint spacing and location, late cutting of control joints, ineffective curing procedures, and uneven subgrade or restraint (Fig. 16.3).

Plastic Shrinkage Cracking. Plastic shrinkage cracking occurs while the concrete is still plastic owing to rapid moisture loss at the surface by evaporation. The cracking pattern will appear as very thin cracks that occur in a polygonal pattern on the surface. Because only the top up to ¼ in. of the surface is affected, this problem is mostly a cosmetic issue. However, if the concrete is exposed to moisture and freezing conditions, plastic shrinkage cracking will allow greater ingress of moisture and increase the vulnerability of the concrete to scaling. Several factors are related to the cause of plastic shrinkage cracking.

Excess bleed water reworked into the surface paste. This elevates the water-to-cement ratio and increases the shrinkage potential and porosity. The potential for plastic shrinkage cracking is greatest when concrete is placed directly on an impervious subgrade, which will initiate all bleed water toward the surface.

Adding water to the surface to aid finishing. This also can produce plastic shrinkage cracking. This will increase the porosity of the surface paste and reduce the

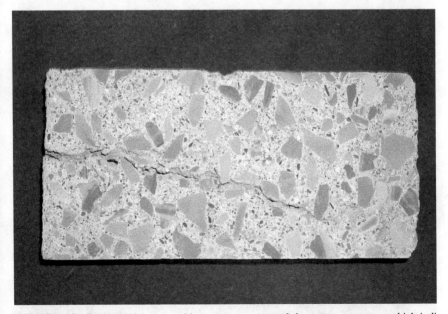

FIGURE 16.3 Drying shrinkage cracking propagates around the coarse aggregate, which indicates that cracking developed in the early hours after placement.

amount of air entrainment. Exterior slabs that have been "blessed" during finishing will be more vulnerable to freeze-thaw deterioration.

Hot, dry, and windy conditions. These will increase the likelihood of moisture loss during placement and curing significantly and lead to plastic shrinkage cracking at the surface.

Tearing

Concrete placements requiring a troweled surface need to be protected in hot and windy conditions with low humidity. If the surface dries out prematurely while the bulk of the concrete is still plastic, tearing can occur during troweling of the surface (Fig. 16.4).

Surface Delamination

A common problem with interior slabs on grade occurs when they are placed on impervious subgrades. A plane of subhorizontal void space is created when rising bleed water coalesces just beneath a dense, impervious troweled surface paste. This plane of weakness usually is restricted to the upper ⅛ to ½ in. The cause is *premature finishing* before bleed water has had a chance to reach the surface. However, there are several mitigating circumstances that will have a direct impact on the cause and severity of surface delamination.

High water-to-cement ratio concrete. Such mixtures will have more bleed water that rises to the surface. The higher the water content, the greater is the potential for delamination.

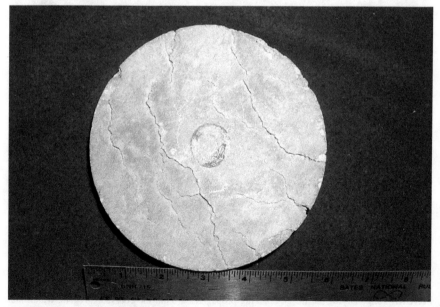

FIGURE 16.4 Surface cracks of tearing can occur during troweling if the surface paste has dried out while the concrete below is still plastic.

An impervious subgrade. This will direct most or all of the bleed water to the surface, increasing the likelihood of delamination. High water table, clayey soils, and placement directly onto wood, metal, concrete, or plastic are examples of impervious substrates. Core samples obtained from slabs on grade with surface delamination often were placed on polyethylene used as a vapor retarder.

Air entrainment. This is perhaps the key factor that affects surface delamination in concrete slabs. Rising bleed water will be delayed significantly by entrained air. The higher the entrained-air content, the slower the bleed water will rise to the surface. Delaminating concrete slabs must be examined carefully to document all parameters that may be involved. What may appear initially to be a contractor issue (premature finishing) actually may be related to the material supplied.

Blisters

Blisters are produced essentially the same way as surface delamination, with the difference being that the trapped bleed water is isolated in circular pockets usually a few inches across. The troweled top surface is still plastic enough to be deflected slightly by the trapped bleed water to create a small but noticeable mound on the slab. Blisters develop most often when the subgrade is cool and the concrete in the bottom sets slowly. Under traffic, the raised surface paste breaks apart, creating holes on the surface (Fig. 16.5).

Topping Delamination

Delamination of toppings on floors and decks along cold-joint contacts is very common and is affected by several different factors. Proper surface preparation prior to the application of a topping is critical. The following two items are important for success:

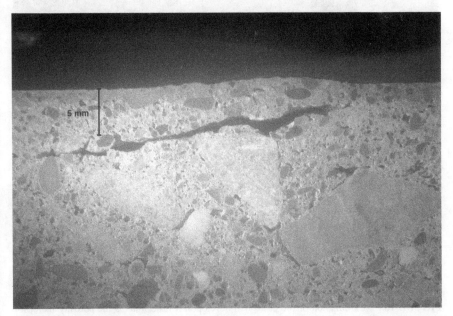

FIGURE 16.5 Beneath a surface blister is a void space that was created by rising bleedwater that became trapped directly under a dense, troweled surface paste.

A *good mechanical bond.* This is necessary on a concrete surface to allow a bonding agent and/or the topping material to have something to physically "grab onto." Dense and smooth troweled surfaces make it especially hard to achieve good adhesion. Sometimes the surface will need to be roughened to aid in achieving an effective bond.

Incompatibility of pH. This will adversely affect the adhesion of many bonding agents applied to carbonated surfaces of the base concrete. Since the bonding agents are designed to work in the high pH environment of fresh concrete, their effectiveness is compromised when they are applied to carbonated paste with a significantly lower pH. These surfaces will need to be physically removed down to sound, uncarbonated concrete (Fig. 16.6).

Scaling

The most common surface distress observed in exterior concrete that is exposed to moisture and freezing conditions is scaling. There are a couple of different types of scaling and several causes for this type of distress.

Disintegration scaling occurs where the top up to approximately ¼ in. of the surface paste breaks up into very small pieces owing to repeated freeze-thaw cycles when saturated. Some people have described this type of scaling while sweeping their sidewalk or driveway; the debris they swept up was scaled concrete.

Sheeting and delamination scaling occurs in highly vulnerable slabs and larger structures such as bridge decks and parking structures. The deterioration occurs as subhorizontal cracking that increases in frequency near the surface. The crack spacing ranges from a few millimeters to greater than 1 in. apart in severe cases (Fig. 16.7).

FIGURE 16.6   Carbonation of the paste will develop at a faster rate as the porosity increases. Noncarbonated paste with a higher pH turns pink when exposed to phenolphthalein indicator solution.

FIGURE 16.7   Severe subhorizontal cracking that was caused by freeze-thaw damage in a highly porous non-air-entrained concrete.

The causes of scaling concrete can be divided into four categories. A qualified and experienced petrographer should be able to identify the cause(s) during laboratory analysis:

Placement with high water content. This will increase the porosity of the surface paste and allow ingress of a greater amount of moisture. As the water-to-cement ratio increases, so does the potential for scaling. For exterior placements, a water-to-cement ratio of 0.45 or lower is recommended. Reality, especially in residential flatwork, is often quite different. It is not unusual to see the water-to-cement ratio approach 0.60 and above. This high-porosity paste is especially vulnerable and likely will experience scaling. In many cases water is added to the concrete after it is batched into the drum of the ready-mix truck. Evidence that water has been added occurs as darker gray paste areas with concave recesses of the coarse aggregate. These features are called *retempering notches* (Fig. 16.8).

Ineffective air entrainment. This is a common cause of scaling. Many concrete references discuss the recommended parameters for air-content volume in durable concrete. While air-content volume is important, especially entrained-air-content volume, it is the spacing of the entrained-sized air voids that is most critical for durability. The spacing factor required for adequate freeze-thaw durability should be no greater than 0.008 in. If all other parameters of the concrete meet freeze-thaw durability standards, an entrained-air void system with a spacing factor of 0.008 in. or less should provide durable concrete.

Overfinishing. Overfinishing of exterior concrete can produce a significant reduction of entrained air in the surface paste. If the reduction of entrained air reaches the point where the spacing factor exceeds 0.008 in., the cement will be vulnerable to scaling.

FIGURE 16.8 The darker gray paste area in the concave coarse aggregate notch indicates a lower-water-content paste than the lighter paste area and that the concrete was retempered.

Ineffective curing. Ineffective curing of exterior concrete is a very common cause of scaling. Perhaps the most important and least appreciated aspect of durable concrete is curing. Experienced has shown that numerous environmental factors will affect the methods and duration of curing. Factors that affect curing include the following:

Temperature

Hot, dry, and windy conditions

Late fall and winter placements

Use of chemical and pozzolanic admixtures

Water-to-cement ratio

Effective use of curing compounds

Mortar flaking. This is very specific in its appearance and cause. A very thin (typically ¹⁄₃₂ in. up to ⅛ in. thick) section of paste will scale off directly over coarse aggregate particles. This type of scaling can be caused by small quantities of bleed water that coalesce directly above the aggregate particle. The high water-to-cementitious ratio paste creates a weakened plane that can flake off more easily during freezing. Mortar flaking also can occur when the surface paste over aggregates dries out rapidly, retarding hydration. This poorly hydrated paste will be especially vulnerable to freeze-thaw deterioration.

Chlorides

Chlorides present in exterior concrete flatwork through the use of natural deicers (NaCl) play a misunderstood role in surface deterioration. Chloride-based deicers lower the freezing point of water and melt ice. They also increase the number of freeze-thaw cycles that concrete would otherwise experience if they were not present. This certainly will hasten deterioration in concrete that is vulnerable to scaling. However, the true *cause* of the deterioration is not the presence of chlorides. There has to be some other inherent defect in the concrete (high w/cm ratio, an ineffective entrained-air void system, poor finishing/curing, etc.) for the chloride to have any effect. Good, durable concrete placed with effective curing can be exposed to high levels chlorides and not experience scaling. Highway pavements are good examples of concrete exposed to heavy deicer usage with little or no deterioration.

Dusting

Concrete placements that leave a powdery residue to the touch describe a condition called *dusting*. The top surface of the concrete will be relatively soft and friable owing to a chemical reaction that has occurred during the curing period in the early hours after placement. Carbon monoxide (CO) and carbon dioxide (CO_2) gases in the air or dissolved in water will react with calcium hydroxide (CaOH) in the paste to produce calcium carbonate ($CaCO_3$). Calcium carbonate (the mineral calcite) is actually harder and more brittle than the noncarbonated paste. In addition to the chemical and physical changes that occur, there also will be a significant change in the pH of the carbonated paste. Typically, the pH will drop from 12 or 13 to around 8 or 9, which is in the neutral range. This significant drop in pH will create an environment in the concrete that will be conducive to corrosion of embedded reinforcing steel. The rate and depth of carbonation will be affected by two key factors: exposure to heavy gases and the rate of carbonation.

Exposure to heavy gases includes an overall higher concentration of heavy gases in urban versus rural areas, as well as high vehicle traffic areas, such as roads, bridges, and parking structures.

The rate of carbonation will increase as the water-to-cementitious ratio and overall porosity of the concrete increase. Increased porosity will allow greater ingress of moisture and heavy gases to occur. A construction example that could lead to heavy dusting often occurs in northern latitudes during the winter months. Interior concrete slabs on grade often are placed in enclosed structures with manufactured heat. Forced air heaters will produce high levels of these heavy gases that tend to sit on the surface of a freshly placed concrete surface. If the combustion products are not vented properly, the highly vulnerable plastic concrete will carbonate at the surface and potentially will dust.

The American Concrete Institute (ACI) defines *dusting* in ACI 116R as "the development of a powdered material at the surface of hardened concrete." The ACI lists the causes of dusting in document ACI 302.1R as

Floating and troweling concrete with bleed water on the surface.

Insufficient cement content (Table 6.2.4).

Excessive clay, dirt, and organic materials in the aggregate.

Use of dry cement as a blotter to speed up finishing.

Water applied to the surface to facilitate finishing.

Carbonation of the surface during winter concreting caused by unvented heaters.

Inadequate curing, allowing rapid drying of the surface, especially in hot, dry, and windy weather.

Freezing of the surface.

Dusting Repair. The Portland Cement Association's Document 1S177.03T, *Concrete Slab Surface Defects: Causes, Prevention, Repair,* lists two remedial measures to employ when dusting occurs: grind the surface or apply a surface hardener.

Popouts

Surface defects where the top surface has a relatively shallow, roughly conically shaped pit with a fractured aggregate particle at its base is called a *popout.* Popouts range in size from a few centimeters to 3 to 4 in. in diameter. It is important to distinguish if the offending particle is fractured. An undamaged aggregate particle may indicate that mortar flaking has occurred. There are two distinct types of popouts that are produced by completely different mechanisms of expansion: mechanical popouts and chemical popouts.

Mechanical popouts. These are produced by highly porous aggregate particles located near the surface that absorb moisture and expand outward on freezing. The surface defect produced will be two to several times larger than the offending particle. Highly porous particles vary in composition and include iron oxide, shale, mud stones, silt stones, sandstones, porous carbonates, and various high-grade metamorphic rock types high in mica content (Fig. 16.9).

Chemical popouts. These are produced by aggregates containing very fine-grained or amorphous silica that reacts and expands volumetrically with a highly saturated alkaline-rich paste. Chemical popouts usually occur shortly after placement when the concrete contains significant available moisture. Many reactive aggregates within a concrete mixture can "pop" within a few days after placement.

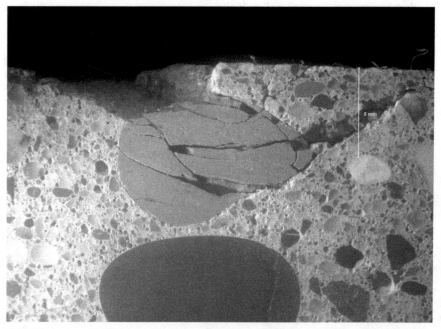

FIGURE 16.9 A highly porous shale particle exhibits extensive cracking and has fractured the surface, producing a popout. Expansion of the particle occurred on freezing when saturated.

However, chemical popouts can occur over a long period of time after placement if moisture is present. Common reactive popout-producing aggregates include chert, and shale, as well as igneous and metamorphic rocks containing strained quartz.

"D" Cracking

A formidable deterioration mechanism commonly seen in pavements in the northern latitudes is a phenomenon called *"D" cracking*. Porous aggregate in a high enough percentage that freezes under saturated conditions will produce a wholesale volumetric expansion of the concrete and exhibit a diagnostic cracking pattern. The crack pattern at the surface tends to run parallel to both the longitudinal and transverse joints. When cracking parallels the intersection of these joints, it forms a pattern shaped like the letter *D*, hence the name. Close inspection of core samples obtained from affected concrete reveals cracking that propagates through the paste and coarse aggregate. The cracking at depth in the concrete is also aligned roughly parallel to the subgrade. The frequency of cracking increases toward the top surface. There are, however, no secondary deposits within the cracks or air voids (Fig. 16.10).

Alkali-Aggregate Reaction (AAR)

One of the most destructive agents that affect concrete structures is reactive coarse aggregates that produce a wholesale volumetric expansion. Much like popouts, the mechanism of expansion can be related to highly porous or chemically reactive aggregates. Instead of isolated surface defects, AAR affects the entire structure. The

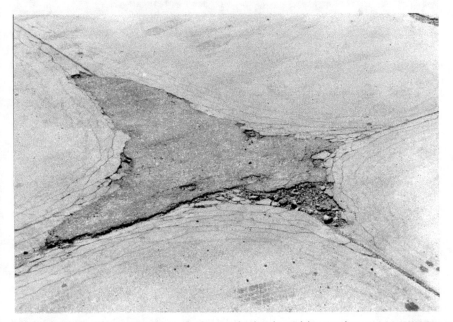

FIGURE 16.10 This previously repaired, severely deteriorated intersecting saw-cut pavement joint was produced by "D" cracking. The expansion is produced from the freezing of water absorbed by highly porous coarse aggregate.

reaction is driven by moisture; thus exterior concrete structures exposed to the elements are the most vulnerable. Large-scale volumetric expansion–type cracking at the surface often will exhibit secondary products leaching from the cracks, and this is a common clue. Reactive aggregates can be classified into two categories: alkali-silica reaction and alkali-carbonate reaction.

Alkali-Silica Reaction (ASR). Concrete with destructive ASR will exhibit several characteristics that are diagnostic:

- Cracking at the surface in a pattern that is consistent with a volumetric expansion (Fig. 16.11).
- Cracking that propagates through both the paste and reactive aggregates (Fig. 16.12).

A white-colored to clear gel will be observed lining and filling fractures and adjacent air voids. The gel is a by-product of the reaction between the siliceous component of the aggregate and the sodium (Na) and potassium (K) alkalis from the cement in the presence of moisture. The gel continues to imbibe moisture and swells, causing the expansion cracking. ASR is a progressive expansion problem as long as there continues to be reactive silica present and available moisture.

Alkali-Carbonate Reaction (ACR). ACR is a rarely seen form of volumetric expansion–type cracking in concrete. Dolomite, a magnesium-rich carbonate ($CaMgCO_3$) aggregate, can undergo an expansive chemical reaction in the presence of moisture.

FIGURE 16.11   A highly deteriorated concrete wall in a parking structure that is experiencing alkali-silica reaction (ASR).

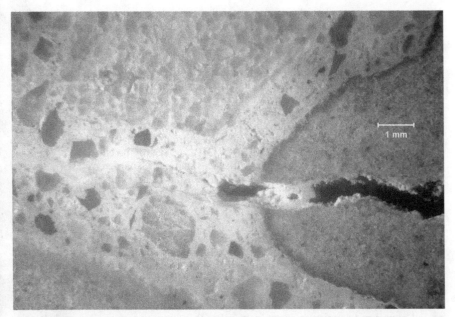

FIGURE 16.12 Cracking propagates through both the paste and the reactive quartzite coarse aggregate particles. White silica gel fills cracking adjacent to the aggregate particles.

Dolomite will decompose into calcite ($CaCO_3$) and brucite ($Mg[OH]_2$). The water addition or hydration that occurs in the formation of brucite produces the expansion.

Ettringite Distress

A relatively new area of concrete distress that has challenged the minds of forensic material scientists is understanding the sometimes destructive nature of the mineral ettringite. A calcium sulfoaluminate mineral ($Ca_6A_{13}[SO_4]_3[OH]_{12} \times 26H_2O$), ettringite is one of the important minerals formed during the hydration of portland cement. In recent years, the quantities of ettringite observed in samples of concrete have increased steadily. Coincidently, these elevated quantities of ettringite are found in concrete experiencing varying degrees of distress. After a 10-year period of research and debate, a consensus was reached about the general destructiveness of excessive quantities of ettringite in concrete. The exact mechanism of harmful expansion associated with ettringite is not agreed on by the experts. However, it is my opinion that the key to ettringite distress is the mineral's ability to behave like a gel and swell through the addition of water in a manner similar to ASR. It is believed that the crystal lattice of ettringite can take on and give up water molecules, thereby changing its physical state and volume.

The theory goes like this. When concrete contains excess sulfur, and moisture is introduced, an ettringite gel is formed within the cement-paste pore system. Whether this gel is truly amorphous (noncrystalline) or cryptocrystalline is unimportant. The key is that this solution behaves like a gel and expands. With repeated or constant exposure to moisture, new gelatinous ettringite is formed, and the expansion is progressive. The reaction will cease when the sulfate source is either depleted or cut off.

In severe cases, complete disintegration of the concrete can occur. A number of ettringite-related types of distress have been identified: deicer distress, delayed ettringite formation, and external sulfate attack.

Deicer Distress. Deicer distress is a very specific form of ettringite distress that is related to concrete exposed to natural deicers. It occurs in northern climates in concrete associated with vehicle-type infrastructures such as pavements, bridge decks, median barriers, etc. Generally, natural deicers consist of the mineral halite (NaCl), with up to 8% impurities of gypsum ($CaSO_4$). These minerals dissolve in solution and migrate into the cement paste, where the sulfate ions combine with constituents (calcium, aluminum) to form ettringite [$Ca_6Al2(SO_4)_3(OH)_{12} \times H_2O$]. The ettringite crystallizes within air voids immediately adjacent to surfaces that allow solution access (joints and cracks). The ettringite-filled entrained-air voids in this zone render the concrete vulnerable to freeze-thaw deterioration. The deterioration will be progressive as long as natural deicer use continues.

Delayed Ettringite Formation (DEF). DEF is different from deicer distress in that the source of the sulfate is internal as opposed to a specific external source. Internal sources of gypsum contamination include the fine aggregate, an oversulfonated cement, and high-sulfur admixture or pozzolans. In the presence of water, the sulfate goes into solution and combines with calcium and aluminum to form ettringite. The destructive expansion is produced when an ettringite gel is formed with associated swelling. Repeated exposure to moisture forms new gel with progressive expansion and deterioration as long as the sulfate source is not depleted. Recent research indicates that elevated-temperature-cured concrete (>165°F) will produce a monosulfate form of ettringite that becomes unstable once the product cures and cools. Subsequent exposure to moisture will allow the unstable form to recrystallize into ettringite by the addition of water molecules with subsequent volumetric expansion (Fig. 16.13).

External Sulfate Attack. Ettringite deterioration where the sulfate is derived from an external source is called *sulfate attack*. Sulfate-rich soils and seawater are the two most common external sources of sulfate. Some of the most aggressive concrete deterioration involves these kinds of sulfate attack (Fig. 16.14).

Environmental Effects

Fire Damage. The damage to concrete from exposure to fire can be quantified and mapped. The most important factors that determine the severity of damage to concrete is the maximum temperature reached and the duration of exposure to heat. Fire damage can be quite severe, as was documented by recent work performed on the Pentagon in Washington, D.C., in the wake of the terrorist attacks on September 11, 2001. Because of the extreme temperatures reached by the combustion of jet fuel from the aircraft that crashed into the building, much was learned from the affected concrete. This information was used to evaluate damage to structural concrete elements during the reconstruction. These data also will be used to aid in future design of critical concrete structures that may be at risk to exposure to fire with extreme heat (Figs. 16.15 through 16.18).

Forensic evaluation of fire-damaged concrete can document physical and chemical changes that give clear indicators of the temperatures reached. These increments of temperature are called *isotherms* and can be mapped within the affected concrete. The following table outlines the temperatures at which specific observable changes occur within concrete exposed to fire.

FIGURE 16.13   Completely disintegrated concrete in an underwater pier with severely corroded prestressed steel, the end result of ettringite distress in its most severe form.

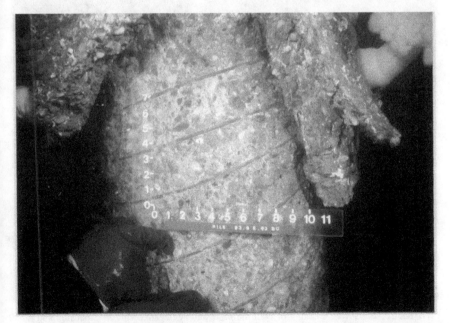

FIGURE 16.14   Severely spalled concrete on an underwater pile experiencing sulfate attack.

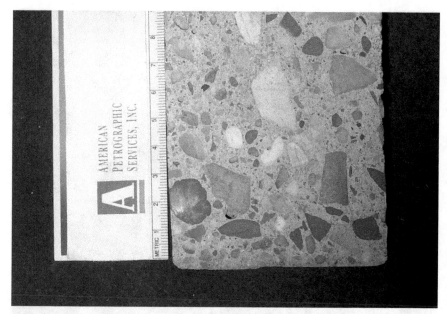

FIGURE 16.15 Iron oxide discoloration within siliceous aggregates from exposure to fire with temperatures in excess of 230°C to a depth of 25 mm.

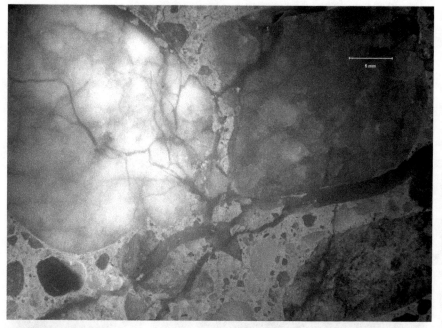

FIGURE 16.16 Severe cracking of both the paste and aggregate owing to expansion and rapid moisture loss within the paste on exposure to fire.

FIGURE 16.17   Rapid loss of water in the paste as it tries to escape produces severe microcracking (100×, plane polarized light).

Approximate Temperature	Effects on Concrete
100°C/212°F	Paste loses free water; craze cracking on the surface begins.
300°C/550°F	Paste and both siliceous and carbonate aggregate color changes from pink to red as temperatures increase. Macrocracking is observed in the paste owing to thermal shock. Cracking will begin to propagate through expanding aggregates as temperature increases.
400°C/750°F	Compressive strength will be reduced by 50%.
570°C/1063°F	Quartz aggregate varieties expand owing to a phase change, producing surface popouts, cracking, and spalling.
600°C/1112°F	Reinforcing steel loses yield strength and buckles under load, producing spalling of the concrete.
900°C/1652°F	Aggregate color changes to buff.

Freezing. Exterior concrete placed in northern latitudes during winter weather runs the risk of freezing if it is not protected properly. If concrete freezes before achieving approximately 500 lb/in.^{2}, the free water within the mixture will form blade-shaped ice crystals in the paste. The ice crystals create void spaces that will crumble once the ice thaws. If the concrete freezes after it has reached 500 lb/in.^{2}, cement hydration will slow and eventually stop. Once the concrete is heated and moisture is present, hydration will continue, and the concrete can reach the specified strength easily. The bladed frost crystals usually are quite small, with the best chance for documentation when reviewed microscopically (Fig. 16.19).

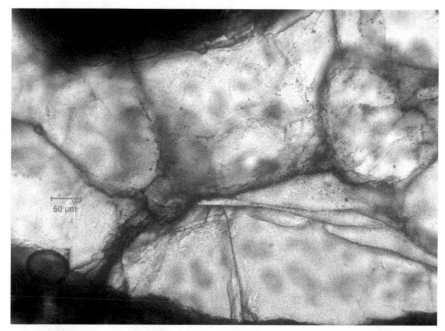

FIGURE 16.18   Red iron oxide staining with cracking of a fine-aggregate-sized quartz particle exposed to heat in excess of 570°C (400×, plane polarized light).

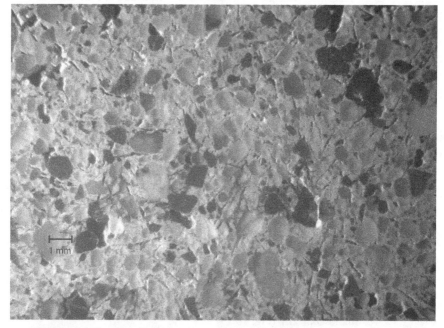

FIGURE 16.19   Blade-shaped frost crystal imprints within a mortar paste that froze.

SAMPLING FOR TEST SPECIMENS

Two key components for the best representation of a particular situation being evaluated are sample location and the number of samples. The number of samples can vary from a single core, section, or cast cylinder to several dozen samples depending on the size and scope of the damage. If a project appears headed for litigation, more samples are recommended. Sample location can be critical for obtaining the best data from testing. The following examples illustrate proper sampling of concrete for petrographic analysis.

Cracking. A sample taken through the cracking is very important. Critical information such as width, depth, and whether cracking propagates around or through the aggregate can only be determined if the crack is present in the sample.

Dusting/blisters/popouts. To best evaluate these types of problems, samples should be obtained with example(s) of the condition being evaluated present in the sample.

Scaling/delamination. Cores should be taken in the area of distress with a portion of the original top surface still remaining. Important evidence to help document the cause of the deterioration may be present in the surface paste.

Slabs on grade. When possible, cores taken to the subgrade are desirable. Full-depth samples help to understand important aspects of subgrade types, bleeding rates, slab thickness, and the potential for restraint.

TESTING PROCEDURES

Consultation with the appropriate experts should result in an investigation plan using noninvasive and invasive testing.

Noninvasive Testing

Rebound Hammer. The Schmidt rebound hammer is used commonly to give an approximation of compressive strength of concrete by testing the surface hardness. (ASTM C805, *Standard Test Method for Rebound Number of Hardened Concrete*) It consists of a spring-controlled hammer that slides along a plunger within a tube-shaped housing. The spring forces the hammer against the surface of the concrete, and the distance of rebound is measured on a scale (Fig. 16.20).

Pulse Velocity. Pulse velocity testing measures the speed that sound travels through concrete (ASTM C597, *Standard Test Method for Pulse Velocity through Concrete*). The pulse velocity through sound concrete travels at approximately 12,000 to 13,000 ft/sec. Cracks and voids will slow the sound wave significantly as it travels around these voids (Fig. 16.21). The following list compares general concrete quality with velocity readings:

Excellent:	Above 15,000 ft/sec
Good:	12,000–15,000 ft/sec
Marginal:	12,000–10,000 ft/sec
Poor:	7,000–10,000 ft/sec
Very poor:	Below 7000 ft/sec

Impact-Echo. A nondestructive method for testing concrete and masonry structures uses impact-generated sound waves that are deflected by internal flaws and exterior

FIGURE 16.20 A Schmidt rebound hammer being used to indicate the compressive strength of an interior slab on grade.

FIGURE 16.21 Pulse-velocity equipment being used to test for cracking, delaminations, and void spaces in an interior reaction wall.

surfaces. Impact-echo can be used to document the location and extent of cracks, delaminations, voids, honeycombing, and debonding. This method has been used with great success in locating flaws and defects in highway pavements, bridges, buildings, tunnels, dams, piers, sea walls, and many other types of structures.

Ground-Penetrating Radar (GPR). GPR is a nondestructive testing tool that can document slab thickness or locate buried objects in the ground or things such as voids, delaminations, conduits, piping, and reinforcing steel in concrete. Data are collected and processed continuously by a portable computer module generating real-time results (Fig. 16.22).

Half-Cell Testing. This test method estimates the electrical half-cell potential of non-coated reinforcing steel to determine the corrosive activity of the reinforcing steel. (ASTM C876, *Standard Method for Half-Cell Potentials of Uncoated Reinforcing Steel in Concrete*).

Invasive Testing

Compressive Strength. Concrete cylinders cast during placement and cores obtained from existing structures are tested routinely for compressive strength (ASTM C39, *Standard Test Method for Compressive Strength of Cylindrical Concrete Specimens,* and C42/C42M, *Standard Test Method for Obtaining and Testing Drilled Cores and Sawed Beams of Concrete*).

Splitting Tensile Strength. This test determines the splitting tensile strength of cylindrical concrete molded cylinders or drilled cores (ASTM C496, *Standard Test*

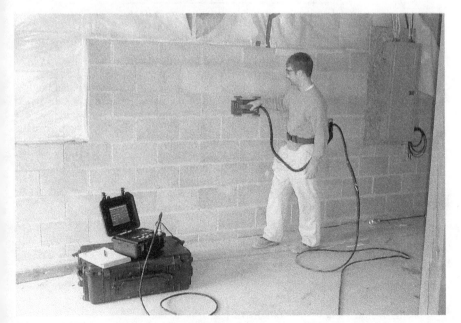

FIGURE 16.22   A concrete engineer performs a ground-penetrating radar survey of a concrete block wall to locate void spaces and steel reinforcement.

Method for Splitting Tensile Strength of Cylindrical Concrete Specimens). A diametral compressive force is applied along the length of the sample until tensile failure occurs.

Flexural Strength. Flexural strength is determined by the use of a simple beam with third-point loading (ASTM C78, *Standard Test Method for Flexural Strength of Concrete*). Reported as the modulus of rupture, the results are used to determine compliance with specifications in slabs and pavements.

Chloride Ion Content. The chloride ion content in concrete is documented using wet chemistry procedures. There are several both water-soluble and acid-soluble methods for determining chloride ion content. Chlorides in excess of 300 to 400 ppm are considered in the range where embedded steel reinforcement will begin to be vulnerable to corrosion. Chlorides measured above these levels will lower the freezing point of water and result in an increase in the number of freeze-thaw cycles in exterior concrete. High chloride levels also will enable water to better permeate exterior concrete at the surface.

Rapid Chloride Ion Permeability. The relative porosity of and rate of potential chloride ion transmission are documented using this test procedure (ASTM C1202, *Standard Test Method for Electrical Indication of Concrete's Ability to Resist Chloride Ion Penetration*). Three 2-in.-thick by 4-in.-diameter pucks cut from two concrete cylinders or cores are used. The test measures the rate in coulombs at which an electric charge is transmitted through the concrete. The higher the coulomb value recorded, the higher the relative porosity of the concrete. Three test values are reported along with the average. The relative permeability ratings are listed below:

Rating	Coulombs
Excellent	<1000
Good	1000–2000
Fair	2000–4000
Poor	>4000

Hardened Air Content Analysis. The hardened air content of concrete can be documented using the modified point-count method or the linear traverse method (ASTM C457, *Standard Test Method for Microscopical Determination of Parameters of the Air-Void System in Hardened Concrete*). The linear traverse method is more accurate than the modified point-count method owing to the significantly higher number of data points that are recorded. This leads to more accurate air content data. Important parameters documented include the entrapped and entrained air content, spacing factor, specific surface, or the total surface area of the air-void system, and the number of air voids per inch. Durable concrete that will be exposed to moisture and freezing conditions should meet the following air-void system parameters:

Total air content, %	5 to 7
Entrained air content, %	3 to 6
Entrapped air content, %	0.5 to 2
Air voids per inch	7.5 to 15
Specific surface, in^2/in^3	>600
Spacing factor, inches	0.008 in. or less

Petrographic Analysis. The most comprehensive testing that can be performed to document the physical and chemical parameters of concrete is petrographic analysis

(ASTM C856, *Standard Practice for Petrographic Examination of Hardened Concrete*). An experienced and competent petrographer can record important parameters such as air content, water-to-cement ratio, degree of hydration, pozzolan presence and amounts, carbonation, aggregate type and quality, surface conditions, and many others. Concrete quality and problems that may arise are commonly addressed through petrographic analysis.

X-Ray Diffraction Analysis (XRD). Minerals that are too small to identify optically can be documented using XRD. X-rays will be diffracted by the internal crystal structure of unknown minerals and measured by a detector that measures angles through a 150 degree arc. The exact angle(s) of diffraction will be diagnostic of a particular mineral and identify it. XRD is used often to identify secondary deposits that occur on concrete surfaces that transmit moisture.

Scanning Electron Microscopy (SEM). A scanning electron microscope can be used to review concrete samples at magnifications of up to 1,000,000 times. High-resolution images and the geochemistry of cementitious compounds can be generated and studied.

Pull-Out Testing. This test uses a special ram to measure the force required to pull out a steel rod cast up to 3 in. deep into the concrete. The concrete is simultaneously put in tension and shear, but the force exerted can be related to compressive strength or help to determine when forms can be removed safely (Fig. 16.23).

Probe Penetration. A Windsor probe is the apparatus used most commonly in the field to perform penetration testing into hardened concrete. The equipment consists of a powder-actuated gun or driver with hardened alloy probes (generally 0.25 in. in diameter by 3.125 in. long). The probe is driven into the concrete by a precision powder charge. The depth the probe penetrates will give an indication of compressive

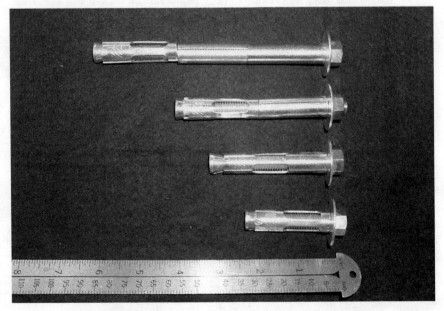

FIGURE 16.23   Various sized steel rods used for pull-out testing.

00 LAB 001 Petrographic Examination of Hardened Concrete
ASTM: C-856

Job #:   Date: 12-9-03/12-11-03
Sample Identification: Truss with Concrete, Base Concrete   Performed by:

I. General Observations

1. Sample Dimensions: Our analysis was performed on an approximately 44 mm (1-3/4") thick base concrete portion of a 178 mm (7") x 89 mm (3-1/2") x 89 mm (3-1/2") and a 178 mm (7") x 95 mm (3-3/4") x 89 mm (3-1/2") thick polished section that was cut from the original 89 mm (3-1/2") x 178 mm (7") x 178 mm (7") long section.

2. Surface Conditions:

    Top: Rough, unfinished surface covered by air void spaces
    Bottom: Rough, irregular, fractured surface

3. Reinforcement: 5 mm (3/16") diameter steelmesh truss impression was observed approximately 35 mm (1-3/8") depth from the top surface on the fractured bottom surface. Also, 5 mm (3/16") diameter steelmesh truss impression was observed on the fractured sides. Moderate corrosion product was observed.

4. General Physical Conditions: The sample consists of an at least 44 mm (1-3/4") thick base concrete with an at least 44 mm (1-3/4") thick concrete topping loosely held within a steelmesh truss. The topping and base concrete were no longer bonded to the truss. Also, the topping is no longer bonded to the base concrete. Paste rich slurry/froth with a significant amount of air voids was observed on the top surface of the base concrete. Several tine or rake marks, approximately 4 mm (5/32") wide, proceed up to 10 mm (3/8") depth from the top surface of the base concrete. Few subvertical drying shrinkage microcracks proceed up to 5 mm (3/16") depth from the top surface. Microcracking was observed in various orientations proximate to the fractured bottom surface of the base concrete. Much of the coarse aggregate in cross section was segregated at least 9 mm (3/8") depth from the bottom surface of the base concrete. The concrete was purposefully air entrained and contains an air void system considered freeze-thaw resistant only under moderate exposure conditions. The purposeful air entrainment was only fairly distributed throughout the sample with common concentration along paste aggregate boundaries. Darker colored, denser, and non-air entrained paste observed in several concave coarse aggregate notches suggests the sample may have been retempered or underwent multiple stage batching. Poor overall condition.

II. Aggregate

1. Coarse: 19 mm (3/4") maximum sized crushed carbonate and argillite. Fairly well graded with fair to poor overall distribution.

2. Fine: Natural, siliceous sand that was fairly well graded. The grains were mostly sub-angular. Good overall uniform distribution.

III. Cementitious Properties

1. Air Content: 3.7% total, 0.3% entrapped, 3.4% entrained
2. Depth of carbonation: Ranged from 1 mm (1/32") up to 6 mm (1/4")
3. Pozzolan presence: None observed
4. Paste/aggregate bond: Fair
5. Paste color: Medium gray becoming darker in the top approximately 2 mm (1/16") of the sample
6. Paste hardness: Medium
7. Paste proportions: 25% to 27%
8. Microcracking: Few subvertical drying shrinkage microcracks proceed up to 5 mm (3/16") depth from the top surface of the base concrete. Microcracking was observed in various orientations in the paste proximate to the fractured bottom surface of the base concrete.
9. Secondary deposits: None observed
10. Slump: Estimated, medium (2" - 4")
11. Water/cement ratio: Estimated at between 0.40 to 0.45 with approximately 10-12% unhydrated or residual portland cement clinker particles.
12. Cement hydration: Alites-well to fully; Belites-low

IV. Conclusions

The general overall quality of the concrete was fair.

FIGURE 16.24   An example of a petrographic analysis data sheet that includes the minimum amount of information that should be documented from the examination.

strength. The instrument should be calibrated for the type of concrete, as well as the type and size of the aggregate.

INVESTIGATIVE SUMMARY

Once all the field and laboratory work has been completed, the information needs to be compiled and analyzed. Consultation between the field and laboratory personnel is required to reach a full understanding of the implications of the data. After these results are assessed properly, a final report should be written that draws the most accurate conclusions and provides appropriate recommendations. Every report should be written as if it will end up in litigation, with clear conclusions that include photographs. Photographic documentation is the best way to illustrate important facts that are relevant to the problem being investigated.

Typical Petrographic Analysis Report

If a petrographer is hired to evaluate concrete, mortar, aggregate, or any other type of cementitious material, a report should be generated on completion of the examination that includes the following sections: introduction, conclusions, sample identification, test results, test procedures, and remarks. The report also should include a detailed petrographic analysis data sheet and photographs that illustrate pertinent aspects documented in the examination that bear directly on the conclusions (Fig. 16.24). The petrographer also should be prepared to discuss the results with the client so that the client has a complete understanding of the work performed.

BIBLIOGRAPHY

ACI 104-71 (97) to ACI 223-98. *ACI Manual of Concrete Practice 2002,* Part 1. American Concrete Institute, Farmington Hills, MI.

ASTM (2002). *Annual Book of ASTM Standards 2002,* Section 4: "Construction; "Concrete and Aggregates," Vol. 04.02. American Society for Testing Materials, Farmington Hills, MI.

Campbell, D. H. (1999). *Microscopical Examination and Interpretation of Portland Cement and Clinker,* 2d ed., Portand Cement Association, Skokie, IL. Farmington Hills, MI.

Kosmatka, S. H., B. Kerkhoff, and W. C. Panarese (2002). *Design and Control of Concrete Mixtures,* 14th ed. Portland Cement Association, Skokie, IL.

St. John, D. A., A. W. Poole, and I. Sims (1998). *Concrete Petrography.* Wiley, New York.

Wolter, S. F. (1997). *Ettringite: Cancer of Concrete.* Burgess International Group, Inc., Minneapolis.

Steel

ROBERT S. VECCHIO, PH.D., P.E.

INTRODUCTION

The safe and reliable operation of structures always has been a major engineering priority. Condition assessment or fitness-for-service evaluation, therefore, plays a fundamental role in ensuring the integrity of such structures. Fitness for service (FFS) is a methodology that accounts for actual service stresses, flaw or damage zone sizes, as-built material properties, and operating environment in determining the actual safety or reliability of a structure in its present condition or expected future state. That is, FFS is a common-sense engineering approach to evaluating the suitability of a structure for its intended service. Moreover, an FFS assessment can provide the foundation for establishing the remaining life of a structure and thereby provide a basis for making capital-based decisions for repair or replacement.

Steel in particular, because of its extensive use in a wide assortment of structures (e.g., bridges, buildings, power plants, etc.) and its ability to be fabricated easily by welding, has become the most widely evaluated material with respect to FFS. In fact, the use of welding was the primary driving force behind the development of most of today's FFS procedures for steel structures. Welded joints inherently contain flaws or discontinuities such as cracks, incomplete penetration, slag, and/or porosity, as well as exhibiting altered base metal mechanical and metallurgical properties. Consequently, an FFS assessment of steel structures can encompass original design review, as-built stress analyses, material property evaluation, quality control, nondestructive examination (NDE), probabilistic risk assessment and, most important, fracture mechanics.

Traditional design of steel structures accounts for weld-joint discontinuities and altered metallurgical properties of the base metal by conservatively limiting the size of discontinuities permitted in fabricated joints and applying cautious factors of safety to joint stresses. In effect, traditional design assumes that no significant flaws or service-induced damage (i.e., fatigue cracking, corrosion, fire-induced distortion) exists in the structure. In contrast, the FFS approach acknowledges the presence of flaws, damage, and/or degraded material properties and assesses a structure in its existing or anticipated condition. Thus the FFS approach is a quantitative assessment that is based on rational evaluation of stresses, material properties, nondestructive examination, and fracture mechanics.

To this end, this chapter provides a general overview of the FFS and condition assessment of steel structures, particularly structures that have sustained damage, such as in-service cracking, corrosion, or fire-induced distortion. In addition, the methodologies and testing procedures generally required in the FFS assessment of steel structures are discussed.

FAILURE-DAMAGE MODES IN STEEL STRUCTURES

In order to assess properly the condition or fitness of a steel structure for continued service, it is important to recognize the potential failure modes, as well as likely types of flaws and damage mechanisms. Generally, failure modes are either instantaneous or progressive in nature; that is, failure occurs with little or no warning, or it may be preceded by a considerable amount of detectable crack growth, deformation, or corrosion deterioration. Flaws and damage mechanisms (other forms of structural steel damage) can develop during service or be preexisting. For example, a 4-in.-long fatigue crack at the toe of a cover-plate-to-flange fillet weld arises owing to in-service cyclic loading, whereas a weld-fabrication flaw, such as incomplete penetration, would be considered a preexisting cracklike flaw. It is essential, therefore, that the damage mechanism be assessed accurately, along with potential failure modes from such damage, so that proper condition and remaining-life assessments can be performed.

The primary failure-damage modes for steel structures are

- Ductile failure
- Brittle fracture
- Buckling/instability
- Corrosion and stress-corrosion cracking
- Fatigue
- Creep
- Metallurgical degradation

Each of these failure-damage mechanisms exhibits unique characteristics such that different design criteria and material behavior must be considered for each in order to preclude failure. The most common aspects of these steel failure modes are discussed in the following sections. Detailed discussions of failure modes and mechanisms can be found in several of the references.

Ductile Failure

Of all the failure modes that can occur in steel structures, ductile failure or excessive distortion (i.e., inelastic/nonrecoverable deformation/strain) occurs least often, yet a significant share of almost every design code for steel that uses yield strength as the critical design parameter (e.g., *AISC Steel Construction Manual,*[1] *ASME Boiler and Pressure Vessel Code,* Sec. VIII, Div. 1[2]) is devoted to the prevention of ductile failure. Ductile failure usually is characterized by excessive inelastic (nonrecoverable) deformation prior to attaining the steel's ultimate or tensile strength. A beneficial aspect of ductile behavior is the large amount of energy absorption that occurs prior to failure. In many instances, therefore, extreme or unanticipated loads can be redistributed within the structure without the consequence of failure. Frequently, excessive deformation occurs over a long enough period of time that the structure can be stabilized or abandoned before unacceptable property damage or injuries occur.

Ductile fractures generally are irregular in appearance and exhibit shear lips and localized "necking down" (thinning of the cross section). An example of ductile fracture in a bridge wind chord is shown in Fig. 17.1. On a microscopic level, ductile fractures exhibit a fracture morphology known as *microvoid coalescence,* as shown in Fig. 17.2.

More often than not, ductile fractures also can be identified by the appearance of the adjacent painted surface or mill scale. If the steel structure has sustained inelastic deformation or applied loads in excess of the yield strength, the paint and/or mill scale—owing to its brittle nature—cracks or crazes in bands perpendicular to the direction of the local principal stress.

However, the deformation capacity of a steel structure can be affected by the service temperature, loading rate, and level of constraint. As the temperature decreases and/ or the loading rate and constraint increase, the capacity for deformation, particularly

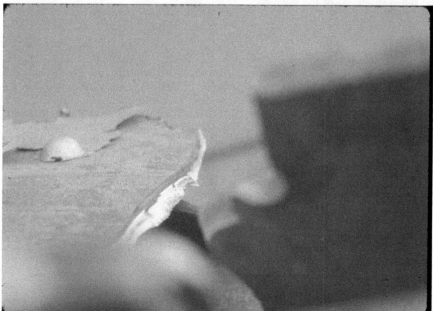

FIGURE 17.1   Ductile deformation and fracture of a bridge wind chord. Note the spalled paint in the vicinity of the deformation.

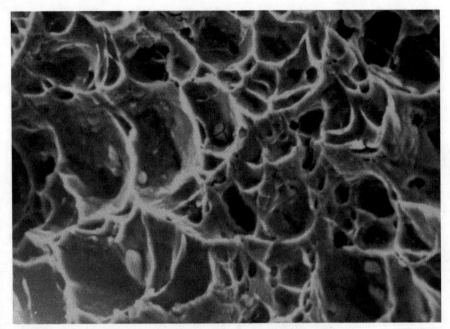

FIGURE 17.2   Scanning electron micrograph showing microvoid coalescence on the surface of a ductile facture.

at notches (stress concentrations), decreases, and the structure can undergo a transition from ductile to brittle behavior.

Brittle Fracture

In contrast to ductile failures, brittle fractures occur with little or no deformation and, therefore, with little or no warning. Brittle fractures initiate and propagate through steel structures at very high speeds (approaching the speed of sound in steel) often at stress levels below the yield strength or even below design allowable stress levels. The rapidity of brittle fracture propagation precludes the intervention of mitigating measures. Consequently, many brittle fractures result in significant structural damage, catastrophic failure, and/or serious injury.

Brittle fracture typically is characterized by flat fractures with little or no associated inelastic deformation, as shown in Fig. 17.3. Additionally, brittle-fracture surfaces exhibit features known as *chevron marks,* which by virtue of their orientation can be used to identify the direction of fracture propagation and, more important, the site of fracture initiation. On a microscopic level, brittle fractures exhibit cleavage facets, shown in Fig. 17.4, a fracture mechanism that requires little energy to propagate cracks through the steel crystal structure.

Most brittle fractures initiate from notches (stress concentrations) such as weld flaws in steels exhibiting low fracture toughness. It should be noted, however, that when local constraint is severe, brittle-like fractures can occur in steel structures even though they posses good ductility and toughness. For example, such brittle fractures have occurred in gas pipelines and ships where thick intersecting plates have been welded together. The intersection of heavy plates creates a local region of high constraint—

FIGURE 17.3 Brittle, low-energy fracture of a jumbo wide flange section used in the tension chord of a roof truss.

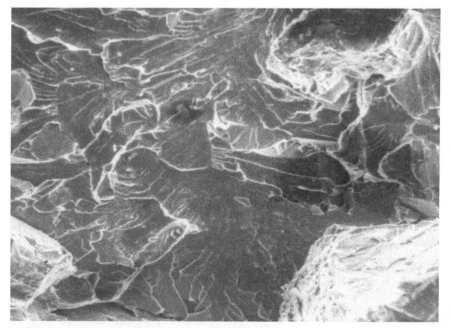

FIGURE 17.4 Scanning electron micrograph showing cleavage on the surface of a brittle, low-energy facture.

that is, a triaxial stress state wherein the three principal stresses are approximately equal—that limits the deformation capacity of the steel at the plate intersection.[3] In fact, the level of constraint at the notch tip in general governs the brittle behavior of steels. The apparent reduction in notch fracture toughness at low temperatures and/or high loading rates is related directly to the effects these parameters have on the yield strength of steel. Low temperatures and high loading rates tend to increase steel's yield strength. Higher yield-strength levels, in turn, generally result in smaller inelastic damage zones at the notch tip and, therefore, apparent lower fracture toughness and brittle behavior. Fortunately, many currently produced steels—by virtue of improved manufacturing techniques (e.g., grain refinement)—do not exhibit such notch sensitivity at low temperatures and/or high loading rates.

Buckling/Instability

Buckling/instability of steel structures is the only failure mode that depends primarily on the structure's geometry and secondarily on the properties of the steel. That is, buckling is essentially an instability phenomenon rather than the actual separation or fracture of material. Moreover, it is the only failure mode that occurs principally under local compressive loading, whereas most other failure modes can occur largely under tensile loading. In general, column buckling behavior can be predicted by any of a number of Euler-like buckling relationships,[4] such as the following tangent modulus equation[5]:

$$\frac{P_{cr}}{A} = F_{cr} = \pi^2 \frac{EL_e}{(L)^2 A} \tag{17.1}$$

EI_e is the effective bending rigidity of the column.

Regardless of the type of structure, buckling occurs almost exclusively in long, slender members, particularly long columns and plates with large width-to-thickness ratios. In addition, buckling can occur globally over an entire member such as a column or locally in a stiffener or web of a plate girder. In all instances, buckling generally is characterized by a rapid change in geometry and excessive distortion. Inelastic deformation usually follows the initial elastic instability.

It has been shown[4] that the dominant variable affecting the buckling strength of a column, other than its slenderness and eccentricity/distortion, is the magnitude and distribution of fabrication-induced residual stresses. The net effect of residual stresses is that they create a stress distribution that can be asymmetric, thereby creating eccentricity in the column and eventual instability under compressive loading.

Fatigue

Ductile failure, brittle fracture, and buckling/instability are failure mechanisms that usually occur under static loading conditions or a single-load event wherein the applied load increases until it exceeds a critical load. In contrast, most structures are subjected to repeating loads of varying magnitude, which are most often below yield strength and design stress levels. Such repeated, or fatigue, loading occurs in bridges, buildings, nuclear power plants, aircraft, ships, railcars, trucks, and medical devices, to name a few.

Fatigue failure is characterized by the initiation and growth of a crack or cracks owing to the repeated loading of the structure. Repeated loading generates microscopic inelastic damage at regions of local stress concentration (i.e., fillet-weld toe, lack-of-penetration flaw, flame-cut penetration, bolt holes, etc.). If sufficient inelastic damage accumulates, then a small crack develops, which then propagates through the structure in a direction that is usually perpendicular to the principal stress. Frequently, fatigue-crack growth will continue until the crack attains a critical size, wherein failure of the entire structure occurs. Initiation and subsequent growth of a fatigue crack are a function of the applied stresses and number of repeated loading cycles. Fortunately, fatigue is a progressive damage mechanism and often is identified before significant structural damage arises.

Fatigue failures typically are characterized by flat fractures, little or no associated macroscopic inelastic deformation, and crack-growth bands (beach marks) on the fracture surface, as shown in Fig. 17.5. In addition, fatigue fractures frequently exhibit markings, referred to as *ratchet marks,* that can identify the site of crack initiation. Since fatigue damage accumulates at regions of stress concentration, fatigue cracks initiate most frequently from the more severe stress concentrations, which in steel structures occur most often at welds (fillet-weld toes, lack-of-fusion flaws, etc.) and penetrations such as bolt and rivet holes. In this regard, recent reviews[6,7] revealed the vast majority of fatigue failures to have occurred at welded connections. In fact, none of the investigated fatigue failures occurred in base metal that did not contain a weld, weld repair, or significant stress concentration associated with the fatigue-crack initiation site.

Corrosion and Stress-Corrosion Cracking

The most common failure mechanism in steel structures is damage due to corrosion. Since steel oxidizes when exposed to oxygen, particularly in moist environments, the opportunities for corrosion of steel structures are endless. Corrosion damage can occur under both static and repeated loading conditions. Under static loading or no-load

FIGURE 17.5   Fatigue fracture of a bolt, characterized by a smooth, dull appearance and faint beach marks; cracking initiated from the thread root.

conditions, such damage is referred to as *general corrosion* or *wastage,* with specific localized forms of corrosion such as pitting, crevice corrosion, and stress-corrosion cracking. Under repeated loading, corrosion damage is usually referred to as *corrosion fatigue*. Corrosion fatigue is similar to general fatigue, except that the time to initiate and/or propagate fatigue cracking is decreased significantly owing to the presence of an aggressive environment.

General corrosion is characterized by broad thinning over a significant region of the structure in an aggressive environment and the formation of an oxide (e.g., rust). The variation in metal loss for general corrosion usually does not vary by more than a factor of 2 or 3 over the surface area of the structure that has sustained the damage. Rusting due to weather (atmospheric) exposure is a prime example of general corrosion, as shown in Fig. 17.6. Localized corrosion encompasses a wide range of attack mechanisms, all of which are limited to relatively small, confined sites. Examples of localized corrosion, shown in Fig. 17.7, include pitting, crevice corrosion, and stress-corrosion cracking. Generally, it is the localized attack that presents the greatest potential for damage, as well as being the most difficult for detection/inspection and analysis of fitness for service.

Creep

As a result of thermal activation at elevated temperatures, plain carbon and alloy steels slowly deform inelastically under constant load even if the service stress is substantially below the yield strength. This damage accumulation, known as *creep,* is time-dependent and is a function of temperature and applied stress. For structures that are exposed to a constant load (e.g., boiler tubes, turbine blades, process piping) and can

FIGURE 17.6 Severe generalized corrosion wastage of a structural steel member web.

FIGURE 17.7   Localized pitting corrosion at a gasketed pipe coupling.

undergo unrestricted deformation, creep results in the continuous accumulation of in-elastic deformation and eventual fracture. Creep generally consists of three stages: primary, secondary, and tertiary.

Initially, creep damage consists of the development and accumulation of microvoids along grain boundaries primarily owing to sliding of grains against each other. Next, the microvoids merge into intergranular cracks, which then grow forming macrocracks. In plain carbon steels, microvoids are evident at the beginning to midpoint of second-ary creep, and microcracks, at the end of secondary creep. In austenitic stainless steels and particularly superalloys, voids and cracks are evident even at earlier stages of creep damage.

Metallurgical Degradation

Metallurgical degradation arises from changes in a steel's microstructure owing to elevated temperature exposure. Changes in microstructure, such as unwanted precipi-tation or excessive grain growth, can affect strength, fracture toughness, ductility, weld-ability, and corrosion resistance. For most steels, thermal aging occurs either during elevated-temperature service, such as steam-cooled piping in a power plant, or during welding, as might occur during the fabrication of a heavy-section truss.

FITNESS-FOR-SERVICE ASSESSMENT PROCEDURES

In general, the concept that a structure with flaws or damage is fit for continued service is not new. Rather, many structures, both old and new, have sustained cracking or damage and continued to operate safely and reliably. Until recently, however, specific guidelines have not been available for structural condition assessment. Recently, a number of codes have been developed to address the FFS of existing steel structures, the most comprehensive of which are the American Petroleum Institute's (API) *Fitness-for-Service Recommended Practice 579*,[8] the American Society of Mechanical Engineers' (ASME) *Pressure Vessel and Boiler Code,* Section XI, "In-Service Inspection of Nuclear Power Plant Components,"[9] the American Society of Civil Engineers' (ASCE) *Guideline for Structural Condition Assessment of Existing Buildings,*[10] and the British Standard BS 7910, *Guide on Methods for Assessing the Acceptability of Flaws in Fusion-Welded Structures.*[11] Currently, the ASME Pressure Vessel and Boiler Code's Post Construction Division is developing an FFS code similar to API 579.

Generally, the results of an FFS assessment should indicate the suitability of a structure for continued service under its current operating conditions. Additionally, an FFS assessment may be required to predict the remaining life of the structure, as well as the behavior of the structure under operating conditions different from those currently in affect. To this end, this section describes the basic elements and procedures required to conduct an effective FFS assessment of an existing steel structure. The essential elements of a FFS assessment are

- Operating history and original design review
- Flaw and damage assessment
- Stress analyses
- Critical damage size/remaining life
- Repair replacement
- In-service monitoring
- Analysis limitations

Each of these FFS elements is discussed below.

Operating History and Original Design Review

Prior to performing an FFS evaluation, all efforts should be made to identify and collect all relevant background and operating history data. This should include but not be limited to original design and as-built drawings and calculations, required margins of safety, load/hydro test results, material manufacturing certificates or test reports, fabrication procedures (weld and/or bolting), operating and maintenance histories, repair records, current operating conditions (both environmental and loading), anticipated operating conditions (both environmental and loading), nondestructive examination results from original fabrication and current conditions, and current and future life requirements.

Flaw and Damage Assessment

In order to evaluate an existing structure properly for continued service, it is of utmost importance that the nature and extent of existing damage be quantified. If this is not done properly, then the FFS assessment is likely to be flawed and nonconservative relative to the structure's remaining life. In this regard, it is important to identify the damage mechanism properly (e.g., fatigue or crevice corrosion) and the root cause of the damage. Once the damage mechanism root cause is established, then the driving force (e.g., unanticipated cyclic loading) can be accounted for properly in assessing remaining life and affecting the most suitable remediation scheme.

Stress Analyses

Virtually all FFS assessments require some level of stress analyses. Such analyses can vary from a review of the original design calculations to detailed strain gage (SGA) and/or finite-element (FEA) analyses. Strain gage analyses are particularly useful for cyclically loaded structures because actual service loading data are far better than any assumed loading or results from an FEA, which often are inaccurate for cyclically loaded structures.

Critical Damage Size/Remaining Life

The most important part of an FFS assessment is calculation of the critical damage size, i.e., the largest acceptable crack size or maximum amount of metal loss tolerable for the structure when incorporating proper safety margins and loading conditions. Once the critical damage size is determined, then all other requirements, such as remaining life or repair requirements, are readily established. For example, the critical crack size in a bridge member can be determined using the fracture mechanics–based failure assessment diagram (FAD), which incorporates unstable, low-energy (brittle) fracture at one extreme and ductile or limit-load failure at the other extreme—failures intermediate to these extremes are enveloped by an FAD failure curve.

The remaining life of a structure is established from the type of damage it has sustained, as well as the anticipated loading. For example, if a weld-related crack has been identified in a pressure boundary component, the remaining life is likely to be governed by the time required to propagate this crack through the structure's thickness to create a leak. In this case, a fracture mechanics fatigue-crack growth assessment would be required to predict the time required to grow this preexisting crack through the thickness or to its critical size.

Repair/Replacement

Inasmuch as an FFS assessment indicates the suitability of a structure for continued service and its remaining life, such results are only applicable provided that some form of in-service inspection or monitoring can be performed. However, if the rate of corrosion attack or crack growth is not known with sufficient confidence or the FFS assessment indicates that the structure is not suitable for continued service, then repair or replacement of the structure is likely to be necessary. If it is determined that a structure should be replaced, then no further FFS analyses are required, except for improving the new design with the results developed from the FFS assessment.

Alternatively, if it is determined that the structure can be repaired, then the repair method, its effect on other parts of the structure, and required inspections must be established relative to the required remaining life. Any repair must return the structure to a safe operating condition with suitable margins of safety.

For cracked structures, the cracks can be removed and welded so as to attain the original configuration. However, this approach must be employed with great caution because cracking certainly will recur and, in all likelihood, much sooner than the original cracking developed unless the crack driving force or environment responsible for cracking is mitigated. Cracking also can be impeded by incorporating a structural detail that provides an alternate load path around the crack or acts as a crack arrestor (e.g., a drilled hole or stiffener). Obviously, a new detail can be designed; however, the impact of the new design should be evaluated thoroughly relative to other parts of the structure.

Corrosion-damaged structures can be repaired by reducing the operating stress or, conversely, restoring their thickness. For example, a corrosion-thinned flange can be reinforced with a cover plate, or applying a weld overlay can restore the thickness of a thinned pressure-vessel wall. For such repair schemes to be effective, however, it is essential that the corrosion attack is diminished or eliminated.

In-Service Monitoring

Regardless of the outcome from an FFS assessment, it is important that an in-service monitoring or inspection program be implemented to ensure the suitability of the structure for its intended function. Such monitoring is performed for several reasons: (1) to verify assumptions made in the FFS assessment, (2) to assess the rate of crack growth or continued corrosion-induced deterioration, (3) to identify new cracking or corrosion damage before it becomes critical in nature, (4) to make certain that changes in recommended operating procedures are being carried out, and (5) to identify significant changes in the operating environment.

The first part of the in-service monitoring program should consist of a baseline survey of the critical locations in the structure so that new damage can be distinguished readily from existing damage. Thereafter, inspections should be performed at the FFS recommended intervals, preferably by the same personnel. Use of the same personnel will ensure that indications are not misinterpreted. The FFS engineer should review results of the in-service monitoring program so that actual damage rates can be compared with predicted rates and changes to the inspection interval can be made, if necessary.

Analysis Limitations

Although most FFS methodologies share many common features, each has unique limitations and applicability requirements. Clearly, it is important that the FFS engineer be familiar with these limitations to ensure that the FFS assessment procedures are applied appropriately and are applicable to the structure under consideration. For example, API 579 is intended for use in the assessment of pressure-boundary components originally designed to a nationally recognized design/construction code. Additionally, when evaluating cracklike flaws in steel pressure components, API 579 requires that the component does not operate in the creep regime, that no significant dynamic loading is imposed, and that the material must be a plain carbon steel with a yield strength of 40 ksi or less.

Clearly, the API 579 requirements are very specific and are intended to ensure that use of the fracture mechanics–based failure assessment diagram (FAD) is applicable. As with any FFS assessment, it is the responsibility of the engineer to ensure that the all procedures and principles are applied in an appropriate manner.

EVALUATING STEEL STRUCTURES WITH CRACKLIKE FLAWS

FFS evaluation of steel structures with cracklike flaws is accomplished most reliably using fracture mechanics.[3,8,11] Cracklike flaws are structural and/or metallurgic discontinuities that are planar in nature, i.e., flaws that are characterized by their length and depth, with a sharp root radius along their boundaries. Typical planar cracklike flaws include cracks, sharp notches and weld discontinuities such as undercut, incomplete fusion, and incomplete penetration. Discontinuities such as porosity and slag inclusions are considered volumetric and are less severe relative to fracture initiation.

Since cracklike flaws induce local stress fields that are not readily characterized by customary stress analysis techniques (including the local stress concentration approach), fracture mechanics has evolved as the only acceptable methodology for predicting the behavior of cracklike flaws. Within the fracture mechanics framework, the failure assessment diagram (FAD) method[8,9,11–13] is the most comprehensive approach.

This method is particularly useful when evaluating structures that potentially can fail by the interaction of rapid-unstable fracture, ductile tearing, and/or plastic collapse— the modes of failure encountered most frequently in steel structures with cracklike flaws. A commonly used FAD failure locus is described by the following relationship in terms of the brittle fracture ratio K_r and ductile (reference) stress ratio (L_T) and is shown in Fig. 17.8:

$$K_r = [1.0 - 0.14(L_r)^2]\{0.3 + 0.7 \exp[-0.65(L_r)^6]\} \quad \text{for } L_r \leq L_{r(\max)} \quad (17.2)$$

where $L_{r(\max)}$ is the ductile fracture cutoff limit and varies from 1.25 for structural steels to 1.80 for austenitic stainless steels.

A FAD analysis requires calculation of the cracklike flaw assessment point $L'_r - K'_r$, where L'_r is the limit-load element, defined as the ratio of the applied stress to the corresponding stress at the limit load, and K'_r is the unstable (brittle) fracture constituent, defined as the ratio of the applied stress intensity factor K to the fracture toughness K_{1c}. Failure is indicated when the assessment point $L'_r - K'_r$, intersects or exceeds the FAD failure locus. Margins of safety also are incorporated easily in the calculation of $L'_r - K'_r$, which allows for the consequence of a failure, the importance of various failure modes, and variations in material properties to be incorporated into the analyses. A typical FAD diagram with $L'_r - K'_r$ assessment points is shown in Fig. 17.9.

For a cracked structure, therefore, stress intensity factor solutions K_r and reference stress solutions L_r for the same flaw configuration must be developed. Such solutions can be obtained from existing compilations[8] or derived using the finite-element

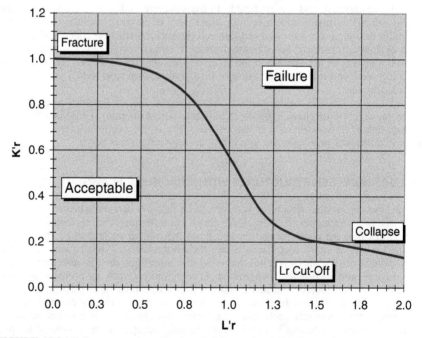

FIGURE 17.8 Failure assessment diagram showing regions of failure and acceptable behavior for a cracked structure.

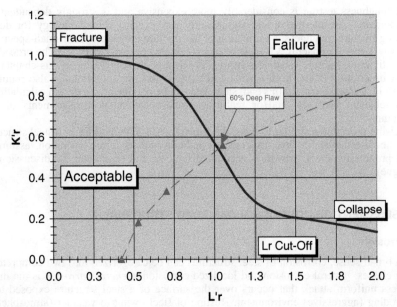

FIGURE 17.9 Failure assessment diagram showing the assessment points for a longitudinally oriented surface flaw in a cylinder subjected to internal pressure.

method. For example, stress intensity and reference stress solutions for a plate with a through-wall crack are as follows:

$$K = (\sigma_m + M_b\sigma_b)(f_w)\sqrt{\pi c} \tag{17.3}$$

where c = crack length
  W = plate width
  σ_m = membrane stress
  M_b = membrane magnification factor
  σ_b = bending stress
  f_w = crack length to plate width factor

and

$$L = \frac{P_b + (P_b^2 + 9P_m^2)^{0.5}}{3(1 - \alpha)} \tag{17.4}$$

$$\alpha = \frac{c}{W}$$

where P_m = primary membrane stress
  P_b = primary bending stress

In addition to these solutions, the fracture toughness and tensile properties for the structure are required. In this regard, coupons often are removed from noncritical locations for material testing. While tensile testing (ASTM A370[14] or E8[15]) is relatively inexpensive and tensile properties are only secondarily dependent on temperature, fracture toughness testing is considerably more expensive and is strongly dependent on temperature. An alternative and significantly less expensive methodology for determining fracture toughness is estimation of the fracture toughness from small-specimen Charpy V-notch impact (CVN) test results conducted per ASTM E23.[16] The procedure for estimating fracture toughness from CVN data is described later in this chapter.

A description of the both normal and extreme loading conditions is also required for an FFS assessment. To this end, such data can be obtained from design calculations, finite-element analyses, and/or real-time strain/acceleration measurements of the structure.

Using procedures similar to those described in API 579, the FFS of a steel structure can be determined for both its current condition and its future operation, assuming that progressive crack growth is accounted for (see end of chapter for discussion of fatigue).

ASSESSMENT OF CORROSION-DAMAGED STRUCTURES

Corrosion

In a broad sense, corrosion of steel structures can be categorized into two comprehensive classes: general corrosion and localized corrosion. *General corrosion* is the more or less uniform attack that occurs over the surface of a steel structure exposed to a corroding (aggressive) environment. Rusting of steel owing to weather (atmospheric) exposure is a prime example of general corrosion. *Localized corrosion* encompasses a variety of attack mechanisms, all of which are limited to relatively small, confined sites. Examples of localized corrosion include pitting, crevice corrosion, galvanic corrosion, and stress-corrosion cracking. Generally, it is the localized attack that presents the greatest potential for damage, as well as the most difficult for detection/inspection and analysis of service fitness.

As with most FFS assessments, the initial stage of a corrosion damage assessment is a triage-type analysis. Accordingly, more advanced corrosion damage is given more rigorous treatment when the consequence of failure is severe. Less severe corrosion damage can be assessed with simpler analyses. However, the simpler analyses incorporate a notable safety margin that can result in excessive conservatism, which, in turn, frequently requires a more rigorous analysis. The most in-depth analyses typically require detailed modeling of the actual geometry, service loading, and material properties. API 579[8] provides an excellent methodology for the assessment of corrosion damage with several assessment levels.

Inspection

As for any FFS assessment, the full extent of corrosion damage must be appraised by inspection, documentation, and evaluation. Most important, a thorough overall visual inspection is necessary. If access from to both sides of the corroded member is not available, then ultrasonic thickness surveying is strongly suggested. In this regard, it is important to identify the region of maximum corrosion attack, i.e., the minimum remaining thickness. If the component being inspected is too large to inspect 100% economically, then a sampling plan should be devised that monitors all regions of expected accelerated corrosion attack and highest stresses. Moreover, the inspection should include regions of stagnation or other areas where moisture and corrosive fluids and/or vapors can accumulate, such as corners, crevices, areas of wear or fretting, regions where coatings are damaged, low points, etc.

Pitting can be particularly difficult to detect because corrosion products from general corrosion may mask pits. A small pit may not be detected simple by visual examination, and often the damage potential is underestimated. ASTM: G46[17] includes inspection methods for evaluating pits. These include ultrasonic, electromagnetic, and dye-penetrant techniques. Crevice corrosion and stress-corrosion cracking damage can be detected using similar techniques. However, the detection of crevice corrosion also may require the disassembly of various parts of the structure because such corrosion generally occurs between faying surfaces and therefore may not be detectable without exposing these surfaces.

Simplified Corrosion Assessment

If it is clear that the consequence of a structural failure owing to corrosion is not a life-safety issue, the economic impact of the failure is low, and if it can be ensured that no cracklike flaws are present, then a simplified corrosion assessment can be performed. Quite simply, this approach assumes that the amount of metal removal by corrosive attack over the entire component is equal to the maximum metal loss detected in the component inspection. As indicated earlier, this generally requires ultrasonic thickness surveying. Thickness measurements by mechanical means are acceptable, provided that adequate access is available to allow accurate measurements. Using the appropriate code requirements, the stress in the component in question can be calculated as a function of time at extended service. The rate of cross-sectional reduction of the load-carrying member is based on the maximum attack and the time of exposure to date. Hence the thickness t as a function of time during future service in the load-bearing member is equal to the current minimum detected thickness, with an adjustment made to account for thinning occurring at the maximum rate (detected during the current inspection) and the time T in future service as follows:

$$t = t_c - (\Delta t_{max}/T_o)(T) \qquad (17.5)$$

where
t_c = current thickness
Δt_{max} = maximum attack found in inspection
T_o = time over which Δt_{max} occurred
T = time in future service

The *limit of remaining life* is the point at which the calculated stress reaches the code allowable design stress or, alternatively, when the thickness reaches the minimum design thickness. Thus the remaining useful life is estimated by assuming that the corrosive attack continues at the maximum rate observed under the service conditions assessed to date. This methodology also assumes that the original design loading is known and is still valid.

The assumption of constant corrosion rates generally is conservative. Typically, the initial corrosion attack proceeds rapidly and decreases to a slower steady-state rate thereafter. Accordingly, when the initial corrosion rate is included in the average, the result is an average rate that is somewhat higher than the actual steady-state rate. If the corrosive attack is in the form of deep pits, the assumption of a constant overall corrosion rate probably will be too conservative, and more rigorous FEA modeling of the corrosion attack geometry may be required.

Complex Corrosion Assessment

When insufficient information is available for a simple analysis, or when a simple analysis proves to be too conservative, corrosion damage can modeled using more rigorous and appropriate analytic techniques, such as FEAs. Appendix B of API 579 provides guidelines for performing FEAs for FFS evaluations.

Accurate structural dimension and geometry data, extent of corrosion attack, actual service loading, and material mechanical properties must be known to fully model the structure. If the loading is not well established, actual in-service loading using in situ measurement techniques, such as strain gages and/or accelerometers, will be required. If the material mechanical properties are not documented accurately, then testing of removed samples may be necessary (see the end of this chapter for more details on assessing steel material properties). However, whenever sample material is removed for property determination, it should be established that the sample is truly representative of the structure in question. It is also important to ensure, if material must be replaced, that the welding, bolting, or other joining procedures used are appropriate and qualified.

If it cannot be guaranteed that cracklike flaws do not exist in the corroded regions, then it is necessary to treat these situations as if such flaws are present. For example, a conservative approach for corrosion damage characterized by pitting would be to assume that the pits are cracklike flaws. In this case, a crack model would be developed that circumscribes the pit(s), and a cracklike flaw analysis as described earlier would be performed.

Remediation

Inasmuch as there are numerous assumptions implicit in all corrosion damage analyses, it is recommended that consideration be given to some form of corrosion remediation or perhaps restoration. With an understanding of the corrosive attack mechanism, attack locations, etc., it may be possible either to modify the design (e.g., add additional protective coatings) or to change the operating conditions to affect a significant reduction in corrosion attack rates. Finally, as with any FFS assessment, long-term monitoring is always beneficial and, in many circumstances, may be the most cost-effective remediation scheme.

ELEVATED-TEMPERATURE DAMAGE ASSESSMENT

Material Degradation

For steels, like all metals, the diffusion rate (mobility) of atoms and crystallographic (internal) defects such as vacancies and dislocations increase at elevated temperatures. As a result, steels can become structurally unstable, and their original microstructure and mechanical properties can change by one of the following metallurgical mechanisms:

- Temper embrittlement
- Graphitization and/or carbide spheroidization
- Precipitation and coalescence of intermetallic phases
- Recovery and recrystallization
- Creep and relaxation
- Creep fatigue and thermal fatigue

This section briefly discusses these elevated-temperature steel damage mechanisms, as well as creep and high-temperature fatigue. General techniques for the detection of elevated-temperature damage are also discussed.

Temper Embrittlement

In the intermediate temperature range of 650 to 1000°F, carbon and low-alloy steels are sensitive to the precipitation of phosphorous, arsenic, antimony, and several other elements along grain boundaries. Concentration of these elements along grain boundaries can occur during slow cooling from high temperatures or tempering for extended time in this temperature range. Such material degradation can result in temper embrittlement and manifests as premature intergranular fracture when a structure is loaded. A substantial shift in the fracture-toughness transition curve to higher temperatures is also observed. As such, the tolerable or critical flaw size in embrittled steel may be reduced substantially. It should be noted that room-temperature strength and ductility are not affected by temper embrittlement, and as such, standard hardness or tensile testing will not identify temper embrittlement in steel. Temper embrittlement can be avoided by heat treatment above this range followed by rapid cooling and/or by a reduction of the concentration of deleterious elements (e.g., phosphorous, antimony, etc.) that contribute to temper embrittlement. Many power-plant components operate in the critical temperature range and therefore are susceptible to temper embrittlement.[18]

Charpy V-notch impact fracture toughness testing is the most effective method for assessing if steel has become embrittled. Moreover, appropriate heat treatment can reverse the detrimental effects of temper embrittlement.

Graphitization and Spheroidization

Graphitization is a microstructural change in plain carbon or low-alloy steels that: (1) contain chromium at levels less than 1.25% and (2) are exposed to elevated temperatures (below 1025°F) for long periods of time. Graphitization results from the decomposition of pearlite (iron and iron carbide) into ferrite (iron) and carbon (graphite). This decomposition, in effect, embrittles (weakens) the steel, particularly when the graphite particles form continuous zones perpendicular to the applied load in a load-carrying member. Graphite particles that are randomly distributed throughout the microstructure cause only a moderate decrease in strength.

Carbide spheroidization is a process of degradation of pearlite and the formation of coalesced or spherical carbides. Carbide spheroidization results in modest strength reduction and is the usual mode of pearlite decomposition at temperatures above 1025°F. In contrast, graphitization is observed below this temperature. Therefore, carbon steel piping and other components that are exposed to temperatures from about 800 to 1025°F for several thousand hours are likely to be embrittled by graphitization.[19]

The effects of graphitization can be determined by measuring the hardness or tensile strength of the affected steel. Alternatively, a microstructural evaluation also can be performed to assess the extent of graphitization and/or spheriodization. Field metallography, using replication techniques, can be used to identify such damage in existing structures without removing samples.

Precipitation of Intermetallic Phases

Degradation of alloy steels at elevated temperatures usually is associated with precipitation and coalescence of intermetallic phases. Austenitic stainless steels, which are used widely in the fabrication of power-plant superheaters and other elevated-temperature structures, operate at 1100°F and sometimes higher. With long-term service, precipitation of carbides, intermetallics, and other phases known as *sigma* and *laves* usually is observed. The presence of such phases can reduce the fracture toughness and increase the propensity for stress corrosion cracking[20] substantially in austenitic stainless steels.

Microstructural evaluation is the most effective method for assessing the extent of precipitation of unwanted intermetallic phases.

Recovery and Recrystallization

Steels often are used in the cold-worked (work-hardened) condition in order to obtain higher strength levels and develop unique geometries such as curved beams. At welded joints, the cold-worked microstructure sustains recovery or even recrystallization that greatly reduces the original strength and therefore can be considered as material degradation. Hardness testing is the simplest and most convenient method to determine the severity of such deterioration.[21] Since these softened zones can be very narrow, microhardness techniques such as Vickers (diamond pyramid hardness) or Knoop are best for detecting such softening.

Creep and Relaxation

As a result of thermal activation at elevated temperatures, plain carbon and alloy steels slowly deform inelastically under constant load even if the service stress is substantially below the yield strength. This damage accumulation, known as *creep,* is time-dependent and is a function of temperature and applied stress. Generally, creep consists of three stages: primary, secondary, and tertiary.[22]

During primary creep, microvoids accumulate initially along grain boundaries, principally owing to sliding of grains against each other, and then the microvoids merge into intergranular cracks, which then merge into larger macrocracks. In plain carbon steels, microvoids are evident at the beginning to midpoint of secondary creep, and microcracks, at the end of secondary creep.[23] In austenitic stainless steels and particularly superalloys, voids and cracks are evident even earlier in creep life.

The primary method of creep damage assessment is microscopic examination of specimens or metallurgical surface replicas. These methods are well developed and also can be used to predict the remaining life of creep-damaged structures. However, other methods, such as creep-stress rupture testing, as well as hardness and oxide scale thickness measurements, also can be employed to predict remaining life.

Stress relaxation is creep damage in structures such as bolts that are exposed to elevated temperatures (e.g., building fire), wherein the overall length of the structure remains essentially constant. Owing to the constant length, accumulation of creep deformation in bolts, for example, results in a reduction in the originally induced preload, which, in turn, results in bolt loosening or creep fracture. Inspection of such structures by torque testing can assess the extent of relaxation damage.

Creep Fatigue and Thermal Fatigue

Creep fatigue is damage owing to the combined effects of inelastic creep deformation and mechanical fatigue associated with cyclic loading at elevated temperatures. Analytic methods have been developed, based on the linear damage-summation approach, to predict the accumulated damage and remaining life for the fatigue-creep phenomenon.[18]

Thermal fatigue occurs when steel is subjected to cyclic temperatures simultaneously with or without cyclic mechanical loading. The exposed steel of a retractable roof is an example of a structure that can be subjected to thermal fatigue because it frequently sustains daily changes in temperature. Life-prediction techniques for thermal fatigue also have been developed from fatigue-creep analyses, often using finite-element analysis.[18]

ASSESSMENT OF FIRE DAMAGE

Behavior of Structural Steels Exposed to Fire

Common structural steel members (e.g., beams, columns, and plates) generally are fabricated from plain low-carbon steel and furnished in the hot-rolled or normalized condition with a ferrite-pearlite microstructure. Accordingly, the strength of these steels does not change significantly when they are exposed to temperatures up to approximately 800°F. On the other hand, typical structural steels exhibit a dramatic loss in strength and modulus at temperatures above 1000 to 1200°F,[24,25] as shown in Fig. 17.10. More important, however, the tensile properties of these steels change only marginally after cooling if the steels are heated to less than approximately 1200°F for a reasonably short time (i.e., several hours). As such, typical structural steels exposed to fire do not exhibit significant changes in tensile properties following the fire.

In contrast, high-strength carbon and alloy heat-treated steels are more susceptible to metallurgical degradation when they are exposed to the heat of a fire. However, such steels are seldom used in general structural applications except for fasteners. High-strength fasteners (ASTM A325 and A490) usually are fabricated by quenching and subsequent tempering in the range of 800 to 1200°F, which develops a microstructure of tempered martensite. Therefore, high-strength fasteners can sustain a substantial reduction in strength following expose to temperatures above about 800°F.

Fire Damage Evaluation

Assessment of damaged structural steel members following a fire is performed most effectively using a multilevel approach. That is, the assessment is performed in steps

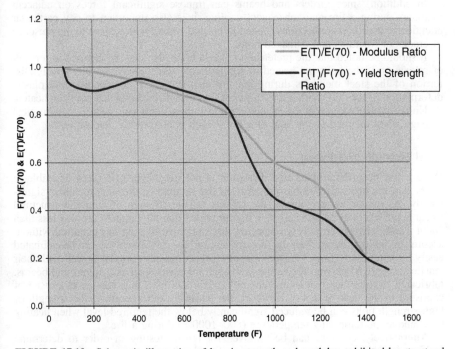

FIGURE 17.10 Schematic illustration of loss in strength and modulus exhibited by structural steels as a function of temperature.

such that the extent of damage is categorized according to overall distortion, followed by subsequent detailed evaluations, as necessary. For example, initial visual examination should document carefully the extent of global distortion, such as camber and sweep, as well as the extent of local deformation, such as stiffener or flange buckling.

A site evaluation of the fire-damaged structure also should be performed as quickly after the fire as possible in order to facilitate identification of the most highly heated locations.[25] In this regard, fire-induced temperatures at various locations can be appraised visually based on the following damaged material characteristics:

- Wood and paper ignite at approximately 450°F; plastics melt or burn between 180 and 350°F.
- Concrete changes color at approximately 550°F and becomes deep red at approximately 1100°F.
- Coatings, markers, and paints usually change color, blister, or spall above 600°F.
- Mill scale on the steel members starts to spall with the associated development of coarse surface texture above 1200°F.

More often than not, the most severe damage generally sustained by structural-steel members exposed to fire is excessive distortion. As such, it is common practice to sort fire-exposed structural steel members into three categories based on the extent of distortion as follows:

1. Visually unaffected
2. Somewhat deformed and economically repairable
3. Severely deformed and replacement is required

In addition, since girders and beams can impose significant forces on adjacent members during a fire and subsequent cooling, it is also important to pay particular attention to their end connections (bolted joints and welds) with respect to the presence of cracks or fractures and damaged bolts.

It should be noted that the pretension force in bolts also may be lowered as a result of relaxation if the bolts are heated to temperatures above approximately 950°F. Distortion of the steel members during the fire also can cause overloading and plastic deformation of the bolts with a loss of pretension even if the bolts had not been heated.

Unlike typical structural steels, austenitic stainless steels can sustain metallurgical damage when exposed to the heat of a fire, which is not readily detectable.

Steel Strength (Hardness)

Once it has been determined that distortion is not excessive, it is often desirable to verify that the heat of a fire has not affected the strength of the steel members significantly. As mentioned previously, the postfire tensile strength of common structural steels is not likely to be affected by a fire, provided that the temperature was less than about 1200°F. However, if it is suspected that excessive heating has occurred without accompanying distortion, then the in-place strength of steel members can be estimated easily by hardness testing. The hardness of steels correlates reasonably well with their tensile strength. Moreover, the hardness of steel members such as columns and beams, fabricated from low-carbon steel, generally exhibits little, if any, change as a result of exposure to the heat of a fire. However, the hardness and, therefore, the strength of high-strength bolts can be reduced significantly below the required level when heating to or above the tempering temperature (800–1000°F) during a fire.

Alternatively, coupons can be removed for tensile testing in order to determine yield and tensile strength, as well as ductility. However, care must be taken when selecting coupon locations so as to avoid adversely affecting the load-carrying capacity of the member. Moreover, it is important to ensure that following removal of the coupon the member cutout is ground and radiused so as reduce any stress-concentrating effects owing to cutting the member.

Residual Stresses

When steel members sustain distortion (permanent inelastic deformation), they develop internal residual stresses, as well as impose loads on adjacent members. As such, it may be necessary to establish the level and distribution of such residual stresses, particularly if the steel exhibits low fracture toughness. However, it is important to note that the presence of residual stresses does not affect the ultimate moment capacity of a member, provided that the steel exhibits reasonable fracture toughness and ductility. Nonetheless, if measurement of residual stresses is required, then such measurements are performed most accurately by placing strain gages at critical locations and subsequently removing coupons containing the strain gages. It should be noted that the change in strain following removal of the coupon is a measure of the total load in the member. Accordingly, the dead and live loads must be subtracted from the measured strains in order to estimate the residual stresses. It is also worth noting that the estimation of residual stresses owing to the heat from a fire can be further complicated by the presence of residual stresses developed during fabrication and construction. In addition, any distorted nearby members can impose stresses in the member being measured.

Metallographic Analysis

The microstructure of rolled structural members fabricated from plain carbon steel consists of ferrite (iron phase) and pearlite (mixture of iron and iron carbide phases). For typical fire exposure times (i.e., several hours or less), pearlite will decompose partially at approximately 1100°F or decompose completely with the formation of a ferrite spheroidized carbide microstructure at 1200 to 1300°F. In the temperature range of 1350 to 1650°F, partially retransformed ferrite-pearlite microstructure forms. Therefore, if the duration of a fire is known, then the temperature to which the steel had been exposed can be estimated from the change in microstructure, if any.

In order to evaluate the microstructure and, therefore the temperature, in-situ metallographic replication is usually employed. The surface of a structural member is ground and polished at selected locations and then etched to reveal the steel microstructure. The microstructure can be examined using a portable microscope, or replicas of the microstructure can be prepared for subsequent laboratory examination (see end of chapter for replication details).

ASSESSING STEEL MATERIAL PROPERTIES

A fitness-for-service assessment of existing steel structures generally requires knowledge of the mechanical and metallurgic behavior of the structure's materials.

Mechanical behavior is described most often by the material's yield strength σ_y and ultimate tensile strength σ_u. Additional mechanical properties, such as fracture toughness K_c, fatigue-crack initiation and crack growth parameters, stress-corrosion crack growth parameters, etc., also characterize the material mechanical behavior and often are necessary for FFS assessment.

Tensile Properties

The most useful tensile property data for existing steel structures are test data obtained directly from specimens removed from the structure. It should be noted that samples for tensile specimens should be removed from the structure so as not to weaken the structure's strength. For example, samples from moment connected beams should be removed from the midheight of the web. For simply supported beams, samples should be removed from the end of the flange that is not connected to the column. Tensile testing of the specimens should be performed in accordance with ASTM specifications A370[14] or E8.[15]

If it is not possible to cut tensile specimens from the structure, field hardness measurements can be used to estimate ultimate tensile strength. Hardness measurements can be performed on existing structures without inflicting any damage to the structure. A number of portable hardness measurement devices are available that provide reliable results if used properly.

A structure's measured hardness levels can be converted to ultimate tensile strength using standard conversion charts for steel. Unfortunately, there are no direct correlations between hardness and yield strength. There are, however, several approximate correlations between ultimate tensile strength σ_u and yield strength σ_y. For non-cold-worked steels, for example, the yield strength can be estimated from the following relationship[26]:

$$\sigma_y = 1.05\sigma_u - 30,000 \ (\text{lb/in.}^2) \tag{17.6}$$

Therefore, nondestructive hardness measurements of a structural member can be used to estimate yield and ultimate tensile strengths, provided, of course, that these correlations are used within the ranges for which they are valid. Nominal mechanical properties of a structural member can be also obtained from the manufacturer's certificate, if available.

Fracture Toughness

Fracture toughness (K_{1c} and/or K_{1d}), a measure of a material's resistance to fracture in the presence of a cracklike flaw, is a necessary material property for FFS evaluations that use the FAD or other fracture mechanics–based procedures. A material's fracture toughness can be obtained by direct testing of specimens removed from the structure. However, fracture toughness tests are very expensive and, in general, frequently are not necessary for structural-steel applications.

An alternative and significantly less expensive methodology for determining fracture toughness is estimation of fracture toughness from small-specimen Charpy V-notch impact (CVN) test results conducted per ASTM E23.[16]

K_{1c} and K_{1d} can be estimated from CVN test results according to the following procedure[3]:

1. Perform standard CVN testing over the transition temperature region.
2. For each test temperature, calculate K_{1d} using the following empirical correlation:

$$K_{1d}^2/E = 5(\text{CVN}) \tag{17.7}$$

where K_{1d} is the dynamic fracture toughness (lb/in.$^2\sqrt{\text{in.}}$), E is the modulus of elasticity (lb/in.^{2}), and CVN is the absorbed energy (ft · lb) from the CVN tests.
3. Determine the temperature shift (T_{shift}, °F) between K_{1d} and K_{1c} using the following correlations:

$$T_{\text{shift}} = 215 - 1.5 \cdot \sigma_y \quad \text{for 36 kips/in.}^2 \leq \sigma_y \leq 140 \text{ kips/in.}^2 \tag{17.8}$$

$$T_{\text{shift}} = 0 \quad \text{for } \sigma_y > 140 \text{ kips/in.}^2$$

where σ_y = material yield strength, kips/in.^{2}

4. Determine K_{1c} as a function of temperature by shifting K_{1d} values at each temperature obtained in step 3.

This procedure is reasonably conservative and limited to the lower end of CVN transition curve, where CVN values in foot pounds are less than about one-half the yield strength in kips per square inch.

Since notch acuity and loading rate do not affect fracture toughness significantly in the upper shelf and upper transition regimes, the following CVN–K_{1c} relationship[3] can be used to estimate static fracture toughness K_{1c} directly from CVN results without using a temperature shift:

$$\left(\frac{K_{1c}}{\sigma_y}\right)^2 = 5\left(\frac{CVN}{\sigma_y} - 0.05\right) \tag{17.9}$$

It should be noted that CVN specimens should be cut so as not to weaken the existing structure. In addition, the notch in the CVN specimens should be oriented to the most probable crack direction in the structure.

Fatigue Behavior

Traditionally, the resistance of a material to cyclic loading (fatigue) has been estimated using the endurance limit S_f or stress amplitude S versus cycles-to-failure N life diagrams (S-N curves) developed from the testing of smooth base metal specimens. As a first approximation, the fatigue strength or endurance limit for structural steels, such as ASTM A36, can be estimated from the following equation[27]:

$$S_f = (0.40 \text{ to } 0.55)\sigma_u \tag{17.10}$$

Using this approach, the fatigue life of a structure was believed to be infinite, provided that the applied cyclic stress was less than the endurance limit. Unfortunately, for real structures, which are fabricated by welding, bolting, and/or riveting, the use of smooth-specimen endurance limits frequently is nonconservative.

Rather, the vast majority of fatigue failures occur at component connections[6,7] (e.g., welds, bolt holes) because of the stress-concentrating effects of inherent discontinuities such as undercut, incomplete penetration, porosity, etc. These stress concentrations induce locally high cyclic stresses sufficient to initiate fatigue cracking, even though the global or nominal cyclic stress may be significantly less than the smooth-specimen endurance limit.

Accordingly, the accumulated fatigue damage and remaining fatigue life of a welded-steel structure can be determined using the AASHTO or BS7608 S-N fatigue-of-weldments approach for steel structural members.[28–30] Three principal variables have been identified that govern the fatigue life of welded and bolted connections.[28] In order of importance, they are the applied stress range ($\Delta\sigma = \sigma_{max} - \sigma_{min}$), connection detail type, and number of loading cycles. Depending on the severity of the connection stress concentration, fatigue life will vary significantly for a given loading profile. Hence a description of component loading—i.e., cyclic stress range and frequency of occurrence—is required to evaluate fatigue life. A significant number of experimental studies have been performed in which the fatigue behavior of a large number of different welded and bolted connections has been quantified.[28–30] These studies have resulted in the classification of different connection types into unique fatigue life categories (e.g., AASHTO categories A through E', with E' being the most severe fatigue condition). Statistical analysis of the fatigue data within each of these categories has resulted in a set of lower-bound fatigue life curves, shown in Fig. 17.11, that form the basis of the current AASHTO *Guide Specifications for Fatigue Design of Steel Bridges*[31] and similar curves for British Standard BS7608, *Fatigue Design and Assessment of Steel Structures.*[30] These lower-bound welded and bolted connection fatigue curves correspond to a 2.3% probability of failure at a given stress range (i.e., 2 standard deviations below the mean). Further, these fatigue curves are applicable to both constant-amplitude and variable-amplitude (random) fatigue-loading conditions.

Most dynamically loaded structures are subjected to variable (random) rather than constant-amplitude loads (stresses). Consequently, a suitable cumulative damage criterion must be applied in order to predict fatigue life. As discussed earlier, it has been shown that the AASHTO weldment fatigue curves can be used to predict both constant-amplitude and variable-amplitude (random) amplitude loading. In this regard, variable-amplitude loading is accounted for using the Miner's effective-stress-range method to calculate the appropriate effective-stress-range ($\Delta\sigma_{eff}$) as follows:

$$\Delta\sigma_{eff} = (\Sigma\gamma_i\sigma_i^n)^{1/n} \tag{17.11}$$

where γ_i is the ratio of cycles at the ith load sequence to the total number of cycles, σ_i is the stress range for the ith load sequence, and n is equal to 3.0 when Miner's cumulative damage rule is used. It also should be noted that the AASHTO steel fatigue curves have been constructed with a constant slope of -3.0 to facilitate fatigue life calculations of structures subjected to variable-amplitude loading.

The total fatigue life N, corresponding to a 2.3% probability of failure of a given welded or bolted connection, can be calculated according the following relationship:

$$\log(N) = A - 3 \log(\Delta\sigma_{eff}) \tag{17.12}$$

where $\Delta\sigma_{eff}$ is the effective stress range, and A is the coefficient for a given detail type and is given below.

Steel Weldement Fatigue Life Coefficient A

Category	Coefficient A
A	10.40
B	10.08
C	9.65
D	9.34
E	9.03
E'	8.59

A brief, simplified description of typical structural steel members and connections covered by AASHTO/AWS categories A through E' are as follows:

A Plain plate and rolled beams
B Plain welds and welded beams and plate girders
C Stiffeners and short attachments (less than 2 in. long)
D Long attachments
E Cover-plated beams
E' Thick flanges and thick cover plates

Fatigue Crack Growth

For FFS evaluations it is also sometimes necessary to know the actual rate of fatigue crack growth. The rate of fatigue crack growth in steel structures can be characterized by the fracture mechanics stress intensity factor range (ΔK).[3] Fatigue crack growth rate is measured and correlated as a function of ΔK, as shown in Fig. 17.12. As shown, fatigue crack growth exhibits three growth-rate regimes. Region I is referred to as the *threshold regime*, wherein the crack growth rate is vanishingly small and below the threshold ΔK_{th}; cracks theoretically do not propagate under cyclic loading.

For martensitic, bainitic, ferrite-pearlite, and austenitic steels, ΔK_{th} can be estimated by

$$\Delta K_{th} = 6.4(1 - 0.85R) \text{ kips/in.}^2 \cdot \sqrt{\text{in.}} \quad \text{at } R > 0.1 \quad (17.13)$$

where $R = \sigma_{min}/\sigma_{max}$, σ_{min} is the minimum cyclic stress, and σ_{max} is the maximum cyclic stress.

Region II is located above ΔK_{th} and in general can be described by the following equation:

$$\frac{da}{dN} = A(\Delta K)^m \quad (17.14)$$

where da and dN are the incremental changes in crack length and applied loading cycles, and A and m are material constants. Upper-bound crack growth rate relationships for ferritic, martensitic, and austenitic stainless steels are as follows[3]:

$$\frac{da}{dN} = 3.6 \times 10^{-10}(\Delta K)^{3.0} \quad \text{Ferritic}$$

$$\frac{da}{dN} = 0.66 \times 10^{-8}(\Delta K)^{2.25} \quad \text{Martensitic} \quad (17.15)$$

$$\frac{da}{dN} = 3.0 \times 10^{-10}(\Delta K)^{3.25} \quad \text{Austenitic}$$

Crack growth in region III is very rapid and is followed rapidly by fracture. As such, crack growth behavior in this regime usually is not quantified.

Thus a deterministic estimate of fatigue crack growth life can be obtained using fracture mechanics and the appropriate crack growth-rate relationship. For example, the number of cycles N_p required to propagate a crack from an initial size a_i to its critical size a_c in a ferritic steel member can be obtained by integrating Eq. (17.15) as follows:

$$N_p = \int_{a_i}^{a_c} \frac{da}{3.6 \times 10^{-10}(\Delta K)^3} \quad (17.16)$$

It should be noted that Eqs. (17.15) and (17.16) are applicable to both constant-amplitude and variable-amplitude loading. For variable-amplitude loading, ΔK is replaced by ΔK_{eff}, which is calculated using $\Delta\sigma_{eff}$. A cumulative damage criterion must be applied in order to calculate $\Delta\sigma_{eff}$, as discussed earlier.

Metallurgical Properties

In general, metallurgical properties of structural-steel materials such as general macrostructure, general microstructure, grain size, etc. are not necessary for FFS evaluations. However, if discontinuities (flaws) are found in an existing structural member, the nature of the discontinuities (flaws) should be determined and taken into consideration during an FFS evaluation. It is important to identify whether such discontinuities were preexisting, service-induced, or formed during repair, if any.

The most reliable results can be obtained by detailed metallurgical evaluation of a section cut from the structural member containing the discontinuity. The evaluation can include but should not necessarily be limited to

- Visual examination and dimensional analysis
- Stereomicroscope examination
- Scanning electron microscopy
- Compositional analysis
- Determination of material mechanical properties
- Macrostructural examination
- Microscopic metallographic examination

If it is not possible to remove a sample from the structural member, field surface replication can be performed. To this end, surface-replication locations should be rough ground, fine ground, and polished using field metallographic techniques. Then, in order to reveal the structural member material microstructure, the replication locations can be etched. Replicas are taken by applying cellulose acetate tape softened with acetone. After allowing the tape to harden, the replica backs are darkened and then removed carefully. After that, the replicas are examined using light optical microscopy.

In order to determine the fracture morphology, surface replication can be applied to a fracture surface. Thereafter, the replicas can be coated and examined in a scanning electron microscope.

REFERENCES

1. AISC (1989). *Manual of Steel Construction—Allowable Stress Design,* 9th ed. American Institute of Steel Construction (AISC), Chicago.
2. ASME (2000). *Pressure Vessel and Boiler Code,* Sec. VIII, Div. 1. American Society of Mechanical Engineers (ASME), New York.
3. Barsom, J. M., and S. T. Rolfe, (1987). *Fracture and Fatigue Control in Structures: Applications of Fracture Mechanics,* 2nd ed. Prentice-Hall, Englewood Cliffs, NJ.
4. Tall, L. (ed.) (1986). *Structural Steel Design,* 2nd ed. Krieger Publishing, Malabar, FL.
5. Salmon, E. H. (1921). *Columns: A Treatise on the Strength and Design of Compression Members.* Oxford Technical Publications, London.
6. Vecchio, R. S., (1996). Lucius Pitkin, Inc., in-house review of failures.
7. Barsom, J. M., (1996). United States Steel, in-house review of failures.
8. API 579 (2000). *Fitness-for-Service,* 1st ed. American Petroleum Institute, Washington, D.C.
9. ASME (2000). *Pressure Vessel and Boiler Code,* Sec. XI. American Society of Mechanical Engineers (ASME), New York.
10. ASCE (2000). *Guideline for Structural Condition Assessment of Existing Buildings.* American Society of Civil Engineers (ASCE), Reston, VA.
11. British Standard BS7910 (1999). *Guide on Methods for Assessing the Acceptability of Flaws in Fusion-Welded Structures.* British Standards Institute, London.
12. Dowling, A. R., and C. H. A. Townley, (1975). "The Effects of Defects on Structural Failure: A Two-Criteria Approach," *International Journal of Pressure Vessels and Piping* **3**: 77–137.

13. Anderson, T. L., (1995). *Fracture Mechanics: Fundamentals and Applications,* 2nd ed. CRC Press, Boca Raton, FL.

14. ASTM A370 (2000). *Test Methods and Definitions for Mechanical Testing of Steel.* American Society for Testing and Materials, West Conshohocken, PA.

15. ASTM E8 (2000). *Test Methods for Tension Testing of Metallic Materials.* American Society for Testing and Materials, West Conshohocken, PA.

16. ASTM E23 (2000). *Test Methods for Notched Bar Impact Testing of Metallic Materials.* American Society for Testing and Materials, West Conshohocken, PA.

17. ASTM G46 (2000). *Guide for Examination and Evaluation of Pitting Corrosion.* American Society for Testing and Materials, West Conshohocken, PA.

18. Viswanathan, R. (1989). *Damage Mechanisms and Life Assessment of High-Temperature Components.* ASM International, Metals Park, OH.

19. ASM International (1975). *Metals Handbook: Failure Analysis and Prevention,* Vol. 10, 8th ed. ASM International, Metals Park, OH.

20. *Stainless Steel Handbook.* Allegheny Ludlum Steel Corporation, Pittsburg, PA, 1956.

21. *Welding Handbook,* Vol. 5, 7th ed., American Welding Society, Miami, FL, 1984.

22. Boyer, H. E. (ed.) (1988). *Atlas of Creep and Stress-Rupture Curves.* ASM International, Metals Park, OH.

23. Kerzher, G., I. Sprung, and V. Zilberstein, (1988). "Field Metallography for Life Extension Studies of High-Temperature Components," presented at the American Society for Nondestructive Testing Spring Conference, Orlando, FL.

24. NFPA (1990). *Fire Protection Handbook,* 16th ed., National Fire Protection Association, Quincy, MA.

25. Tide, R. H. R., (1998). "Integrity of Structural Steel After Exposure to Fire," *Engineering Journal of American Institute of Steel Construction* **35**: 26–38.

26. Datsko, J., (1996). *Material Properties and Manufacturing Processes.* Wiley, New York.

27. Timoshenko, S. P., (1956). *Strength of Materials.* Van Nostrand Company, Princeton, NJ.

28. Fisher, J. W., (1977). *Bridge Fatigue Guide.* American Institute of Steel Construction, Chicago.

29. Barsom, J., and R. S., Vecchio, (1997). *Fatigue of Welded Components.* Welding Research Council Bulletin 422, New York.

30. British Standard BS7608, (1993). *Fatigue Design and Assessment of Steel Structures.* British Standards Institute, London.

31. AASHTO LRFD (1994). *Guide Specifications for Fatigue Design of Steel Bridges.* American Association of State Highway and Transportation Officials, Washington, D.C.

Masonry

DAVID TRANSUE, P.E.

INTRODUCTION

The condition evaluation of masonry includes field and laboratory investigation of the units, mortar, and masonry assemblage. Nondestructive and semidestructive test methods are available for the determination of properties of in-place masonry materials, and some of these are standardized. Building codes provide requirements and guidance for the use of testing to determine engineering properties for structural evaluation and for quality assurance. In lieu of testing, guidelines are available for the estimation of material properties. Standard test methods for new construction materials can be used to measure properties of materials sampled from structures; however, tests on building materials that have been in service should not be used to verify compliance with specifications because environmental exposure may change material properties.

Building performance is defined in terms of functional qualities such as resistance to weather, load-bearing capabilities, and appearance. Occupants and evaluators typically observe these properties, but performance characteristics also can be measured and used as quantitative descriptors for building evaluation.

CRITICAL ENGINEERING PROPERTIES

Brick

Property requirements for brick are specified in the American Society for Testing and Materials (ASTM) C216, *Standard Specification for Facing Brick (Solid Masonry Units Made from Clay or Shale)*, and in ASTM C652, *Standard Specification for Hollow Brick (Hollow Masonry Units Made from Clay or Shale)*. Bricks are graded as *moderate weathering* (MW) or *severe weathering* (SW) based on their strength and absorption properties. These property requirements are shown in Table 18.1.

Compressive Strength of Brick. The compressive strength of brick ranges from less than 1000 lb/in.² for old, soft-fired bricks to greater than 15,000 lb/in.² for modern, hard-fired bricks. The test method for measuring compressive strength is included in ASTM C67, *Standard Test Methods for Sampling and Testing Brick and Structural Clay Tile*. This method is designed for new bricks, but it is also useful for bricks removed from service.

Initial Rate of Absorption of Brick. The *initial rate of absorption* (IRA) is defined as the weight of water absorbed by the bedding surface of the brick when submerged to a depth of ⅛ in. for 1 minute. It is expressed in units of grams per minute per 30 square inches, and the test method is described in ASTM C67. In note 5 of ASTM C216, it is stated that bricks with an IRA greater than 30 g/min/30 in.² should be wetted prior to laying. In contrast to this permissive language, wetting of bricks with an IRA greater than 30 g/min/30 in.² is required by the Masonry Standards Joint Committee (MSJC), *Specification for Masonry Structures* (ACI 530.1/ASCE 6/TMS

TABLE 18.1   Physical Requirements of ASTM C216, *Standard Specification for Facing Brick,* for Grade SW (Severe Weathering) and Grade MW (Moderate Weathering) Face Bricks (Average of Five Measurements), and Typical Measured Properties of Pre-1900 Brick and Post-1900 Brick

Brick	Compressive Strength (lb/in.²)	Absorption (5-Hour Boiling, Percent)	Saturation Coefficient
ASTM C216 grade SW	3000 (minimum)	17 (maximum)	0.78 (maximum)
ASTM C216 grade MW	2500 (minimum)	22 (maximum)	0.88 (maximum)
Pre-1900 brick	800–1600	12–22	< 0.78
Post-1900 brick	>3000	<17	< 0.78

602). An approximate measure of brick absorption can be determined in the field by drawing a quarter-size circle on a brick using a wax crayon and dropping water into the circle using an eyedropper. If the brick absorbs 20 drops or more in 90 seconds, it is recommended to wet the brick prior to use (Brick Industry Association, Technical Note 39).

Cold Water and Boiling Water Absorption and Saturation Coefficient. The 24-hour cold water and 5-hour boiling water absorptions are measurements of the weight of water absorbed by a brick under these conditions. These simple tests, described in ASTM C67, are used to describe brick porosity. The ratio of the 24-hour cold water absorption to the 5-hour boiling water absorption is the *saturation coefficient.* A brick with a saturation coefficient of 1 will be completely saturated at or below room temperature. A brick with a saturation coefficient of less than 1 will have unfilled air voids at or below room temperature. Unfilled air voids provide resistance to freezing-thawing damage because the air voids provide space for water to occupy because the water increases in volume by approximately 9% on freezing. The completely saturated brick is more likely to be damaged by freezing because there are no voids to accommodate the freezing water. In general, bricks with a low saturation coefficient are more durable than bricks with a high saturation coefficient because they are more likely to have unfilled air voids. Absorption and saturation coefficient property requirements of ASTM C216 are shown in Table 18.1.

Tensile Strength of Bricks. The tensile capacity of bricks can be measured by the modulus of rupture test, as described in ASTM C67, *Standard Test Methods for Sampling and Testing Brick and Structural Clay Tile,* and shown in Fig. 18.1, or by the splitting tensile strength, as described in ASTM C1006, *Standard Test Method for Splitting Tensile Strength of Masonry Units,* and shown in Fig. 18.2. The modulus of rupture of bricks is used occasionally in specifications of bricks for restoration projects, but otherwise brick tensile strength tests rarely are required.

Brick Moisture Expansion. Clay bricks experience reversible and irreversible volume changes with variation in moisture content. The irreversible volume change is the expansion of brick that occurs soon after manufacture. Bricks are at their smallest when they leave the kiln and become larger as time passes. Fifty percent of this expansion occurs in the first 6 months, and then the rate of growth slows, as shown in Fig. 18.3 (Brick Industry Association, Technical Note 18). Typical values for brick

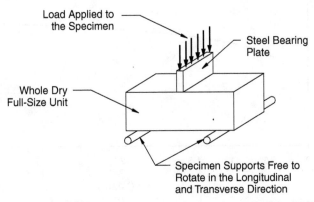

FIGURE 18.1   Modulus of rupture test arrangement.

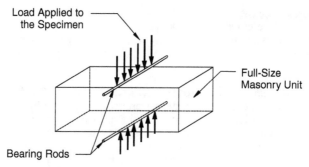

FIGURE 18.2 Splitting tensile strength test arrangement.

expansion are in the range of 0.0002 to 0.0009 in./in. The Brick Industry Association (BIA) recommends the use of a design value of 0.0003 in./in. for composite walls and 0.0005 in./in. for veneer walls where an upper bound of movement is estimated.

Terra Cotta

Terra cotta is a fired-clay product that was used extensively as exterior architectural detail units on buildings from the late nineteenth century until the 1930s. There are no standard specifications for historic terra cotta replacement material. Tindall[36] reported compressive strength and absorption test results from early-twentieth-century tests of terra cotta, including tests by the National Bureau of Standards in 1918 of terra cotta from eight different plants. Compressive strengths ranged from 2200 to 15,885 lb/in.^{2}, and absorption from 7.5% to 12.2%, as shown in Fig. 18.4. The Architectural Terra Cotta Institute issued a set of specifications in 1961 that included physical requirements for terra cotta.[36] These requirements, similar to those for bricks, are shown in Table 18.2.

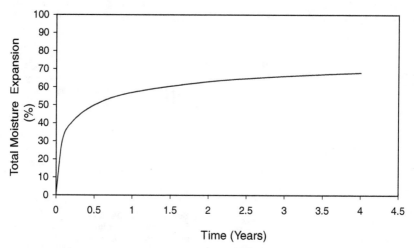

FIGURE 18.3 Expansion of brick over time. (*From BIA Technical Note 18.*)

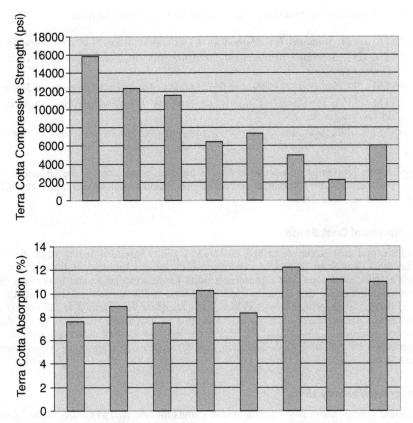

FIGURE 18.4   Absorption and compressive strength test results for terra cotta from 8 plants, tested by the National Bureau of Standards in 1918. (*Data from Tindall, 1989*).

Stone

Stone typically is characterized by compressive strength, absorption, density, and flexural strength. Stone properties vary widely, and stone usually is selected based on appearance, durability, and workability criteria. ASTM publishes standards for dimension stone according to stone type, including granite (ASTM C615), limestone (ASTM C568), marble (ASTM C503), quartz-based stone (ASTM C616), and slate (ASTM C629).

TABLE 18.2   Physical Requirements of Terra Cotta[36] (Average of Five Measurements), Architectural Terra Cotta Institute, 1961

Unit	Compressive Strength (lb/in.²)	Absorption (5-Hour Boiling, Percent)	Saturation Coefficient
Machine-extruded and mold-pressed ceramic veneer	5000 (minimum)	16 (maximum)	0.8 (maximum)

ASTM standards pertinent to evaluation of building stone include:

ASTM C170, *Standard Test Method for Compressive Strength of Dimension Stone*

ASTM C97, *Standard Test Method for Absorption and Bulk Specific Gravity of Dimension Stone*

ASTM C99, *Standard Test Method for Modulus of Rupture of Dimension Stone*

ASTM C1354, *Standard Test Method for Strength of Individual Stone Anchorages in Dimension Stone*

ASTM C1496, *Standard Guide for Assessment and Maintenance of Exterior Dimension Stone Masonry Walls and Facades*

Engineering properties of a wide variety of building stone used in the United States have been published by Robinson and colleagues,[26] and a summary of these properties is provided in Table 18.3.

Architectural Cast Stone

Architectural cast stone, described in ASTM C1364, *Standard Specification for Architectural Cast Stone,* is precast concrete cast in shapes of masonry units or elements such as sills, wall caps, and replacement units for existing stone and terra cotta elements. Architectural cast stone is required to have a 28-day compressive strength of 6500 lb/in.^{2}, as determined by ASTM C1194, *Standard Test Method for Compressive Strength of Architectural Cast Stone.* Absorption of cast stone is limited to less than 6% for cold water absorption and less then 10% for boiling water absorption, as determined by ASTM C1195, *Standard Test Method for Absorption of Architectural Cast Stone.*

Concrete Masonry Units

Standard specifications for concrete masonry units include ASTM C90, *Standard Specification for Loadbearing Concrete Masonry Units,* and ASTM C1372, *Standard Specification for Segmental Retaining Wall Units.* ASTM C90 includes dimensional tolerances, strength and absorption requirements, and shrinkage limits. Three classes of concrete units are described in ASTM C90: light weight, normal weight, and medium weight. The designations *type I, moisture-controlled units,* and *type II, nonmoisture-controlled units,* no longer exist.

Concrete Masonry Unit Compressive Strength. Units are required by ASTM C90 to have compressive strength of 1900 lb/in.^{2} for an average of three units and not less than 1700 lb/in.^{2} for any one unit based on net cross-sectional area. Typical values for concrete masonry units are in the range of 2000 to 3000 lb/in.^{2}. ASTM C140, *Standard Test Methods for Sampling and Testing Concrete Masonry Units and Related Units,* provides the test method for compressive strength.

Concrete Masonry Unit Tensile Strength. Like concrete, concrete masonry units usually have tensile strength in the range of 10% to 20% of their compressive strength. Tensile tests of hollow concrete masonry blocks are complicated by their geometry. Modulus of rupture tests can be conducted on coupons cut from whole blocks, but this procedure is not common. A modulus of rupture test for roof pavers is described in ASTM C140.

Absorption of Concrete Masonry Units. Water absorption requirements are provided in ASTM C90 for light weight, medium weight, and normal weight concrete masonry units. Absorption testing is described in ASTM C140.

TABLE 18.3 Ranges of Values for Physical Properties of Building Stones

Stone	Bulk Density (lb/ft³)	Porosity* (%)	$\alpha^{\dagger}$	Compressive Strength (lb/in.²)	Modulus of Rupture (lb/in.²)	Modulus of Elasticity (kips/in.²)	Mohs' Hardness
Granite	156–170	0.1–4	2.8–6.1	11,600–47,850	1450–10,150	4350–8700	5–7
Rhyolite	137–156	4–15	2.8–5.0	8700–31,900	145–10,150	8700–10,150	5–6.5
Quartzite	156–170	0.3–3	5.6–6.7	15,950–52,200	1450–14,500	8700–14,500	4–7
Marble	150–169	0.4–5	2.8–5.0	5800–27,550	580–4350	2900–10,150	2–4
Sandstone	125–162	1–30	4.4–6.7	2900–36,250	145–5800	435–11,600	2–7
Limestone	112–168	0.3–30	2.2–6.7	2900–34,800	145–7250	1450–10,150	2–3
Travertine	125–169	0.5–5	3.3–5.6	1450–21,750	290–1450	1450–8700	2–3

* Porosity is the ratio of pore volume to bulk volume.

† Coefficient of lineal thermal expansion $\times 10^{-6}$ strain per degree F.

Source: Adapted with permision from Robinson.[26]

Concrete Masonry Unit Shrinkage. Like clay bricks, concrete masonry units have reversible and irreversible volume changes that are related to their moisture content. Unlike clay bricks, the irreversible volume change in concrete masonry units is shrinkage. NCMA TEK 10-1A, *Crack Control in Concrete Masonry Walls,* states that typical drying shrinkage coefficients range from 0.0002 to 0.00045 in./in. The maximum allowable linear drying shrinkage per ASTM C90 is 0.00065 in./in. from a 100% saturated condition, as measured per ASTM C140.

Calcium Silicate Units

Calcium silicate masonry units are manufactured by combining sand, lime, and water, pressing into the desired shape, and then autoclaving the formed units. ASTM C73, *Standard Specification for Calcium Silicate Brick (Sand-Lime Brick),* provides requirements for calcium silicate units. Similar to concrete masonry, calcium silicate units undergo irreversible shrinkage after manufacture. Their properties are measured using ASTM C140, *Standard Test Methods for Sampling and Testing Concrete Masonry Units and Related Units.*

Mortar

Masonry mortars are specified by either proportion or properties in ASTM C270, *Standard Specification for Mortar for Unit Masonry.* The letters *M, S, N, O,* and *K* (from the words *MaSoN wOrK*) distinguish the five mortar types. Table 18.4 shows proportions, required minimum compressive strength, and average measured compressive strength[14] for each of the five mortar types. For some mortars, such as historic masonry replacement mortars, vapor transmission and stiffness are more important performance indicators than compressive strength. To date, however, these properties and the test methods for determining them are not standardized or specified in the United States.

Nonstandardized test methods exist for in-place evaluation of hardened mortar, including pendulum hammer hardness tests[37] (shown in Fig. 18.5), drilling resistance tests,[33] anchor pullout tests, and probe penetration tests.[13] The in-place shear test is a good indicator of mortar structural performance, and the *International Existing Building Code* (IEBC), Appendix A, Chapter A1, "Seismic Strengthening Provisions for Unreinforced Masonry Bearing Wall Buildings," states that "the quality of mortar in all masonry walls shall be determined by performing in-place shear tests according to the following. . . ." The IEBC provides a one-paragraph description of the test and requirements for the number of tests needed. The in-place shear test is described detail

TABLE 18.4   Mortar Classification Chart

Mortar Type	Proportion by Volume			Compressive Strength	
	Cement	Lime	Sand	Minimum (lb/in.²)	Typical* (lb/in.²)
M	1	1/4	2¼ to 3 times the sum of	2500	6400
S	1	1/2	the volumes of cement	1800	3625
N	1	1	and lime	750	1850
O	1	2		350	540
K	1	3		75	190

*Typical strengths are from Frey[14] for mortars prepared and tested in the laboratory.
Source: Developed from ASTM C270, *Standard Specification for Mortar for Unit Masonry.*

FIGURE 18.5   The pendulum hammer hardness test of mortar.

in ASTM C1531, *Standard Test Methods for in Situ Measurement of Masonry Mortar Joint Shear Strength Index*.

The constituent proportions of hardened mortar can be determined using ASTM C1324, *Standard Method for Examination and Analysis of Hardened Masonry Mortar*. This method also is used to diagnose certain mortar conditions, such as frozen paste, and segregation of components, and certain workmanship problems.

Mortar Compressive Strength. The effect of mortar strength on masonry assemblage compressive strength is minimal, especially for mortars with strengths over 1000 lb/in.^{2}.[12] Increasing mortar compressive strength above the 1000 lb/in.^{2} range does not result in significant increases in prism strength. Mortar strength tests commonly are specified for quality control in new construction, using the procedures of ASTM C109/C109M, *Standard Test Method for Compressive Strength of Hydraulic Cement Mortars (Using 2-in. or [50-mm] Cube Specimens)*.

Other Mortar Properties. More important than the compressive strength of mortar are the contribution of the mortar to the flexural bond strength, the condition of the mortar-unit interface, and the durability of hardened mortar. These properties will determine the water resistance capabilities of the wall and thus will affect the long-term performance of the masonry substantially.

Chemical Analysis, Petrography, and Sand Gradation. ASTM C1324, *Standard Test Method for Examination and Analysis of Hardened Masonry Mortar*, provides procedures for determining the proportions of masonry mortars using petrography and analytic chemistry. Goins[15] provided a detailed method for determining constituents of historic mortars using petrographic techniques. For older mortars with lime binder, the aggregate-to-binder ratio can be determined by dissolving an oven-dry sample in hydrochloric acid and measuring the weights of the soluble and insoluble material. Cliver[9] has published detailed acid digestion procedures. During digestion of mortar in acid, vigorous bubbling and amber color indicate the presence of lime, whereas weak bubbling and a murky green color indicate the presence of portland cement. The results of the analysis can be converted from weight proportions to volumetric proportions using measured bulk density values or bulk density values reported in ASTM C270, which are provided here in Table 18.5. The quality of the results obtained from an acid digestion of mortar depends on the nature of the aggregate present in the mortar. The presence of calcareous aggregates or impurities in the mortar can cause misleading results.

Jedrzejewska[17] described an economical procedure for the volumetric determination of carbon dioxide evolved during the acid digestion of a mortar. This procedure was developed for a comparative study of hundreds of mortars from Polish buildings destroyed during World War II.

Aggregate remaining from acid digestion can be cleaned, dried, and passed through sieves to measure the aggregate gradation. The procedure for sieve analysis of masonry sands is provided in ASTM C136, *Standard Test Method for Sieve Analysis of Fine and Coarse Aggregates*. The allowable range for aggregate gradation for masonry mortar is specified in ASTM C144, *Standard Specification for Aggregate for Masonry Mortar*. Figure 18.6 shows a gradation curve of a sieve analysis of two mortar samples and the fine and coarse gradation limits described in ASTM C144. Mortar from structures built in the early nineteenth century and earlier frequently contain aggregate-size gradations outside the limits set in ASTM C144, as is the case for the aggregates shown in Fig. 18.6.

Grout

Fine and coarse grout properties are described in ASTM C476, *Standard Specification for Grout for Masonry*. The two types of grout are distinguished by aggregate gradation. Fine grout, intended for spaces 2 in. wide and less, has no aggregate larger than ¼ in.; coarse grout, intended for spaces greater than 2 in. wide, has no aggregate larger than ⅜ in. ASTM C404, *Standard Specification for Aggregates for Masonry Grout*, provides the required gradation for fine and coarse grout aggregates. Unlike concrete, workability, rather than water-cement ratio, is the main consideration for the mix design. A flowable consistency is necessary to ensure that the grout fills all voids in the masonry assembly. The high water content of the grout is reduced automatically on placement by suction of masonry units.

Grout is specified by either proportions or strength. Grout proportions are provided for both fine and coarse grout in ASTM C404 and are shown in Table 18.6. When

TABLE 18.5 **Bulk Density of Mortar Components for Converting Between Volume Proportions and Batch Weights**

Material	Bulk Density, lb/ft^3 (kg/m^3)
Portland cement	94 (1,505)
Hydrated lime	40 (640)
Lime putty	80 (1,280)
Sand	80 (1,280)

Source: From ASTM C270, *Standard Specification for Mortar for Unit Masonry.*

| ASTM C 144: Natural sand | | Test Data |
| % passing | % passing | % passing |
Sieve # Upper bound	Lower bound	Average of samples 1-3
4	100.00	100.00	95.7
8	95.00	100.00	85.4
16	70.00	100.00	70.7
30	40.00	75.00	50.9
50	10.00	35.00	29.1
100	2.00	15.00	15.8
200	0.00	5.00	7.8

Aggregate grain size distribution curves for three mortar samples from a structure. The gradation range specified for masonry mortar in ASTM C-144 is indicated in dashed lines.

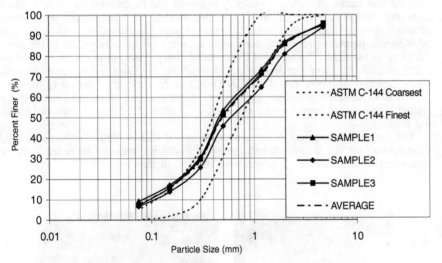

FIGURE 18.6   Sand gradation curves of mortar aggregate.

TABLE 18.6   Grout Proportions and Typical Strength

| Grout Type | Proportion by Volume | | | Maximum Aggregate Size (in.) | Typical Compressive Strength (lb/in.²) |
	Cement	Lime	Sand		
Fine	1	0–1/10	2¼ to 3 times the sum of the	<¼	3000–4000
Coarse	1	0–1/10	volumes of cement and lime	<³⁄₈	

Source: From ASTM C476, *Standard Specification for Grout6 for Masonry.*

grout is specified by compressive strength, the minimum strength is 2000 lb/in.^{2}. Grout samples for compression testing are formed using masonry units in order to simulate the water absorption that occurs in place. ASTM C1019, *Standard Test Method for Sampling and Testing Grout,* describes the forming and testing of grout specimens. Alternative methods of obtaining grout specimens are allowed, including coring of grout-filled masonry units, which is useful for verification of grout strength in existing structures.

The shear bond between grout and masonry units can be measured using the grout shear bond test, which is shown in Fig. 18.7. A standard method for this test is in development by ASTM Committee C12.

Masonry Assemblage

Masonry Compressive Strength. In design, the required strength of a masonry assemblage can be specified in two ways: by specifying the properties of the mortar and masonry units or simply by requiring a minimum masonry strength. The compressive strength and elastic modulus of masonry materials can be determined using in-place tests and laboratory tests. ASTM C1314, *Test Method for Constructing and Testing Masonry Prisms Used to Determine Compliance with Specified Compressive Strength of Masonry,* provides procedures for testing masonry prisms constructed in the laboratory or in the field. The National Concrete Masonry Association (NCMA) provides a test method, *Standard Test Method for Determining the Compressive Strength of In-Place Concrete Masonry,* found in TEK 18-9, that describes the removal of masonry prisms from service for testing. Verification of compressive strength of masonry by testing of prisms cut from existing construction is allowed by the *International Building Code.*

The Federal Emergency Management Agency's FEMA 356, *Pre-Standard and Commentary for the Seismic Rehabilitation of Buildings,* provides three methods for measurement of the expected compressive strength of masonry:

FIGURE 18.7 The grout shear bond test.

1. Extraction of test prisms from an existing wall and testing per Section 1.4.B.3 of the MSJC *Specification for Masonry Structures*
2. Fabrication of prisms using extracted masonry units and a surrogate mortar designed using the results of a chemical analysis of mortar samples.
3. Measurement of peak compressive stress using flatjack testing per ASTM C1196, *Standard Test Method for in Situ Compressive Stress within Solid Unit Masonry Estimated Using Flatjack Measurements* (Note: The referenced test should be C1197, *Standard Test Method for in Situ Measurement of Masonry Deformability Properties Using the Flatjack Method*).

In lieu of testing, FEMA 356 provides default values for estimation of the masonry compressive strength. Table 18.7 shows these values. The compressive strength of recently constructed masonry also can be estimated using design tables found in the *Uniform Building Code* (UBC), the *International Building Code* (IBC), and the MSJC *Specification for Masonry Structures*. Values estimated in this way likely will be conservatively low, however, because the strength of modern masonry usually exceeds these minimum design values. Table 18.8 shows values from the MSJC *Specification for Masonry Structures* for selected prism compressive strength as a function of mortar type and unit strength.

In many retrofit and restoration projects, the bearing stress applied to masonry by beams and other members is the limiting design factor. The MSJC *Building Code Requirements for Masonry Structures* (ACI 530/ASCE 5/TMS 402) limits the bearing stress applied to masonry to 25% of the specified or estimated compressive stress of the masonry.

Modulus of Elasticity of Masonry. The modulus of elasticity can be determined during the course of the prism strength test by examining the stress-strain curve. The UBC specifies that the modulus of elasticity should be determined by the secant method, wherein the slope of a line from $0.05f'_m$ to $0.33f'_m$ is taken as the modulus. The test specimen height-to-thickness ratio must exceed $2:1$ for this procedure to be valid (UBC 2107.1.8.2).

The modulus of elasticity can be estimated as a function of the compressive strength of the masonry, and such estimates are found in building codes for use in design. A database of masonry prism compression tests compiled and analyzed by Atkinson and Yan[2] shows that the ratio of elastic modulus to compressive strength is near to 550, indicating that building codes substantially overestimate the modulus of elasticity of masonry. FEMA 356, on the other hand, follows the work of Atkinson and Yan[2] and

TABLE 18.7 Default Material Property Values from FEMA 356 for Existing Masonry for Use in Lieu of Testing

Masonry Condition	Compressive Strength (lb/in.²)	Flexural Tensile Strength (lb/in.²)*	Shear Strength (lb/in.²)†
Poor	300	0	13
Fair	600	10	20
Good	900	20	27

Note: These values are conservative, and actual tests typically will provide strengths in excess of the listed values.

* For unreinforced masonry.

† For running bond only.

TABLE 18.8 Compressive Strength of Masonry Based on the Compressive Strength of Masonry Units and Mortar Type

Unit	Net Area Compressive Strength of Masonry Units, lb/in.^{2} (MPa) Type M or S Mortar	Type N Mortar	Net Area Compressive Strength of Masonry, lb/in.^{2} (MPa)
Brick	1700 (11.72)	2100 (14.48)	1000 (6.9)
	4950 (34.13)	6200 (42.75)	2000 (13.79)
	8250 (56.88)	10,300 (71.02)	3000 (20.69)
CMU*	1250 (8.62)	1300 (8.96)	1000 (6.9)
	2800 (19.31)	3050 (21.03)	2000 (13.79)
	4800 (33.10)	5250 (36.2)	3000 (20.69)

*Concrete masonry unit.
Source: Values selected from MSJC *Specification for Masonry Structures.*

uses a ratio of 550. Ratios of the modulus of elasticity to compressive strength of clay and concrete masonry from these sources are compiled in Table 18.9.

Masonry Tensile and Flexural Strength. The flexural tensile strength of the masonry assemblage is a good indicator of masonry performance in both resisting loads and in functioning as a durable weather barrier. The tensile strength is indicative of the quality of the bond between mortar and masonry unit. The unit-mortar interface substantially defines the water resistance capability of masonry because delaminations at joints provide paths for water ingress.

The MSJC *Building Code Requirements for Masonry Structures* provides allowable masonry flexural tensile strength values for use in design of unreinforced masonry, as shown in Table 18.10. These values can be used as an upper-bound estimate of the flexural tensile strength of existing masonry. Flexural tensile loads in reinforced masonry are assumed to be carried by the reinforcing steel alone.

Flexural tensile strength can be measured in the field and in the laboratory. Two laboratory procedures are described in ASTM C952, *Standard Test Method for Bond Strength of Mortar to Masonry Units:* a direct tension test for brick masonry and a flexural tension test for concrete masonry. ASTM C1072, *Standard Test Method for Measurement of Masonry Flexural Bond Strength,* describes a laboratory method using a bond-wrench apparatus for measuring the flexural strength of brick or concrete masonry. The bond-wrench procedure of ASTM C1072 has been adapted for field use.[32] Finally, ASTM E518, *Standard Test Methods for Flexural Bond Strength of Masonry,* describes two laboratory tests for measuring bond strength of brick or concrete masonry prisms constructed as beams: one method using third-point loading of a simply supported beam and one method using a uniform loading of a simply supported beam.

TABLE 18.9 Modulus of Elasticity of Masonry as a Function of Masonry Compressive Stress

Source	Clay Masonry, E_m/f'_m	Concrete Masonry, E_m/f'_m
UBC	750	750
MSJC	700	900
Atkinson and Yan^2	Approximately 550	Approximately 550
FEMA 356	550	550

TABLE 18.10   Allowable Flexural Tension for Clay and Concrete Masonry, MSJC
Building code for Masonry Structures, **psi, (kPa)**

	Mortar Types			
	Portland Cement/Lime or Mortar Cement		Masonry Cement or Air Entrained Portland Cement/Lime	
Masonry	M or S	N	M or S	N
Normal to bed joints				
Solid units	40 (276)	30 (207)	24 (166)	15 (103)
Hollow units, ungrouted	25 (172)	19 (131)	15 (103)	9 (62)
Hollow units, fully grouted	65 (448)	63 (434)	61 (420)	58 (400)
Parallel to bed joints—running bond				
Solid units	80 (552)	60 (414)	48 (331)	30 (207)
Hollow units, ungrouted	50 (345)	38 (262)	30 (207)	19 (131)
Hollow units, fully grouted	80 (552)	60 (414)	48 (331)	30 (207)

In lieu of testing, FEMA 356 provides recommended values for use in evaluation, and they are provided here in Table 18.7.

Masonry Shear Strength. For new construction, the shear strength of the masonry assemblage is considered separately for unreinforced and reinforced masonry in the MSJC building code. For reinforced masonry, it is important to consider the shear resistance of the reinforcing steel in comparison with the shear resistance of the masonry because brittle failure modes of masonry shear walls are to be avoided. Over-reinforcing can encourage brittle failure modes. Drysdale, Hamid, and Baker[12] provide a thorough description of this issue in their Chap. 10 on shear walls.

For projects involving seismic strengthening of existing unreinforced masonry, the *International Existing Building Code* (IEBC) requires that shear strength be determined by in-place shear tests. The IEBC describes the test in one concise paragraph and also includes requirements for the number of tests. The in-place shear test is described detail in ASTM C1531, *Standard Test Methods for in Situ Measurement of Masonry Mortar Joint Shear Strength Index.* FEMA 356 recommends that shear strength be determined by in-place testing and provides recommended values for use in lieu of testing, as shown in Table 18.7. These values are not for reinforced masonry. Shear strength of specimens removed from service can be tested per ASTM E519, *Standard Test Method for Diagonal Tension (Shear) in Masonry Assemblages.*

Comparative Properties: Thermal Expansion Coefficients. Values for the coefficient of thermal expansion for various materials are shown in Table 18.11. These values are useful when calculating differential strains in connected components of different materials.

CODES AND STANDARDS

Codes

Among the applicable building codes are the *International Building Code* (IBC), the *Uniform Building Code* (UBC), the *Uniform Code for Building Conservation* (UCBC), the *International Existing Building Code* (IEBC), the Masonry Standards Joint Com-

TABLE 18.11 Coefficient of Thermal Expansion for Various Materials

Material	Average Coefficient of Lineal Thermal Expansion $\times$ 10^{-6} Strain per Degree F
Dense aggregate concrete masonry	5.2
Light aggregate concrete masonry	4.3
Clay or shale brick masonry	3.6
Granite	2.8–6.1
Limestone	2.2–6.7
Marble	2.8–5.0
Gravel aggregate concrete	6.0
Lightweight structural concrete	4.5
Aluminum	12.8
Structural steel	6.7
Stainless Steel	9.6
Gypsum plaster	7.6
Perlite plaster	5.2
Pine, parallel to fiber	3.0
Pine, perpendicular to fiber	19.0

mittee's (MSJC) *Building Code Requirements for Masonry Structures* (ACI 530/ASCE 5/TMS 402) and *Specification for Masonry Structures* (ACI 530.1/ASCE 6/TMS 602), and *The EUROCODE*. The IBC is the new, renamed version of the UBC, and the IEBC is the new, renamed version of the UCBC. In general, with the exception of the IEBC, building codes are concerned with new structures and do not specifically address evaluation of existing structures. The IEBC Appendix A, "Seismic Strengthening Provisions for Unreinforced Masonry Bearing Wall Buildings," Chapter A1, includes requirements for the materials and configuration of unreinforced masonry, as well as allowable stress values for existing masonry and allowable load values for new materials used in conjunction with existing masonry. This resource, although intended for seismic retrofit projects, is very useful for any project involving engineering of older masonry structures.

Test methods, guidelines, and allowable property values for new materials can be adapted for evaluation and analysis of materials in service. However, exposure to service will change the properties of the materials, so it is not necessarily valid to use results of tests on materials that have been in service to verify compliance with specifications or code. In general, local building officials are responsible for interpreting and enforcing building codes, so applicable codes are locality-specific for each evaluation project. Also, local building officials frequently are knowledgeable and helpful, and contacting an official early in a project can save much time.

Standards

Among the applicable standard-writing organizations are the American Society for Testing and Materials (ASTM), the American National Standards Institute (ANSI), the German Institute for Standards (DIN), the International Union of Laboratories and Experts in Construction Materials, Systems and Structures (RILEM), and the standards of the *Uniform Building Code* (UBC).

The ASTM provides standard test methods, practices, and guidelines for evaluation and testing of building materials, including new and existing masonry. ASTM Committee C12 on mortars and grouts for unit masonry, Committee C15 on manufactured masonry units, and C18 on dimension stone are responsible for producing and maintaining standards relating to masonry. These committees produce the standards that are used most commonly for masonry materials in the United States.

RILEM is a nonprofit, nongovernmental organization based in Paris, France, that promotes the advancement and sharing of knowledge related to buildings. Recently, RILEM Technical Committee TC 127-MS published a series of recommended tests for masonry evaluation in RILEM's journal, *Materials and Structures.*[39] These tests are

RILEM MS-A.1, "Determination of the Resistance of Wallettes against Sulphates and Chlorides"

RILEM MS-A.2, "Uni-Directional Salt Crystallization Test for Masonry Units"

RILEM MS-A.3, "Uni-Directional Freeze-Thaw Test for Masonry Units and Wallettes"

RILEM MS-A.4, "Determination of the Durability of Hardened Mortar"

RILEM MS-A.6, "Method for Triaxial Compression Tests on Mortar Specimens Taken from Bed Joints"

RILEM MS-B.1, "Freeze-Thaw Test of Masonry Panels"

RILEM MS-B.2, "Measurement of the Shear Strength Index of Bed Joints"

RILEM MS-B.3, "Bond Strength of Reinforcement in Bed Joints"

RILEM MS-B.4, "Determination of Shear Strength Index for Masonry Unit/Mortar Junction"

RILEM MS-B.5, "Determination of the Damage to Wallettes Caused by Acid Rain"

RILEM MS-D.1, "Measurement of Mechanical Pulse Velocity for Masonry"

RILEM MS-D.2, "Determination of Masonry Rebound Hardness"

RILEM MS-D.3, "Radar Investigation of Masonry"

RILEM MS-D.5, "Measurement of Ultrasonic Pulse Velocity of Masonry Units and Wallettes"

RILEM MS-D.6, "In-Situ Measurement of Mortar Bed Joint Shear Strength"

RILEM MS-D.7, "Determination of Pointing Mortar Hardness by the Pendulum Hammer"

RILEM MS-D.9, "Determination of Mortar Strength by the Screw (Helix) Pull-out Method"

RILEM MS-D.10, "Measurement of Moisture Content by Drilling"

Technical Notes, Recommended Practices, and Guidelines

Important guideline documents include those of the National Concrete Masonry Association (NCMA), the Brick Industry Association (BIA), the Federal Emergency Management Agency (FEMA), the American Society of Civil Engineers (ASCE), and the American National Standards Institute (ANSI).

The Structural Engineering Institute (SEI) of the ASCE Committee on Structural Condition Assessment and Rehabilitation of Buildings prepares standards on three areas: structural condition assessment, assessment of the building envelope, and assessment of buildings for seismic considerations. This committee has completed and maintains the important document SEI/ASCE 11, *Guidelines for Structural Condition Assessment of Existing Buildings.* The SEI also has worked with the FEMA to publish FEMA 356, *Pre-Standard and Commentary for the Seismic Rehabilitation of Buildings.* This document contains detailed guidelines for determining engineering properties of existing masonry.

Another useful document from FEMA is FEMA 306, *Evaluation of Earthquake Damaged Concrete and Masonry Wall Buildings, Basic Procedures Manual*. The manual presents test and investigation guidelines for nondestructive testing methods, including visual inspection, sounding, rebound hammer testing, reinforcing bar detection, ultrasonic pulse-velocity testing, impact-echo testing, spectral analysis of surface waves, radiography, ground-penetrating radar testing, and others.

The NCMA *TEK Manual for Concrete Masonry Design and Construction* is an invaluable series of notes providing comprehensive technical details of concrete masonry design, detailing, construction, and related topics. The BIA's *Technical Notes on Brick Construction* is an invaluable resource for information on the design, detailing, and construction of brick masonry.

COMMON TYPES, CAUSES, AND CONSEQUENCES OF DEGRADATIONS AND DEFECTS

Water is the single most important cause of deterioration of masonry materials and systems. The control of water at the building site and the maintenance of water-control elements are items of interest in every building evaluation.

Moisture Resistance of Wall System

Non-moisture-resistant masonry tends to deteriorate rapidly if subject to wetting and drying cycles. This type of deterioration increases in severity over time because the water penetration causes material deterioration, which allows further water penetration. Consistently damp walls also encourage growth of fungus and mold, accelerate corrosion of metal components, dampen insulation, and damage interior finishes. The resistance of a wall system to moisture penetration can be measured using ASTM E514, *Standard Test Method for Water Penetration and Leakage Through Masonry,* which simulates the effects of a driving rainstorm on the wall. Also available are spray tests, including one using a spray rack apparatus[20] for repeatable water penetration testing.

Poor mortar-unit bond is a common cause of water penetration. Poor bond can be the result of poor workmanship, low-quality or poorly mixed mortar, mortar that has frozen during construction, mortar that has dried out during construction, mortar that has lost water to suction by absorptive brick that was not prewetted, the presence of impurities in the mortar or on the units, differential movement or volume change between mortar and units, and cracking from a variety of causes. Bond quality can be evaluated visually by examining the mortar-unit interface. Even narrow delaminations at the bond line between units and mortar can result in significant leakage. However, results of tests by Birkeland and Svendsen[5] indicate that cracks of 0.004 in. and less will not be a source of water penetration from wind-driven rain. Strength measurements of the mortar-unit bond such as flexural bond strength and in-place shear can classify bond quality indirectly. Remedies for water penetration owing to poor bond include repointing, control of water, and grout injection. Walls that rely on coatings to resist water penetration require periodic inspection and reapplication of the coating. Sealants around openings and flexible-joint sealants also require periodic replacement to prevent water infiltration.

In cavity walls, flashing details around openings, roof level, floor levels, and the base of the wall require end dams and weepholes in order to function properly. Interior details can be examined by drilling holes and observing the interior with a fiberoptic borescope[30] or by removing units.

Site Characteristics

Measuring the slope of the ground surface adjacent to a structure is useful for determining whether runoff is directed toward the structure. Observations of the interior below grade and exterior near grade will reveal problems with groundwater, and detailed moisture content mapping of masonry surfaces can aid in these diagnoses. Excessive moisture transfer through a wall often is evidenced by effluorescence, biologic growth, and deterioration. *Rising damp* is a condition that occurs when water rises by capillary action to a certain height in a wall and then evaporates at the surface. This phenomenon can cause damage to the masonry at the evaporation zone, where the evaporating water deposits harmful soluble salts.

Finally, lack of or poor maintenance of water-control systems such as gutters, scuppers, and roof drains is responsible for the majority of deterioration problems observed in practice. Windows and window accessories that are damaged or functioning poorly also cause water infiltration and masonry deterioration problems. Observation of a building during heavy rain is recommended to help diagnose these systems.

Deterioration Processes in Bricks

Freeze-thaw damage and damage from the action of soluble salts are the two most common brick deterioration mechanisms. Other modes of brick deterioration include acid damage from pollutants and cleaning compounds, mechanical damage including loss of material from sandblasting and high-pressure water sprays, structural damage owing to overstress conditions, foundation movements, and corrosion of embedded metals.

Deterioration Processes in Concrete Masonry Units

Concrete masonry units deteriorate in a manner similar to that of concrete. Freeze-thaw deterioration and soluble salt crystallization are the main agents of deterioration, and acid damage from pollutants and cleaning compounds is also possible. ASTM C90 provides strength requirements and absorption limits for concrete masonry units. Durability properties are not addressed directly in the specification. A thorough discussion of freeze-thaw resistance of concrete can be found in the American Concrete Institute's *Guide to Durable Concrete* (ACI 201.2R-92, 1997).

Freeze-Thaw

Damage from freezing of saturated masonry is presumably due to the pressure caused by the 9% increase in the volume of water on freezing. Thomson[35] provides a review of theory and tests related to the freeze-thaw resistance of mortars in historic masonry. ASTM C216 indirectly addresses the effects of freeze-thaw damage in bricks by specifying maximum absorption and saturation coefficient criteria. These criteria are based on the work of McBurney[23] and others at the National Institute of Standards and Technology (NIST), formerly the National Bureau of Standards. McBurney and colleagues conducted strength, absorption, saturation, and freeze-thaw tests on 3368 bricks from 209 manufacturers during the 1930s.

Robinson and colleagues[27] compiled a set of test results on 5217 commercially marketed bricks and found that 23% of the bricks determined to be nondurable bricks passed the ASTM C216 criteria for "severe weathering" brick. Kung[21] determined that the critical saturation coefficient is a function of the materials and firing techniques used in production and that underfired, nondurable bricks can pass the test. Winslow and colleagues[38] found that many bricks that were failing in service passed the ASTM

C216 criterion for bricks graded for severe weathering regions. Although these researchers have established that the saturation coefficient criterion allows some failing brick to be approved, the criterion does screen many poor brick, and the test is inexpensive and relatively fast when compared with freeze-thaw testing.

Pore Size Distribution

Mercury intrusion porosimetry consists of submerging an evacuated piece of material in mercury and forcing the mercury into the sample under pressure. The volume of mercury intruded into the sample at various pressure ranges is used to describe the distribution of the sizes of the pores in the sample. Although there are no American standardized methods for mercury intrusion porosimetry of masonry materials, ASTM D4404, *Standard Test Method for Determination of Pore Volume and Pore Volume Distribution of Soil and Rock by Mercury Intrusion Porosimetry*, provides applicable technical details.

Maage[22] developed a "durability factor" to describe the durability of bricks based on mercury porosimetry testing. Winslow and colleagues[38] evaluated the durability factor as a measure to predict brick durability and compared it with the performance of bricks in service. The advent of scanning electron microscopy (SEM) has made the direct observation of pore structure practical. Bortz and colleagues[6] recommend that SEM be incorporated as a regular procedure in future durability testing.

Effluorescence and Cryptofluorescence

White, powdery deposits seen frequently on masonry surfaces are visually distracting and also can be damaging to masonry. These deposits are crystallized salts and are referred to as *effluorescence* when they occur on the surface of masonry materials and *cryptofluorescence* when the crystallization occurs below the surface of the material.

Effluorescence is caused by salt molecules that are transported in solution to the masonry surface or near to the masonry surface where the water evaporates, leaving the salts behind. In some masonry materials, the pore structure is such that water can evaporate from a certain depth below the surface, traveling as a gas through the material to the surface, where it escapes to the atmosphere.[1] This evaporation zone can remain wet while the surface appears dry. This cryptofluorescence mechanism can be particularly damaging because salts collect in this zone and create the potential for the face of the material to separate from the whole. This occurrence is common in sandstones and in certain porous bricks.

The most common fluorescing salts are sulfates of calcium, sodium, and magnesium. More rarely, salts of chlorides, nitrates, carbonates, sulfites, vanadium, and molybdenum cause effluorescence.[16] Sulfate salts are particularly damaging as cryptofluorescence because these salts can exist in many states of hydration. For example, sodium sulfate, Na_2SO_4, can incorporate as many as 10 water molecules per Na_2SO_4 molecule into its crystalline structure. The preferred hydration state depends on temperature and humidity, and when water is incorporated into the crystalline structure, the volume change exerts pressure on the confining material, causing damage. For a state-of-the-art account of effluorescence mechanisms, see Charola.[8] Salts are present in various degrees in stones, bricks, concrete masonry units, mortar, and grout and are also introduced from external sources, including groundwater. ASTM C67 includes a test method to determine whether bricks have effluorescing potential.

Salt damage is common on the inside face of masonry foundations and in exterior masonry surfaces that are prone to frequent wetting, such as from contact with soil or poor water-control details. Efforts to seal the affected surfaces usually are ineffective and sometimes exacerbate the damage by trapping salts and moisture behind the sealant layer. An effective solution is to control the water source, such as by improving drainage, adding a dampproof course, or repairing gutters and downspouts to prevent chronic wetting of masonry.

Effluorescence can be cleaned simply by brushing it off the masonry surface (NCMA TEK 8-3A). Washing effluorescing surfaces with water is not necessarily effective because introduction of the wash water can cause further effluorescence.

Shrinkage and Expansion

A common cause of distress in concrete masonry structures is tensile cracking owing to shrinkage of restrained or partially restrained masonry. Figure 18.8 shows an instance of shrinkage cracking. Brick masonry expands over time and causes cracking at the ends of brick panels and corners where expansion joints are absent or infrequent. Figure 18.9 shows expansion cracking of brick masonry. Expansion of bricks can cause cracking and bowing of masonry walls when the brick masonry is confined. This problem is exacerbated when the confining structure is composed of materials that tend to shrink or creep, such as concrete or concrete masonry. Tall reinforced concrete frame buildings with brick veneer sometimes suffer from excessive compressive stresses in veneer panels between floors owing to shrinkage and creep of the frame. Composite walls of brick and concrete masonry sometimes develop curvature owing to the combined effects of brick expansion and concrete shrinkage.

TYPES AND CONDUCT OF TESTS

In Situ Testing

Water Penetration Testing. The resistance of a wall to water penetration can be measured quantitatively using a modified version of ASTM E514, *Standard Test*

FIGURE 18.8   An instance of shrinkage cracking in concrete masonry.

FIGURE 18.9   An instance of expansion cracking in brick masonry.

Method for Water Penetration and Leakage Through Masonry. A test chamber is clamped to the wall to simulate a driving rain on an area. The chamber is sealed to the wall surface and pressurized with air, and water is sprayed on the masonry continuously from a spray bar at the top of the chamber. The water runs in a closed system from a drain port on the bottom of the chamber, to a reservoir, and back to the spray bar. The water lost to the wall is measured by recording the volume of water in the reservoir at regular time intervals. A test in progress is shown in Fig. 18.10. An example of quantitative interpretation of the results of ASTM E514 testing by Raths[24] is shown in Table 18.12. However, the test method is intended for comparative purposes, e.g., for the evaluation of different masonry coatings in the laboratory. There is disagreement concerning the use of field ASTM E514 results as a quantitative measure of wall performance,[7] and the ASTM is developing a new test method for the field determination of water penetration of masonry wall surfaces. The ASTM E514 modified field test undoubtedly is useful for comparing water penetration through masonry walls in the same structure in different areas and in the same area before and after treatment procedures such as the application of water repellents, joint pointing, or grout injection.

Karsten Tube. The Karsten tube can be used to measure the rate of water absorption of masonry units or of mortar joints. This test uses a simple tube with a reservoir that is attached the wall with putty, as shown in Fig. 18.11. Tubes are available to test both horizontal and vertical masonry surfaces. The tube is filled with water, and the water level is recorded on a regular time interval chosen depending on the material; the range of 5 to 15 minutes is reasonable for most materials. By plotting the volume of water versus time, the rate of absorption per unit area is obtained. Information about the Karsten tube and its use can be found in RILEM Test Method II.4, *Measurement of Water Absorption under Low Pressure.*

Spray Test. Spray tests are useful for detecting leaks around windows, movement joints, and other discontinuities in masonry surfaces. It is advisable to perform the test

FIGURE 18.10   Field-adapted ASTM E514 water penetration testing in progress.

using a calibrated nozzle and specific water pressure. The Architectural Aluminum Manufacturers Association's *Field Check of Metal Curtain Walls for Water Leakage* (AAMA 501.2-83) contains a useful procedure that can be readily adapted for use on masonry walls. This procedure, designed to test seams in aluminum curtain walls, uses a calibrated nozzle and a pressure gauge and is performed by spraying the test area from a distance of 1 ft. During the test, the interior is monitored for leaks. Figure 18.12 shows the spray test in progress using a nozzle and gauge per AAMA 501.2-83.

A spray test described by Krogstad and Weber[20] is a modified version of ASTM E1105, *Standard Test Method for Field Determination of Water Penetration of Installed Exterior Windows, Curtain Walls, and Doors by Uniform or Cyclic Static Air Pressure Difference.* This test uses a spray rack with a controlled flow rate to evaluate water

TABLE 18.12   Guidelines for the Interpretation of Results of Field Modified ASTM E514, *Standard Test Method for Water Penetration and Leakage Through Masonry.* **There is Disagreement Concerning Quantitative Interpretation of ASTM E514 Results**

Leakage Rate	Rating of Wall
Less than ½ gallon per hour	Good performance
½ to 1 gallon per hour	Questionable performance
More than 1 gallon per hour	Serious performance problems

Source: Used with permission from Rath (1985).

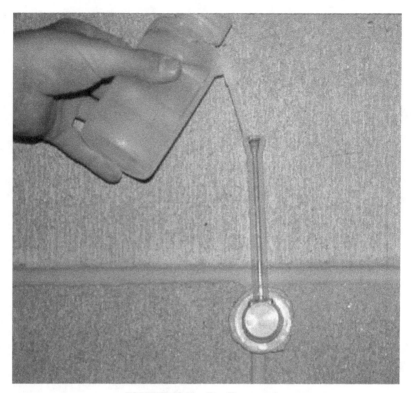

FIGURE 18.11 The Karsten tube.

resistance of wall components and interfaces. The spray rack is left in place for several hours while the interior is monitored for leaks. This test is useful for identifying the source of leaks in a building exterior.

Moisture Meters. Several varieties of moisture meters are available to determine the moisture content of masonry materials. These include sensors that are left in place to measure changes in moisture over time and meters that are used to measure the moisture content of material from the surface. Different technologies include capacitance-based measurements and electric permeance measurements. It is often useful to map moisture content of masonry areas to diagnose water problems, as shown in Fig. 18.13.

Flatjack Testing: Measurement of Masonry Deformability Properties. The modulus of elasticity can be measured in place using the procedure of ASTM C1197, *Standard Test Method for in Situ Measurement of Masonry Deformability Properties Using the Flatjack Method.* This test is a compression test conducted on an isolated portion of masonry. Inflatable steel bladders, or "flatjacks," are placed in slots cut into the mortar bed joints above and below the test area, as shown in Figs. 18.14 and 18.15. The test area is instrumented to measure strain, and the flatjacks are pressurized to induce compressive stress between the two jacks. The stress and strain are measured during the loading cycles. Results from a field test using this method are shown in Fig. 18.16. In addition to deformation characteristics, the compressive strength of the masonry can be measured using this test arrangement by applying load until failure.

FIGURE 18.12   The spray test per AAMA 501.2, *Field Check of Metal Curtain Walls for Water Leakage.*

Care must be taken to ensure that loads are in fact applied axially and that local flexural stresses do not influence the test. It is also important to choose a test site that has sufficient overburden load to resist the upward reaction forces that will be applied during the test. Limitations of this test include a pressure limit of 1000 lb/in.^{2} for commercially available flatjacks and that the test procedure is not suitable for hollow masonry units.

Flatjack Testing: Measurement of in Situ Compressive Stress.   The state of stress of masonry in service can be measured using ASTM C1196, *In Situ Compressive Stress within Solid Unit Masonry Estimated Using Flatjack Measurements.* This test is useful for identifying overstressed veneers or bending moments in walls and columns. Gauge points are installed above and below the test site, and the distance between them is measured. A horizontal saw cut is then made in a mortar joint between the gauge points. After the cut, the gauge points move closer together as the masonry deforms, and this distance is measured. A flatjack is then placed in the cut and pressurized, forcing the cut mortar joint to enlarge until the distance between the gauge points returns to the value measured before the saw cut. The pressure required to return the gauge points to their original position is recorded. This pressure value, modified with correction factors described in ASTM C1196, is a measure of the stress in the wall at the test area prior to the saw cut.

In-Place Shear Testing.   The shear strength at the mortar-unit interface can be measured directly by removing a masonry unit and using a hydraulic ram to test an adjacent unit. Alternatively, the head joints on either side of a unit can be removed and the load applied by a small flatjack inserted in one of the joints. ASTM C1531, *Standard*

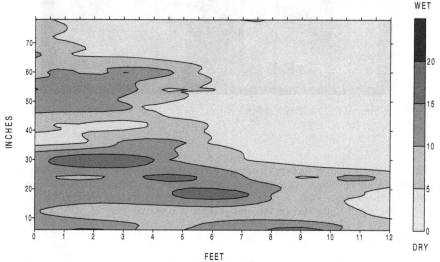

FIGURE 18.13 A moisture-content map of a masonry area used to diagnose a moisture-infiltration problem.

Test Methods for in Situ Measurement of Masonry Mortar Joint Shear Strength Index, thoroughly describes these test methods, and UBC 21-6, *In Place Masonry Shear Tests,* outlines the test method. Figure 18.17 shows an in-place mortar joint shear test in progress. When using this procedure with concrete masonry or brick masonry with strong mortar, it is important to choose a test site away from openings or corners to avoid causing wall failure during the test.

The *Uniform Code for Building Conservation* (UCBC) requires that for seismic strengthening projects, mortar quality in masonry walls be determined using in-place shear tests per UBC 21-6. A minimum test result of 30 lb/in.² is required, where the result is the maximum shear stress recorded in the test minus the estimated in situ compressive stress, or overburden stress.

Anchor Bolt Pullout and Shear. The strength of masonry anchors can be determined by testing in direct tension, shear, or torsion. ASTM E488, *Standard Test Methods for*

FIGURE 18.14 Photograph of test setup for masonry deformability testing using ASTM C1197-92, *Standard Test Method for in Situ Measurement of Masonry Deformability Properties Using the Flatjack Method.*

Strength of Anchors in Concrete and Masonry Elements, describes the conduct of tension and shear tests. UBC Standard 21-7, *Tests of Anchors in Unreinforced Masonry Walls,* describes direct tension testing, shear testing, and torsion testing. The torque test is the simplest anchor bolt test because the only equipment required is a torque wrench, and the load is applied manually. The bolt passes the test if it can withstand the applied torque: 40 ft · lb for ½-in.-diameter bolts, 50 ft · lb for ⅝-in.-diameter bolts, and 60 ft · lb for ¾-in.-diameter bolts.

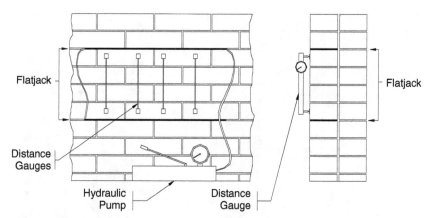

FIGURE 18.15 Schematic of test setup for ASTM C1197-92, *Standard Test Method for in Situ Measurement of Masonry Deformability Properties Using the Flatjack Method.*

The tension test is conducted using a hydraulic ram and a reaction frame to distribute the load over an area of the wall. The shear test similarly uses a hydraulic ram, and the anchor frame must be able to resist the shear force applied to the anchor bolt. Figure 18.18 shows an anchor bolt shear test in progress. Depending on the purpose of the test, a maximum deflection or minimum load criterion is used, or the anchor bolt is tested to failure.

Testing for the Presence Reinforcing Steel and Grout. Detecting reinforcing steel and grout in reinforced masonry can be accomplished using metal detection devices,

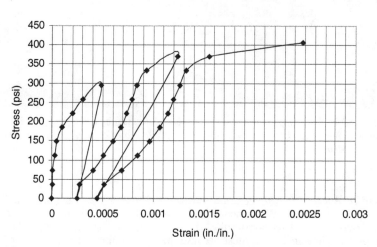

FIGURE 18.16 Stress and deformability plot for three loading cycles of ASTM C1197, *Standard Test Method for in Situ Measurement of Masonry Deformability Properties Using the Flatjack Method.*

FIGURE 18.17 An in-place shear test in progress.

FIGURE 18.18 An anchor bolt shear test configuration.

sounding, and drilling to verify conditions. Pachometers for the detection of reinforcing bars in concrete are ideal for use with reinforced masonry. Most pachometers have a sensing depth of approximately 6 in. and can be used to determine the depth of a reinforcing bar given its size or the size of the bar given its depth. When both size and depth are unknown, it is up to the operator to interpret the results. It is difficult to identify or characterize lap splices with a pachometer. In addition, wire joint reinforcing steel will be sensed.

To determine if a cell is grouted, the simplest technique is to "sound" the cell with a small hammer. Empty cells, when knocked with a small hammer, will produce a sharp "thok" sound because the face shell vibrates as a free plate. Grouted cells produce a dull, lower-frequency "thump" sound because the face shell is bound to the grout and cannot vibrate freely. The quality of sound produced is a function of the material and weight of the hammer head and the length and flexibility of the handle. It is beneficial to sound the masonry with several hammers and choose one that produces discernible sound differences. Figure 18.19 shows an assortment of sounding hammers.

The presence of grout should be verified by drilling into the masonry. The drilling resistance of the cell will indicate the presence or absence of grout, and observing the wall interior with a fiberoptic borescope will further define the condition.[30]

In addition to sounding, there are several technologically advanced methods for sensing the presence of grout in a cell, including ultrasonic pulse velocity, impact-echo, radar reflections, and thermographic imaging with infrared cameras. These techniques can be efficient for inspection of large areas of wall. For many projects, however, sounding with hammers and the use of metal detectors are sufficient to provide the required information.

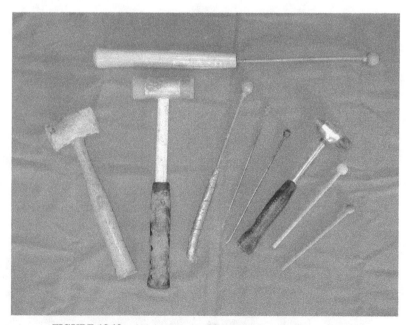

FIGURE 18.19   An assortment of hammers for sounding masonry.

Nondestructive Evaluation

Sonic and Ultrasonic Pulse Velocity. Measurement of ultrasonic and sonic and wave velocity in masonry materials can be used to determine material properties and infer interior characteristics such as wall solidity and masonry configuration. Suprenant and Schuller[33] provided detailed descriptions of equipment and data-processing techniques for sonic and ultrasonic testing of masonry. Krautkrämer and Krautkrämer[19] provided full theoretical and technical details of ultrasonic testing for all materials. Ultrasonic testing is well developed in the field of concrete evaluation, and the standard method for concrete, ASTM C597, *Standard Test Method for Pulse Velocity through Concrete,* is useful for masonry.

Ultrasonic pulses provide greater accuracy and resolution, but sonic pulses travel longer distances because lower-frequency waves attenuate less. Typically, sonic testing is used when ultrasonic testing is not possible, i.e., when ultrasonic waves attenuate to unmeasureable levels because of long path length or nondense material.

Three types of stress waves occur in solids: compression or longitudinal waves, shear waves, and surface waves. In masonry testing, compression waves typically are measured. Compression wave speed in a material is the same regardless of frequency and is a function of the density, elastic modulus, and Poisson's ratio of the material:

$$c = \sqrt{\frac{E}{\rho} \frac{(1 - \nu)}{(1 + \nu)(1 - 2\nu)}}$$

where c = wave speed
E = modulus of elasticity
ν = Poisson's ratio
ρ = density

Wave speeds of various materials are provided in Table 18.13.

TABLE 18.13 Stress Wave Speed in Various Materials[19,26,31,34]

Material	Compression Wave Speed, Meters per Second
Aluminum	Approximately 5000–6300
Cast iron	Approximately 3500–5800
Steel	Approximately 5900
Porcelain	Approximately 5600–6200
Concrete	Approximately 2500–5000
Old brick or stone masonry*	Approximately 900–3000
Building stones	Approximately 1000–6000
Marble	Approximately 5000
Ice	3980
Water (25°C)	1493
Sea water (25°C)	1533
Air (0°C)	331
Air (20°C)	343

*Atkinson and Noland & Associates.

Sonic and ultrasonic testing of masonry has been shown to be useful in determining masonry solidity,[18] locating voids in masonry walls,[33] measuring the efficacy of strengthening by grout injection,[3] and as a measure of deterioration. Tomographic imaging techniques have been used to create maps of the interior properties of masonry elements from large data sets of sonic and ultrasonic measurements.[25] Figure 18.20 shows the results of an ultrasonic investigation of a historic masonry construction.

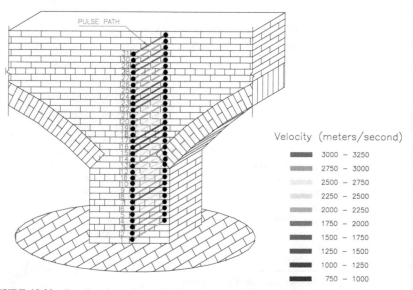

FIGURE 18.20 Results of an ultrasonic investigation of a historic masonry construction.

TABLE 18.14   Relative Dielectric Permittivity of Various Materials[10,11]

Material	Relative Dielectric Permittivity
Sandstone (dry)	2–3
Sandstone (wet)	5–10
Granite (dry)	5
Granite (wet)	7
Limestone (dry)	7
Limestone (wet)	8
Concrete (dry)	4–10
Concrete (wet)	10–20
Ice	3–4
Water	80
Sea water	81–88
Air	1

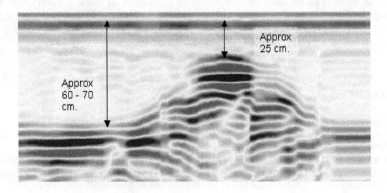

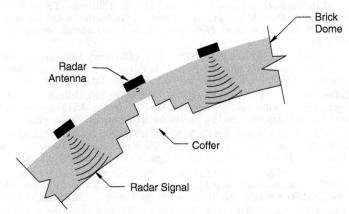

FIGURE 18.21   Results of a surface-penetrating radar investigation of a historic masonry construction.

Surface-Penetrating Radar. Measurements of the reflections of electromagnetic waves can be used to characterize subsurface features of masonry. RILEM Committee 127-MS has developed a standard method for radar investigation of historic masonry, RILEM MS-D.3, *Radar Investigation of Masonry,* that details equipment requirements, procedures, and limitations. Daniels[11] presented theoretical and technical details of surface-penetrating radar testing in general. Nondestructive testing of masonry and other construction materials typically is conducted using microwave radar devices that use frequencies of 300 to 2000 MHz. The most common equipment configuration is to use one antenna to both generate the impulse and measure the time of arrival and amplitude of reflections from the object.

As with visible light, radar waves are reflected at interfaces between materials that have different wave transmission properties. The property that defines the speed of the wave in the material and affects the magnitude of the reflection at the interface between two materials is the *relative dielectric permittivity.* The speed of radar waves in a material is equal to the speed of light in a vacuum divided by the square root of the relative dielectric permitivvity.[10] Table 18.14 shows the relative dielectric permittivity of various materials.

As with stress waves, higher-frequency radar waves provide better resolution at the cost of depth of penetration owing to the greater attenuation and dispersion of higher-frequency waves. The speed and depth of penetration of radar waves are substantially reduced by the presence of water, and the waves reflect perfectly from metal surfaces.

Radar investigations have proven useful in evaluation of stone masonry,[4] in determining configurations of massive historic masonry elements,[30] and in locating reinforcing bars and grout in concrete masonry. Figure 18.21 shows the results of a surface-penetrating radar investigation of a historic masonry construction.

REFERENCES

1. Amoroso, G. G., and V. Fassina, (1983). *Stone Decay and Conservation—Atmospheric Pollution, Cleaning, Consolidation and Protection.* Materials Science Monographs 11, Elsevier, New York.
2. Atkinson, R. H., and G. G. Yan, (1990). *A Statistical Study of Masonry Deformability.* Research Report, Part of National Science Foundation Grant No. CES-8806180, "Development of a Database for Compressive Stress-Strain Behavior of Masonry," Washington, D.C.
3. Berra, M., L. Binda, L. Anti, and A Fatticcioni, (1992). "Utilization of Sonic Tests to Evaluate Damaged and Repaired Masonries," in *Proceedings, Nondestructive Evaluation of Civil Structures and Materials, University of Colorado at Boulder.* Atkinson-Noland and Associates, Boulder, CO.
4. Binda, L., G. Lenzi, and A. Saisi, (1997). "NDE of Masonry Structures: Use of Radar Test for the Characterization of Stone Masonries," in *Proceedings of Structural Faults and Repair Conference, Edinburgh, UK,* Engineering Technics Press, Edinburg, UK, pp. 505–514.
5. Birkeland, O., and S. D. Svendsen, (1962). "Norwegian Test Methods for Rain Penetration through Masonry Walls," in *Symposium on Masonry Testing.* ASTM Special Technical Publication No. 320, American Society for Testing and Materials, Philadelphia.
6. Bortz, S. A., S. L. Marusin, and C. B. Monk, Jr., (1990). "A Critical Review of Masonry Durability Standards," in *Proceedings of the Fifth North American Masonry Conference,* Vol. IV. University of Illinois at Urbana–Champaign, June 3–6, Johnson Publishing, Boulder, CO.
7. Brown, M. T., (1990). "A Critical Review of Field Adapting ASTM E514 Water Permeability Test Method," in Matthys, J. H. (ed.): *Masonry: Components to Assemblages.* ASTM Special Technical Publication 1063, American Society for Testing and Materials, Philadelphia.
8. Charola, A. E., (2000). "Salts in the Deterioration of Porous Materials: An Overview," *Journal of the American Institute of Conservation of Historic and Artistic Works* **39**: pp. 327–343.

9. Cliver, E. B., (1974). "Tests for Analysis of Mortar Samples," *Bulletin of the Association for Preservation Technology* **4**(1): 68–73.

10. Conyers, L. B., and D. Goodman, (1997). *Ground Penetrating Radar: An Introduction for Archaeologists.* Altamira Press, Walnut Creek, CA.

11. Daniels, D. J., (1996). *Surface-Penetrating Radar.* IEEE Radar, Sonar, Navigation and Avionics Series 6, Institution of Electrical Engineers, London.

12. Drysdale, R. G., A. A. Hamid, and L. R. Baker, (1999). *Masonry Structures: Behavior and Design.* Masonry Society, Boulder, CO.

13. Felicetti, R., and N. Gattesco, (1998). "A Penetration Test to Study the Mechanical Response of Mortar in Ancient Masonry Buildings," *Materials and Structures* **31**: pp. 350–356.

14. Frey, D. J., (1975). "Effect of Constituent Proportions on Uniaxial Compressive Strength of 2-in. Cube Specimens of Masonry Mortars," Master of science thesis, Department of Civil and Environmental Engineering, University of Colorado, Boulder, CO.

15. Goins, E. S., (2002). "Standard Practice for Determining the Components of Historic Cementitious Materials," project report for the National Center for Preservation Technology and Training, National Park Service, Grant Agreement No. MT-2210-9-NC-29, PTT Publications No. 2002-20, Natchitoches, LA.

16. Grimm, C. T., (1985). *Durability of Brick Masonry: A Review of the Literature.* Special Technical Testing Publication 871, American Society for Testing and Materials, Philadelphia.

17. Jedrzejewska, H., (1960). "Old Mortars in Poland: A New Method of Investigation: Studies in Conservation," *Journal of the International Institute for Conservation of Historic and Artistic Works* **5**(4): pp. 132–138.

18. Kingsley, G., J. Noland, and R. Atkinson, (1987). "Nondestructive Evaluation of Masonry Structures Using Sonic and Ultrasonic Pulse Velocity Techniques," in *Proceedings of the Fourth North American Masonry Conference, University of California, Los Angeles.*

19. Krautkrämer, J., and H. Krautkrämer, (1990). Ultrasonic Testing of Materials, 4th ed., Springer-Verlag, Berlin.

20. Krogstad, N., and R. A. Weber, (1993). "Using Modified ASTM E1105 to Evaluate Resistance of Masonry Barrier, Mass, and Skin Walls to Rain," in *Masonry: Design and Construction, Problems and Repair.* ASTM Special Technical Publication 1180, American Society for Testing and Materials, Philadelphia.

21. Kung, J. H., "Frost Durability of Canadian Clay Bricks," in *Proceedings of the 7th International Brick Masonry Conference, Melbourne, Australia.*

22. Maage, M., "Frost Resistance and Pore Size Distribution in Bricks," Institut for Bygningsmaterialleare Rapportunummer, BML 80.201, University of Trondheim, Norway.

23. McBurney, J. W., and J. C. Richmond, (1942). *Strength, Absorption and Resistance to Laboratory Freezing and Thawing of Building Bricks Produced in the United States.* Building Materials and Structures Report BMS60, U.S. Government Printing Office, Washington D.C.

24. Raths, C. H., (1985). "Brick Masonry Wall Nonperformance Causes," in Grogan, J. C., and J. T. Conway (eds.): *Masonry: Research Application, and Problems.* Special Technical Publication 871, American Society for Testing and Materials, Philadelphia.

25. Rens, K. L., D. J. Transue and M. P. Schuller, (2000). "Acoustic Tomographic Imaging of Concrete Infrastructure," *Journal of Infrastructure Systems,* American Society of Civil Engineers, (1982). **6**(1): 15–23.

26. Robinson, E. C., (1982). "Physical Properties of Building Stone," in *Conservation of Historic Stone Buildings and Monuments.* Report of the Committee on Conservation of Historic Stone Buildings and Monuments, National Materials Advisory Board, National Research Council, National Academy Press, Washington, D.C.

27. Robinson, G. C., J. R. Holman, and J. F. Edwards, (1997). "Relation Between Physical Properties and Durability of Commercially Marketed Brick," *American Ceramic Society Bulletin* **56**(12): 1071–1076.

28. Schuller, M. P., R. H. Atkinson, and G. R. Kingsley, (1995). "Nondestructive Evaluation of Masonry Structures: Standardization and Current Practice," in Colville J., and Amde, A. M. (eds.): *Proceedings of Research Transformed into Practice, Implementation of NSF Research.* ASCE Press, New York.

29. Schuller, M. P., R. H. Atkinson, and J. L. Noland, (1997). "Structural Evaluation of Historic Masonry Buildings," *APT Bulletin, The Journal of Preservation Technology* **26**(2–3): 17–27.

30. Schuller, M. P., (2003). "Nondestructive Testing and Damage Assessment of Masonry Structures," in *Progress in Structural Engineering and Materials* **5**(4): 239–251.

31. Serway, R. A., (1992). *Physics for Scientists and Engineers,* 3d ed., Saunders College Publishing, Philadelphia.

32. Shrive, N. G., and D. Tilleman, (1992). "A Simple Apparatus and Methods for Measuring On-Site Flexural Bond Strength," in *Proceedings of the 6th Canadian Masonry Symposium, University of Saskatchewan, Saskatoon, Canada.*

33. Suprenant, B. A. and M. P. Schuller, (1994). *Nondestructive Evaluation and Testing of Masonry Structures.* The Aberdeen Group, Addison, IL.

34. Telford, W. M., L. P. Geldart and R. E. Sheriff, (1990). *Applied Geophysics, 2d ed.* Cambridge University Press, Cambridge, England.

35. Thomson, M., G. T. Suter, and L. Fontaine, (1995). "Freeze-Thaw Durability of Mortars in Historic Masonry: Test Methodologies," in *Proceedings of the Seventh Canadian Masonry Symposium, Hamilton, Ontario,* Department of Civil Engineering, McMaster University.

36. Tindall, S., (1989)."How to Prepare Project-Specific Terra-Cotta Specifications," *Bulletin of the Association for Preservation Technology* **21**(1): 26–36.

37. Transue, D. J., M. P. Schuller, and K. L. Rens, (1999). "Use of the Pendulum Hammer Hardness Test for Mortar Evaluation," presented at the Eighth North American Masonry Conference, Austin, Texas.

38. Winslow, D. N., C. L. Kilgour, and R. W. Crooks (1988). "Predicting the Durability of Bricks," *Journal of Testing and Evaluation* **16**(6): 527–531.

39. RILEM Technical Committee 127-MS, (1996–1998). "Recommended Test Methods," *Materials and Structures* **29–31.**

USEFUL STANDARDS, CODES, AND GUIDELINES

The Brick Industry Association (BIA) Technical Notes on Brick Construction, Brick Industry Association, 11490 Commerce Park Drive, Reston, VA.

NCMA TEK Manual for Concrete Masonry Design and Construction, National Concrete Masonry Association (NCMA), 13750 Sunrise Valley Drive, Herndon, VA.

ACI 201.2R-92, *Guide to Durable Concrete, American Concrete Institute's Manual of Concrete Practice,* American Concrete Institute, Detroit, MI.

MSJC Building Code Requirements for Masonry Structures (ACI 530/ASCE 5/TMS 402) and *Specification for Masonry Structures* (ACI 530.1/ASCE 6/TMS 602), *Masonry Standards Joint Committee.*

SEI/ASCE 11-99, *Guideline for Structural Condition Assessment of Existing Buildings,* American Society of Civil Engineers, New York, 1999.

International Existing Building Code, International Code Council, Inc.

International Building Code, International Code Council, Inc.

FEMA 356, *Pre-Standard and Commentary for the Seismic Rehabilitation of Buildings,* the Federal Emergency Management Agency, Washington, D.C.

UBC Standard 21-7, *Tests of Anchors in Unreinforced Masonry Walls.* Whittier, CA

RILEM Test Method II.4, *Measurement of Water Absorption under Low Pressure.* Paris.

AAMA 501.2, *Field Check of Metal Curtain Walls for Water Leakage,* American Architectural Manufacturers Association, Des Planes, IL.

ASTM C67, *Standard Test Methods for Sampling and Testing Brick and Structural Clay Tile.*

ASTM C73, *Standard Specification for Calcium Silicate Brick (Sand-Lime Brick).*

ASTM C90, *Standard Specification for Loadbearing Concrete Masonry Units.*

ASTM C97, *Standard Test Methods for Absorption and Bulk Specific Gravity of Dimension Stone.*

ASTM C109/C109M, *Standard Test Method for Compressive Strength of Hydraulic Cement Mortars (Using 2-in. or [50-mm] Cube Specimens).*

ASTM C136, *Standard Test Method for Sieve Analysis of Fine and Coarse Aggregates.*

ASTM C140, *Standard Test Methods for Sampling and Testing Concrete Masonry Units and Related Units.*

ASTM C144, *Standard Specification for Aggregate for Masonry Mortar.*

ASTM C170, *Standard Test Method for Compressive Strength of Dimension Stone.*

ASTM C216, *Standard Specification for Facing Brick (Solid Masonry Units Made from Clay or Shale).*

ASTM C270, *Standard Specification for Mortar for Unit Masonry.*

ASTM C404, *Standard Specification for Aggregates for Masonry Grout.*

ASTM C426, *Standard Test Method for Linear Drying Shrinkage of Concrete Masonry Units.*

ASTM C476, *Standard Specification for Grout for Masonry.*

ASTM C503, Standard Specification for Marble Dimension Stone (*Exterior*).

ASTM C568, *Standard Specification for Limestone Dimension Stone.*

ASTM C615, *Standard Specification for Granite Dimension Stone.*

ASTM C616, *Standard Specification for Quartz-Based Dimension Stone.*

ASTM C629, Standard Specification for Slate Dimension Stone.

ASTM C652, *Standard Specification for Hollow Brick (Hollow Masonry Units Made from Clay or Shale).*

ASTM C952, *Standard Test Method for Bond Strength of Mortar to Masonry Units.*

ASTM C1006, *Standard Test Method for Splitting Tensile Strength of Masonry Units.*

ASTM C1019, *Standard Test Method for Sampling and Testing Grout.*

ASTM C1072, *Test Method for Measurement of Masonry Flexural Bond Strength.*

ASTM C1194, *Standard Test Method for Compressive Strength of Architectural Cast Stone.*

ASTM C1195, *Standard Test Method for Absorption of Architectural Cast Stone.*

ASTM C1196, *Standard Test Method for in Situ Compressive Stress within Solid Unit Masonry Estimated Using Flatjack Measurements.*

ASTM C1197, *Standard Test Method for in Situ Measurement of Masonry Deformability Properties Using the Flatjack Method.*

ASTM C1314, *Test Method for Constructing and Testing Masonry Prisms Used to Determine Compliance with Specified Compressive Strength of Masonry.*

ASTM C1324, *Standard Test Method for Examination and Analysis of Hardened Masonry Mortar.*

ASTM C1354, *Standard Test Method for Strength of Individual Stone Anchorages in Dimension Stone.*

ASTM C1364, *Standard Specification for Architectural Cast Stone.*

ASTM C1372, *Standard Specification for Segmental Retaining Wall Units..*

ASTM C1496, *Standard Guide for Assessment and Maintenance of Exterior Dimension Stone Masonry Walls and Facades.*

ASTM C1531, *Standard Test Methods for in Situ Measurement of Masonry Mortar Joint Shear Strength Index.*

ASTM D4404, *Standard Test Method for Determination of Pore Volume and Pore Volume Distribution of Soil and Rock by Mercury Intrusion Porosimetry.*

ASTM E96, *Standard Test Methods for Water Vapor Transmission of Materials.*

ASTM E122, *Standard Practice for Choice of Sample Size to Estimate a Measure of Quality for a Lot or a Process.*

ASTM E447, *Compressive Strength of Masonry Assemblages.*

ASTM E488, *Standard Test Methods for Strength of Anchors in Concrete and Masonry Elements.*

ASTM E514, *Standard Test Method for Water Penetration and Leakage through Masonry.*

ASTM E518, *Standard Test Methods for Flexural Bond Strength of Masonry.*

ASTM E519, *Standard Test Method for Diagonal Tension (Shear) in Masonry Assemblages..*

ASTM E1105, *Standard Test Method for Field Determination of Water Penetration of Installed Exterior Windows, Curtain Walls, and Doors by Uniform or Cyclic Static Air Pressure Difference.*

▰ CHAPTER 19

Timber

DONALD W. NEAL, P.E., S.E.

INTRODUCTION

There is a strong demand for structural assessment that is accurate and based on sound scientific and engineering principles. The client requires reliable test data devoid of speculation for repair and maintenance issues, to guide decisions on rehabilitation of aging infrastructure, and for adapting structures to differing usage. The collection of raw test data sufficient to render a structural conclusion involves hands-on use of procedures and test equipment and may involve challenging site conditions. This chapter discusses common types of degradation and explains test procedures for assessing the condition of structural lumber, timbers, and engineered wood products.

Wood is a natural-growth orthotropic (anisotropic) material whose engineering properties depend on wood grain orientation within a member—longitudinal, tangential, or radial. Since the radial and tangential wood ring orientation in a sawn or glulam wood structural member is random and their difference in engineering properties is small, the engineer need only consider properties parallel or perpendicular to grain. Distinctly different engineering properties are tabulated for tension and compression depending on stress parallel or perpendicular to grain.

ENGINEERING PROPERTIES OF WOOD

Bending

Most timber structural members are used primarily in bending, so bending is often considered the defining design property. Basic bending stress is often used to label the grade for both glulam members and mechanically graded dimension lumber. Within the elastic range, the flexural stress block is considered to be triangular, and extreme fiber bending stresses are calculated as M_C/I. The *modulus of rupture* (MOR), by definition, is the value of M_C/I calculated using the bending moment M at ultimate load. The MOR is not the real maximum fiber stress at failure because the stress block beyond the elastic range is not triangular. However, the MOR is a useful index for comparison of test results.

Tension Parallel to Grain

Until recent years, test equipment was not readily available to pull full-sized pieces of lumber to failure in tension, so only small strip-tension tests were performed. Older tabulated allowable design values for both solid sawn timbers and glulam grouped bending and tension parallel to grain together on the theory that pure tension stress would behave the same as the tension portion of bending stress. The 1968, *National Design Specification* (NDS)[1] was the first to provide separate allowable stresses with different design values for bending and tension.

Tension Perpendicular to Grain

Tension perpendicular to grain, also called *cross-grain tension,* describes tension stress perpendicular to the long axis of the wood fiber. This may result from gravity load applied at the lower edge of a member, from truss web forces perpendicular to grain at chord, restrained shrinkage across the grain at mechanical connections, or lateral load from a diaphragm applied to the top of a ledger where the lower portion of the ledger is fastened securely to a wall.

When tension perpendicular to grain is induced by an increase in radius, straightening the arc of a curved glulam member, tension perpendicular to grain commonly is termed *radial tension.* Cross-grain tension is the least understood mechanical property of timber. Cross-grain tension is not presently included in allowable design tables.

Allowable values for radial tension are covered in the body of the NDS and in the American Institute of Timber Construction (AITC) 117 design specification.[2] The basic radial tension allowable stress is 15 lb/in.[2] for Douglas fir–larch, Douglas fir–south, hem-fir, western woods, and Canadian softwood species subjected to gravity loading. The allowable radial tension stress is one-third the allowable horizontal shear stress for the preceding species subjected to lateral loading and for southern pine subjected to all types of loading. Low wood strength perpendicular to grain is a function of the bond between earlywood and latewood portions of the annual wood growth ring. The low value of 15 lb/in.[2] for radial tension was assigned on the basis of judgment and field experience. Curved glulam beams of western softwood species require steel reinforcement parallel to grain where radial tension stress exceeds the allowable value. Since the allowable value is so low, this includes most curved glulam beams. The only disadvantage of steel reinforcing is that the glulam member is restrained against shrinkage, and when changes in moisture content occur, microchecking in the wood member occurs in the vicinity of the reinforcing.

Compression Parallel to Grain (Short Compression)

This is termed *crushing strength* under the provisions of American Society for Testing and Materials (ASTM) Standard D198[3] and is defined as the maximum stress sustained in compression parallel to the grain by a specimen having a ratio of length to least dimension of less than 11. Compression parallel to grain tests are often possible to obtain during a condition assessment because a relatively small test specimen is required, and the test setup is simple. From this test, compressive modulus of elasticity (MOE), proportional limit, and ultimate compression data may be obtained. While useful as comparative data in evaluation, allowable stresses in compression parallel to grain are of limited value to design engineers because most compression member capacity is limited by buckling.

Buckling (Long Compression)

Buckling strength of timber members in compression is defined by the Euler critical buckling stress and is governed by member stiffness and the slenderness ratio l/d, where l is the distance between points of lateral support, d is least cross-sectional dimension, and E is modulus of elasticity (MOE). From the 1950s until the 1977 NDS, the allowable column stress was the lesser of the crushing strength or the following Euler buckling equation, which includes a factor of safety:

$$F_c = \frac{0.3E}{(l/d)^2}$$

From the 1977 NDS until the 1991 NDS, columns were classified as short, intermediate, and long.

Short columns. Columns with an l/d ratio of 11 or less used an allowable stress based on crushing strength.

Long columns. Columns with an l/d ratio greater than a K factor, defined in the specification, used the adjusted Euler buckling stress design equation.

Intermediate columns. Columns with an l/d ratio between 11 and K used a formula representing the interaction between crushing and the adjusted Euler buckling stress equation.

The 1991 NDS incorporated the single-continuous-column formula, often referred to as the *Ylinen formula,* for all ranges of slenderness ratio with interaction between

crushing and Euler buckling failure modes. The continuous-column formula is a refinement of the Euler equation with different input constants depending on the type
of timber member [glulam, visually graded lumber, machine-evaluated lumber (MEL),
poles and piles etc.]:

$$F_C' = F_C^* \, C_P$$

$$= F_c^* \left[\frac{1 + (F_{CE}/F_C^*)}{2c} - \sqrt{\left[\frac{1 + (F_{CE}/F_C^*)}{2c} \right]^2 - \frac{F_{CE}/F_C^*}{c}} \right]$$

where F_C^* = tabulated compression design value multiplied by all applicable adjustment factors except C_P
F_{CE} = critical buckling design value for compression members
= $K_{CE} E' / (l_e/d)^2$
K_{CE} = Euler buckling coefficient for columns
= 0.3 for visually graded lumber
= 0.419 for glulam and machine-stress-rated (MSR) lumber
c = 0.8 for sawn lumber
= 0.9 for glulam or structural composite lumber
l_e = effective column length

Compression Perpendicular to Grain

Allowable compression perpendicular to grain stress is used in design primarily to size
the required bearing area of beams over supports or under concentrated loads. Failure
in compression perpendicular to the grain cannot be as well defined as with other
modes of failure. Excessive compression loading perpendicular to grain causes wood
cells to collapse and crush but rarely causes total member collapse or catastrophic
failure. Prior to the 1982 NDS, allowable compression perpendicular to grain stresses
were based on a proportional limit stress. The proportional limit stress is that stress at
which the stress-strain behavior of the material first deviates from linearity. Starting
with the 1982 NDS, the basis for the allowable design value in compression perpendicular to the grain was revised as the mean stress at a stated deformation. The time-
dependent load duration factor C_D is not applicable to an allowable value defined by
the deformation limit and is not appropriate for use with compression perpendicular
to grain calculations.

Shear

Tabulated allowable shear stresses are for the shear plane parallel to the faces of
finished lumber and approximately parallel to wood fiber orientation. Since most wood
members are beams oriented horizontally, this is commonly termed *horizontal shear*.
For the most part, allowable shear stress is unchanged between grades of a given
species. Data obtained by shear testing of glulam per ASTM D198 resulted in increased
allowable shear stresses being adopted in 1994 plus additional increases in 1998 and
2003.

Stiffness

Young's modulus, or the modulus of elasticity (MOE), is the slope of the stress-strain
curve and may be measured in the testing laboratory when members are loaded axially
or in flexure. The tabulated MOE included in tables of allowable design stresses is an
average value. MOE in bending is included with all design property tables for dimen-

sion lumber, timbers, and glulam. Glulam design tables include MOE in bending for both axes of the member plus MOE in axial stress. Machine-stress-rated (MSR) grades of dimension lumber have MOE measured for each piece at the point of manufacture. The MOE rating and an accompanying allowable bending stress are used for member identification of MSR grades, and tabular values are listed in the NDS supplement.

Density

The density of wood varies substantially both within and between species. Balsa has a specific gravity of about 0.16, whereas greenheart and purpleheart often exceed a specific gravity of 1.00. Wood is used within a wide range of climatic conditions and has a wide range of moisture content values in use. Moisture makes up a portion of the weight of any wood product. Wood density values reflect the combined weight of the basic wood material plus retained moisture. *The Wood Handbook—Wood as an Engineering Material*[4] provides a table for determining density of wood as a function of specific gravity and moisture content. It is common knowledge that wood strength, in general, is a function of density. Higher-density softwoods such as Douglas fir and southern pine have higher strength than lower-density softwoods such as cedar or redwood. Higher values of wood density and/or specific gravity generally result in higher allowable material stresses, stiffness, and connector values.

Moisture Content

Wood is a natural-growth organic material with a cellular structure that retains water. Water is a living tree's best friend but the worst enemy of finished wood products. Wood moisture content is expressed as a percentage of the weight of water present to the dry weight of wood. The fiber saturation point of most wood species occurs in the middle 20% range, with Douglas fir about average at 25% to 26%. Additional moisture above the fiber saturation point may be retained as free water within the cells. Since water softens the cellular wall, strength and stiffness increase as moisture content is reduced below the fiber saturation point. Wood in use and not subject to free water will come to equilibrium at a moisture content as a function of temperature and humidity. Design values for solid sawn softwood lumber and timbers require a wet-service factor C_M adjustment to be applied where moisture content exceeds 19% for an extended period of time. Glulam is usually produced at a moisture content range of 12% to 15%. Design values for glulam are based on a wood moisture content averaging 12%, with a wet-service factor C_M adjustment to be applied where moisture content exceeds 16%.

CODES AND STANDARDS

Model Building Codes

Most building codes in the United States have been based on one of the three regional model codes: the *Uniform Building Code*[5] (UBC), published by the International Conference of Building Officials (ICBO); the *National Building Code,*[6] published by Building Officials and Code Administrators International (BOCA); and the *Standard Building Code,*[7] published by the Southern Building Code Congress International (SBCCI). The UBC has been used predominately in the West, BOCA in the Midwest and Northeast, and standard in the South with some overlap. Most building jurisdictions adopt a model code and write a supplement or specialty code tailoring it to specific climatic or site conditions such as permafrost, hurricanes, expansive soils, or snow. In 1995, the three regional model codes jointly formed the International Code

Council (ICC) for the creation of a single national building code. Recently the National Fire Prevention Association (NFPA) developed the *NFPA 5000 Building Code* in competition with the ICC code. It is imperative in structural evaluation to know the specific code and code revisions against which a structure must be evaluated. The building code adopted by the authority having jurisdiction is the legal document for structural compliance, and if differences exist between the code and other specifications or standards, the code prevails. The code is a minimum standard, and the prudent engineer may incorporate more conservative criteria of which he or she has knowledge.

ASTM

The American Society for Testing and Materials (ASTM) is a nonprofit organization that provides a forum for interested parties to write standards for materials, products, systems, and services. The standards relating to wood are contained in Vol. 4.10 under Sec. 4, "Construction." Structural condition assessment invariably involves testing either materials or structural systems. Where possible, the use of established testing laboratories following accepted test standards, such as ASTM, is highly recommended. ASTM standard tests mentioned here are listed in the References at the end of this chapter.

ANSI/AITC Standard 190.1 and AITC 200 Inspection Manual[8]

ANSI/AITC Standard 190.1 for wood products, *Structural Glued Laminated Timber,*[8] is a consensus standard concerned primarily with the nuts and bolts of glulam manufacture. The standard is revised periodically. The current edition is ANSI/AITC 190.1-2002. The standard is developed under provisions of the American National Standards Institute (ANSI). Development of consensus for the standard was accomplished by the Canvass method. The standard contains requirements for the manufacture and quality control of structural glued laminated timber. *The AITC 200 Inspection Manual*[9] is based on the ANSI/AITC 190.1 standard. The manual covers production quality control, testing and inspection requirements, test procedures, and standard AITC tests.

AITC Standards

American Institute of Timber Construction (AITC) publishes the following standards, designated by AITC number and title:

104 *Typical Construction Details*
108 *Standard for Heavy Timber Construction*
109 *Standard for Preservative Treatment of Structural Glued Laminated Timber*
110 *Standard Appearance Grades for Structural Glued Laminated Timber*
111 *Recommended Practice for Protection of Structural Glued Laminated Timber During Transit, Storage and Erection*
112 *Standard for Tongue-and-Groove Heavy Timber Roof Decking*
113 *Standard for Dimensions of Glued Laminated Structural Members*
114 *Structural Glued Laminated Timbers for Electric Utility Framing and Crossarms*
117 *Standard Specifications for Structural Glued Laminated Timber of Softwood Species*
119 *Standard Specifications for Hardwood Glued Laminated Timber*

All the standards are potentially valuable for assessment work. The most commonly used of the standards and the source document for design with softwood glulam timber, is Specification 117, *Design,* which provides allowable design values and design examples. The latest edition available is AITC 117-2004. Specification 117 typically has been issued in two documents, Design and Manufacturing. The manufacturing document showing lamstock grade requirements and layup zones for the various glulam combinations and is of interest primarily to those in glulam manufacture.

Grading Rules: Lumber and Timber

Rules-writing agencies have evolved that develop lumber grades for specific species within a geographic area and publish grading rules. Lumber mills producing graded lumber join an appropriate grading agency, and their products carry that agency's stamp. The following is a listing of rules-writing agencies certified by the American Lumber Standards Committee for inspection and grading of untreated lumber.

NELMA—Northeastern Lumber Manufacturers Association, Cumberland Center, ME

RIS—Redwood Inspection Service, Novato, CA

SPIB—Southern Pine Inspection Bureau, Pensacola, FL

WCLIB—West Coast Lumber Inspection Bureau, Portland, OR

WWPA—Western Wood Products Association, Portland, OR

NLGA—National Lumber Grades Authority, New Westminster, British Columbia, Canada

Voluntary Product Standard PS 2[10]

The Engineered Wood Association has promulgated a series of specifications and standards for wood-based panel products that include plywood. Perhaps the most useful of these documents is Voluntary Product Standard PS 2-92, *Performance Standard for Wood-Based Structural-Use Panels* (U.S. Department of Commerce, National Institute of Standards and Technology, Gaithersburg, MD, 1992).

COMMON TYPES, CAUSES, AND CONSEQUENCES OF DISTRESS, DEGRADATIONS, AND DEFECTS

Decay

Decay is a potential hazard in wood structural members. Decay occurs when, in conditions favorable to their growth, wood-destroying fungi use wood as their food source. In addition to the food source, these organisms require air, moisture, and a favorable temperature to survive. Lacking air, continuously submerged wood will not decay owing to attack from most wood-eating organisms, although in saltwater it may be attacked by marine borers. The wood-eating organisms are dormant at temperatures below approximately 4°C. While working in the cold environment of the Aleutian Islands, I observed 40-year-old untreated bridge timbers in good condition where the steel bolts had deteriorated to rust traces owing to corrosion from sea salt. Wood decay in the wet environment progressed very slowly owing to the low temperatures. The simplest of these three factors to control is moisture content. In the normal environment where people live, air, a favorable temperature, and wood-eating organisms are always present, requiring only moisture as the catalyst to induce decay. Pressure preservative treatment of wood members at the time of manufacture is the best defense against wood decay in service. Under the most favorable of conditions, the thickness of the treatment envelope is a bit less than 1 in., with the inner core of wood untreated. Penetration of this envelope by boring or saw cuts exposes untreated wood to decay hazard. Paint-on treatment products have limited value because little penetration is achieved. One after-market treatment involves injection of a toxic fumigant into a bored hole, which is then plugged. Most utility poles are so treated to extend their life. This treatment is not approved for interior use because of the toxicity of the treatment chemical. Impel rods are solid boron cylinders, like candles, that may be used as after-market treatment and are safe for interior use. Impel rods are inserted into bored holes and are activated by elevated wood moisture content. The disadvantage of impel rods is that when the treatment chemical is activated by moisture, the chemical is dissipated.

Insect Infestation and Marine Borers

Several types of insects bore into and damage wood, termites being one of the most destructive. Most wood-eating insects require moist wood, although a few insects will attack dry wood. Damage from termites and other wood-eating insects is a function of temperature, and such damage is more prevalent in warmer climates. Termites are virtually nonexistent in the northern heartland away from the tempering effect of oceans. Subterranean termites maintain ground colonies with mud tunnels to the wood structure above. The same preservative treatments used to control decay usually are effective against insects.

Wood exposed to saltwater environments is subject to a family of wormlike marine borers that, in conditions favorable to them, may cause rapid destruction of pilings and submerged wood members. As with termites, their activity is a function of temperature. Heavy treatment with creosote is very effective in marine environments. A dual treatment of waterborne salts followed by creosote treatment is most effective to resist *Limnoria* and other marine borers. *Limnoria* are invertebrate crustaceans that produce a lacework of burrows just below the surface of the wood. Wave action in the tidal zone often erodes this thin covering over the burrows as the borers go progressively deeper into the wood. *Teredo* burrow deep within the wood and may completely honeycomb a wood piling with little visible surface evidence. The U.S. Army Corps of Engineers has restricted the use of creosote in certain waterways such as the Columbia River estuary, creating a dilemma for designing effective preservative treatment of marine structures where these restrictions apply.

Notching

Notching of timber structural members at the tension side creates a critical combination of section loss, stress-riser concentrations at the notch, and interaction of tension perpendicular to grain and shear stresses. This creates a complex situation for analysis. Knowledge of fracture mechanics and stress concentrations beyond that possessed by most practicing structural engineers is required to analyze the situation created by notching timber flexural members. The following formula for shear stress at a tension-face notch has been used since early versions of the NDS:

$$f_v = (3V/2bd_n)(d/d_n)$$

The formula for shear stress at a tension-face notch was revised in the 2001 NDS as follows:

$$f_v = (3V/2bd_n)\,(d/d_n)^2$$

where V = shear at notch location
         b = member width
         d = full member depth
         d_n = member depth at notch

The original formula is based on testing of short, relatively deep beams with a span-to-depth ratio of 9 and addresses only shear stress. Caution should be used in reliance on this formula, which does not address stress risers, characteristics and sharpness of the notch, existing microchecks, separations, or low strength perpendicular to grain in the shear plane. Over the years, owing to service problems with notched beams, additional restrictions have been added, including squaring the d/d_n term in the notched-beam formula. The NDS limits notch depth at tension face for solid sawn member to $d/4$ at a support, and to $d/6$ inside supports, and no notches are permitted in the middle third of span. The tension side of bending members of 3½-in. and greater net thickness should not be notched, except at bearing points over a support. The AITC *Timber Construction Manual*[11] limits notch depth at the tension face to $d/10$ for glulam beams. In-service problems suggest that beams tapered or notched at the tension face should be avoided.

Chemical Deterioration

Wood is naturally resistant to many chemicals and is the preferred construction material in environments that corrode steel, such as fertilizer storage buildings. As with decay, the heartwood is more resistant to chemical deterioration than the sapwood. In the cross section of a living tree, sapwood consists of the outermost recent growth rings. The sapwood contains both living and dead cells, transports moisture and sap, and stores food. Heartwood is the wood at the core of a tree and extending from the pith center to sapwood, the cells of which no longer participate in life processes of the tree. Gums, resins, and other materials make the heartwood darker in color and more decay-resistant than sapwood. In many wood species, such as Douglas fir and western red cedar, the dark-colored heartwood is easily distinguished from the light-colored sapwood. Acids and acid salts cause permanent strength loss owing to chemical changes in the cell structure. Alkalines cause permanent strength loss owing to destruction of lignins, which bind wood fiber together. Fire-retardant salts impregnated into wood in high concentrations have a detrimental effect on strength. Testing by myself of fire-retardant treated timber members from World War II era aircraft hangars showed strength loss in the 10% range with no loss in stiffness.

Heat

Wood strength decreases with higher temperature and increases with lower temperature. However, the effects are small enough in the normal environment where people live that these effects generally may be disregarded. In my experience, heat was a factor in structural failures at a paper mill, over coolers in a food processing plant, and at attic trusses in a desert southwest U.S. location. I used stress adjustment for temperature in the design of a storage warehouse in Saudi Arabia. The temperature adjustment factor C_t, with values from 1.0 to 0.5, is presented in tabular form in the NDS. Tabulated design values must be multiplied by C_t for structural wood members with sustained exposure to elevated temperatures.

Joint Eccentricity

Older bowstring trusses often were constructed using double chords with single webs in the space between chords. In this case, the webs could not be connected concentrically because of web-to-web interference. Bowstring trusses with small web forces usually were able to tolerate this eccentricity. Older precomputer structural analyses by graphics or the method of joints assume pinned concentric joints, and web-to-chord eccentricity often was ignored. Truss heel eccentricity where centerline intersection of heel members is eccentric with the truss reaction induces a torque into the heel connection that increases member bending and may create distress in trusses not designed for this condition. Internal truss member eccentricity, where centerline intersections of adjacent web members are eccentric with chords, creates the same condition. For bowstring trusses, distress owing to heel eccentricity is more likely than distress owing to internal web eccentricity because the member forces at the heel are much higher (Fig. 19.1).

Lateral Buckling

Columns, arches, and truss members stressed in compression usually are controlled by buckling or the interaction of buckling and compression rather than by pure compres-

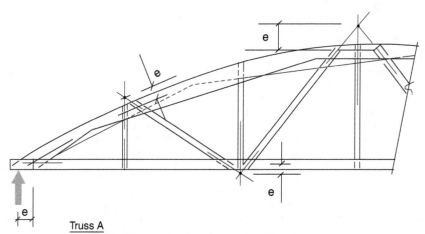

Truss A
1. Illustrating bowstring truss where center-line intersection of heel members are eccentric with truss reaction.
2. Illustrating bowstring truss where center-line intersection of adjacent web members are eccentric with chords.

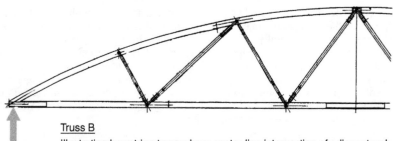

Truss B
Illustrating bowstring truss where center-line intersection of adjacent web members are concentric with chords and where center-line intersection of heel members are concentric with truss reaction.

FIGURE 19.1   Truss member eccentricity.

sion. Trusses and arches are rarely stable alone until secured into a diaphragm or some type of lateral bracing system; therefore, buckling problems with these structural systems are most likely to occur during construction. Inadequate bracing has been the primary cause of light truss problems investigated by me. Causes include inadequate temporary bracing, diaphragm construction lagging behind truss or arch placement, and economic pressure to release rented cranes.

Unbalanced Load

Structural systems with continuity across connections or across points of support that carry uniform loads should be checked for unbalanced loading. For multiple-span or cantilever-beam systems, this means full dead load plus alternate span live or snow load. For truss design, balanced uniform loading will control chords, but unbalanced live or snow load may control web and web-to-chord connections depending on web configuration. Full unbalanced truss loading usually means full dead load plus live or snow load at half span only. For three-hinged arch design, half-span unbalanced live or snow load will control shear at hinge. For reticulated spherical dome design, the tension ring and most primary members will be controlled by the uniform-load condition, but unbalanced live or snow load at alternate radial or tangential bays will cause load reversal to tension at some compression members, and this must be addressed both in design of members and in connections. The model building codes all have some language requiring consideration of unbalanced loading where it creates a more severe design situation than balanced loading. The 1997 *Uniform Building Code* (UBC) in Sec. 1607 requires unbalanced loading where such loading will result in larger members or connections, but an exception is included where alternate span loading need not be considered if the uniform roof live load is 20 lb/ft^2 or more. However, most cantilever-beam systems will be controlled by unbalanced live or snow load at alternate spans. My design review experience has shown that unbalanced loading often is not considered in the design of cantilever-beam systems, and when considered, cost savings from the cantilever-beam system tend to vanish.

Ponding

Water ponding on flat or nearly flat roofs may be caused by drainage obstructions, or deflection owing to lack of stiffness in the roof framing members. Ponding problems are more common in areas where roof live loads, rather than snow loading, control the design. Under the UBC, the roof live load in non-snow-load areas may be reduced to 12 lb/ft^2 for flat roofs with a tributary loaded area of 600 ft^2 or more. This leaves little reserve for water ponding compared with northern areas requiring roof design for heavier snow loading. Water ponding is a progressive-deflection problem. Weight of water accumulation causes roof beams to deflect. Roof beam deflection creates a deeper pond with greater water weight, which causes additional deflection, greater water weight, and additional deflection. This progressive deflection continues until beam bending moment is balanced by beam resistance, the water escapes, or the beam fails. A common thread in roof failures caused by water ponding is more than one of the following conditions:

- Flat roof with less than ¼ in. per foot of residual roof slope at all roof framing members following long-term member deflection
- Lack of or inoperable roof drains
- Design for 12 lb/ft^2 live load
- Flexible beam framing system
- Lack of overflow scuppers or scuppers more than 2 in. above primary roof drains

Secondary Effects

Secondary effects are the stresses induced in a structure by deformations from the primary loading. Secondary effects, and particularly the effect of nodal rotation on trusses, should be considered in design or evaluation. In an ideal world, web-to-chord truss joints should be as designed—i.e., welded if designed fixed or frictionless pins if designed pinned. Most web connections in timber trusses are designed as pinned but have partial fixity owing to restraints in the connection. If long, rigid steel side plates (designed to transfer axial loads only) are used for web-to-chord connections and the truss is flexible with considerable nodal rotation, then the connectors will resist that rotation by creation of moment couples in the web members. In severe cases these couples create a cross-grain tension force that can split the truss member. In my experience, a roof truss failure at a Seattle library is a case in point. The outer truss web split owing to a combination of nodal rotation and truss heel eccentricity. Long-welded steel web-to-chord connector plates rotated and moment couples resisting the rotation split the outer web member (Figs. 19.2 and 19.3). In another truss failure case I investigated, nodal rotation at a wood chord–wood web truss with glued connections failed owing to cross-grain tension stresses induced at the web-to-chord connections.

TYPES AND CONDUCT OF TESTS

The ASTM and AITC have developed standard tests for basic material properties, chemical retention, moisture content, adhesive viability, etc. Controversy and/or litigation may follow a structural assessment report depending on financial responsibility for remediation and other factors. The assessment report is strengthened when documented test procedures in accordance with standard tests are used to support test conclusions. In some instances, specimens from a distressed member or from similar adjacent members may be removed for laboratory testing. More commonly in assessment and evaluation work, the only options are testing in place or demolition. For this reason, nondestructive testing (NDT) on structural elements is among the more valuable tools in the engineer's toolbox. All the AITC tests listed below are intended primarily for plant qualification or plant quality control purposes. However, this does not preclude their value as tests for evaluation and assessment purposes.

Destructive Laboratory Tests

ASTM D198, Standard Test Methods of Static Tests of Lumber in Structural Sizes. This standard uses test specimens of lumber in structural sizes for the following tests:

Flexure
Compression (short column)
Compression (long column)

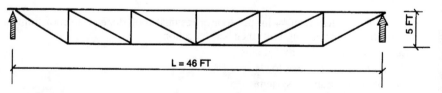

FIGURE 19.2   Truss profile, Rainier Beach Library, Seattle, Washington.

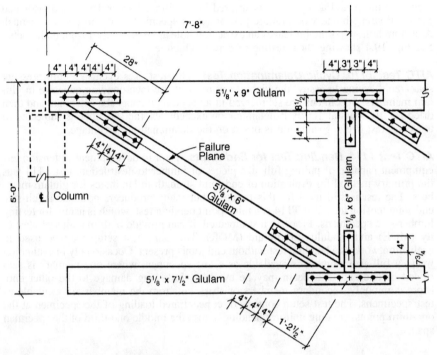

FIGURE 19.3   Truss heel detail. Failure occurred at outer diagonal web. Rainier Beach Library, Seattle, Washington.

Tension

Torsion

Shear modulus

ASTM D4761, Standard Test Methods for Mechanical Properties of Lumber and Wood-Base Structural Material.[12] This standard uses test specimens of lumber and timber in structural sizes. It has evolved from ASTM D198 but is more versatile in that nonlaboratory testing is possible and loading is not required to failure. The standard applies to the following tests:

- Bending edge-wise
- Bending flat-wise
- Tension parallel to grain
- Compression parallel to grain

AITC Test T107, Shear Test. This test is used primarily to evaluate face-bond glue-line strength in glulam members by measuring the adhesive shear strength at failure and estimation of wood failure percentage. Test T107 permits either the *stepped-block* specimen or the *core* specimen. The stepped-block specimen uses a block cut from a glulam member. The core specimen is used widely owing to simplicity of specimen extraction and testing. A cylindrical core 1-in. in diameter is extracted from a glulam member perpendicular to the wide face of laminations so that gluelines are perpen-

dicular to the core. Each glueline is sheared in a calibrated shear device and evaluated per shear strength and wood failure percentage as specified by the inspection manual. A custom hollow-coring bit leaves an extraction hole approximately 1⅜ in. in diameter. See Fig. 19.4 showing shear testing of core at glueline.

AITC Test T110, Cyclic Delamination Test. This test uses cyclic vacuum pressure soak/rapid drying to assess the viability of the wet-use face-bonding adhesive in glulam members. Specimens are submerged in a vacuum chamber for 2½ hours and then dried for 10 hours at 160°F. If delamination exceeds 5% the specimen is subjected to a second test cycle. Evaluation is based on the delamination percentage.

AITC Test T114, Bending Test for End Joints. Prior to development of tension test equipment capable of pulling full size pieces of lumber to destruction, test T114 was the primary method for evaluation of end-joint strength in laminates for glulam members. The test is still used for this purpose, but many producers now use a full-size end-joint tension test. Test T114 is a third-point bending test, which is useful for testing lumber-size specimens. Properly instrumented, it can provide both modulus of elasticity (MOE) and modulus of rupture (MOR). The same test setup may be used in evaluation work as a bending test without end joints present. Occasionally in evaluation work, a full-size member is available as a test specimen, where one member is sacrificed or one end of a beam may be available following damage to the other end. More commonly, secondary members or strips cut from the member are available as test specimens. The test setup consists of concentrated loading of the specimen at the one-third points, creating uniform bending across the middle one-third of the specimen span.

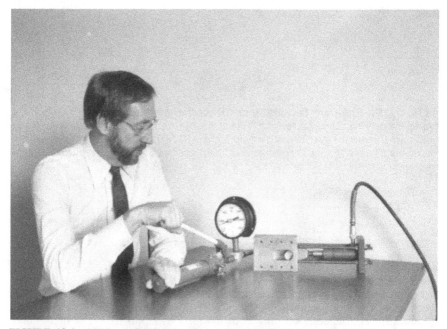

FIGURE 19.4 AITC test T107—shearing at adhesive line in core extracted from glulam to measure strength of face-bonding adhesive.

AITC Test T119, Full-Size End-Joint Tension Test. Laminates are surfaced at the narrow face to the approximate width of finished glulam sizes. When evaluating end joints cut from an existing glulam specimen, the laminates containing an end joint may be sawn apart and tested per AITC T119. Also, the full laminate may be tested in tension without an end joint using the same test procedure.

Nondestructive Tests

Visual Inspection. Visual inspection is the starting point for all condition assessment evaluations. The more advanced stages of decay may be detected visually, and the inspector notes areas for further, more detailed investigation. Crushed wood and depressions on the face of wood members often indicate a thin shell of sound wood on the surface with decay voids inside. Other visual red flags indicating wood deterioration are fruiting bodies, discoloration, insect activity, and areas of high moisture content.

AITC Core Shear Test. AITC test T107 for evaluating glueline quality in glulam members was covered earlier as a production shear test. If the core is field extracted from an existing structure with engineering control, it may be considered a nondestructive test. This requires access to design calculations, location of the core away from maximum stress locations, and a design check to ensure the 1⅜-in. hole does not overstress the member. If the integrity of the face-bond adhesive is suspect, or if a party to the controversy challenges the adhesive bond, a nondestructive core shear test is the simplest method to obtain quantifiable data. A glued plug is recommended when the core is extracted from an in-place structure.

Sonic Stress-Wave Testing. A *stress wave* is a disturbance or excitation of the molecular particles in a material such as a structural member. An impact device creating a wave in the sonic frequency spectrum induces the stress wave. The terms *sonic* and *ultrasonic* refer only to the frequency of excitation used to impart a wave into the member. Wave velocity through the member depends solely on material characteristic of the member and is unrelated to whether the wave is induced in sonic or ultrasonic frequencies. Most stress-wave instruments use a hammer or pendulum to induce the stress wave. Instruments are available commercially with accelerometer sensors and timers that can measure elapsed time of travel of a stress wave induced into the material, and they are battery powered and portable for field use. Figure 19.5 shows stress-wave testing on an existing structure. Instrumentation necessary to measure elapsed time of travel of the stress wave may be visualized as a type of expensive stopwatch. A start sensor turns the timer on as the wave is induced, and a stop sensor turns it off as the wave passes the stop sensor. Wave velocity may be calculated knowing elapsed time and the distance between sensors. Dynamic MOE may be determined by the following formula if both density and wave velocity can be measured:

$$\text{MOE} = \frac{DV^2}{g}$$

where
D = material density
V = wave velocity through material
g = constant of gravity, 32.2 ft/sec^2

A calibrating factor with static MOE measured by physical testing and dynamic MOE measured by stress-wave testing of the same specimen is necessary to estimate

FIGURE 19.5   Stress-wave testing on an existing structure.

static MOE using the stress-wave procedure. When stress-wave testing is used to estimate MOE, the stress wave is induced parallel to grain (Fig. 19.6).

A valuable procedure for detecting decayed wood is stress-wave testing with the wave induced perpendicular to grain (Fig. 19.7). Decayed wood slows the wave velocity, and the location of deteriorated wood may be mapped using a series of stress-wave tests in a grid pattern.

The digital output from stress-wave readings is elapsed time, usually in microseconds. The output readings will vary depending on the characteristics of the impact,

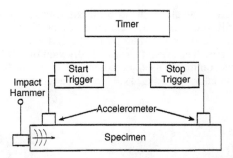

FIGURE 19.6   Schematic of stress-wave testing in wood parallel to grain.

the extent of microchecking in the specimen, and coupling. *Coupling* is the mechanics of getting the stress-wave signal into the specimen with the impact device and getting the signal out with the stop device. The shape and surface texture of the specimen can affect coupling. Interpretation of stress-wave test output requires a high level of expertise. A lack of understanding as to how these factors affect instrument output easily can lead to erroneous conclusions. Disadvantages of stress-wave testing include poor signal coupling at weathered specimen surfaces, an increase in wave transmission time from microfractures, the need to access both sides of a test specimen, and the difficulty of obtaining repeatable readings from in situ testing. An advantage of stress-wave testing is that quantitative rather than comparative data are generated.

Ultrasonic Testing.   Instruments are available for testing wood members by monitoring a stress wave induced in the ultrasonic frequency spectrum by a pulse generator. The wave transmission time and characteristics of the test specimen are used to calculate specimen MOE, the same as with sonic stress-wave testing. Ultrasonic testing has limited use for timber testing owing to the large transducers required and is prone to coupling problems. Coupling works well under water, making it suitable for testing submerged timber members.

Resistance Drill Testing.   Coring or drilling of timber members using small drill bits has long been a crude but common procedure for timber inspection. Drill resistance can be detected and shavings examined. Resistance drilling is a more refined version

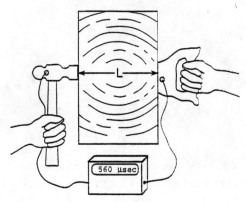

FIGURE 19.7   Schematic of stress-wave testing in wood perpendicular to grain.

of this same procedure. A small needle drill bit is power fed at a constant rate into a wood member as drill torque is plotted continuously on a graph. Instruments for resistance drilling were developed in Europe primarily for use by arborists detecting deterioration in live trees. The system works equally well on lumber and timber members. The instruments commonly used are compact enough for field use and are powered by cordless drill motors—see Fig. 19.8 showing a resistance drill in use for testing

FIGURE 19.8 Resistance drill in use for testing an in-place structure.

an in-place structure. The instrumentation is sensitive enough to show the density differences within each annual wood growth ring on the test graph. Checking shows up as a short downward spike on the graph. Decayed wood appears on the graph as a low, flat line. See Fig. 19.9 for a test graph showing a check and Fig. 19.10 for a test graph showing decay at test locations. Positive attributes of resistance drilling include an accurate archive record of the test, a graph easily understood by the layperson, an instrument that is simple to use, and the ability to conduct the test with access to only one side of the test specimen. A disadvantage of Resistance drill testing is that data on the test graph are comparative rather than quantitative.

Wood Moisture Content Determination. A precise measurement of wood moisture content is a destructive test in which a specimen is weighed before and after moisture is driven off in an oven per ASTM D4442.[13] A nondestructive approximation of wood moisture content may be obtained easily by using a calibrated electrical resistance moisture meter, where the electrical resistance between two insulated needles driven into the wood is measured parallel to grain per ASTM D4444.[14] Wood moisture content gradient is valuable evaluation data and may be obtained by recording moisture content at several penetration depths. Probes are commonly available with 1- or 3-in. needles on resistance-type moisture meters. A faster dielectric type moisture meter does not require needles but uses a sensor on the wood surface. Elevated wood moisture content is a catalyst for decay, and most strength and stiffness properties are adjusted downward for elevated wood moisture content. As such, a wood moisture meter should be available for any assessment evaluation of structural timber. Most wood moisture meters are calibrated for the Douglas fir species. Species and temperature adjustments to the meter reading must be made from data supplied for the particular meter model. Figure 19.11 shows an electrical resistance–type wood moisture meter in use.

Slope-of-Grain Determination. Wood fibers are formed in the living tree with their long dimension approximately parallel to the axis of the tree trunk. Wood grain direction is the wood fiber orientation. For engineering purposes, design properties are referred to as parallel or perpendicular to the grain. This means *approximately* parallel or perpendicular to the faces of finished lumber or timbers. The wood fiber is not precisely parallel or perpendicular to the faces of finished lumber because of natural growth characteristics or sawmill tolerances. The angle between wood fiber orientation and the faces of finished lumber is termed the *slope of grain*, and its impact on strength and stiffness is addressed in the lumber-grading rules. The annual rings are formed radially owing to wood density differences during the growth seasons. In a finished piece of lumber, the orientation of annual ring lines is not necessarily parallel to the wood fiber or grain orientation. Checking that develops as wood dries will be parallel to the grain orientation. A simple needle stylus that may be carried easily in the pocket is all that is necessary for field checking slope of grain in wood members. The needle follows the wood fiber when pulled by hand in a general parallel-to-grain direction with the needle embedded slightly in the wood. Electronic instruments have been developed that can detect the slope of grain when placed against the finished face of lumber or timbers.

Chemical Testing. Wood is more resistant to chemical action than most construction materials but is not free from chemical deterioration. Acids, acid salts, and alkalines may cause permanent strength loss in wood. Fire-retardant treatment (FRT) chemicals are commonly borates, ammonium sulfate, and ammonium phosphate, all of which cause wood degradation by acid hydrolysis. During an evaluation, questions often arise regarding the presence of preservative treatment. Testing for treatment chemicals is a common and relatively inexpensive procedure. Standard tests for wood treatments are

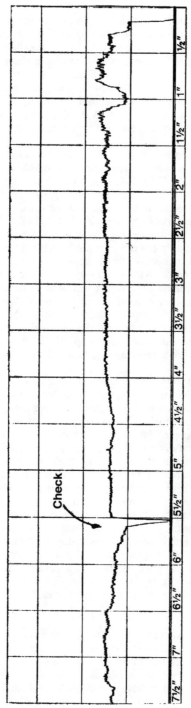

FIGURE 19.9 Resistance drill graph showing sound wood with a check 5½ in. from drill entry face.

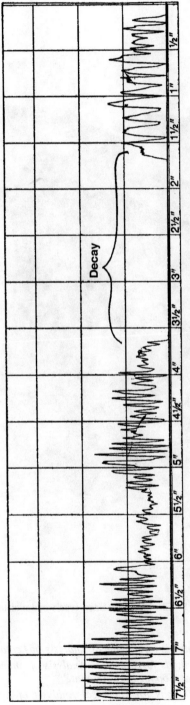

FIGURE 19.10 Resistance drill graph showing decayed wood between 1¾ and 3⅝ in. from drill entry face.

FIGURE 19.11 Electrical resistance–type wood moisture meter.

ASTM D1274, *Test Methods for Chemical Analysis of Pentachlorophenol*[15]

ASTM D1326, *Test Methods for Chemical Analysis of Ammoniacal Copper Arsenate and Ammoniacal Copper Zinc Arsenate*[16]

ASTM D1627, *Test Methods for Chemical Analysis of Acid Copper Chromate (ACC)*[17]

ASTM D1628, *Test Methods for Chemical Analysis of Chromated Copper Arsenate (CCA)*[18]

A spectrometer analysis may be used to determine the elements present from retained wood treatment chemicals. Penetration depth of the chemical in wood is important, and a chemical indicator applied to the specimen is used commonly for this purpose. Some chemicals, such as those used for fire-retardant purposes, are more detrimental near the wood surface owing to their interaction with atmospheric moisture. Testing for the presence of chemicals may be classified as nondestructive because very small test specimens are required.

Species and Grade Identification. Some woods, such as white oak or western red cedar, are distinctive in look, color, and odor. Others, such as Douglas fir and western larch, have distinctive foliage, but the wood is so similar in appearance it is graded together, and the difference is difficult to distinguish by visual examination. The surface of structural wood members in service is often weathered and coated with paint, varnish, or grime. Lacking shop drawings or other source documents, the services of a wood technologist may be required to identify species by microscopic examination of the wood fiber. In order to estimate the grade of material in evaluation work, species identification is required. Material grade may be estimated using lumber grading rules, but if a positive lumber grade determination is required, the services of a certified lumber grader will be necessary.

Screw Probe. The device for this test is a power-driven type of screw often termed a *drywall screw.* In this test, a relatively long screw is used, preferably 5 to 6 in. in length. The head is ground off, and the smooth shank is chucked into a drill motor. In sound wood, the drill operator cannot restrain the screw from auguring into the wood when the drill motor is running clockwise. In decayed wood, the operator is able to hold the running drill motor without the screw auguring into the wood.

Icepick or Awl Probe. Simple probing with an icepick can locate areas of wood decay by relative resistance to probe penetration. Crude but fast and effective, the icepick probe can quickly locate questionable areas that may be inspected more thoroughly with additional techniques. Embedding the pick ⅛-in. or so and prying out a wood splinter will fracture the splinter. A splinter breaking "long" or away from the pick indicates sound wood, whereas splinters breaking "short" or at or near the pick indicate brashy or decayed wood.

Hammer Sounding. Striking a test specimen at various locations is a procedure out of antiquity, but it is fast and a good starting procedure for locating wood deterioration. Easier to accomplish than to describe, the tone quality will indicate the general condition of the wood. Severely decayed wood will sound like a drum when struck by a hammer. Sounding is most effective for dimension lumber size specimens of 4-in. nominal or less that are difficult or impractical to test with stress wave or resistance drill techniques.

Increment Boring and Drilling. An increment boring tool is a handheld device designed for use by foresters to extract a small core from live trees. The tool is slow to use because it must be hand augured into the test specimen. It works best in the damp wood of live trees because the small cores are often difficult to extract when boring into dry wood. Another common evaluation technique is drilling with an auger or high-speed drill bit and roughly evaluating the wood condition by drill resistance plus the look and smell of drill shavings. These traditional techniques continue to have limited value, but I prefer resistance drill testing to either increment boring or drilling because it produces a more definite test result with less structural or cosmetic damage to the test specimen.

REFERENCES

1. American Forest and Paper Association, (2001). *National Design Specification for Wood Construction (NDS)*, ANSI-AF&PA NDS, AFPA, Washington, D.C.
2. AITC 117 (2001). *Standard Specifications for Structural Glued Laminated Timber of Softwood Species*. American Institute of Timber Construction, Englewood, CO.
3. ASTM D 198–99. *Standard Methods of Static Tests of Lumber in Structural Sizes*. American Society for Testing and Materials, West Conshohocken, PA.
4. *Wood Handbook: Wood as an Engineering Material—1999*. Reprinted from Forest Products Laboratory, General Technology Report FPL-GTR-113, Forest Products Society, Madison, WI, 1999.
5. *Uniform Building Code*, Vol. 2. 1997 International Conference of Building Officials (ICBO), Whittier, CA.
6. *National Building Code*, Building Officials and Code Administrators International (BOCA), Chicago, IL.
7. *Standard Building Code*. Southern Building Code Congress International (SBCCI), Birmingham, AL.
8. *ANSI/AITC Standard A 190.1-2002 for Wood Products—Structural Glued Laminated Timber.* American Institute of Timber Construction, Englewood, CO, 2002.
9. *Inspection Manual AITC 200-92 for Structural Glued Laminated Timber.* American Institute of Timber Construction, Englewood, CO, 1992.
10. Voluntary Product Standard PS-2, *Performance Standard for Wood-Based Structural Use Panels*. U.S. Department of Commerce, National Institute of Standards and Technology, Gaithersburg, MD, 1993.
11. American Institute of Timber Construction (AITC) (1994). *Timber Construction Manual*. Wiley, New York.
12. ASTM D4761–96 *Standard Test Methods for Mechanical Properties of Lumber and Wood-Base Structural Material*. American Society for Testing and Materials, West Conshohocken, PA.
13. ASTM D4442–92 *Standard Test Methods for Direct Moisture Content Measurement of Wood and Wood-Base Material*. American Society for Testing and Materials, West Conshohocken, PA.
14. ASTM D4444–92 *Standard Test Methods for Use and Calibration of Hand-Held Moisture Meters*. American Society for Testing and Materials, West Conshohocken, PA.
15. ASTM D1274–95 *Test Methods for Chemical Analysis of Pentachlorophenol*. American Society for Testing and Materials, West Conshohocken, PA.
16. ASTM D1326–94 *Test Methods for Chenmical Analysis of Ammoniacal Copper Aarsenate and Ammoniacal Copper Zinc Arsenate*. American Society for Testing and Materials, West Conshohocken, PA.
17. ASTM D1627–94 *Test Methods for Chemical Analysis of Acid Copper Chromate (ACC)*. American Society for Testing and Materials, West Conshohocken, PA.
18. ASTM D1628–94 *Test Methods for Chemical Analysis of Chromated Copper Arsenate (CCA)*. American Society for Testing and Materials, West Conshohocken, PA.

CHAPTER 20

Fabric

WESLEY R. TERRY, P.E., and TIAN-FANG JING, P.E.

CRITICAL ENGINEERING PROPERTIES

Geometric Properties

Fabric membrane materials behave differently structurally than other typical building materials. Conventional building materials carry the applied loading usually in bending and shear or, in the case of an axial member, in tension or compression. Fabric membranes behave as a pure membrane with only the in-plane forces available to resist the applied loadings.

Most fabric membrane structures are of woven fabric made from a woven substrate of glass or polyester threads. Most commonly, the glass are coated with a Teflon fluorinated ethylene propylene (FEP) and polytetrafluoroethylene (PTFE) fluoropolymer resin, whereas the polyester threads are coated with a polyvinyl chloride (PVC) polymer coating. Fabric, being usually a woven material, has two principal membrane directions: the warp direction of threads and, 90° to the warp, the weft (also called *fill*) direction of threads. Most fabric membranes behave as orthotropic materials with different stiffnesses in the two principal directions.

A prestressed fabric membrane typically will have two principal directions of curvature, one convex and one concave (referred to as *direction 1* and *direction 2* in Fig. 20.1). The fabric generally is oriented so that the yarn fibers are parallel to these principal directions. In order to produce a stable surface, the fabric membrane surface must have double curvature in the two principal directions and must originate on opposite sides of the surface. This is known as *anticlastic curvature.* The basic shape is defined mathematically as a hyperbolic paraboloid.

Uniaxial Strength Properties

Three main uniaxial strength properties are of importance in fabric membrane structures. These are *uniaxial strip tensile strengths* in the warp and fill directions, *trapezoidal tear strengths* in the warp and fill directions, and *coating adhesion strength.* At a minimum, tests for these three strength properties for the base fabric membrane rolls prior to fabrication and installation of a fabric membrane should be conducted. For testing and evaluation of existing structures, fabric membrane samples from the existing structure can be taken and compared against the virgin roll material.

The uniaxial strip tensile strength is a measurement of the breaking force in lbf/in. of width (kN/m) of the fabric in both the warp and fill directions. The basic set of tests consists of preparing several 1-in.-wide test samples in the warp and fill directions using a constant-rate elongation testing machine operated at 2.0 in./min elongation, applying the load and recording the ultimate breaking strength of the fabric material in the two primary directions. This is a uniaxial test of the fabric.

The trapezoidal tear strength is a measurement of the individual thread breaking force in lbf (kN) of the fabric used in the warp and fill directions. The basic test consists of taking a rectangular test sample with a partial slit in the middle parallel to either the warp or fill directions and tearing the sample at the tip of the slit using a constant rate elongation testing machine operated at 12.0 in./min elongation. The average tear strength from the five highest peak forces is then recorded in the warp and fill directions. The trapezoidal tear strength gives a measurement of the potential

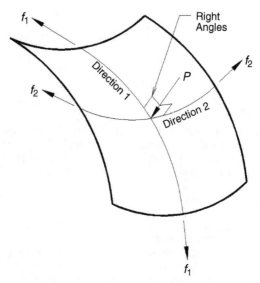

FIGURE 20.1 Principal direction of fabric.

for the fabric membrane to tear at corners and areas of discontinuity in the fabric membrane structure.

The coating adhesion strength is a measurement of the adhesion of the coating to the base fabric substrate in lbf/in. of width (kN/m). The basic test specimen consists of sealing two pieces of fabric together and cutting into 1-in.-wide samples. A constant rate elongation testing machine operated at 2.0 in./min pulls in opposite directions on the two sealed pieces to tear at the adhesion surface. The average adhesion strength is then recorded from the five highest peak forces. The adhesion strength is a measurement of the bond between the coating and the base substrate and is indirectly related to the seam strength between two layers of fabric membrane.

Biaxial Behavior Properties

The strength of fabric membrane materials usually is characterized by the uniaxial tensile strength, but the uniaxial tests do not describe the actual ultimate strength of the material adequately. Most fabric materials made from a woven substrate demonstrate orthotropic material properties owing to the interaction between the warp and fill threads, resulting in a decrease in ultimate strength in a biaxial stress loading (Fig. 20.2). This is due to the interaction of the opposing threads bearing against each other and causing a decrease in strength.

In testing done by Owen-Corning Fiberglass in the 1980s, biaxial strength results for Structo-Fab 450 fabric material, which is a Teflon-coated fiberglass fabric, were compared with their uniaxial tensile strengths. Three biaxial loading procedures were compared against the uniaxial tensile strengths, as shown in Table 20.1. A biaxial cylinder sample was loaded in a monotonic manner until rupture under a 1:1 biaxial stress ratio between the warp and fill directions. A step loading of a similar cylinder sample in 30-minute, 50-pli increments to rupture was applied. A stress-rupture test also was performed to determine the long-term performance of the fabric under sustained loading. The stress-rupture test was developed to determine truly long term properties. Some of these test specimens were under load for up to 2½ years.

From the preceding results it is obvious that biaxial loading of the fabric membrane and the duration of the loading have an effect on the load capacity of the fabric membrane. This is one of the reasons that large safety factors on uniaxial tensile test results are incorporated into the design of fabric membranes. Usually a safety factor of 4 is used against wind loads, and a safety factor of 5 is used for long-term loads such as live or snow loads.

Appearance and Maintenance Properties

All fabric materials exhibit degradation with time owing to ultraviolet (UV) exposure, water exposure, and weathering due to wind, rain, snow, and other environmental effects. Appearance and maintenance properties are somewhat of a subjective nature. There are some qualitative measurements that can be made to aid in determining the

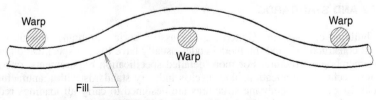

FIGURE 20.2   Woven fabric substrate.

TABLE 20.1 Biaxial Strength Results for Structo-Fab 450

Uniaxial strengths:	
Warp	1100 pli
Fill	900 pli
Biaxial strengths:	
Monotonic loading	600 pli
Step loading	500 pli
Stress rupture	400 pli

appearance and maintenance properties. Teflon-coated fiberglass fabric materials tend to perform best in retention of material properties. This is due to the materials that are used in the manufacture of the fabric. The fiberglass fabric substrate is inert to UV degradation, and of the available coating materials, Teflon is the most chemically inert, having the lowest coefficient of surface friction, which creates a surface that is readily cleanable and having a temperature range of -450 to $+550°F$. Test data show that the Teflon-coated fiberglass materials can expect retention of nominal strengths of 90% after 10 years, 80% after 20 years, and 70% after 30 years of normal service.

The PVC-coated polyester fabric materials perform at a lesser level than the Teflon-coated fiberglass fabrics. This is due to the inherent degradation properties of the base materials. The polyester fabric substrate is not inert to UV degradation and therefore must have UV blockers in the coating. The PVC coating material itself does not perform well in outdoor exposure and soon begins to degrade and weather. Therefore, almost all PVC-coated architectural applications have an additional top finish to improve the degradation properties. The three main top finishes are an *acrylic topcoat,* which is a thin liquid coating applied to the surface; a *polyvinylidene fluoride topcoat* (PVDF), which is also a thin liquid applied to the surface; and is a Tedlar PVF film, which is not a coating but a film layer that is bonded to the PVC coating in the manufacturing process and also must be removed to be heat weldable. Two quality levels of PVDF are available; the lowest level of protection is a blended coating of acrylic and PVDF that is heat weldable, whereas the higher level of protection is an essentially pure topcoating of PVDF that must be removed to be heat weldable.

The various topcoats offer varying levels of protection to PVC-coated fabric. The acrylic-coated and blended PVDFs do not perform well, and significant surface appearance degradation can be expected within 5 years. The pure PVDF topcoating and the Teldar PVF film coating retain their surface appearance significantly better than the acrylic and blended PVDF films but not as well as the Teflon-coated materials. Experience has shown that with the pure PVDF topcoating and the Tedlar PVF film coating, adequate surface appearance and protection to the base material can be expected for 10 to 15 years, although little test data exist beyond 10 years for the various PVC coated materials.

CODES AND STANDARDS

Local building codes in the United States classify fabric membrane structures under "special construction," and the model codes usually have limited specification criteria for the membrane structure. For more detailed specification requirements, one needs to review technical publications and discuss industry standards with manufactures in the field. In general, membrane structures are designed to carry all loadings required in the model building codes for live load, wind, snow, and other environmental loadings. Steel frames and cables are designed by the code provisions or specifications developed by trade associations and engineering societies such as AISC and ASCE. The fabric membrane should be designed with an adequate safety factor per industry standards and technical trade association publications, taking in to account long-term degradation, seam strength, biaxial tension, and load duration. This is most important because fabric membranes lose strength with time at varying rates depending on the material of construction and loading conditions. Following is a partial list of standards and codes that are used in the industry:

ASCE, *Air Supported Structures*

ASCE, *Tensioned Fabric Structures: A Practical Guide*

International Building Code

Uniform Building Code

ASTM D4851, *Tested Methods for Coated and Laminated Fabrics for Architectural Use*

ASTM E84, *Tested Method for Surface Burning Characteristics of Building Materials*

ASTM E108, *Tested Method for Fire Resistance of Coverings*

ASTM E136, *Tested Method for Behavior of Materials in Vertical Tube Furnace at 75°C*

COMMON TYPES, CAUSES, AND CONSEQUENCES OF DEGRADATION AND DEFECTS

Types of Failures

In general, there are three types of common failures that occur in fabric membrane structures. The first is a *tensile* or *bursting failure* of the membrane. This is caused by an overloading of the membrane to failure and is usually catastrophic. This failure mode can happen, for example, when improper drainage has been provided on the membrane, and an accumulation of water causing ponding or an accumulation of snow occurs to a point of reaching ultimate tensile failure. The second type of failure is a *tear failure,* which occurs at an open edge at the perimeter or at an exposed cut or rip in the fabric membrane, usually caused by something external cutting the fabric or flying debris being blown through the fabric membrane during a windstorm or hurricane. The third type of failure is a *seam failure,* where the seam between two fabric panels opens up, exposing the fabric membrane to a loss of prestress and change of shape and subjecting the fabric membrane to tear failures. The usual causes of seam failures are improper sealing of seams in the shop during manufacturing or in the field during construction.

Loss of Structural Strength

Loss of structural strength usually is a consequence of weathering or degradation of the material with time. This can be caused by UV degradation, loss of coating owing to abrasion, and water absorption and migration into the base substrate.

UV degradation is very important with PVC-coated polyester fabrics because both the base materials exhibit UV degradation with time. Therefore, these types of materials usually will have a topcoat of acrylic, PVDF, or Tedlar to protect the PVC coating and to provide additional UV protection as well. The Teflon-coated materials do not exhibit this loss of strength owing to UV degradation because their base components of Teflon and glass fibers generally are UV-resistant.

Loss of coating owing to weathering and abrasion is an important consideration for both Teflon-coated and PVC-coated fabrics. Loss of coating owing to weathering eventually exposes the base fabric to UV degradation and consequent loss of strength owing to water absorption and migration. The Teflon-coated materials are less susceptible to weathering and abrasion owing to the inherently harder surface but still will exhibit loss of coating over time. The PVC-coated materials have a softer surface than Teflon and will weather quicker over time unless a harder topcoat surface is provided such as PVDF of Tedlar.

Water absorption and migration in the base cloth are caused by the porosity of the coating materials and by loss of coating owing to weathering or nicks and abrasions that can be caused by external sources. The water absorption and migration tend to reduce the strength of the fabrics and can become a concern with loss of coating protection as time goes on.

Loss of Biaxial Properties

Fabric membrane structures usually are installed under a uniform prestress that is the same in the warp and fill directions. This initial prestress is nominally 20 to 40 pli, which is approximately 5% of the ultimate uniaxial strip tensile strength. The purpose of the initial prestress is to provide a smooth, curved surface for aesthetic reasons and, second, to provide a stable, curved shape to resist wind and snow loadings without excessive deformation or loss of prestress causing wrinkles, bagginess, and potential water ponding.

The loss of prestress owing to creep in the fabric materials usually is not an issue with Teflon-coated fiberglass fabrics because the glass fabric exhibits little creep over time with prestress levels at 20 to 40 pli. Some of the PVC-coated polyester fabrics can exhibit creep with time and, in general, should have adjustable attachments that would allow periodic retensioning of the membrane to maintain a stable stressed shape.

Loss of Aesthetics and Physical Appearance

This is a somewhat subjective matter in that what is physically appealing to one person may not be the same to another. In general, the physical appearance of fabric membrane structures can be measured by color fastness of the fabric and dirt accumulation as compared with the original virgin roll fabric before installation.

Teflon-coated fabrics tend to retain the same appearance with time. They should appear the same, regarding dirt accumulation, in 20 years as they did in the first year owing to the inherent properties of the Teflon coating. Excessive dirt accumulation on Teflon-coated fabrics is an indication of extreme air pollution or a failure in the coating material.

PVC-coated fabrics show a wide range of dirt accumulation depending on the coating composition and the topcoating. In general, untopped PVC-coated membranes and PVC-coated membranes that are acrylic or blended PVDF topcoated, perform poorest and can exhibit severe dirt accumulation in 1 to 5 years. Pure PVDF-topcoated vinyls and Tedlar-topcoated vinyl fabrics perform much better but do not retain the initial appearance as well as Teflon-coated fabrics do.

It should be noted that physical appearance of the fabric with regard to dirt retention and other aesthetic measurements is not a measurement of the physical strength of the fabric. Fabric membrane structures that may have lost their life from an aesthetic viewpoint may still have retained 100% of their physical properties.

TYPES AND CONDUCT OF TESTS

Structural Strength Tests

The most common procedure for testing of fabric membrane materials used for architectural purposes is that specified by ASTM D4851, *Standard Test Methods for Coated and Laminated Fabrics for Architectural Use*.[1] The following tests are addressed in ASTM D4851:

- Fabric count
- Mass per unit area
- Fabric thickness
- Fabric length
- Fabric bow
- Adhesion of coating to fabric
- Uniaxial elongation under static load
- Fabric breaking force
- Breaking strength after crease fold
- Elongation at break
- Fabric trapezoid tear force
- Resistance to accelerated weathering
- Solar optical properties

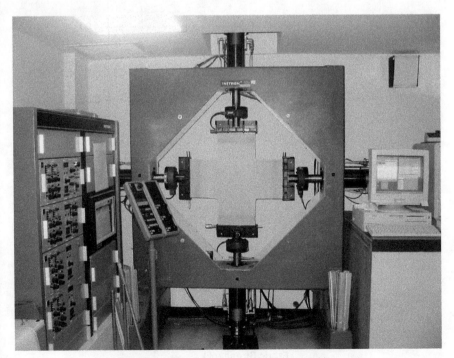

FIGURE 20.3   Biaxial strength testing machine.

- Fabric flame resistance
- Noise reduction coefficient

Biaxial Property Test

There are no standardized test procedures, such as by ASTM, for biaxial stress strength testing, so fabricators in the industry have developed biaxial loading machines for their own use. These machines are usually of a cruciform shape, and the fabric is cut into

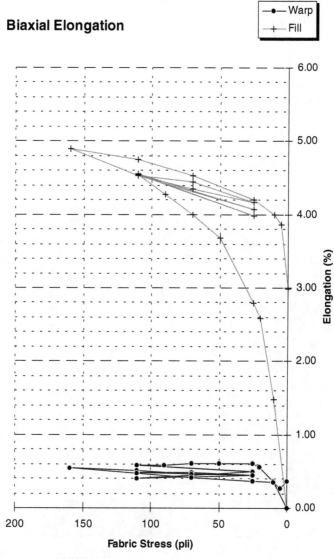

FIGURE 20.4   Typical biaxial stress-strain curve.

a cruciform aligning with the warp and fill directions and loaded by hydraulic cylinders in the warp and fill directions. (Fig. 20.3)

The biaxial test usually is done under a 1:1 biaxial stress ratio between the warp and fill directions. Figure 20.4 shows a typical biaxial stress verses elongation graph for Teflon coated fiberglass material loaded within its working stress range. It is readily recognizable from this graph that the material properties are orthotropic and that in the initial load cycle there is considerable creep elongation that occurs until the warp-thread/fill-thread crimp interchange is taken out. After the initial cycle, the material approximates a linear elastic material behavior in the working stress range.

Physical Appearance Tests

As with the biaxial tests, there are no standardized test procedures for physical appearance that are followed for architectural fabrics. Manufactures have testing data for their products such as gloss retention, color change, and dirt accumulation, but test procedures vary from manufacturer to manufacturer and cannot always be relied on to draw qualitative comparisons.

REFERENCES

1. ASTM D4851, (1997). *Standard Test Methods for Coated and Laminated Fabrics for Architectural Use.* ASTM, West Conshohocken, PA.

2. Mott, R., (n.d.). *Fabric Structure Design Methodology Based on Finite Element Membrane and Biaxial Elastic Property Testing.* Owens Corning Fiberglass Technical Center.

3. Dupont Public Literature, (1999). *A Comparative Study: Architectural Fabric Top Finish Performance.* Dupont.

4. Chemfab Public Literature, (1990). *Flexible Advanced Materials.* Chemfab.

5. Ferrari Public Literature, Textile Architecture. *Fluotop T 1002.* Ferrari.

CHAPTER 21

Field Load Testing

DONALD O. DUSENBERRY, P.E.

INTRODUCTION

Structural analyses and materials tests alone may not be sufficient to reveal the behavior of structural components. The composition of structural elements, the effects of deterioration, and uncertainties surrounding the interaction of various elements within a system sometimes dictate field testing as a means to resolve performance questions. Testing, together with an adequate understanding of materials characteristics and validating analytical studies, can provide the highest level of confidence in the load-carrying capacity of structural components and structural systems (Fig. 21.1).

Even when we have detailed design drawings and as-built drawings, there can be uncertainties about key elements of a building's construction. For example, we often

FIGURE 21.1   Full-scale load test of fire-damaged precast concrete tee. (*Courtesy of MACTEC Engineering and Consulting.*)

cannot be certain about the actual position of reinforcing steel in a reinforced concrete member. Further, we often do not have a basis to evaluate the expected performance of a member that lacks the detailing features that are the underlying bases for conventional design assumptions. Can we be certain about the lap length of reinforcing bars in a splice? Can we be certain of the actual strength of a splice that does not have code-prescribed length? What should we expect of a composite beam that does not have sufficient shear studs to fully engage the slab? Does a heavy timber member with knots that exceed grading limitations still have adequate strength for a particular purpose?

Deterioration introduces uncertainty in the load-carrying capacity of structural elements. Detailed inspections can be performed to reveal the nature and extent of the deterioration, but it often is not possible to determine the residual strength by analytic studies alone. There may be no reference available to serve as a guide for evaluation of the strength of a member that has lost a component of its strength to deterioration or mechanical damage (e.g., Can the strength of a beam with a deformed flange be calculated reliably?). Spalling on the underside of a concrete beam may damage the bond in a way that is difficult to restore.

An engineer may repair corrosion on a steel member by adding cover plates, but the critical nature of the repaired beam still may warrant postrepair testing as a means to validate the adequacy of the repair. Field testing often is the only way to verify that a structural repair has restored the required strength to a damaged or deteriorated element.

Sometimes there simply are no analytic tools that will allow an engineer to obtain the level of precision needed for accurate studies. The natural frequency of a structural system often cannot be determined to high precision because nonstructural elements—drywall, non-load-bearing partitions, curtain walls, etc.—add measurable but uncertain stiffness. Also, connections that are designed as pinned usually can carry moment during low-amplitude vibrations. Analyses for these conditions can be very costly, and they do little to increase the confidence in the results of studies that require high precision.

While field testing often can be used to address construction uncertainties, effects of deterioration, and performance of detailing features, full-scale testing also is a means to validate analytic models. When it is important to understand the implications of changes to a structure under conditions when the performance of detailing is uncertain, field testing can be used as a means to calibrate an analytic model that then can be used to extrapolate for the evaluation of modifications. If an engineer wants to evaluate the influence of bracing on interstory drift, considering the contribution of behaviors not commonly considered in design, it might be prudent to test the accuracy of the original model by comparing its prediction of performance with actual data that can be measured in the field. If correlation is good, then the level of confidence in the modeling assumptions is raised.

Field testing can answer some questions unequivocally. Testing can silence critics and convince doubters in a way that analytic studies can never do. However, field testing sometimes cannot answer all questions. Further, often it is expensive, it usually is difficult to perform in occupied buildings, and testing sometimes leads to structural damage that is difficult to repair. For these reasons, the decision to test existing structures needs to be considered carefully.

Owners, building officials, and other stakeholders need to be educated about the risks, the benefits, and the shortfalls that often accompany full-scale testing. Engineers who undertake tests need to be certain about the validity and usefulness of the data that will be acquired.

PROCEDURES AND STANDARDS

Field testing often is performed for one of two reasons: (1) to establish that a specific construction has adequate strength for a certain purpose without necessarily establishing its actual strength or (2) to establish the reliable strength of a detail or a component of construction for the purpose of sizing or specifying the detail. In the first case, it is preferable that the test be performed to verify the adequacy of the construction without inducing damage that leads to unnecessary repairs. Hence tests of this character need to be performed with careful attention to load application and to monitoring the test element to detect evidence of impending damage. This form of testing generally is referred to as *acceptance testing by attributes,* with the primary conclusion of the effort being a determination of whether specific construction is or is not suitable for its intended purpose.

When the goal is to establish the reliable strength of a detail, it often is necessary to test to failure a substantial number of similar installations of the detail. An example might be testing to establish the safe design load for a particular type of anchor that is to be used to connect architectural panels to a masonry support structure. In this case, an engineer might consider testing exemplary installations of the anchors to generate data that will allow a determination of the safe load for the anchor–support structure combination. This form of testing, which is referred to as *acceptance testing by variables,* normally produces failures of the test elements.

Regardless of the type and purpose of the testing effort, it is important to develop and document a test procedure prior to execution. The procedure will define the process for the test and help to formulate the goals and the means to verify the critical performance. In many cases, engineers responsible for field tests will need to develop specific custom procedures to address unique field conditions or to verify certain performance criteria. The engineer considering testing must understand the behavior that is to be validated, the salient structural features that must be evaluated and monitored, the means for load application, the limitations for the test, data processing, facility protection during the test, and safety issues.

Procedures for Acceptance Testing by Attributes

Several recognized standards define procedures for in situ load testing for some specific conditions. In some cases, these procedures are mandatory under local law. Even if they are not mandatory, to the extent that these standards provide procedures that are directly applicable to the testing need, their use can enhance credibility in the outcome of the tests. When the standard procedures are not directly relevant, they still can be useful bases for development of custom procedures for verification of unique conditions.

There is strong precedent and long history for testing as a means to resolve uncertainties in construction. Many building codes allow building officials to accept the results of competently performed tests as proof of structural performance. For instance, Section 1708.1 of the *Massachusetts State Building Code*[1] states

> Where proposed construction is not capable of being designed by approved engineering analysis, or where proposed construction design method does not comply with the applicable material design standard . . . , the system of construction or the structural unit and the connections shall be subjected to the tests prescribed in [this code]. The code official shall accept certified reports of such tests conducted by an approved testing agency. . . .

In addition to certain state building codes, there are loads standards, materials standards, and model codes that set forth load test procedures. The provisions in some of these references are summarized below.

International Code Council, International Building Code (IBC 2003). Section 1713 of the *International Building Code* (IBC 2003)[2] establishes a procedure for load testing when needed to verify structural capacity. This section of IBC 2003 states

> Whenever there is a reasonable doubt as to the stability or load-bearing capacity of a completed building, structure or portion thereof for the expected loads, an engineering assessment shall be required. The engineering assessment shall involve either a structural analysis or an in-situ load test, or both. The structural analysis shall be based on actual material properties and other as-built conditions that affect stability or load-bearing capacity, and shall be conducted in accordance with the applicable design standard. If the structural assessment determines that the load-bearing capacity is less than that required by the code, load tests shall be conducted in accordance with [this code].

IBC 2003 first refers the user to applicable material standards for load test procedures. When appropriate procedures are published in the applicable material standard, they are to be used for structural verification. However, IBC 2003 provides guidelines for a test procedure if an applicable procedure is not contained in the relevant material standards.

Under the provisions of this standard, the test procedure, which must be developed by a registered design profession and approved by the building official, must simulate loads and conditions of application that are expected to exist in the completed structure when in normal use.

Basic elements of the required test procedure include

- For components that are not part of the seismic load-resisting system, the test load is to be twice the unfactored design loads.
- The test load is to be left in place for 24 hours.
- Under the design load, the structural deflection must not exceed deflection limits published in the "Serviceability" subsection of the IBC 2003 section on "General Design Requirements." (This subsection of IBC 2003 refers the user to deflection limits published in materials design standards.)
- Within 24 hours after removal of the test load, the structure must recover not less than 75% of the maximum test deflection.
- During and after the test, the structure must not show evidence of failure.

Massachusetts State Building Code. Some building codes maintained by states and other jurisdictions contain load test procedures. As an example, the *Massachusetts State Building Code*[1] contains a procedure that is very similar to the IBC 2003 procedure. When there is reasonable doubt about the stability or load-bearing capacity of a completed building, engineering assessments consisting of analyses or load tests or both are required. When analyses determine that the load-bearing capacity is less than that required by the code, then load tests are required.

As with IBC 2003, the *Massachusetts State Building Code* refers the engineer to test procedures in appropriate materials standards. When no such test procedures exist, a registered professional engineer must develop and administer the test program.

The basic elements of the test procedure are the same in the *Massachusetts State Building Code* as they are in IBC 2003, except that the *Massachusetts State Building Code* allows the Structural Engineer of Record (SER) to set deflection limits during the test rather than defaulting to serviceability limits in materials design standards.

Other states and jurisdictions may have different requirements to verify structural adequacy. Engineers responsible for acceptance testing should consult with local building officials.

Structural Engineering Institute of the American Society of Civil Engineers, Minimum Design Loads for Buildings and Other Structures (SEI/ASCE 7). The reference standard *Minimum Design Loads for Buildings and Other Structures* (SEI/ASCE 7-02)[3] is a reputable reference that often finds legitimacy in the resolution of uncertainties or disputes about expected performance of structures. This standard specifically refers to testing as a means to resolve uncertainties about strengths. Section 1.7 of SEI/ASCE 7-02 states, "A load test of any construction shall be conducted when required by the authority having jurisdiction whenever there is reason to question its safety for the intended occupancy or use."

SEI/ASCE 7-02 does not specify protocol or acceptance criteria for load tests. Section C1.7 of the Commentary to this standard notes, "No specific method of test for completed construction has been given in this standard since it may be found advisable to vary the procedure according to conditions." For specific guidance on procedure, this standard cites other references.

American Concrete Institute, Building Code Requirements for Structural Concrete and Commentary (ACI 318). The American Concrete Institute (ACI) has developed test protocol specifically for validation of the adequacy of concrete elements. Chapter 20 of the *Building Code Requirements for Reinforced Concrete Structures* (ACI 318-02)[4] prescribes the procedure, data acquisition, and data analyses to a level of detail that ensures some uniformity in expectation and repeatability of field tests on reinforced-concrete flexural elements. The intent of the prescribed procedure is to cause reinforced-concrete elements to be loaded to near their failure loads and to verify performance by measuring the tested element's ability to sustain the load and recover its pretest configuration after the test load is removed.

The Commentary to the load testing chapter of ACI 318-02 states, "Provisions of Chapter 20 may be used to evaluate whether a structure or a portion of a structure satisfies the safety requirements of this code. A strength evaluation may be required if the materials are considered to be deficient in quality, if there is evidence indicating faulty construction, if a structure has deteriorated, if a building will be used for a new function, or if, for any reason, a structure or a portion of it does not appear to satisfy the requirements of the code." Hence the load testing provisions of ACI 318-02 are intended to address uncertainties in existing buildings; they are not intended for the verification of new materials or design approaches.

ACI 318-02 further differentiates between occasions when tests and analyses are appropriate. This reference states that analyses are sufficient to resolve questions about adequacy when the effect of a strength deficiency is well understood and it is feasible to obtain dimensions and materials properties necessary for analyses. Tests are appropriate when the effects of the strength deficiency are not well understood or when it is not feasible to establish the dimensional information or materials properties necessary for analyses. Among the circumstances when tests are appropriate, according to ACI 318-02, are occasions when deterioration raises questions about safety.

ACI 318-02 further cautions that analyses should be performed, when possible, to support the results of tests.

Some specific features of the test procedure in ACI 318-02 are

- Patterns of loading should be such that the critical sections are exposed to maximum deflections and stresses.
- When a single pattern of loads cannot test all relevant behaviors, additional tests should be perform to capture the relevant responses.
- Test loads should be to 85% of the design load (specifically to $0.85(1.4DL + 1.7LL)$, where DL is dead load and LL is live load).
- Loads should be applied in at least four approximately equal increments.
- Response should be monitored after each load increment and 24 hours after application of the full test load.
- The test load should be removed immediately after the last recording of response.
- Structural recovery should be monitored 24 hours after load removal.

Regarding acceptance criteria, ACI 318-02 requires the following:

- No portion of the structure should show evidence of failure (e.g., spalling or crushing of concrete).
- The maximum deflection under full test load should not exceed $l_t^2/20,000h$, where l_t and h are the span and thickness, respectively, of the element being tested.
- On load removal, the residual defection should not exceed one-quarter of the maximum deflection under full test load.
- If the maximum deflection and the residual deflection do not remain within these limits, it is permissible to retest the element.
- In any retest, the residual deflection must not exceed one-fifth of the new maximum deflection under the full retest load.

American Institute of Steel Construction, Load and Resistance Factor Design Specification for Structural Steel Buildings (AISC). A test procedure for structural steel, published in Chapter N of the American Institute of Steel Construction (AISC)

Load and Resistance Factor Design Specification for Structural Steel Buildings,[5] provides a testing procedure to establish the live-load rating of an existing steel floor or roof structure. While less detailed in its specific requirements, the AISC procedure contains many of the same elements as are in ACI 318-02.

The AISC procedure allows for testing an entire structure or, when not feasible, a segment or zone of not less than one complete bay that is representative of the most critical conditions. The AISC procedure establishes that the test load should be the sum of the load applied during the test and the effects of the in-situ dead load. Load should be applied incrementally, and inspections should be performed at each loading increment to detect signs of distress or imminent failure.

The live-load rating is determined based on the maximum test load using the following formula:

$$P_t = 1.2DL + 1.6LL$$

where P_t is the maximum test load.

Hence the live-load rating, based on a test in accordance with the AISC procedure, is

$$LL = (P_t - 1.2DL)/1.6$$

However, the maximum nominal live-load rating of the floor structure cannot exceed that which can be calculated using the applicable provisions of the specification.[5] For roof structures, consideration needs to be given to load combinations that involve snow or rain loads.

Unlike in the ACI procedure for concrete construction, there is no presumption that the failure load is any particular multiple of the test load. The live-load rating is derived directly from the test load. Hence tests following the AISC procedure should be halted on observation of damaging distress or when the desired load rating is verified.

The AISC procedure requires the maximum load to be held for 1 hour and that the maximum displacement does not increase by more than 10% during that holding period. Should the increase in displacement exceed this limit, it is permissible under the AISC procedure to repeat the test. Displacements also are to be measured 24 hours after the test load is removed in order to determine the amount of permanent set. However, no particular limit is placed on the permissible amount of permanent set, based on the observation that the amount of acceptable permanent deformation depends on the specific structure.

Materials Other than Concrete or Steel. The materials design standards for concrete and steel provide guidance on test procedures to verify structural adequacy. However, there are no universally accepted parallel procedures for the field testing of systems with wood or other materials other than concrete and steel. Engineers generally will find themselves without authoritative guidance for the testing of such systems and will need to develop appropriate procedures within the testing requirements, if any, of the governing building code.

Special consideration is required when developing test procedures for materials such as wood, which have high variability in strength properties from element to element, time-dependent strength relationships, significant creep deflections, and other relevant features such as strengths that depend on moisture content. Further, one might question whether subjecting wood systems to near their failure loads will induce damage that will reduce future service strengths.

Suter[6] suggests a two-phase test for timber beams based on work performed to evaluate aged structures. The first phase verifies the deflection performance of the element, whereas the second phase establishes the strength limit.

For deflection performance, the test load under Suter's procedure is calculated to produce the required deflection limit (e.g., live-load deflection in service not exceeding $l/360$, where l is the span length) and checked against the deflection under the intended service live load. The test is then performed for either or both of the calculated load to produce the deflection limit or the service live load, and the deflection at full test load is verified.

Suter suggests taking an initial deflection reading after application of all dead loads, including a superimposed load to account for anticipated dead load that is not in place at the time of the test. Test load, representing live load, is then applied in four approximately equal stages, with deflection readings taken approximately 15 minutes after each application. The full test load remains in place for 24 hours, with deflections recorded at sufficient intervals through that period to demonstrate that deflections have stabilized.

Test load is then removed in a single stage, with deflections measured approximately 15 minutes after the entire test load is removed but with the superimposed dead load, if any, still in place. If 90% (a conservative value, as acknowledged by Suter) of the maximum test deflection is not recovered, then deflection measurements should be repeated 24 hours after removal of test load to determine if 90% recovery occurs after stabilization.

If the maximum test deflection or the deflection recovery does not satisfy the acceptance criteria, then the test can be repeated for a lower load to establish the superimposed live load that corresponds to the required live-load deflection limit.

Suter recommends that the second phase of the test, to verify strength, be performed for a test load that is twice the deflection test load. The procedure for application of the test load is similar to that for the deflection test, except that test load is applied in at least six stages so that careful attention can be applied to signs of weakness or impending failure, such as cracking or splitting. In addition, when under full test load, more frequent deflection measurements should be made to allow characterization of creep behavior and to identify potential weaknesses.

Acceptance criteria include successful support of the strength test load without damage that impairs future strength of the element and recovery of at least 75% of the test deflection within 24 hours after test load removal. Suter suggests repeating the test if the deflection recovery criterion is not met. After the retest, the deflection recovery should be at least 85% of the maximum test deflection.

Suter does not explicitly discuss adjustments to account for load duration, moisture content, or other factors. However, he does recommended recording this information and providing in the report recommendations for design loads for future service of the tested element. Since the test is performed in a matter of a day or so in ambient conditions and the service loads normally would be applied for longer durations and perhaps under different conditions, references such as *National Design Specification for Wood Construction* (NDS)[7] can be used to adjust loads for expected service conditions.

For instance, if the purpose of the test is to verify the flexural strength of an assembly, and the responsible engineer wishes to adjust the short-term test load to account for live-load application over the life of the structure, the load duration factor C_D specified in NDS can be applied by multiplying the test load by the ratio of values of C_D for the expected application duration in service (say, 1.0 for live-load cumulative duration of 10 years) to the actual duration of the test load (say, 1.33 for a test duration of approximately 1 day). Hence the long-term service load could be estimated by this approach to be approximately 75% of the test load.

Similar adjustments can be made for other factors (e.g., temperature and moisture) that affect service strength. Unfortunately, the values of the coefficients for these properties are not always the same for all mechanical properties of interest (e.g., C_D does not apply to compression perpendicular to the grain), and the influence of moisture content can vary among wood species.

Standard practices of the American Society for Testing and Materials (ASTM) adjust test results to establish allowable service stresses in wood. ASTM D245-00, *Standard Practice for Establishing Structural Grades and Related Allowable Properties for Visually Graded Lumber,*[8] and ASTM D2555-98, *Standard Test Methods for Establishing Clear Wood Strength Values,*[9] establish procedures for testing clear wood specimens and determining from those tests allowable stresses for design of wood elements. The results of flexural tests performed under ASTM D2555-98 are divided by 2.1, which accounts for duration and a factor of safety, to establish allowable loads. For flexural strength tests, which take 5 to 10 minutes under these ASTM standards, the duration factor is taken as 1.6 to adjust for load duration in normal (live-load) service.

The potential to induce damage to wood structures during acceptance testing needs careful consideration. The allowable stresses for wood members are developed on a statistical base, with the allowable stresses based on a 5% exclusion factor. Therefore, testing wood structures to full test load, calculated rigorously based on documented wood strengths, should be expected to cause failures in 1 in 20 tests of "in grade" lumber. Further, it should be recognized that safety margins for various mechanical properties are not all the same. Finally, wood can sustain damage in one-time loading that diminishes its strength to support subsequent loads.

For these reasons, the engineer responsible for tests of wood structures needs to make careful observations about the potential for damage during the test, use appropriate judgment about terminating the test when indications appear, and apply the observations to conclusions and recommendations at the end of the test. (Madsen[10] reported that "in grade" tests, intended to determine grading of wood elements as they are used in service, generally do not reduce subsequent strength. However, in-grade tests usually do not stress wood elements to near their failure loads, so it is not applicable generally to extrapolate such observations to strength-verification tests.) Presently, it might not be possible to develop a truly rigorous field load test for wood structural assemblies that are not simple (i.e., when tests must verify several components or mechanical properties simultaneously) or when the margin is small between the required strength and the true safe strength of the structure.

Similar observations apply to materials other than wood. Brittle materials that rely on uncracked conditions for strength have the potential to develop damage that can go undetected during the test. When this happens, the posttest strength of the assembly can be diminished to a level that voids conclusions based on the test.

Summary of Procedures for Acceptance Testing by Attributes. Several standard procedures are available for reference. In some cases, features of test procedures will be dictated by code or industry standard. In other cases, the responsible engineer will need to develop unique procedures. In any event, in most circumstances, acceptance testing procedures will have several common features when applied to components in existing buildings:

- Test load is near to the theoretical failure load of the test element.
- Test load is applied in increments so that performance can be monitored and the test can be terminated if evidence of serious damage appears.
- Test load is held for a period of time so that deflections can stabilize and the test element can demonstrate its ability to support the load.
- Deflection recovery is measured after the test load is removed to determine whether there has been permanent damage to the test element.

- The test element is deemed to have passed the test if it supported the load without evidence of serious damage and it recovered most of its maximum deflection on completion of the test.

These basic features, at a minimum, should be incorporated into all acceptance tests of components in existing buildings.

With all tests, and especially with tests of materials or structures that are brittle or that have widely varying intrinsic strength properties, the responsible engineer must give careful consideration to the potential that the test will produce erroneous results. Causes of such an outcome can be that the test, although carefully developed based on available information, failed to verify an adequate margin of safety in a critical property of the tested element or induced damage that is not easily detected.

Procedures for Acceptance Testing by Variables

Acceptance testing by variables usually requires repetitive testing of a significant number of similar test elements so that the acquired data provide a statistically valid base for assessment of the reliable strength of the subject detail. Since there usually is little concern about the test inducing damage to the subject detail (the intent is to load the test elements to failure), performance determination testing often can be conducted with relatively little monitoring of the condition of the test element during load application. Of course, if the engineer wishes to fully understand the performance of the construction, particularly if there is a need to know serviceability limits that could control designs using the subject detail, then systematic observations should be made throughout the test.

It usually is not necessary to apply load in increments. A single, continuous application of load at a quasi-static rate suffices for most strength-determination tests. Nevertheless, it is important to establish a baseline starting point, with apparatus firmly set and recording devices zeroed.

Test Procedure Development

When mandated by code or when dictated by a building official, the nature and many of the details of the test procedure will be prescribed. However, the engineer with responsibility for performance evaluation still assumes the responsibility for the test protocol and for acquiring sufficient data to be able to render appropriately substantiated opinions about the adequacy of the questioned structural element. Further, when faced with the desire to conduct meaningful tests when there is no universally accepted guidance on procedures, the engineer must take necessary steps to ensure that the outcome of custom tests will fulfill the purpose.

First in the process for developing test protocol is justification of the need for tests and prediction of the validity of the outcome. The cost of tests and the potential for disruption or structural harm often require careful consideration of alternative approaches. It should be demonstrated that analysis alone cannot be used to justify performance or that verification by tests will be less costly than by analysis. There should be an expectation that tests will verify that performance is acceptable rather than that they will prove a deficiency. Otherwise, consideration should be given to application of the costs for tests toward the installation of upgrades that will remediate the potential deficiencies when such alternatives are feasible.

Common Expectations

There should be concurrence among the stakeholders that tests will lead to meaningful results. Those directly affected during testing must understand the impact on their operations or interests so that they can adjust and cooperate. This act of acceptance usually is the most important step in the process, for if there is no specific concurrence on the need for and validity of testing, or if the impact on stakeholders and other affected parties is too large or poorly understood, field testing can fail to serve a useful function in the verification of structural performance. As a worse outcome, field testing can harm structural performance or firmly held interests in a way that more than negates the usefulness of the data collected.

Load tests often are performed under an atmosphere of contention. The economic burden for the correction of a suspected deficiency often shifts from one party to another based on the outcome of load tests. To the extent that all parties with economic interests can agree on the need, protocol, and meaning of tests before tests are performed, posttest disputes related to the efficacy of the tests can be tempered. In contentious situations, parties with interest should witness the tests for their own protection and to ensure that accepted protocol is implemented.

Once the stakeholders and affected parties are educated about the importance of tests, the engineer in charge must implement a meaningful protocol that will verify the critical performance features while maintaining the safety of the structure, its occupants, and testing personnel. The engineer must consider sampling techniques, equipment for load application, safety, shoring and bracing, measurement, data acquisition, and data reduction. Finally, it will be important to be able to report the findings in a persuasive fashion that conveys both the knowledge gained through tests and the significance of any uncertainties and limitations that remain. A substantial amount of advance planning is needed even before a specific test protocol is developed.

SAMPLING

Sampling techniques vary depending on the purpose of the tests, the desired reliability of the conclusions that will result, and the size of the population of similar elements that need to be verified. Guidance is presented below for sampling approaches for acceptance testing by attributes and by variables.

Acceptance Test Sampling

Acceptance testing usually requires the answer to a straightforward question: Is the in-place construction adequate for its intended purpose? The sampling decision to address this question also is straightforward when there is only one or are just a few installations of the questionable construction. However, when the population of suspect installations is large, then a program for selecting the elements for testing becomes more complicated.

Acceptance Test Sampling for Isolated Elements. The sampling technique depends on the specific purpose for testing. In cases where there is one element that is suspect, or when a few similar elements have uncertain performance characteristics, the sampling technique can be resolved relatively easily. Full resolution is gained if all suspect elements are tested. Often when there are relatively few similar elements in question, testing the one or the several that are representative of the group will suffice. This approach holds when it is possible to verify that all elements in the test group share the features that lead to uncertainty and that might affect test outcome and that the tested elements can be shown to be worst-case examples of the offending conditions. If a population's group is to be validated by testing a small sample, confidence can be gained by testing individual elements that have characteristics that make them, among all the elements in the population group, the ones most susceptible to failure.

Acceptance Sampling by Attributes. When there are many essentially identical elements that must be verified by testing, random sampling techniques can give confidence to the performance of the entire population by testing a relatively small fraction of the suspect elements. The level of confidence one gains from sampling techniques depends on the acceptance criteria, the number in the test group, and the expectation on the randomness of the relevant conditions within the full population of suspect elements. For conditions (both in the potential for deficiencies in the population of elements and in the method for sampling from the population) that are truly random, standard sampling theories[11] dictate the number of elements in the test group needed to ensure the reliability of the decision that follows from testing.

To develop a sampling plan, one must determine the acceptance criteria and the confidence with which the engineer wishes to know the accuracy of the findings. Clearly, it is never possible to be 100% sure about the suitability of a population of elements that could contain deficiencies without testing every element in the population. Hence, if an engineer is to develop an opinion about the suitability of a population of elements without testing them all, the engineer must determine the required level of confidence in the results based on testing a sample. Commonly, engineers might decide that the confidence level must be 90%, 95%, or another similar order that the results of the sample tests represent the strengths of the elements in the entire population. This acknowledgment that we can never be sure without testing the entire population leads to the ability to select a manageable sample for testing.

When it is important to know that a particular structural detail is adequate for its intended purpose without knowing its failure strength, the sampling and testing can be performed in accordance with theories for *acceptance testing by attributes*. In this approach, each test item is evaluated as being either "good" or "bad," and decisions are made on the basis of the number of "bad" test items found in the sample. Acceptance testing by attributes criterion normally is expressed, when testing a sample of elements from a larger population, as the number of deficient elements permissible in a certain population size. For example, an engineer might conclude that it is acceptable to allow a 5% deficiency rate for floor joists in a group, on the theory that floor joists are repetitive elements that share load and, therefore, provide redundant load paths.

An engineer might, therefore, conclude that the test program must determine with 90% confidence that no more than 5 joists per 100 in the population are deficient. Given a sample size n and a deficiency rate p' within that sample, then the probability of accepting a sample that has a certain number of failures c during testing is

$$L(p') = \sum_{d=0}^{c} \binom{n}{d}(p')^{d}(1 - p')^{n-d}$$

where $\binom{n}{d} = \dfrac{n!}{d!(n - d)!}$

It often is cumbersome to determine sample sizes from this relationship. Fortunately, assuming a Poisson approximation for the distribution yields a rapid solution.[11] Tabular results are provided in many texts on statistics. In addition, references on statistics[12] publish operating characteristic (OC) curves that plot the confidence levels as a function of various sample sizes, deficiency rates in the full population, and numbers of "bad" test items in the sample.

In the example assuming that 5 deficient joists per 100 would represent acceptable performance, to have a confidence level of 90% that the deficiency rate in the full population of joists does not exceed this rate, it would be necessary to test 78 joists (regardless of the size of the population, as long as the population size is much larger then 78) and find no more than 1 to be deficient.

In the interest of cost control, it is possible to conduct a *two-phase sampling procedure*. In the first phase, relatively few elements are selected for testing from a large population, but the acceptance criteria are selected to be tight. If the elements in the first set of tests do not fail the acceptance criteria, then the entire population can be judged to conform. However, if the first sample contains failures exceeding the acceptance criteria, then a second phase of tests can be performed to further verify the performance expectation for the entire population.

Continuing with the example introduced earlier, it is possible to achieve a confidence level of 90% by testing only 46 joists if none of the tested joists proves to be deficient. On this basis, one might develop a two-phase testing program that has, as its first phase, the random selection and testing of 46 joists to determine if they are "good" or "bad." If no deficiencies are found, then the engineer can conclude with a 90% confidence level that the entire population does not have a deficiency rate exceeding 5%. If, in the sample of 46 joists, 2 or more test items fail, then the engineer might conclude that the full population will not meet performance expectations. However, if only 1 joist in the original sample of 46 is found to be deficient, it might prove cost-efficient to select (at random) an additional 32 joists to bring the total number of tested joists to 78. Then, if no additional deficiencies are found in the second sample, the combined samples provide a basis to conclude, with a 90% confidence level, that the entire population does not have a deficiency rate exceeding 5%.

A two-phase approach also can be advantageous when high uniformity of the design and construction of structural components can be demonstrated and the test load in the first phase can be set to exceed, by an appropriate margin, the essential load necessary for adequate performance (i.e., the rated load). Under this condition, a relatively few elements can be tested to the augmented test load. Should all support the test load without failing acceptance criteria, then engineers sometimes can determine that the probable range in variability of expected performance throughout the population of elements will not result in individual critical failures, should the structure be subjected to the rated load.

In such a two-phase approach, when failures occur at loads that are less than the rated load, engineers can establish with only few tests that the population of elements does not have adequate strength. When there are failures at loads that are above the rated load, then engineers can decide to perform more extensive tests to loads approximating the rated load to quantify the adequacy of the elements in the population.

This approach is best suited for noncritical elements and when there is redundancy in the support of loads so that single untested elements that might fail in service, because their performance is at the low end of the entire population's performance, do not lead to catastrophic failures or safety risks.

Sampling techniques and acceptance criteria must be developed generally for conservatism but without inappropriately high performance expectations. A test designed to avoid acceptance of deficient elements by establishing unrealistically high acceptance criteria for the tested group will fail in its purpose. Engineers can unwittingly design test protocols that the population of suspect elements cannot survive, even though the elements in the population are fully adequate for the intended purpose. This can lead to flawed remediation decisions and unnecessary damage to construction.

Acceptance Sampling by Variables. When it is important to establish reliable estimates of specific properties (e.g., strength or stiffness) for a population of elements, then the sampling and testing can be performed in accordance with theories for *acceptance testing by variables*. In this approach, it is necessary to use random sampling techniques to identify test items in the full population or in exemplary construction and to test each to failure.

Resources to determine the sample size and the means to process the data are found in references on statistics[12] and in ASTM standards. ASTM E122-00, *Standard Practice for Calculating Sample Size to Estimate, With a Specific Tolerable Error, the*

Average for a Characteristic of a Lot or Process,[13] instructs on determination of the sample size. The process involves first making an assumption about the variability of the data by estimating the standard deviation of the values in the population and assuming a maximum allowable error between the value found through the sample testing and the value that would be determined if every element in the population were tested.

A process needs to be developed to ensure that the selection of samples is truly random. This is best accomplished by a method such as assigning "serial numbers" to the elements in the population and generating random numbers that select the serial numbers of the test elements without regard for factors that might incidentally affect the distribution of strengths within the sampled elements.

As an example of the procedures presented in the ASTM reference, given the estimate of the standard deviation and the assumption of allowable error, the required sample size in populations that have values that are assumed to be symmetric about the mean is

$$n = (K\sigma_o/E)^2$$

where σ_o = the advance estimate of the standard deviation.
          E = maximum allowable error resulting from the sample size.
          K = a factor that establishes the certainty with which the actual error will not exceed the chosen value of E.

K ranges from 3 when the intent is to be "practically certain" (probability of exceeding E equal to approximately 3 in 1000) to 1.64 when the intent is to achieve rather approximate values (probability of exceeding E equal to approximately 1 in 10). On completion of testing, the standard deviation of the data can be compared with σ_o to test the validity of the initial assumptions.

ASTM E141-91, *Acceptance of Evidence Based on the Results of Probability Sampling,*[14] provides procedures to test the validity of the data acquired during testing and to express the level of confidence in the properties established from the test data.

Normally, test data will be averaged to determine the value that is taken to be the average of the value of the property in the full population. This calculation is represented by

$$\theta = \bar{y} = \sum_{i=1}^{n} y_i/n$$

where θ = average of the entire population
          $\bar{y}$ = average of the sample
          y_i = individual measurements
          n = size of the sample

To establish the level of confidence that $\bar{y}$ is a valid estimate of θ, one must calculate the *standard error* of the data:

$$se(\bar{y}) = \sqrt{\sum (y_i - \bar{y})^2/n(n-1)}$$

where $se(\bar{y})$ is the standard error of the data.

Then the actual value of the average in the population can be bounded with confidence using probability distribution values of the Student's t distribution (commonly called simply a t *distribution*). The upper and lower bounds on the actual average value are

$$\theta_{\text{Upper}} = \bar{y} + t_{\alpha/2}(\nu)se(\bar{y})$$

$$\theta_{\text{Lower}} = \bar{y} - t_{\alpha/2}(\nu)se(\bar{y})$$

where $t_{\alpha/2}(\nu)$ = t distribution value for a specific confidence level and number of degrees of freedom in the data (Values of t distributions can be found in most references on statistics.[11,15])

α = referred to herein as the *exclusion limit,* meaning that $100 \times \alpha\%$ of the values will fall outside the limits

ν = number of degrees of freedom in the data and is equal to $n - 1$

As an example of the application of these ASTM approaches, assume that an engineer wants high certainty that tests will reveal an average fastener strength that is within 2 lb of the actual average strength of all fasteners in the population. Assume that prior test data suggest that the standard deviation for the strength of this type of fastener also is 2 lb. The formula in ASTM E122-00 yields the following sample size:

$$n = (3 \times 2/2)^2 = 9 \text{ tests}$$

Then assume that nine tests are performed, with the following results: 103.2, 97.7, 101.3, 99.6, 98.3, 101.9, 99.0, 101.7, and 102.0 lb. The average of these values is 100.5 lb. The standard deviation, at 1.91 lb for this hypothetical test program, compares favorably with the initial assumption. Hence the sample size was approximately correct to achieve the desired result, assuming a normal distribution for strength properties and that the standard deviation of the test data is the same as the standard deviation of the full population.

Employing the approach in ASTM E141-91, the standard error of the data is 0.638 lb. Assume that it is important to calculate the bounds on the actual average fastener strength with a confidence level of 95% (i.e., 5% exclusion limit, meaning $\alpha = 0.05$). From references on statistics, one can find that

$$t_{\alpha/2}(\nu) = t_{0.025}(8) = 2.306$$

On this basis, the data reveal that the true average of the fastener strengths is known, with 95% confidence, to be between

$$\theta_{\text{Upper}} = 100.5 + 2.306 \times 0.638 = 102.0 \text{ lb}$$

$$\theta_{\text{Lower}} = 100.5 - 2.306 \times 0.638 = 99.0 \text{ lb}$$

Hence the product of the application of the approaches presented in these ASTM procedures allows the selection of an initial sample size and a test of the validity of the sample size. Then the estimate of the value of the property can be expressed together with the standard error and the sample size (or number of degrees of freedom in the data). Finally, by selecting the desired confidence level, bounds can be placed on the true average value of the property.

The ASTM references should be consulted for a full presentation of the procedure to select sample sizes and to express the test results in meaningful terms.

Interacting Elements

Whether all elements in a population are to be tested or a sample is to be examined by test, it is important that the samples be selected and isolated appropriately to allow tests to verify the critical performance characteristics. Imagine that there is a need to test the adequacy of joists in a floor system and that the deck that spans between the joists has substantial strength and rigidity. Under this condition, it is inadequate to apply a tributary load to a single joist and expect the performance of the joist during the test to represent performance in service. Unless the single test joist is isolated from adjacent joists, it is entirely possible that the adjacent joists will share in the support of the test load, and performance expectations will not be met.

Consideration must be given to the potential that nearby nonstructural elements will influence the outcome of load tests. Sometimes interior partitions can support substantial loads. Curtain walls can influence the performance of spandrel beams. To the extent that these interactions are artificial to the true performance of the test specimen and detract from the accuracy of the results, they should be prevented from influencing the tests.

To raise the probability that the salient features of an indeterminate structural system will be tested, it often is necessary to isolate the test joist (such as by cutting the deck on both sides—but providing appropriate supplemental bracing) or to load several joists at the same time (Fig. 21.2). If the test joist and a sufficient number of joists on both sides of the test joist are loaded simultaneously, load transfer from the test joist to adjacent joists will be minimized, and performance of the test joist can be judged accurately. Often it is important to perform analyses that demonstrate the extent of interaction among adjacent elements so that appropriate test protocol can be developed. For instance, in the preceding example it normally would be important to anticipate the relative stiffness of the primary load-carrying elements (the joists) and the transverse load-carrying element (the deck) to identify how many joists need to be loaded to ensure that the center, tested joist receives its full complement of load.

Even when adjacent, similar elements do not share load with the test specimen, there sometimes are performance characteristics that must be adjusted to achieve valid results. Continuous beams sometimes are designed for application of pattern loads over several spans. In order to validate important performance characteristics in such cases, it sometimes is necessary to apply the appropriate pattern loads. Alternatively, the test span in such instances may need to be loaded in a fashion that does not directly conform to the expected load in service but simulates the effects by achieving element stresses that correspond to those of the in-service conditions. For instance, when midspan moment capacity of a continuous beam is suspect and it is not feasible to place

FIGURE 21.2 Wood bracing installed to prevent buckling of metal joist disconnected from deck. (*Courtesy of Simpson Gumpertz & Heger Inc.*)

pattern loads on all spans that contribute measurably to the forces in the test specimen, it could be appropriate to place an "overload" on the tested span. The amount of the overload would be calculated to achieve the same moment at midspan by loading the single test span as would occur with several spans loaded in patterns. The calculations need to be based on the theoretical behavior of the beam, acknowledging the relative stiffnesses of interconnected beams and columns.

Of course, under these circumstances, it is essential to know that the applied overload will not impose damaging forces that normally would not be experienced in service. Such damaging forces could be unrealistically high shears at the end of the tested beam or high column moments in adjacent columns.

Sampling, therefore, also must consider the independence of the test specimen and the influences that are derived from primary and secondary structural behaviors that originate outside the specific test zone. Sampling and specimen preparation sometimes cannot be separated fully from loads.

EQUIPMENT

Test equipment needs to be selected carefully and maintained properly so that tests can be performed safely and efficiently, and the results will be reliable.

Load Application

Almost without exception, tests to evaluate the strength of a structural system involve the application of load or the enforcement of displacement. In most cases, the magnitude of the test load required to verify performance is large enough that substantial load must be applied in increments and in a controlled fashion that does not overload adjacent areas of a structure or inappropriately influence the results of the test. For these reasons, the selection of the means to apply load and the procedure to transport weights to the test location are critical.

When tests are performed to evaluate performance under gravity loads, weights can be applied to the test structure to represent loads in service. Relatively high-density materials that can be handled in modest amounts serve best because they can be delivered to the test location under control. Drums filled with water can serve well when test loads are on the order of 150 to 200 lb/ft^2. For higher loads, sand, solid brick or block, or pallets of metal can be used. In all cases, it is helpful if the materials selected are readily available near the site and can be assembled into units of loads that can be applied and remove incrementally.

Consideration must be given to the independence of the structural behavior from the method of transfer of loads to the structure. The engineer must consider whether relatively strong and stiff objects, such as steel plates of substantial length compared with the length of the test element, transfer load appropriately to the test element or, instead, span as structural elements themselves. In the latter case, the strength of the test element is not evaluated appropriately because transfer of the test load does not simulate the load in service.

In many cases, the logistics related to handling the loading materials influence selection of the materials. Often structural systems cannot support forklifts or other machines for handling weights. Even when machines normally could be supported on the structural system, it rarely is possible to use them in the vicinity of a test specimen once much of the test load has been applied. In these instances, load application involves hand work.

Some difficulties associated with handling weights for load tests can be alleviated by applying loads from below the test element (Figs. 21.3 and 21.4). When test loads

FIGURE 21.3   Suspended platforms to support loads for overhead test. (*Courtesy of Simpson Gumpertz & Heger Inc.*)

FIGURE 21.4   Metal weights suspended from overhead structure. (*Courtesy of Simpson Gumpertz & Heger Inc.*)

can be suspended, weights can be brought to the test site with minimal influence on the structural components that are being tested. Obviously, care still is required to ensure that the loads can be supported by the structure below the test element.

There are cases when jacks (Fig. 21.5), hoists, or winches can be used to apply loads. Such devices often are necessary when test loads must be applied horizontally. This equipment requires reaction surfaces, so there must be a nearby structure of sufficient strength and stiffness to allow the test element to be loaded fully during the test without adversely affecting the outcome of the test.

Dynamic testing often requires a device that can excite the structure near its natural frequency or at some other desired frequency. This can be done by installing a shaker that can apply oscillating loads at various frequencies. Shakers often have weights mounted on a sliding ram or eccentrically on a rotating shaft. The driving shafts are rotated through a range of frequencies to encompass the expected frequency of the structure. When the frequency of the shaker approaches the frequency of the structure, the response is amplified. Care must be taken to control the amount of energy that is introduced to avoid resonance that could cause unwanted damage. Further, to minimize the potential for damage, consideration must be given to the implications of the induced movements in all directions.

An alternative for measuring the frequency of the lowest modes of structural vibration involves "plucking" the structure and measuring its response. The concept is much like bouncing on the end of a diving board and noting the frequency of vibration. The process works well for components but is complicated for a full-scale structure; sufficient energy must be imparted to induce a vibration without creating a risk for the structure or personnel. Dropped weights sometimes are used in structures (but the

FIGURE 21.5   Hydraulic jack used to test anchors in concrete. (*Courtesy of Simpson Gumpertz & Heger Inc.*)

weight must be included in the calculation of the response). Under very controlled conditions, cables sometimes can be used to induce a deflection and can be released suddenly to start a structural vibration that then can be monitored. Unfortunately, it often is difficult to develop a safe process for stressing the cable and suddenly releasing it without causing a potentially dangerous slingshot response in the cable.

When the goal is to monitor structural response to natural stimuli, it usually is unnecessary to apply artificial loads. However, the natural stimulus, whether it is wind, earthquake, moisture, temperature, or other influences, must be monitored so that effective load magnitudes can be assessed.

Measurement

The response of the structure to the application of loads must be monitored. The means to obtain the measurements without influencing the test, the level of precision, and the frequency of observation all must be considered during planning phases.

Clearly, is it not appropriate to use a measurement method that cannot be observed without influencing the reading. For instance, for relatively flexible structures and light loads, devices that need to be read by an observer who must walk onto the test structure can give erroneous results. Environmental sensors need to be clear of external influences; thermometers should not be near radiators, and anemometers should be in position to accurately measure wind speed at the place of interest (or, alternatively, at standard height in an exposure that can be characterized unambiguously). Therefore, when designing test protocols, the engineer must select instruments in consideration of the physical limitations of the site and access to measurement locations.

Deflections of most common structures under service or close to ultimate loads can be monitored without difficulty to precision on the order of $\frac{1}{16}$ in. However, unusual structures (e.g., long-span structures) may not require this level of precision. On the other hand, to capture slow, time-dependent phenomena (e.g., settlement) in relatively short test periods, much more refined precision may be necessary. Of course, the more refined the needed precision becomes, the greater is the care that is necessary to obtain accurate and repeatable measurements over extended time periods. To capture settlement trends in a relatively short time, it may be necessary to commit very-high-precision surveying techniques that have accuracy that is at least two orders of magnitude more precise than is common for conventional surveying techniques used to lay out a structure for construction.

Vibration studies usually require measurements to relatively high precision. Deflections of interest often are small. Furthermore, data collected by accelerometers, often used for such studies, must be integrated to extract information about amplitudes of motion. As such, precision in data collection during such tests is critically important.

Observations need to be made frequently enough to be able to plot trends in deformation or, in the case of vibration studies, to extract meaningful interpretations. During strength tests, it could be very valuable to establish the load at which behavior ceases to be linear. The transition from linear to nonlinear response may signal the onset of yielding or another form of nonreversible damage.

In some instances, onset of nonlinear response may be irrelevant. However, depending on the load at which the transition occurs and the implications of the residuals from nonlinear response, it might be advisable to terminate a test when nonlinear response is detected. In any event, for strength tests, it is prudent to measure deformation at least at load increments not exceeding one-quarter of the total test load.

Deformation can be measured with several devices. The simplest is a ruler that an observer places against a test element and compares its position with a nearby fixed object. While not very elegant, this method can be very effective and cost-efficient in appropriate circumstances. Perhaps the most familiar instrument to measure deformations is a dial gauge with a spring-loaded probe (Fig. 21.6). The body of the gauge

FIGURE 21.6   Dial gauges installed to measure deflections. (*Courtesy of Robert T. Ratay Consulting Engineer.*)

is attached to a fixed object, and the probe presses against the test element (or vice versa). As the test element moves, the probe slides and registers the movement on the dial (or digital display) in the body.

Dial gauges are attractive because they are easy to install and use. However, the probes on dial gauges have limited travel. Although gauges are available to measure deformation up to several inches, they still are limited in the amount of deformation they can record. When displacement exceeds the limit of the gauge, sometimes it is possible to reset the gauge during the test by recording the movement at the end of a load increment, removing and resetting the gauge with a new "zero" value, and summing the measurements to calculate total deformation after the gauge is reset.

Deformations also can be measured using a linear variable-displacement transducer (LVDT). This is an electronic device that functions much like a dial gauge. Generally, LVDTs have probes the slide in a cylinder and produce an electric signal that can be interpreted to indicate the amount of movement (Fig. 21.7). LVDTs generally have output resolution that is very refined and can be obtained in standard construction to measure movements greater than 1 ft.

Optical methods also can be used to measure deformation. The range of possibilities again starts with the ruler. In this case, a ruler can be affixed to a test element and read with a surveyor's level from a distance. There are optical comparators, which have two parts—one with a grid and another with crosshairs (Fig. 21.8). The two parts are separately mounted to components that move relative to each other, with the motion registered as changes in position of the crosshairs on the grid field. Optical methods using lasers and fiberoptics also can record elongations that can be interpreted either as relative movement or average strain over a distance.

Clearly, surveyor's tools—theodolites, rangers, levels, and tapes—can measure movements through comparisons of a set of initial readings with readings after loads are applied. Direct measurements and determinations through triangulation apply, de-

FIGURE 21.7   LVDTs installed on frame inside to measure pipe deformations. (*Courtesy of Simpson Gumpertz & Heger Inc.*)

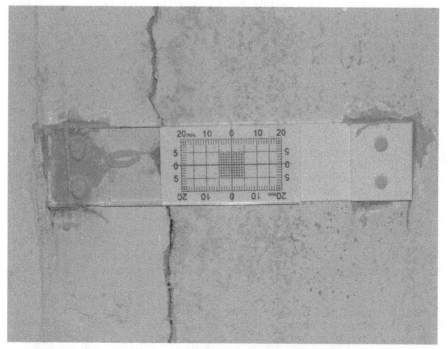

FIGURE 21.8   Optical comparator to monitor movements of across a crack. (*Courtesy of Simpson Gumpertz & Heger Inc.*)

pending on the circumstances. For measurements such as are necessary to detect slowly changing deformations, high-precision instruments and repetitive readings often are necessary for accurate detection of small movements.

Tiltmeters (Fig. 21.9) can record slope and change in slope. They operate on several principles and can be acquired to operate over various ranges of tilt and to differing levels of precision. They can be affixed to a surface or simply placed on a flat surface. Some have digital displays of slope for direct reading; others produce an electric signal that can be interpreted and displayed or recorded for later processing.

Loads applied through jacks or similar devices can be measured with in-line load cells (Fig. 21.10). Load cells most frequently contain a deforming internal component that is instrumented to measure its strain. The stain, in turn, is calibrated for reporting as an applied load. Load cells are available to measure tension loads and compression loads. They can be designed to be placed in line with a cable or rod in tension, between an object and a jack in compression, and for a wide range of other configurations. Load cells can be ring-shaped to allow a rod or shaft to pass through the middle. Similar instruments can measure torque.

Jacks that are operated with hydraulic fluid can be monitored by observing the pressure in the fluid and calculating the resulting load based on the known surface area of the piston in the cylinder. To achieve uniform load on a series of jacks, the single monitored source of hydraulic pressure can have a manifold that accommodates several jacks. In this configuration, jacks with the same piston surface area will carry the same load at the same time.

Self-weights of the test specimen and any hardware for execution of the test need to be considered when measuring test loads. These weights, as appropriate, should be considered in the first load application cycle.

Strain gauges (Fig. 21.11) can be used to determine stresses in loaded elements. Most commonly, strain gauges are fine, looped wires that change electrical resistance as they are stretched or compressed. A variation is a vibrating-wire strain gauge (Fig.

FIGURE 21.9   Tiltmeter mounted on wood bulkhead. (*Courtesy of Simpson Gumpertz & Heger Inc.*)

FIGURE 21.10   Load cells on frame, attached by suction, to test glazing anchorage. (*Courtesy of Simpson Gumpertz & Heger Inc.*)

21.12), which senses changes in elongation by changes in tension in a wire. When such gauges are affixed to the surface of an element to be loaded, they can register strain changes in the base material. With this information, stress and force in the loaded element can be calculated. Strain gauges can be adhered to a separate bar or plate that can be clamped to a test element. As long as the bar or plate is sufficiently compliant compared with the test element and the clamps are sufficiently secure to enforce the strain gauge to be deformed along with the test element, this approach allows for rapid installation and complete removal of the test instrument.

Strain gauges using principles of fiberoptics also are available. These gauges tend to be longer and, therefore, are useful for determining average strain over relatively long distances.

Dynamic response can be measured either by continuously recording changes in instruments that measure deformation and processing that data for frequency content and amplitude or by accelerometers. Accelerometers sense motion (velocity or acceleration), the signal for which can be processed to obtain frequency content and amplitude.

Data Acquisition

Data can be acquired and recorded by hand or through automated processes (Fig. 21.13). Electronic instruments can be connected to dataloggers that process and retain signal information from the instruments. Dataloggers can be used to display output for concurrent observation and interpretation, or they can be used for postprocessing. Coupled with a modem or Internet connection, dataloggers can be accessed remotely. Also, dataloggers can be equipped with a call-out feature that will alert remote observers

FIGURE 21.11   Strain gauges to record strains for stress calculations. (*Courtesy of Simpson Gumpertz & Heger Inc.*)

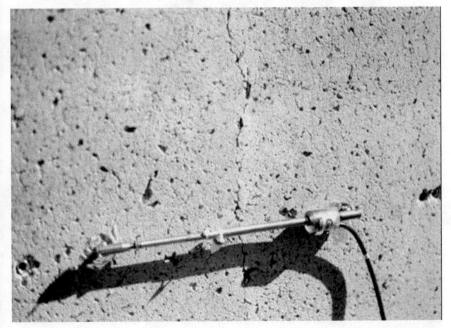

FIGURE 21.12   Vibrating-wire strain gauge. (*Courtesy of Simpson Gumpertz & Heger Inc.*)

FIGURE 21.13 Data logging equipment in cabinet for long-term monitoring. (*Courtesy of Simpson Gumpertz & Heger Inc.*)

when preset data thresholds are exceeded. Dataloggers so equipped can be constant monitors that can collect data and check status of a structure over a long term without human presence.

Calibration and Reliability of Data

All equipment should be calibrated for accuracy. The frequency and method of calibration depend on the instrument and the needs of the testing program. The greater the required accuracy in the results, the more important it is to be sure that instruments are calibrated properly at the time of the test. In the most critical circumstances, it might be advisable to order a professional calibration immediately before and immediately after a sequence of tests.

Some instruments can be calibrated and certified by external testing agencies that perform rigorous procedures to verify the operation of test instruments. Other instruments can be calibrated simply by the user. For example, a user can apply a known weight to a load cell to determine whether the load cell is working properly. Once one load cell is calibrated, it can be used in series with others to verify that they are calibrated as well. Depending on the required precision, the function of a device such as a tiltmeter can be verified in part simply by placing it once on a flat surface, noting the slope, and then rotating the instrument 180 degrees and placing it again on the same surface. A properly calibrated tiltmeter should indicate the same slope when placed both ways.

Sometimes engineers have only one opportunity to perform a test on a structure. The probability of a successful test is increased when redundancy is built into the data-acquisition procedure and means are provided to maintain continuity in data should monitoring devices be disturbed (e.g., the tripod supporting a level is bumped). This is especially true for long-term monitoring programs during which there could be failures of data-acquisition instruments. Redundancy can be accomplished by providing more instruments than are minimally necessary to record important data and by providing independent monitoring programs. Independent systems can provide backup data should the primary acquisition system fail, and they can be used for calibration and quality checks on the primary system.

It often is advisable to document tests with photographs and videotape recordings. In addition to providing useful images to convey the nature of the work to the client and other interested parties, these records can help with interpretation of results by documenting such features as sequence of load application and the appearance of evidence of structural distress.

SAFETY

Field load testing can be hazardous, particularly when large components are involved. Often tests require handling heavy weights on a structural frame, sometimes above testing personnel or building occupants. Almost by definition, the adequacy of the structure is suspect, and hence its ability to support a heavy load is uncertain. There is potential for a failure during the test.

Tests usually induce high strain energy in the structure and in test apparatus. Cables transferring forces to the structure can rupture, and failures of the test structure can cause sudden releases of this energy, with flying parts creating hazards to persons nearby. In addition to parts of the structure itself, pieces of the test apparatus can be propelled away by the failure of the test component. For these reasons, planning and execution of tests need to be performed with rigorous commitment to the safety of testing personnel and the general public.

In some instances, barriers should be installed to contain pieces that might be ejected when an element fails. Observers must be cognizant of the potential for failure at any time, keep their hands away from pinch areas, wear safety helmets and goggles when appropriate, and keep away from the test area except when it is safe and essential to approach.

Tests should be performed in strict conformance with the requirements of the Occupational Safety and Health Administration 29 CFR 1910, *Occupational Safety and Health Standards*,[16] and 29 CFR 1926, *Safety and Health Regulations for Construction*,[17] as appropriate, and other appropriate federal, state, local government, and industry safety standards.

SHORING AND BRACING

Field load tests often are intended to expose a suspect structural system to loads that approach the ultimate capacity. By definition, it is to be expected that a structural element could fail during a test, suddenly releasing its stored strain energy and allowing superimposed and suspended weights to fall or stretched cables to go slack. It is critically important to be prepared for these occurrences by providing adequate shoring and barriers to minimize the potential for damage. Additionally, it often is beneficial to provide a means for rapid reduction in test load should an impending failure be detected during a test.

Shoring (Fig. 21.14) should be placed below any element that will be loaded. To account for deformations that will occur during the test, it will be necessary to install the shoring to near the underside of the element but not in contact over the test region. The engineer will need to anticipate the amount of the test deflection and provide sufficient gap for the test to proceed unimpeded. In some cases, the anticipated deflection could be large enough that the shoring might have to stop several inches below the test element. In this case, it might be advisable to insert removable shims above the shoring at the beginning of the test, to keep the potential drop distance as small as is practical. If this is done, as the test proceeds and deflection occurs, shims can be removed successively to accommodate the structural behavior. Of course, the test observers will need to monitor the size of the gap between the test element and the shoring to prevent the test element from coming to rest on the shoring during an ongoing test.

Shoring normally needs to be designed by a competent person to support the full weight of the test element and the supported loads. In addition, shoring design could consider an additional impact factor to account for sudden failures that can occur in some tests. When sudden failures are anticipated, it is particularly important to keep

FIGURE 21.14   Shoring installed below concrete precast tee. (*Courtesy of MACTEC Engineering and Consulting.*)

gaps as small as practical to minimize the influence of impact. If gaps are kept small, impact factors on the order of 2.0 usually suffice to account for sudden failures.

Shoring must rest on a structure that is able to support the superimposed load and associated impact in the event of a failure. In an elevated structure, this may require shoring to more than one level below the test element so that the load can be shared by several floor structures. Proper consideration must be given to the point loads created by the legs of shoring, particularly on structures that are designed only for distributed loads.

During tests that require structural elements to be isolated artificially from one another (e.g., by cutting a deck to allow testing of a single joist), it may be necessary to add back bracing for the compression regions to prevent premature buckling of test elements. When this is done, it must be accomplished in a manner that replicates the bracing that exists in the as-built structure without voiding the purpose for isolating elements.

PREPARATION

The decision to go forward with a test program must be accompanied by careful planning and disclosure. There are risks involved in any field test. Several stakeholders must be informed and concur with the need and the process. Careful planning of the logistics will minimize disruption and cost.

Most owners and tenants would prefer to have questions about structural performance resolved through nonintrusive means. Many will be reluctant to agree to the disruptive and potentially damaging activities that are necessary for field testing.

The first task for the engineer is to demonstrate to the stakeholders (including the building official in some cases) the need to conduct field tests. This often involves explaining the inadequacies of other options to confirm structural behavior.

Part of this demonstration is an explanation of the risks associated with field tests. In particular, field testing to verify strength has potential to damage the test elements, resulting in repair costs that otherwise might not be needed. The owner needs to accept this risk as a balance against the likelihood that the tests will demonstrate a viable cost-saving approach to address the underlying concern.

The engineer and the stakeholders should agree in advance about the significance of the data that will be generated during field tests. To the extent possible, all interested parties should understand whether tests will be conclusive, whether a second phase of testing could be required, and whether there will remain uncertainties that will need separate resolution.

Sometimes tests are performed to demonstrate to the building official that a structure has sufficient strength and reliability to be left unrepaired, even when it has suspect details or details known to deviate from conventional practice. In these circumstances, it is critically important for the building official to understand the intent of the tests and to agree to the meaning of the outcome. The building official should be informed of any limitations in the expected use of the data and means that likely will be used to address issues that cannot be resolved by tests.

To the extent that there are interfaces with the public during tests, local public safety officials need to be involved. Tests on the outsides of buildings may require sidewalk sheds and partial street closures. Should it be necessary to shut off sprinklers, the fire department needs to be informed and may require a presence at the site. Thought needs to be given to the potential that tests will disturb electric and gas facilities and to the implications should that happen. Of course, the engineer and owner need to consider both the safety of and disruption to any building occupants.

To address these issues, the engineer should prepare a test plan. This plan should address the following items:

- Purpose and scope, including a statement of goals
- Identification of authorizing entity
- List of parties, with contact information, that are affected by the work
- Review of the ability of the structure to support the test loads and potential debris loads, with backup calculations
- Review of shoring design, with backup calculations
- List of required equipment for the test, with the identification of suppliers of that equipment
- List of required on-site resources (e.g., water, electric service, access, etc.)
- List of required instrumentation, with records of calibration or intended calibration procedures
- Statement of the structural preparations that must be made to address demolition and modifications necessary to accommodate the test, with supporting calculations concerning the effects of the modifications
- Statement of restoration plans, identifying elements that are anticipated to need repair and restoration, with supporting calculations for structural restorations that will be needed
- Statement of required safety precautions, with identification of the individuals with responsibility for safety
- Methology for the test, including
  - Test protocol to be applied, including references to relevant standards (e.g., ASTM standard test procedures or ACI load test procedures)
  - Load increments to be applied, with substantiating calculations
  - Procedures for transporting and applying load
  - Procedures to monitor for evidence of damage
  - Data to be collected, with timing for the collection
  - Data reduction approaches
- Schedule for the test
- Schedule for interaction with affected parties
- Table of contents for test report

Once the test plan is developed, the engineer should coordinate with the owner. This communication will resolve issues related to the scope of the test and the expected outcome. At this time, the owner should approve the plan and give authorization to proceed. Responsibilities should be assigned to contact stakeholders in the process, including building occupants, regulatory bodies, and other interested parties. Also at this time, the test schedule should be established so that any contractors and building occupants affected by the work can plan and activate for the test.

Generally, tests require the support of a contractor. Contractors can prepare the structure for the test by performing such tasks as cutting test elements free from adjacent structure, supplying and assembling shoring, and acquiring and delivering materials for weights. During the test, contractors can handle the test weights on the site. Contractors also remove test apparatus and restore the structure to the pretest condition on completion of tests.

At the time of the test, the test area must be isolated and secured. As necessary, tenants need to be relocated temporarily and their possessions protected. Depending on the nature of the test, the isolation of the test area and the associated hazards, barriers may need to be installed to prevent unauthorized persons from approaching the test area. Persons authorized to be in the test area need to understand the hazards. They also need to understand their roles in the process and the timing of their participations. The primary goal should be a safe and efficient operation.

To the extent that the test might generate debris, dirt, dust, or spray water, adjacent areas need to be protected.

EXECUTION OF THE TEST

At test time, there should be ample observers and support personnel to perform the essential functions without adding unnecessary personnel to confuse the test process and potentially create undue risk. Each person needs to understand his or her essential role in the test. Engineers need to conduct the test and direct the acquisition of data. Contractors, in coordination with the lead engineer, need to control the application and removal of test loads and the adjustment of shoring, if necessary. The owner needs to address the interface with tenants and public safety personnel, if necessary.

There should be a clear chain of command. Someone should be charged with the authority to make essential decisions about when to adjust the test protocol or when to discontinue the test should there be an indication that proceeding could create unnecessary damage. Someone should be charged with authority over site safety. That individual needs to be aware of the activities of the observers and contractors on the site so that hazardous conditions can be avoided. When there is need to interface with public safety agencies, the individual in charge of site safety should lead in those contacts.

The test should be scheduled carefully to minimize disruption and to increase efficiency. Ample preparation time should be allowed to include verifications of the means to handle loads and the calibration and function of the instrumentation. Building occupants should be aware of the limitations in access at the time of the test. Specific points of contact need to be established so that communication is unambiguous.

Contingency plans should be developed to address unexpected obstacles. In most instances, it is prudent to have backup instrumentation available in case there is an equipment failure. With the amount of preparation that precedes a field test and the disruption that tests create for building occupants and others, delays caused by a component that fails to operate at the critical time can be unnecessarily costly and annoying. Furthermore, once a test is begun, undue delay might void the validity of the outcome.

Prior to beginning the test, the engineer in charge should verify that critical instruments have been calibrated and that weights to be applied to the structure are true. Throughout the test, collection of data should be verified. It is good practice to query electronic dataloggers periodically to verify that data are being recorded. When observers record data by hand, it often is helpful for a second person to view gauges or, when data are recited by an observer to another person for entry into a log, that the person making the entry repeats the readings back to the observer for verification.

The test should start promptly with everyone fully aware of the protocol and the responsibilities, follow the schedule to the extent possible, and conclude with prompt securing of the site and disassembly of the test apparatus.

In the event that the test induces serious damage in structural components, that damage needs to be assessed on completion of the test. A process to stabilize the damaged structural elements should be initiated promptly so that potential hazards can be removed. If the test does not induce damage, in most cases the original configuration of the structural system should be restored so that the test area can be placed promptly back into service.

CLEANUP

On completion of the test, the test apparatus should be dismantled as soon as reasonably possible. However, shoring should be left in place until damage, if any, to the test element can be assessed fully and appropriate repairs are made. Structural elements that were modified for the test (e.g., floor systems that were severed to establish independence of the test element) should be repaired or restored. Dust, dirt, and debris should be removed. Damaged or removed interior finishes and architectural features should be restored.

REPORTING

On completion of the test, instrumentation should be examined to establish that it did not receive physical damage that might have fouled data. In some cases, instrumentation should be recalibrated to verify that test data remained accurate throughout the test. Dataloggers should be queried to establish that the full data set was recorded. To the extent possible, a preliminary assessment of the validity of data should be performed (such as by plotting load-deflection curves) to test that the data are reasonable.

The engineer should apply a logical format for the presentation of the results. The report should include the essential information in the test plan, and a full reporting of the data and the results of data analyses performed after the test. The report should include interpretation of the data and the essential conclusion about the satisfactory or unsatisfactory performance of the test element, with a description of the bases for the conclusions. If there were deviations from the test plan, or if data were not valid for any reason, these changes from the anticipated outcome of the test protocol should be identified and described.

The test report should identify the parties present, together with a listing of the responsibility of each individual. Specific observers should be associated with each data set, as well as each author and checker for portions of the data and the report.

Finally, the report should include a clear statement about the validity of the test, the conclusions that can be drawn from the test, and recommendations for additional tests, if any. If there needs to be any qualifications for the findings, those statements should appear in the report as well.

REFERENCES

1. Commonwealth of Massachusetts, (1997). *The Massachusetts State Building Code,* 6th ed., Boston.
2. International Code Council, (2002). *International Building Code.* ICC, Country Club Hills, IL.
3. American Society of Civil Engineers, ASCE 7-02, (2002). *Minimum Design Loads for Buildings and Other Structures.* ASCE, Reston, VA.
4. American Concrete Institute, ACI 318, (2002). *Building Code Requirements for Reinforced Concrete.* ACI, Farmington Hills, MI.
5. American Institute of Steel Construction, (1999). *Load and Resistance Factor Design Specification for Structural Steel Buildings.* AISC, Chicago, IL.
6. Suter, G. T. (1982). "Evaluation of in Situ Strength of Aged Timber Beams," in Meyer, R. W. and Kellogg, R. M. (eds.), *Structural Uses of Wood in Adverse Environments.* Society of Wood Science and Technology, Van Nostrand Reinhold, New York.
7. American Forest and Paper Association, (2001). *National Design Specification for Wood Construction.* AFPA, Washington, D.C.
8. American Society for Testing and Materials, (2002). *Standard Practice for Establishing Structural Grades and Related Allowable Properties for Visually Graded Lumber.* ASTM D245-00(2002)e1, ASTM, West Conshohocken, PA.
9. American Society for Testing and Materials, (1998). *Standard Test Methods for Establishing Clear Wood Strength Values.* ASTM D2555-98e1, ASTM, West Conshohocken, PA.

10. Madsen, B. (1992). *Structural Behavior of Timber*. Timber Engineering, North Vancouver, BC, Canada.

11. Bowker, A. H., and G. J. Lieberman, (1959). *Engineering Statistics*. Prentice-Hall, Englewood Cliffs, NJ.

12. Ang, A. H-S., and W. H. Tang, (1995). *Probability Concepts in Engineering Planning and Design*. Wiley, New York.

13. American Society for Testing and Materials, (2000). *Standard Practice for Calculating Sample Size to Estimate, With a Specific Tolerable Error, the Average for a Characteristic of a Lot or Process*. ASTM E122-00, ASTM, West Conshohocken, PA.

14. American Society for Testing and Materials, (1991). *Acceptance of Evidence Based on the Results of Probability Sampling*. ASTM E 141-91, ASTM, West Conshohocken, PA.

15. Benjamin, J. R., and C. A. Cornell, (1970). *Probability, Statistics and Decision for Civil Engineers*. McGraw-Hill, New York.

16. Occupational Safety and Health Administration, (2003). *Occupational Safety and Health Standards*. 29 CFR 1910, U.S. Department of Labor, Washington, D.C.

17. Occupational Safety and Health Administration, (2003). *Safety and Health Regulations for Construction*. 29 CFR 1926, U.S. Department of Labor, Washington, D.C.

Preparation of Reports

ROBERT T. RATAY, PH.D., P.E.

The documentation of a structural condition assessment project is the written report. The work is incomplete, and of questionable value, until a written report is rendered. Depending on the assignment, it may be a simple one-or-two-page letter or it may be a multi-volume set of documents—consistent with the scope and extent of the project. It is advisable to agree with the client, preferably before but at least during the project, on the extent of the report that he or she expects and you intend.

For a walk-through or cursory condition assessment a letter report is often adequate and appropriate. Even a simple letter report should be organized in sections with subtitles so that a reader can easily locate the part(s) he or she is interested in. A reasonable organization of a letter report is the following:

Date
Addressee
Reference
Salutation
Introductory paragraph
Purpose and scope of work
Materials reviewed
Description of the structure
Findings
Conclusions and recommendations
Sign-off and signature
Enclosures, if any

A large project with in-depth investigation may warrant a large, possibly even multi-volume report. It may be advisable to deliver it in phases—or first a preliminary and later a final report—so as not to overwhelm the client and to give him/her the opportunity to see that it addresses all of his/her needs.

A reasonable organization of a large formal report may be the following:

Cover page
Title page
Table of contents

Executive summary

Introduction (background, purpose, general scope of the work, qualification of the team)

Description and history of the structure

Governing codes and standards

Documents and information reviewed and used

Methods and equipment used

Field inspection and observations

Probes

Field and laboratory testing

Structural analyses and results

Evaluation of field and analytical findings

Summary

Conclusions

Recommendations

Disclaimer

Photographs and Figures

Appendixes

The ASCE standard *SEI/ASCE 11, Guideline for Structural Condition Assessment of Existing Buildings,*[1] the CASE *National Practice Guidelines for the Preparation of Structural Engineering Reports for Buildings,*[2] and the FHWA *Bridge Inspector's Training Manual,*[3] are good references for formatting structural condition assessment reports.

Public agencies, multifacility owners, engineering firms and even individuals often have their own report format to which the consultant may have to adhere. Certain structural types, materials, and circumstances may dictate the need for special emphases that are reflected in the organization of the report.

Instead of listing the items that are self-evidently included in the individual sections of the report and are listed in References 1, 2 and 3, the following paragraphs are reminders of items of importance that have to be addressed in appropriate sections of the report.

Compose a descriptive but succinct title. Sometimes a several-word qualifying subtitle is useful after a short main title.

Be sure to date interim reports as well as the final one.

In a cover letter, or at the beginning of the report, write an executive summary of the purpose, scope, conclusions, and recommendations. Recognize that this is all that some readers of the report will read, and those readers may be nontechnical but "important" people.

State the name and address of the structure early in the report.

Note the purpose, date(s) and extent of the condition survey. Be sure to point out any limitations and restrictions on your work, whether by instructions from the client, or on account of restricted time, money, access, or other reason. State the background and experience of the survey team.

Discuss in appropriate detail the date of the structure's construction, type of its construction, its past and present uses, history of alterations and repairs, and abnormal occurrences, if any. All of these have bearings on the assessment.

Cite the codes and standards that were applicable to the structure when it was designed and constructed, and those that are governing now. When reasonable and useful, include applicable quotes from the codes and standards.

List all of the material you have reviewed, including original construction documents, records of violations, maintenance, repairs, renovations, abnormal occurrences, etc. Also list those documents that would be necessary or helpful but are not available. Discuss the methods and equipment used to the extent that they imply the completeness, accuracy, reliability, and limitations of the work.

The "heart" of the assessment work is usually the field inspection and resulting observations. Discuss how it was done. Describe in detail the location, nature, and seriousness of each problem discovered; refer to the item in the checklist (if a checklist was used); supply sketches and/or photographs to illustrate the problem. Make an attempt to identify the cause, and indicate your reasons for your opinion as well as the degree of certainty of that opinion. Itemize and describe in detail those defects that warrant attention. Highlight these so that they are readily noticeable in the text.

If probes were made, explain the reasons for the probes and how their locations were selected. Indicate verbally and, if warranted, graphically the locations of the probes, and describe of what they consisted. Describe whether and how they were repaired. Inform of the availability and whereabouts of saved materials, if any. Discuss the findings from the probes. If considered necessary, document the details of the probing in an appendix.

If field and/or laboratory testing were performed, explain the reasons for doing them. Name the laboratory where the testing was done. Give a description of the techniques and provide literature references. Present the results and their implications. Document the necessary details of the testing in a separate section or in an appendix.

If structural analyses were performed, explain the reasons for doing them. Make literature reference to the method and/or computer program(s) used. List major assumptions in developing the model and indicate the expected accuracy and reliability of the results. Discuss the indications of the results as to the assessment of the structure's condition. Document the details of the analysis in the same or in a separate section or in an appendix.

"Pull together" the observations from the field inspections, findings from the probes, results of the tests, and results of the structural analyses, and review and synthesize them to present an evaluation of the specific conditions of the structure. Provide rationale behind the evaluation. *This part is the essence of the report* in that this is an almost stand-alone discussion of all the foregoing work and of the reasoning and the judgment behind the conclusions.

Present a list, or at least an overall description, of those structural items that were not inspected because of lack of access, time, money, or other reasons; and opine whether special effort should be made to inspect them.

Write a brief summary assessment of the condition of the structure; think of it as "general appearance and overall quality." Compose this part carefully, clearly, and concisely, for *this may become the most-quoted part of the report.* Point out, without detailed discussions, those defective conditions that are particularly serious, require further investigation, and/or are in need urgent repair.

Make recommendations fur further inspection, probing, testing, analysis, emergency and/or permanent repair, or other action on the item(s) needing it. Indicate the level of urgency of the recommendations. If you have no ready solution for mitigating a bad condition, do not be intimidated; it is entirely acceptable to have recognized a problem but to not know immediately the best remedy for it.

It is not unreasonable to write a well-considered and carefully worded disclaimer to limit your liability to the specific intent and content of the report. It may be advisable to alert the client at the outset of the project that the report will have a reasonable disclaimer.

At the end of the report or in the cover letter, indicate your availability for follow-up and further service on this project.

In addition to the above-described material, copies of field logs should be in an appendix, probably under separate cover, or their availability should be indicated.

Depending on the type of client and on the relationship that developed between you and the client, it may be a good idea in the cover letter to make an offer of an oral presentation where the client can ask questions and you can clarify points in the report.

Beware of pronouncing the structure "safe." Safety or the lack of safety of a structure is not inherent but rather it is an opinion based on observations, calculations, and tests, all of which may be performed and judged differently by different investigators. In addition, "safe" for a particular exposure, use, or load that is known may not be "safe" for *another* exposure, use, or load that is either known or unknown.

Keep in mind that the report will be read by all sorts of people. Keep the language of the general sections understandable to the nontechnical reader, but direct the detailed discussions and results to engineers who will review and use them.

Do not treat the report lightly, and do not write it carelessly. Rightly or wrongly, your own competence and the reliability of your report are often judged by the quality of the writing.

Be aware that your condition assessment report may have serious financial and legal consequences to the client! At the same time, however, be mindful of your own professional liabilities as well as your responsibilities for public safety!

REFERENCES

1. *SEI/ASCE 11, Guideline for Structural Condition Assessment of Existing Buildings* (1999), American Society of Civil Engineers, Reston, VA.
2. *National Practice Guidelines for the Preparation of Structural Engineering Reports for Buildings* (1995), Council of American Structural Engineers (CASE), Washington, DC.
3. *Bridge Inspector's Training Manual*, July 1991, FHWA, Washington, DC.

CPSIA information can be obtained
at www.ICGtesting.com
Printed in the USA
LVHW012025220723
753193LV00005B/40